Public Health Ethics Analysis

Volume 5

Series Editor
Michael J. Selgelid,
Centre for Human Bioethics, Monash University, Melbourne, VIC, Australia

During the 21st Century, Public Health Ethics has become one of the fastest growing subdisciplines of bioethics. This is the first Book Series dedicated to the topic of Public Health Ethics. It aims to fill a gap in the existing literature by providing thoroughgoing, book-length treatment of the most important topics in Public Health Ethics—which have otherwise, for the most part, only been partially and/or sporadically addressed in journal articles, book chapters, or sections of volumes concerned with Public Health Ethics. Books in the series will include coverage of central topics in Public Health Ethics from a plurality of disciplinary perspectives including: philosophy (e.g., both ethics and philosophy of science), political science, history, economics, sociology, anthropology, demographics, law, human rights, epidemiology, and other public health sciences. Blending analytically rigorous and empirically informed analyses, the series will address ethical issues associated with the concepts, goals, and methods of public health; individual (e.g., ordinary citizens' and public health workers') decision making and behaviour; and public policy. Inter alia, volumes in the series will be dedicated to topics including: health promotion; disease prevention; paternalism and coercive measures; infectious disease; chronic disease; obesity; smoking and tobacco control; genetics; the environment; public communication/trust; social determinants of health; human rights; and justice. A primary priority is to produce volumes on hitherto neglected topics such as ethical issues associated with public health research and surveillance; vaccination; tuberculosis; malaria; diarrheal disease; lower respiratory infections; drug resistance; chronic disease in developing countries; emergencies/disasters (including bioterrorism); and public health implications of climate change.

More information about this series at http://www.springer.com/series/10067

Euzebiusz Jamrozik • Michael Selgelid
Editors

Ethics and Drug Resistance: Collective Responsibility for Global Public Health

 Springer

Editors
Euzebiusz Jamrozik
Monash Bioethics Centre
Monash University
Melbourne, VIC, Australia

Michael Selgelid
Monash Bioethics Centre
Monash University
Melbourne, VIC, Australia

Department of Medicine, Royal
Melbourne Hospital
University of Melbourne
Melbourne, VIC, Australia

Wellcome Centre for Ethics and the
Humanities and The Ethox Centre,
Nuffield Department of Population Health
University of Oxford
Oxford, UK

ISSN 2211-6680   ISSN 2211-6699   (electronic)
Public Health Ethics Analysis
ISBN 978-3-030-27873-1   ISBN 978-3-030-27874-8   (eBook)
https://doi.org/10.1007/978-3-030-27874-8

This Springer imprint is published by the registered company Springer Nature Switzerland AG.
The registered company address is: Gewerbestrasse 11, 6330 Cham, Switzerland

Introduction: Ethics and Drug Resistance

Background

Drug resistance is widely acknowledged to be one of the greatest threats to global public health in the coming decades. Tedros Adhanom Ghebreyesus, Director General of the World Health Organization, describes antimicrobial resistance (AMR) as 'a global crisis' and 'the perfect example of the complex, multi-sectoral, multi-stakeholder challenges we will increasingly face in the future'.[1] In addition to scientific research, addressing the challenge of drug resistance requires coming to grips with numerous difficult ethical questions. This book thus provides up-to-date ethical analyses, from multiple perspectives, of many aspects of this crucial public health problem.

Infectious diseases cause significant morbidity and mortality worldwide, with a disproportionately high disease burden among disadvantaged populations. Resistance to drugs for important pathogens also frequently tracks disadvantage, meaning that increasing rates of drug-resistant infections threaten to widen global health inequalities. Meanwhile, rising levels of resistance make it harder, or in some cases impossible, to effectively treat common bacterial (and other) infections. This has significant implications for healthcare and seriously jeopardises many of the gains of twentieth century clinical medicine, even in well-resourced settings. Successful surgery, transplantation, care of newborn children, and chemotherapy for cancer, for example, all depend upon effective antibiotics to treat infections that could otherwise be fatal.

AMR occurs when pathogens evolve resistance mechanisms in response to anti-microbial exposure and/or when resistance is spread from one microbe to another. Resistant pathogens arise in both humans and animals, and they can spread between individuals and between species. Antimicrobial treatment, infection control practices, and agricultural policies directly affect resistance patterns. This in turn raises

---

[1] https://www.who.int/antimicrobial-resistance/news/WHO-GAP-AMR-Newsletter-may-2017.pdf?ua=1

inescapable ethical questions about how to make trade-offs between different kinds of risks and benefits.

Increasing rates of increasingly drug-resistant infections might be driving humanity towards a 'post-antibiotic era' – i.e. a future situation analogous to the situation before effective antibiotics were discovered and/or became widely available. This would involve a dramatic increase in harms to patients and the costs of treatment. It could also have significant effects on public health policy and potentially dramatic effects on social life.

At this critical moment in the history of medicine, public health ethics has a key role to play in the shaping of practice and policy. This book provides unprecedented, comprehensive, in-depth analysis of ethical issues associated with drug resistance.

Part I: Ethics and Drug-Resistance in Context

Part I provides an overview of drug resistance in multiple contexts. Chapter 1 begins with a broad survey of the causes and consequences of drug resistance as well as potential policy responses to this problem. Chapters 2, 3, and 4 focus on analyses of drug resistance in the contexts of tuberculosis (TB), HIV/AIDS, and malaria – which cause especially high disease burdens among the worst off, predominantly in low- and middle-income countries (LMICs). Drug resistance threatens to undermine public health programmes to treat and control these diseases and could thereby stall progress in global health and socioeconomic development. Chapters 5, 6, and 7 explore the involvement of different sectors in the development of, and response to, drug resistance. These chapters focus (respectively) on private healthcare providers, the hospital as a nidus of drug resistance, and drug-resistant infections in non-human animals.

Chapter 2 focuses on drug-resistant (TB). In the most severe cases, patients with active multi- and extensively-drug-resistant tuberculosis (MDR- and XDRTB) have few (if any) effective treatment options and face high mortality rates. One TB control strategy involves treating individuals with latent (i.e. asymptomatic) TB infection (LTBI) before they develop active disease, but the treatment of *resistant* LTBI has until recently been a neglected topic in clinical research. Nguyen et al. focus on practical ethical challenges arising in the design and conduct of clinical research on MDR LTBI treatment in Vietnam, including community understanding of LTBI and the acceptance of such research in low-income settings. Such analyses arguably have wider implications, since asymptomatic carriage by otherwise healthy individuals is a significant feature of many other (drug-resistant) pathogens, and these phenomena are often poorly understood and/or neglected by researchers, despite having significant implications for research and public health.

In Chapter 3, Bridget Haire explores ethical issues related to drug-resistant HIV/AIDS. Highly active anti-retroviral therapy (HAART) can be highly effective in suppressing (but not curing) HIV/AIDS, thus significantly reducing rates of disease and/or transmission. Unfortunately, resistance to HAART sometimes develops, particularly where HIV/AIDS patients have difficulty accessing a continuous (life-long) supply of antivirals and/or where patients are (for other reasons) not able to take medicines reliably. As Haire points out, since the development of resistance can lead to treatment failure and higher risks of disease and transmission, and since diagnostic testing for resistance and second line HAART involves significantly increased costs, policymakers face difficult trade-offs. Haire highlights the ethical aspects of priority setting and cost-effectiveness assessments in the context of HIV/AIDS control policy, which can be particularly challenging in low-income settings where HIV/AIDS is most prevalent and public health resources are most constrained.

In Chap. 4, Cheah et al. focus on malaria, a pathogen for which a mid-twentieth century global eradication effort failed, in part due to the evolution of antimalarial resistance. An especially problematic recent development is the rapid emergence of resistance to newer anti-malarials, particularly in South-East Asia (SEA). Though there are strong moral reasons to prevent the spread of resistant malaria, relevant public health interventions pose numerous ethical and practical challenges. *Inter alia*, several strategies would necessarily involve treating, and thereby imposing risks upon, apparently healthy individuals (including some who are asymptomati-cally infected with malaria) – just as tuberculosis interventions sometimes target those with latent infection – raising questions regarding whether, or when, such treatment would be justifiable. Furthermore, preventing the spread of resistant malaria parasites to communities with the highest malaria-related mortality (e.g. in Sub-Saharan Africa) may require intensifying intervention in other communities (e.g. in SEA), thus involving burdens for one group in order to prevent even greater burdens for others.

Despite progress towards increasing access to healthcare in SEA, many individu-als in this region, as in others, rely on private decentralised health providers for access to antimicrobial drugs. Chapter 5, by Liverani and colleagues, provides a rich analysis of the links between under-regulation of the private healthcare sector and the emergence of drug resistance in SEA. Gaps in surveillance systems, high rates of overprescription, and the dispensing of low quality (and/or counterfeit) antimi-crobials are among several causes of increased risks of drug-resistant infections. Liverani et al. demonstrate the complexities of drug resistance as a public health problem across multiple pathogens highlighting tensions between access to antimi-crobials and the excesses of profligate use. These tensions, and the associated chal-lenges in the ethical governance of multiple sectors and countries, are recurring themes in the book.

In Chap. 6, Gilbert and Kerridge consider how and why hospitals have often become epicentres of antibiotic resistant bacteria – and why in-hospital strategies, such as antimicrobial stewardship and infection prevention/control, have often been only partially effective. Part of the problem is that even 'appropriate' (i.e. not just 'inappropriate' and/or 'excessive') use of antibiotics inevitably contributes to the emergence and persistence of resistant strains of bacteria. Ethically salient consequences of antibiotic use in hospitals include direct harms related to resistant bacterial infections (in patients, staff, and – through transmission – the wider community), stigma and other burdens endured by carriers of drug-resistant strains and those who care for them, and the significant costs of reactive (as compared to preventive) infection control interventions. They conclude by outlining moral obligations of individuals and organizations to contribute to reduction of hospital-acquired drug-resistant infections.

Boden and Mellor, in Chap. 7, consider links between drug-resistance in animals and in humans. Although controversy continues to surround questions regarding the degree to which drug resistance in animals contribute to infections in humans, they characterise antibiotic resistance as a typical 'One Health' problem. As such, coherent policy responses are needed across multiple sectors (e.g. human health, animal health, food production, and environmental management). Boden and Mellor argue that international policy-making in particular should take existing socioeconomic inequities in to account, being careful to avoid unnecessarily punitive measures in LMICs which could compromise animal health as well as food production, while still attempting to reduce the incidence and spread of drug-resistant infections (across multiple species).

Part II: Theoretical Approaches to Ethics and Drug Resistance

In Part II, ethical issues associated with drug resistance are analysed via diverse theoretical lenses, appealing to a variety of philosophical and economic concepts including virtue, duty, rights, capabilities, justice, and public goods.

In Chap. 8, Justin Oakley approaches ethical dilemmas in antimicrobial prescription from a virtue ethics perspective. Oakley argues that prescribers should be guided not only by patient-centred virtues, but also by community-centred virtues, including the virtue of justice, in order to strike the right balance between the benefits of antimicrobials for patients and the societal harms of promoting resistant infections. He argues that this is especially important where the expected benefits to the patient from antimicrobial treatment would be low and the wider societal harms are potentially significant. Oakley notes that physicians' decisions are also influenced by a number of cognitive biases and situational factors. A virtuous physician would thus need to cultivate practical wisdom and meticulousness in addressing her own biases. In addition, individual prescribers need the support of healthcare policymakers and institutions to situate them in systems that foster and support virtuous prescribing practices.

In Chap. 9, Giubilini and Savulescu focus on cases where restricting one person's use of antibiotics could be plausibly described as an 'easy rescue'. These are situations in which, at little or no cost, one can consume less antibiotics and thus reduce imposition of risk on others (as well as, perhaps, reducing one's own risk of future resistant infections). The authors argue that individuals have moral obligations to avoid imposing risks on others, including by avoiding profligate use of antibiotics. Thus, policies restricting antibiotic use would have particularly strong ethical justification in situations of 'easy rescue' since (other things being equal) there would be few important countervailing moral considerations. They note that such cases are particularly likely in high-income countries, where individuals have a reliable health system to support them in the event that the cost of not taking antibiotics (which is *ex ante* uncertain) turns out to be more significant (i.e. that the 'rescue' turns out to be less easy than expected). Furthermore, they argue that states have reciprocal duties to individuals whose antimicrobial use is restricted (e.g. by more stringent prescribing policies), which may plausibly include duties to provide various forms of compensation and/or healthcare. Finally, the authors situate their claims in broader notions of collective obligations to contribute to common goods and argue that antimicrobial effectiveness can be conceived of as such a common good – one that is undermined by overuse (as well as therapeutically justified use) of relevant drugs.

In Chap. 10, Shawa and colleagues apply a human rights approach to the problem of highly resistant strains of TB, for which the development of new drugs and wider access to existing treatments are urgently needed – especially in LMICs. In addition to the right to health, the authors argue that there is a right to enjoy the benefits of scientific progress and that states and international agencies have duties to respect, protect, and fulfil both of these rights (among others). Inadequate access to effective TB drugs and the longstanding relative neglect of resistant TB by funders, researchers, and diagnostics/drug developers are therefore framed as failures to fulfil (at least) these two human rights. Shawa and colleagues argue that duties to fulfil the right to enjoy the benefits of scientific progress in particular entail wide-ranging responsibilities, for example, to pursue legislation to promote TB research and expand access to newer TB drugs (even where this would involve overriding or reducing the scope of intellectual property rights such as those implicit in patents that often make such drugs unaffordable). Noting that there has not yet been international agreement on the minimum core obligations entailed by this right, they outline potentially useful ways of specifying these in order to provide more explicit guidance for states to respond appropriately to this urgent public health problem.

Carl Coleman, in Chap. 11, looks specifically at the right to refuse treatment and the conditions under which legislation and policy should endorse isolation and/or non-consensual treatment. He argues that although laws in some jurisdictions permit mandatory treatment for infectious diseases such as tuberculosis, such compulsion is ethically justifiable only in rare cases in which certain conditions are met, for example, where (i) treatment refusal poses grave risk to others, (ii) the imposed treatment is safe, effective, and not overly burdensome, and (iii) less restrictive measures are infeasible. Coleman surveys international human rights documents and laws in multiple jurisdictions regarding non-consensual treatment and/or TB, noting that provisions for compelled (diagnostic testing and) treatment often exist, even if they are rarely invoked. Even in high-risk cases, such as XDR-TB, the public can usually be protected from harm by isolating the patient (whether or not the patient accepts treatment in isolation). Coleman thus argues that there would rarely be adequate justification for enforcing treatment on the isolated individual. In cases where, for example, isolation facilities are overwhelmed (e.g. during a large outbreak) and large numbers of people refuse effective treatment for infection, mandatory treatment might be more justifiable. However, Coleman points out that – in practice, particularly with adequate community engagement from public health authorities – conditions sufficient to justify such non-consensual treatment will seldom, if ever, be met.

In Chap. 12, Jamrozik and Selgelid explore the public health ethics implications of asymptomatic carriage and transmission of drug-resistant bacteria by otherwise healthy people. This chapter first summarises current evidence regarding the widespread carriage of key drug-resistant bacteria, noting important gaps in current data. The authors then analyse potential public health interventions for carriers in light of existing public health ethics frameworks, arguing that the relative burdens imposed by public health measures on healthy carriers (as opposed to sick individuals) warrant careful consideration and should be proportionate to the expected public health benefits in terms of risks averted. Ultimately, more surveillance and research regarding community transmission will be needed in order to clarify relevant risks and design proportionate policies, although community surveillance itself also requires careful ethical consideration.

In Chap. 13, Byskov et al. introduce a capability framework to enrich analysis of the burdens of public health interventions among carriers of multi-drug-resistant organisms. They note that carriers can face stigma and other harms as well as restrictions on their particular opportunities for choices (e.g. regarding freedom to chose where to go, with whom to interact, or which occupation to pursue). Thus carriers are potentially subjected to a wide range of potential burdens and/or harms, depending on the policy in question. The authors argue that examining these adverse effects in terms of reductions in the capabilities and functionings of carriers helps to illuminate the ways public health policies aimed at controlling the spread of resistant pathogens can constrain the lives of those affected. Because adverse effects on carriers' capabilities will be highly context specific, the authors ultimately aim to identify a rich taxonomy of ethically relevant considerations to help policymakers (i) determine the likely burdens of being a carrier and of a given intervention and (ii) weigh these burdens against the public health benefits (and costs) of potential interventions.

In Chap. 14, Michael Millar likewise draws on 'the capabilities approach' to illustrate the ways in which access to effective antibiotics among children is critical to secure normal childhood development and fully flourishing adult capabilities. He notes that there is significant inequality in the distribution of risk of (resistant) infectious disease and access to treatment. Millar argues that this is especially concerning where lifelong capabilities are adversely affected. Uncontrolled (or effectively untreatable) resistant infectious diseases can, furthermore, lead to a clustering of disadvantage in a particular individual or community, amplifying existing health injustices. Millar raises the compelling example of growth stunting among LMIC children. Often caused in part by early childhood infections, stunting results in poor long-term physical and cognitive outcomes. However, proposals for the mass treatment of children with antibiotics raise familiar tensions between assuring good health and promoting the rise of drug resistant infections. Given the global inequalities in the distribution of relevant risks and the potential for resistant pathogens to spread across national borders, international co-operation is required (but can itself be threatened by persistent injustice).

In Chap. 15, Francis and Francis examine the collection and use of information in the measurement of, and response to, infectious diseases with a particular focus on the value of fairness in the context of public health surveillance. Relevant types of information, broadly conceived, include data regarding infected individuals and disease transmission as well as knowledge arising from research. Francis and Francis analyse the use of such information from the interconnected perspectives regarding 'vectors' (those who transmit infection) and 'victims' (those harmed by infection), noting that people may experience both of these states (often simultaneously). The authors emphasize that excessive focus on the vector perspective may lead to unnecessary stigmatization of individuals and punitive policies that can become counterproductive since infected individuals will have strong incentives to conceal their diagnoses. In contrast (simultaneous) concern for individuals as victims may help to foster less burdensome interventions and more support for those infected, but overemphasizing such concerns may lead to, for example, overuse of antibiotics and thus more drug resistance. Francis and Francis argue that the ethical principles of fairness and reciprocity should guide infectious disease policy formation to an appropriate balance of each perspective, especially where difficult trade-offs are required.

In Chap. 16, Lynette Reid examines the links between drug resistance and health inequalities, illustrated by cases such as the rising resistance of many sexually transmitted infections (including among sex workers) and the risks of resistant intestinal parasitic infections. Reid argues that drug resistance undermines global health development narratives because worsening drug resistance may make it impossible to mitigate the persistent large infectious disease burden associated with poverty. Thus, drug resistance is predicted to increase inequality, and a focus on improvements in infectious disease *prevention* (e.g. by addressing the social and economic inequalities that predispose people to infection) would arguably do more to reduce long-term health injustice than expanding access to increasingly ineffective treatments. On the other hand, Reid points out that drug resistance in high-income countries could lead to a 'levelling down' in health equality by undercutting the safety of high-cost interventions (such as complex surgery and immunosuppression). However, as effective antimicrobials become a scarce resource, their use could also be unjustly monopolized by the well-off (in both HICs and LMICs). In any case, Reid argues that policy should go well beyond assuring minimally sufficient access to water, sanitation, and antimicrobials – and address the underlying political and economic forces that result in the persistence of unjust risks of disease among underprivileged individuals.

In Chap. 17, Coast and Smith focus on intersections between economic and ethical analyses of antibiotic resistance. This chapter conceptualizes problems associated with AMR in terms of 'public goods' (a concept related to the idea of common goods invoked above). In economics, public goods are said to be non-rival (i.e. one individual's use/enjoyment of a good does not limit its use/enjoyment by others) and non-excludable (i.e. it is difficult or impossible to prevent access to the good).

Coast and Smith note that the lack of new antibiotics is a predictable consequence of economic forces leading to market failure. They also highlight alternative systems that have been proposed to stimulate useful research and development. At the level of consumers of antibiotics, they note that although many of the benefits of antibiotic use accrue locally and in the present, the harms and costs are (sometimes unfairly) distributed across space and time. Current consumption of antibiotics, for instance, may compromise interests of future people. Ultimately, they argue that economic and ethical considerations will often converge on similar policy recommendations. For example, they note that infection prevention (which is often more cost-effective than providing treatment once an infection becomes clinically apparent) and research into non-antibiotic treatments may be part of solutions aiming to achieve (economically and ethically) optimal improvements in health via a reduction in the burden of infectious diseases.

Part III: Ethics, Regulation, Governance, and Drug Resistance

The chapters in Part III provide unique perspectives on ethical issues associated with policy regarding drug resistance. These analyses draw on concepts related to game theory, collective action, risk limits in research, solidarity, environmental ethics, law, and social policy.

In Chap. 18, Jonathan Anomaly gives an account of the international co-operation urgently needed to regulate the use of antibiotics in animal agriculture. Anomaly starts by asserting that the situation of farmers deciding whether to use antibiotics as a means to sustain animals in crowded conditions is akin to the game theoretical model of a many person prisoners' dilemma. In short, those who opt not to use antibiotics are predicted to lose out economically as factory farmers heavily using antibiotics drive down the price of meat. In the long term, resistance becomes rampant and all are worse off. This predictable 'market failure' to secure the public good of antibiotic effectiveness is one reason in favour of regulating antibiotic use. Regulation needs to reduce the negative effect of 'free riders' (i.e. profligate antibiotic users) and provide assurance for individuals who use antibiotics carefully (i.e. in line with the social optimum) that others will do the same. Anomaly argues in favour of an international treaty and outlines how such a treaty might be designed and implemented.

In Chap. 19, Nichols King analyses an issue raised in many other chapters in this volume in more detail. Given what is known about the complexity of the causal and perpetuating factors involved in the problem of antimicrobial resistance, will 'technological fixes' (e.g. new diagnostics, drugs, vaccines) alone provide an adequate response, or are broader social, behavioural, political, and economic changes more likely to achieve sustainable improvements in public health? Noting that multiple co-ordinated policy responses are likely required, King traces the history of 'technological fixes' and examines the ways in which over-reliance on such approaches has implications for distributive justice.

In Chap. 20, Littmann, Rid, and Buyx explore the concept of 'rational use' of antibiotics. They note that, in some cases, there will be ethical conflicts between patients' interests and the need to preserve effective antibiotics for the future. Littmann et al. draw an analogy from research ethics regarding the acceptable limits to risks to which participants in research may be exposed and provide a framework for policymakers to evaluate potential antibiotic use policies ethically, particularly with respect to the degree of risk to which current patients are exposed due to potential reductions in antibiotic use.

In Chap. 21, Holm and Ploug also approach the problem of the ethical justification for restricting the use of antibiotics in cases where they will provide patients some benefit (but also entail risks of antibiotic resistance). They argue that the concept of solidarity can help to guide physicians and policymakers in such contexts, and also that it can help to promote the support of such policies among the general public. Once persons realize that each could (or anyone could) easily be affected by a drug-resistant infection, it should arguably lead them to act in solidarity with others by giving appropriate weight to the risks related to potential increases in drug resistance. Thus, such individuals would be more inclined to avoid the use of (or prescription of) antibiotics for self-limiting conditions. Furthermore, the authors give an account of how solidarity can inform public health policy, and the ways in which it might be expanded to the global level in the context of the international spread of drug-resistant infections.

In Chap. 22, Nijsingh et al. examine the ethical and evidentiary justification for public health policy responses to drug-resistant infections. Evidentiary justification can be challenging where little high-quality evidence is available and where the underlying causal pathways driving drug resistance are complex and/or poorly understood. In turn, ethical analysis of current policy must often be sensitive to the (sometimes limited) degree of evidence regarding the (cost-)effectiveness of an intervention (or package of interventions). With these complexities in mind, the authors give a thorough analysis of the application and limitations of the Precautionary Principle in the context of antibiotic resistance, as well as a number of other ethical and evidentiary challenges facing policymakers.

In Chap. 23, Anne Schwenkenbecher gives an account of how prospective moral responsibility for one's contribution to antimicrobial resistance – one's 'antimicrobial footprint' – can help to support collective action to reduce the problem. Shwenkenbecher considers arguments that failing to reduce one's use of antimicrobials (where it is possible to do so at acceptable costs) can contribute to unfairness and/or lead to collective harm, concluding that moral reasons to avoid contributing to collective harm (e.g. from drug resistant infection) should support individual action to reduce one's antimicrobial footprint.

The next two chapters address issues related to global health governance in the context of antimicrobial resistance. In Chap. 24, Bennett and Iredell explore the challenges of using existing governance frameworks. They begin by giving an account of the WHO/World Health Assembly 2015 Global Action Plan for Antimicrobial Resistance and the United Nations resolutions that closely followed. The complex problem(s) of antimicrobial resistance have been framed in a number of different ways in different policies, and the authors argue that policymakers will need to overcome conceptual, practical, and political challenges in order to implement coherent and effective policy in this complex area. A key conceptual challenge is the need to define (and achieve international agreement on) what constitutes 'appropriate' antimicrobial use in order to implement effective regulation and accountability measures.

In Chap. 25, Lee and Ho also take the Global Action Plan as a starting point and complement the above analysis by offering a legal and regulatory toolkit to support effective health governance related to drug-resistant infections. The authors argue that an equitable regulatory 'lever' should be one key part of co-ordinated policy responses and will often be required in order to enable other policies. Lee and Ho give a detailed account of how regulation should be used to support (among other important priorities) quality assurance in antimicrobial production, optimum prescribing and dispensing practices, and the assurance of equitable access to antimicrobials.

Finally, in Chap. 26, Littmann, Viens, and Silva describe antimicrobial resistance as a 'super-wicked' problem. As noted in many other chapters, drug resistance is a complex policy area: there are huge numbers of contributors, stakeholders, and causal pathways involved in creating, perpetuating, and responding to the level of drug resistance across a wide range of pathogens, sectors, and settings around the globe. Littmann et al. conclude the volume by highlighting the many distinctive ethical issues arising in relation to drug-resistant infections and the ways in which ethical analysis should inform policy and response activities.

Acknowledgments

Dr Jamrozik worked on this volume during a residency at the Brocher Foundation. His fellowship at the University of Oxford is supported by the Wellcome Trust [216355/Z/19/Z], [203132/Z/16/Z].

Monash Bioethics Centre   Euzebiusz Jamrozik
Monash University
Melbourne, Australia
Department of General Medicine
Royal Melbourne Hospital
Melbourne, Australia
Wellcome Centre for Ethics and the
Humanities and The Ethox Centre,
Nuffield Department of Population Health,
University of Oxford
Oxford, UK

Monash Bioethics Centre   Michael J. Selgelid
Monash University
Melbourne, Australia

Contents

Part I
Ethics and Drug Resistance in Context

Chapter 1
Drug-Resistant Infection: Causes, Consequences, and Responses

Euzebiusz Jamrozik and Michael J. Selgelid

Abstract This chapter provides an overview of the causes and consequences of, and possible policy responses to, the problem of drug resistance. Throughout, we highlight the ways that ethical and conceptual analyses can help to clarify relevant issues and improve policy, especially in public health, broadly conceived. Drug resistant pathogens arise, persist, spread, and produce harm due to a complex set of causes: biological processes (e.g., related to microbial evolution, the transmission of genetic determinants of resistance between microbes, and human host immunity) as well as human behaviors (e.g., antimicrobial use and hygiene practices) and other social factors (e.g., access to clean water, sanitation, healthcare, and antimicrobials). Furthermore, the ethically salient consequences of drug resistance include not only morbidity and mortality from untreatable infections (that are often inequitably distributed), but also broader effects on human freedom, privacy, and well-being. Public health ethicists are ideally placed to identify and weigh the values that might be promoted or compromised by potential policies and/or interventions that aim to address the problem of drug resistance. This chapter concludes by discussing potential policy responses, including those related to surveillance, research, animal and human antimicrobial use, the broader social determinants of health, infection control practices, and vaccination.

Keywords Antimicrobial resistance · Drug resistance · Drug resistant infection · Ethics · Bioethics · Public health ethics

E. Jamrozik (✉)
Monash Bioethics Centre, Monash University, Melbourne, VIC, Australia

Department of Medicine, Royal Melbourne Hospital, University of Melbourne, Melbourne, VIC, Australia

Wellcome Centre for Ethics and the Humanities and The Ethox Centre, Nuffield Department of Population Health, University of Oxford, Oxford, UK
e-mail: zeb.jamrozik@gmail.com

M. J. Selgelid
Monash Bioethics Centre, Monash University, Melbourne, VIC, Australia
e-mail: michael.selgelid@monash.edu

© The Author(s) 2020 3
E. Jamrozik, M. Selgelid (eds.), *Ethics and Drug Resistance: Collective Responsibility for Global Public Health*, Public Health Ethics Analysis 5, https://doi.org/10.1007/978-3-030-27874-8_1

1.1   Introduction

It is widely acknowledged that drug resistance poses one of the greatest threats to global public health during the coming decades. Drug resistance compromises the treatment of infections (that were commonly debilitating and/or fatal before the development of antimicrobial drugs), and thereby undermines many advances in surgery, cancer treatment, and immunosuppression that depend on our ability to treat infections effectively. Microbes – bacteria, mycobacteria, parasites, fungi, and viruses – have, over billions of years and untold numbers of microbial generations, developed mechanisms (via evolutionary processes) to protect themselves from harm and transmit such portective mechanisms to other microbes (of the same or different species) (Holmes et al. 2016). The rapid increase in antimicrobial drug resistance in the twentieth and twenty-first centuries is a result of these powerful evolutionary mechanisms combined with human activities that affect the microbial world, including the widespread production and use of antimicrobial drugs. Resistance is a matter of degree (for example, low levels of resistance can be over-come with a higher dose or longer course of antimicrobials) and its impact is also relative to the availability of alternative treatments (where second line drugs are readily available, resistance to first line agents may initially be less of a concern). Thus, the recent emergence of strains of clinically important pathogens that are highly resistant to all, or nearly all, available therapies (e.g. extensively drug-resistant tuberculosis (TB) and pan-resistant gram negative bacteria) is an urgent challenge for public health (Schwaber et al. 2011; Birgand et al. 2016; World Health Organisation 2017a).

Drug resistance is an important topic for ethical analysis since (i) human actions and inactions are major contributors to the problem, (ii) the consequences for human health and well-being are highly significant and inequitably distributed, and (iii) policies aiming to reduce the rates of resistant pathogens may involve balancing this reduction in risk with other morally salient risks, burdens, and benefits. Thus, this volume aims to provide a timely exploration of many of the ethical aspects of the phenomenon of drug resistance. This first chapter highlights the complex causes and significant consequences of drug resistance, and the ways in which ethical and conceptual analysis can inform and improve relevant policy responses. We link these discussions with other chapters in this volume, as well as gesturing towards future directions for ethicists, empirical scientists, and public health policymakers.

1.2   Causes

1.2.1   Evolution and Transmission of Resistance Genes

The microbial world is ancient, abundant, ubiquitous, and complex. As a result of random mutation over trillions of microbial generations, bacteria have adapted to their environments, in part by developing genes that code for mechanisms of resis-tance to various threats – including, for example, heavy metals, naturally occurring antibacterial compounds (including beta-lactams such as penicillins and carbapenems) and synthetic antimicrobials (e.g. fluroquinolones and sulphonamides) (Holmes et al. 2016; D'Costa et al. 2011). Certain microbes are also able to horizontally transfer genes coding for resistance mechanisms to other microbes (Holmes et al. 2016; Chang et al. 2015). With the dramatic, unprecedented increase in human interventions in the microbial world (especially the widespread use and overuse of antimicrobial agents), strong evolutionary selection pressures have been applied to microbes leading to the emergence, increasing frequency, and persistence of resistant microbes in humans, animals and the environment (Holmes et al. 2016).

1.2.2 Antimicrobial Use in Humans

In his 1945 Nobel Prize speech, Alexander Fleming (who first discovered the antibiotic properties of penicillin) famously noted that 'the ignorant man may easily underdose himself and by exposing his microbes to non-lethal quantities of the drug make them resistant' (Fleming 1945). If antimicrobial treatment is inadequate, that is, then resistant strains of microbes that otherwise would have been killed off may survive and become more strongly established in the absence of microbal competitors, in the environment of a person's body. Resistant pathogen strains thus selected can then be transmitted to other persons. For pathogens like tuberculosis (TB) and HIV, requiring months (or, for HIV and extensively drug-resistant TB, years) of multi-drug therapy, undertreatment (e.g. due to 'noncompliance' of patients, or inadequate access to medicine, etc.) has played a central role in the emergence and persistence of highly resistant strains (see Chaps. 2, 3, 5, 10 and 26). In the case of malaria – where, for various reasons, parasites may be exposed (in human patients) to sporadic and/or sub-therapeutic concentrations of antimalarials, sometimes as a result of partial treatment – underuse of drugs likewise plays a role in the emergence of antimalarial resistance (White 2017) (See Chap. 4).

In the case of antibiotic resistance in common bacterial pathogens, however, *overuse* of antibiotics is far more important than undertreatment. Despite years of rhetoric regarding the need to 'complete the prescribed course' for common uncomplicated bacterial infections, this now appears, except among a subset of pathogens and specific sites of infection, to have been ill-founded[1] and, on balance (when generalised to all infections), harmful advice (Llewelyn et al. 2017). Overuse and 'appropriate' use are much more dominant causes of resistance. This is because the human body (particularly in the digestive and respiratory tracts, and on the skin) contains billions of bacteria, many of which are indiscriminately exposed to an antibiotic used either appropriately to treat one particular pathogen (e.g. bacterial pneumonia) or inappropriately (e.g. a viral infection mistaken for a bacterial

---

[1] In part because, until recently, few trials had addressed the question of whether shorter courses for common infections may be just as effective, with less development of resistance (and less side effects) – the few trials that have now been done generally support the use of shorter courses in uncomplicated infections.

infection) (Llewelyn et al. 2017; Carlet 2012). Exposure to more antibiotics in a given individual is predictive of a higher rate of asymptomatic carriage of resistant pathogens (although this rate does decrease over time) (Nasrin et al. 2002; Bryce et al. 2016). These resistant bacteria, although in usual circumstances causing no harm, can lead to invasive (resistant) disease – for example when a person's skin is cut or incised by a surgeon, or when bowel bacteria spread to other locations in a person's body, or when a person becomes immunosuppressed (Tischendorf et al. 2016; Safdar and Bradley 2008). Those who thus become 'carriers' of resistant bacteria can transmit such pathogens to others (whether the initial carrier is symptomatic or not) (Smith et al. 2004; Lerner et al. 2015; Jamrozik and Selgelid 2019).

Overuse of antibiotics in humans is thus a collective action problem – in some respects a classic 'tragedy of the commons' (Hardin 2009), but complicated, *inter alia*, by the transmission of resistance between pathogens and also the transmission of resistant pathogens between humans.[2] The (simplified) structure of the collective action problem is that each doctor or patient seeks individual benefit of the patient (minimising the risk of severe bacterial infection) by, respectively, prescribing and taking antibiotics even in cases where this may have only marginal expected benefits for the patient; but their decisions/actions collectively (in conjunction with other causal factors) bring about high levels of antibiotic resistance – which, in the long term, is a major threat to all.

Problematic treatment decisions (that contribute to resistance) are sometimes related to diagnostic uncertainty. When a patient has symptoms associated with respiratory illness, for example, there is often no sufficiently rapid and accurate test to determine whether it is caused by a bacterial pathogen. As a result of this uncertainty, combined with risk-aversion among doctors and patients, and a (mis)perception that a course of antibiotics 'does no harm' (or that the risks of side-effects and generating resistance are outweighed by potential benefits[3]), millions of antibiotics each year are taken when they are not required. Solutions that rely on individuals acting in accordance with the social optimum (especially if, in doing so, they take on more risk to themselves) are, at best, incomplete or, at worst, doomed.

The problems of both antimicrobial overuse and underuse are magnified in some low- and middle-income countries where many people lack access to basic diagnostic testing and antimicrobials—and/or where antimicrobials are commonly available without prescription (Holmes et al. 2016; Laxminarayan et al. 2016; Dar et al. 2016) (see Chap. 5). There is an inherent tension between 'access and excess', i.e. many die because they are unable to obtain diagnosis and/or antimicrobial treatment

---

[2] In standard commons tragedies, such as overfishing, individuals collectively deplete a common resource (e.g. by fishing), ultimately leading to its collapse. In the case of drug resistance, the resource (e.g. effective antimicrobials) can be depleted (in a way akin to standard depletion) through use of antimicrobial drugs, leading to resistance, but also, for example by the transmission of drug resistant strains from one person to another (regardless of whether either has recently used antimicrobials) – thus the relationships at play may be more complex than standard commons tragedies.

[3] The actual (as opposed to perceived) risks and benefits are rarely quantified 'at the bedside'.

when it is really needed while, at the same time, antibiotics are used when they are not required, leading to resistance. This, in turn, exacerbates problems of access – because the second line drugs required (after first line drugs have been rendered useless) are more expensive/less affordable (Laxminarayan et al. 2016; Dar et al. 2016). In parallel, the transmission of resistant pathogens is amplified by a lack of access to readily available clean water, sanitation, and well-resourced healthcare institutions. The burden of drug-resistant infections thus tracks poverty and social disadvantage both within countries and internationally (Llewelyn et al. 2017; Bryce et al. 2016; Guh et al. 2015) (See Chap. 16).

1.2.3   Transmission

Drug-resistant microbes are transmitted between human beings just like other non-resistant pathogens – via airborne or droplet transmission, skin contact, the faecal-oral route, sexual transmission, contact with infected bodily fluids, contaminated water and food, vector transmission (e.g. mosquitoes in the case of malaria), and so on. The epidemiological significance of transmission of some pathogens in some contexts is relatively well understood, whereas the transmission of the same pathogens in other settings may be different and/or less well-studied. In the case of drug-resistant bacteria, for example, in-hospital transmission is well-documented. Such transmission often occurs via contamination of the clinical environment and via healthcare workers – especially those who fail to adhere to basic hand hygiene practices (see Chap. 6), although controversy surrounds the optimum infection control policies to prevent transmission (Morgan et al. 2017). However, the transmission of resistant bacteria (and antibiotic resistance mechanisms) in the general community (i.e. outside healthcare facilities) is poorly understood, and much more evidence is needed to guide policy (Holmes et al. 2016; Dar et al. 2016). Transmission in the community is facilitated not only by direct human contact, but also general environmental contamination with resistant pathogens, the mobile genetic elements that confer resistance, and even antibiotics themselves – with polluted water systems being a key link in indirect transmission between human beings, and between animals and humans (Pruden et al. 2013; Martinez 2009). This problem is of greatest concern in low-income settings with poor access to clean water and sanitation, further exacerbating the inequitable distribution of harms from drug-resistant infection (Laxminarayan et al. 2016; Dar et al. 2016).

The global spread of drug resistance is greatly facilitated by modern air travel. Millions of people become colonised (usually without symptoms) with resistant pathogens or other (non-pathogenic) microbes containing genetic determinants of resistance every year in locations with high rates of resistance and then fly to regions where (whether or not those colonised are sick) resistant pathogens and/or resistance determinants are directly or indirectly transmitted to others (Kennedy and Collignon 2010; Östholm-Balkhed et al. 2013).

1.2.4   Antimicrobial Use in Animals and Agriculture

The widespread use of antibiotics in industrial agriculture and aquaculture, either as 'growth promoters' or in other mass prophylactic uses, has lead to pathogens developing resistance to the agents used. This has had consequences for both animal and human health. Consequences for humans occur when clinically significant pathogens (and/or resistance determinants) are transmitted from animals to humans (either directly via animal contact or consumption of animal products, or indirectly via environmental contamination with resistant pathogens and/or resistance determinants), when humans are themselves exposed to antibiotics used in the food chain (either in the products they consume, or because antibiotics and antibiotic residues are released into the environment), or when humans are exposed to pathogens (or other microbes) that become resistant (and/or carry resistance determinants) as a result of exposure to antibiotics in enviroments contaminated by agricultural use (Holmes et al. 2016; Schwaber et al. 2011; Birgand et al. 2016; Chang et al. 2015; Martinez 2009).[4] The links between animal and human health via our shared microbiome are complex, and the relative importance of different causal pathways in a particular setting is often difficult to quantify (Chang et al. 2015) (See Chap. 7). Likewise, although the agricultural industry uses more antibiotics in total tonnage than human healthcare, the relative contribution of agricultural antibiotic use to the epidemiology of resistant bacterial disease in humans is difficult to study, often unknown and likely varies widely in different settings (Holmes et al. 2016).

1.3   Consequences

1.3.1   Direct Harms to Human Beings

The true global burden of death and disease due to resistant infection is unknown, and from both ethical and scientific points of view there is an urgent need for more accurate estimates. One prominent appraisal published in 2015 suggested that at least 700,000 deaths occur each year due to drug-resistant infection worldwide, and that this annual death toll could rise to ten million by the year 2050 (Antimicrobial resistance: Tackling a crisis for the health and wealth of nations 2015). However, this analysis included only 6 pathogens and acknowledged that the true number is probably already far higher, especially given that more of the burden of disease falls on poor communities that often have incomplete disease surveillance systems and limited access to relevant diagnostic technology (Laxminarayan et al. 2016).

---

[4] In (1) microbes become resistant due to their exposure to antibiotics in animals' bodies; in (2) microbes become resistant due to their exposure to antibiotics in people's bodies (resulting from people's exposure to antibiotics in contaminated environments); in (3) microbes become resistant due to their exposure to antibiotics in contaminated environments.

The inequitable distribution of harms from resistant disease mirrors the inequitable distribution of infectious disease burden more generally. In both high- and low-income countries, the heavy burden disproportionately shouldered by impoverished people and communities is largely explained by 'social determinants of health' (e.g. lack of access to clean water, sanitation, wealth, education, access to a robust health system, etc). On many accounts of justice, wealthy individuals and societies have strong moral reasons to improve these basic determinants of public health for all (Selgelid 2008). Furthermore, since resistant pathogens (like other pathogens) spread across borders, the wealthy have increasingly strong self-interested reasons to provide assistance to others and to prevent others from developing resistant disease.

High rates of resistant pathogens (especially common bacterial species) undermine many of the advances of modern medicine – because the successes of surgery, transplantation, cancer treatment, immunosuppression, intensive care, and obstetric and neonatal care are very often contingent on being able to treat and cure infections. Increasing drug resistance thus has widespread implications for health and healthcare. Although patients who are unwell with other comorbidities are at the highest risk (both of carrying resistant pathogens – due to recurrent treatment and hospitalisation – and of invasive disease from these pathogens), even relatively healthy people are, and will increasingly be, harmed by resistant infectious disease.

Before the advent of antibiotics, a simple skin wound could lead to untreatable sepsis, amputation and/or death, and a 'post-antibiotic era' would entail a return to similarly grim prospects. Increasing drug resistance thus severely threatens the entire global population and, in addition, future generations.

1.3.2 Economic Consequences

Along with direct harms, drug resistance has severe economic consequences. Drug-resistant infections are more difficult (sometimes impossible) and more expensive to treat and cure, and they are more likely to result in incapacitation of the patient and significant economic losses for society. One estimate suggested that total global losses due to resistant infection between now and 2050 could total over \$US 100 trillion (O'Neill 2015), meaning that there are powerful economic reasons to devise and implement effective measures to curb the problem (See Chap. 17).

As noted in several chapters in this book, the *availability of effective antimicrobials* has many features of a 'public good' in economic terms. Standard economic models predict (more or less accurately) that a free market in antimicrobials (i.e. with little or no regulation of access apart from price) leads to 'market failure' and the erosion of the good in question (i.e. availability of effective antimicrobials). In most societies, since access to antimicrobials occurs via healthcare practitioners, the 'market' is relatively controlled (as opposed to free). The incentive structures that lead to a collective action problem for doctors and patients (discussed above) nevertheless lead to a similar erosion of antimicrobial effectiveness – albeit at an attenuated rate, depending upon the degree to which regulations on prescriptions succeed. Private healthcare providers, in any case, can be difficult to regulate (see Chap. 5), especially where they are not part of a centralised and/or universal health-care system.

1.3.3 Burdensome Public Health Interventions

The consequences of drug resistance for human beings are more than just matters of physical health and wealth. In many cases, public health surveillance and related public health practices have ethically salient implications for other aspects of well-being, including psychological well-being (e.g. due to experiences of stigma among carriers of resistant microbes – see below), as well as privacy and other freedoms (e.g. which are compromised by mandatory physical/social distancing measures such as isolation and quarantine). Well-designed public health surveillance and research should be conducted in order to clarify the health risks and costs of resistant infection and the risks, benefits, and burdens of potential public health interventions. High quality data would help policymakers determine whether imposing certain burdens on individuals would be justified as a means to improve public health (e.g. by reducing infectious disease due to resistant organisms) (Fairchild and Bayer 2004). Unfortunately, investment in surveillance for resistance has, globally, been very low and is only just starting to be improved, particularly in wealthy settings.

Those identified by surveillance as carriers of resistant pathogens while inpatients in healthcare settings sometimes experience stigma (Rump et al. 2017) as well as a wide range of effects on well-being, which some argue are best understood through a capabilities approach that explores the broader implications of public health policies for the flourishing of individual lives (see Chap. 13). Such an approach may also help to illustrate the broader aspects of human life that are jeopardised by the lack of access to effective antimicrobials, especially among children, for whom early severe infection may impair long term development (see Chap. 14).

Indeed, as more people become aware of resistance, and more community surveillance is conducted, apparently healthy individuals in the community may be increasingly identified as asymptomatic carriers of resistant organisms (See Chap. 12). Furthermore, such individuals might be monitored, offered or required to undergo treatment, and have other liberties (freedom of movement, free choice of occupation) curtailed by public health policy (Houston and Houston 2015). Those who have had recent contact with carriers might be tested and/or quarantined. There is thus frequently a tension between the aim to protect public health (by identifying infected individuals and reducing transmission of (resistant) disease) and the aim to avoid imposing significant burdens (in terms of compromised well-being, privacy and/or liberty) on individuals in order to prevent the spread of disease to others (Viens et al. 2009).

1.4 Responses

1.4.1 New Drugs

For many decades, even where the emergence of drug resistance was recognized, much of the response (or lack thereof) by individual clinicians as well as policymakers was grounded in (apparently unfounded) confidence that new antimicrobial drugs would be discovered and developed, meaning that resistance to older drugs was of limited significance. Despite early warnings of the consequences (Holmes et al. 2016; Honigsbaum 2016) – profligate use continued and indeed accelerated in humans, animals, and agriculture. Meanwhile, the restricted use of new antimicrobials as 'reserve' agents – although it may help to slow the emergence of resistance – means that there are disencentives to (profit-motivated) research and development of new antimicrobials. For this and other reasons, few new antibiotic classes or agents have been developed in recent decades (Norrby et al. 2005). Boosting more relevant research and development may thus require more public effort/funding and/or realignment of pharmaceutical companies' incentive structures (i.e. so that profit making becomes more compatible with developing products that are most important to global public health) (Banerjee et al. 2010).

New drugs (or other means of treatment/prevention) are arguably most urgently needed for infections that have become nearly pan-resistant (e.g. extensively drug-resistant TB, or multi-resistant gram negative bacteria). Vis-à-vis other responses that target particular causal pathways (e.g. restrictions on prescriptions practices or agricultural use) they would provide a cross-cutting solution to the problem of drug resistance – i.e. addressing the problem regardless of the specific mechanisms by which it was brought about.

It is clear, however, that policymakers (and, indeed, patients) cannot rely on new drugs to 'fix' the problem(s) of drug resistance, since (i) the development of new antimicrobial drugs has, in recent decades, been slow and/or relatively unsuccessful, (ii) the challenges underlying difficulties with drug development have thus far proven difficult to overcome, and (iii) without other interventions to curb the increase in drug resistance, we face a never-ending problem of finding new drugs. Thus, drug resistance requires a multi-faceted and global policy response – yet one that is also tailored to the specific problems and mechanisms of resistance in a given microbe and a given context.

1.4.2 Research and Surveillance

Apart from finding new treatments, other kinds of research are urgently needed, including empirical research in diagnostics, microbiology, vaccines (and other non-drug interventions), as well as social science and public health systems research (Dar et al. 2016). As a starting point, there are large gaps in our knowledge of the epidemiology of most resistant pathogens. Improving local and international public health surveillance would help to determine the impact of various resistance mechanisms and evaluate the cost-effectiveness of interventions. Yet this, in turn, requires development of cost-effective and publically acceptable surveillance mechanisms that can be more widely implemented, including in LMICs, and political and scientific agreement on which data will be collected and shared (Tacconelli et al. 2017). It also requires careful attention to the ethical conduct of public health surveillance (World Health Organisation 2017b).

1.4.2.1 Reducing Use in Humans

There is an urgent need to reduce antibiotic use in human patients. The most ethically straightforward cases are those in which a person is prescribed (or purchases without prescription) an antibiotic (with a risk of side-effects and resistance) for a condition for which it will provide no benefit (e.g. a viral illness) or where a shorter course of antibiotics is as effective as a longer course but the latter confers an increased risk of resistance and/or side-effects. In such cases, antibiotic use constitutes a net harm to the individual and, through the risk of transmission of resistant organisms, to others.

One strategy to reduce use is to develop new diagnostics, so that patients and doctors can avoid using antibiotics where they are not required. But in the absence of perfect tests, changes in professional and public culture are also required. For example, doctors should address their own cognitive biases (see Chap. 8) as well as patient concerns about avoiding the complications of infection, and public awareness campaigns must emphasize that antibiotics are often not required and can themselves entail significant risk (to individuals and public health) (World Health Organisation 2015). International data suggest that some countries have made significant progress in reducing antibiotic use without a significant increase in severe infections (Bronzwaer et al. 2002).

Yet, as effective antibiotics become more and more scarce, there are more difficult ethical tradeoffs to be considered, involving greater uncertainty. For example, if antibiotics become reserved for severe and/or complicated infections – so that use is banned or dramatically reduced for patients with simple infections – many patients with a simple bacterial infection (e.g. mild pneumonia) may end up being more unwell for longer, or even at a small risk of severe outcomes (even though the vast majority will ultimately recover without specific curative treatment), because they do not have access to antibiotics.

Policy to reduce use in either the low risk or the higher risk cases cannot rely entirely on individuals, since the structure of the underlying collective action problem leads to strong incentives to 'free-ride' on others' reduction in use. Thus, some form of regulation is required – e.g. through antimicrobial stewardship (which has been a successful but resource intensive approach in healthcare institutions), and/or restrictions on physician prescribing. How best to design and enforce such regulation is an important matter for debate in public health ethics.

1.4.2.2 Reducing Use in Animals and Agriculture

As argued in later chapters of this volume (see Chaps. 7 and 18), antibiotic use and overuse in agriculture must also be reduced. First and foremost, many have argued that the widespread use in animals of agents that are critical to human health should be dramatically reduced and/or prohibited, especially when such agents are employed for non-therapeutic purposes (e.g. as 'growth promoters') (Marshall and Levy 2011; World Health Organisation 2017c). But even other drugs (and non-drug agents) may lead to the co-selection of resistance determinants for critical drugs among bacteria (and fungi) shared between animals and humans. The overuse of antibiotics in agriculture in part represents a palliative for the high rates of infection in crowded 'factory farms', suggesting that reforming farming practices would have the dual benefits of reducing animal cruelty and reducing drug resistant infections. Reducing or eliminating meat consumption would also obviate many of the dangers to humans of resistance in animals (although not all, since companion animals can also transmit drug-resistant infection (Guardabassi et al. 2004)). The fact that some wealthy countries have dramatically reduced or eliminated profligate antibiotic use in farm animals (in part by giving animals more space and improving infection control practices) often at little or no long term economic cost suggests that it is possible to minimise this driver of drug resistance (McEwen et al. 2018). Whether such practices will be successfully implemented in lower income countries remains to be seen (Dar et al. 2016).

1.4.2.3 Addressing Social Determinants of Health

Infectious diseases, in general, are more prevalent among poor people and communities in both high and low-income countries. Many aspects of social organization and the built environment (water and sanitation systems, health systems, etc.) alter the risk of acquiring infection, and the risk of transmission of resistance mechanisms. Historically, improvements in living conditions arguably contributed more to the decline in infectious diseases in developed/industrialised countries than discoveries of vaccines or antimicrobial treatments (McKeown 1976). One way of reducing the prevalence of drug resistant infection would be to reduce its incidence and transmission by addressing these (and other) such social determinants of health.

The rise of untreatable infections provides a new, and urgent, rationale to ensure universal access to the social conditions that enable healthy living. Even basic measures, especially if provided to all, could help minimize the transmission of resistance (e.g. by providing access to clean water and sanitation) and reduce the use of antimicrobial drugs (e.g. by providing access to high quality, and well-regulated, health systems). Since resistant infections routinely spread via international travel, wealthy nations have reasons not only to act locally but also to act globally by assisting others with less resources (see Chap. 21) – to reduce the incidence and international transmission of resistant infection (Selgelid 2008).

1.4.2.4   Infection Control

Infection control involves measures that aim to reduce the transmission of pathogens in an institution or community. In healthcare institutions, this typically involves screening of patients, monitoring of those carrying (and/or suffering disease from) resistant pathogens, use of personal protective equipment, and social distancing measures such as isolation and quarantine. In some cases, it also involves decolonization of patients. Methicillin-resistant *Staphylococcus aureus* (MRSA) decolonization, for example, involves the use of antibacterial solutions on the skin, hair, and nasal membranes. More dramatically, recent studies have reported successful use of faecal transplantation to decolonize those with highly resistant bowel organisms (Freedman and Eppes 2014; Crum-Cianflone et al. 2015). In many cases, screening for resistant pathogens in hospital does not involve/require informed consent of patients, on the grounds that screening and control measures are required in order to prevent harm to others. Such policies should nonetheless be based upon careful ethical justification as well as, where possible, evidence of cost-effectiveness (see Chap. 6), especially insofar as they infringe on the lives of individual carriers (who may or may not be symptomatic) in significant ways (see Chap. 13).

Infection control policies become more complex when they are applied in the general community. As more people in the community are identified as carriers of highly resistant pathogens, it remains to be seen what kinds of restrictions of individual liberty would or shoud be considered justifiable. When, if ever, for example, should travellers be screened on arrival from high-risk regions – and when should those who test positive for resistant organisms be offered (or required to have) decolonization – and/or be subject to monitoring and/or social distancing measures? Important questions such as these need to be considered (and re-evaluated) as more is learned about community and international transmission of drug-resistant organisms.

1.4.2.5   Vaccines

Vaccines are a cross-cutting intervention with the potential to obviate the need to prescribe antimicrobials for a range of pathogens. In some cases, furthermore, vaccines can prevent infections that would otherwise be untreatable/uncurable (e.g., due to high levels of drug resistance). Existing vaccines against tuberculosis and some bacterial infections (e.g. *Streptococcus pneumoniae, Haemophilus influenzae*) reduce the incidence of infections due to these pathogens and thereby reduce the use of relevant antimicrobials (Dar et al. 2016). Vaccines against specific resistant strains may also lead to their replacement (e.g. in a given ecological niche in the human body) by strains that are easier to treat (Dar et al. 2016), although longitudinal surveillance data regarding strain epidemiology and disease burden are needed to confirm whether such effects lead to a long-term net public health benefit.

Even effective vaccines against viral infections (e.g. influenza, common cold viruses) can lead to a marked reduction in antibiotic use since viral infections are often erroneously treated with antibiotics (Neuzil et al. 2000). New vaccines against malaria and typhoid may help to reduce antimicrobial use and resistance relevant to these pathogens. Because there is a great number of resistant pathogens for which we lack vaccines, however, this is an important area in need of further research and development.

Of course, discovering an effective vaccine may lead to the replacement of one collective action problem (antimicrobial overuse) with another (assuring high vaccination rates) – meaning that even this 'technical fix' would have limitations. Like the development of new drugs, in any case, new vaccines would form only one part of the multi-pronged approach needed to control the problem of drug resistance.

1.5 Conclusions

Drug-resistance is widely recognised to be one of the greatest threats to global public health in the coming decades. Its causes are complex, and more work is needed to determine the relative importance of different causes. The consequences for human health are already highly significant, and, if left unchecked, will be even more dramatic in the near future. These harms, taken together, represent a large, likely underestimated, and ethically salient burden of disease that disproportionately affects underprivileged people and communities worldwide. Yet the presence of untreatable and potentially fatal pathogens and the prospect of such infections becoming more common is a threat to all. Despite current uncertainties and urgent gaps in our knowledge regarding drug-resistant infection, there is a pressing need to develop and implement ethically informed policies to reduce rising levels of resistance and thereby mitigate or avert future harms and injustices. We hope that the subsequent chapters of this volume will make a significant contribution to this important area of public health ethics.

References

Banerjee, A., A. Hollis, and T. Pogge. 2010. The Health Impact Fund: Incentives for improving access to medicines. *The Lancet* 375 (9709): 166–169.

Birgand, G., L.S.P. Moore, C. Bourigault, V. Vella, D. Lepelletier, A.H. Holmes, et al. 2016. Measures to eradicate multidrug-resistant organism outbreaks: How much do they cost? *Clinical Microbiology and Infection* 22 (2): 162.e1–162.e9.

Bronzwaer, S.L.A.M., O. Cars, U. Buchholz, S. Mölstad, W. Goettsch, I.K. Veldhuijzen, et al. 2002. The relationship between antimicrobial use and antimicrobial resistance in Europe. *Emerging Infectious Diseases* 8 (3): 278.

Bryce, A., A.D. Hay, I.F. Lane, H.V. Thornton, M. Wootton, and C. Costelloe. 2016. Global prevalence of antibiotic resistance in paediatric urinary tract infections caused by Escherichia coli and association with routine use of antibiotics in primary care: Systematic review and meta-analysis. *BMJ* 352: i939.

Carlet, J. 2012. The gut is the epicentre of antibiotic resistance. *Antimicrobial Resistance and Infection Control* 1 (1): 39.

Chang, Q., W. Wang, G. Regev-Yochay, M. Lipsitch, and W.P. Hanage. 2015. Antibiotics in agriculture and the risk to human health: How worried should we be? *Evolutionary Applications* 8 (3): 240–247.

Crum-Cianflone, N.F., E. Sullivan, and G. Ballon-Landa. 2015. Fecal microbiota transplantation and successful resolution of multidrug-resistant-organism colonization. *Journal of Clinical Microbiology* 53 (6): 1986–1989.

D'Costa, V.M., C.E. King, L. Kalan, M. Morar, W.W.L. Sung, C. Schwarz, et al. 2011. Antibiotic resistance is ancient. *Nature* 477 (7365): 457–461.

Dar, O.A., R. Hasan, J. Schlundt, S. Harbarth, G. Caleo, F.K. Dar, et al. 2016. Exploring the evidence base for national and regional policy interventions to combat resistance. *The Lancet* 387 (10015): 285–295.

Fairchild, A.L., and R. Bayer. 2004. Ethics and the conduct of public health surveillance. *Science* 303 (5658): 631–632.

Fleming A. 1945. *Penicillin. Nobel Prize lecture.*

Freedman, A., and S. Eppes. 2014. 1805 use of stool transplant to clear fecal colonization with Carbapenem-Resistant Enterobacteraciae (CRE): Proof of concept. *Open Forum Infectious Diseases* 1 (Suppl 1): S65. Oxford University Press.

Guardabassi, L., S. Schwarz, and D.H. Lloyd. 2004. Pet animals as reservoirs of antimicrobial-resistant bacteria Review. *Journal of Antimicrobial Chemotherapy* 54 (2): 321–332.

Guh, A.Y., S.N. Bulens, Y. Mu, J.T. Jacob, J. Reno, J. Scott, et al. 2015. Epidemiology of carbapenem-resistant enterobacteriaceae in 7 US communities, 2012–2013. *JAMA* 314 (14): 1479–1487.

Hardin, G. 2009. The tragedy of the commons. *Journal of Natural Resources Policy Research* 1 (3): 243–253.

Holmes, A.H., L.S.P. Moore, A. Sundsfjord, M. Steinbakk, S. Regmi, A. Karkey, et al. 2016. Understanding the mechanisms and drivers of antimicrobial resistance. *The Lancet* 387 (10014): 176–187.

Honigsbaum, M. 2016. Antibiotic antagonist: The curious career of René Dubos. *The Lancet* 387 (10014): 118–119.

Houston, S., and A. Houston. 2015. Screening and treating UN peacekeepers to prevent the introduction of Artemisinin-resistant malaria into Africa. *PLoS Medicine* 12 (5): e1001822.

Jamrozik, E, and M.J. Selgelid. 2019. Surveillance and control of asymptomatic carriers of drug-resistant bacteria. *Bioethics* 33 (7): 766–775.

Kennedy, K., and P. Collignon. 2010. Colonisation with Escherichia coli resistant to "critically important" antibiotics: A high risk for international travellers. *European Journal of Clinical Microbiology & Infectious diseases* 29 (12): 1501–1506.

Laxminarayan, R., P. Matsoso, S. Pant, C. Brower, J.-A. Røttingen, K. Klugman, et al. 2016. Access to effective antimicrobials: A worldwide challenge. *The Lancet* 387 (10014): 168–175.

Lerner, A., A. Adler, J. Abu-Hanna, S.C. Percia, M.K. Matalon, and Y. Carmeli. 2015. Spread of KPC-producing carbapenem-resistant Enterobacteriaceae: The importance of super-spreaders and rectal KPC concentration. *Clinical Microbiology and Infection* 21 (5): 470.e1–470.e7.

Llewelyn, M.J., J.M. Fitzpatrick, E. Darwin, C. Gorton, J. Paul, T.E.A. Peto, et al. 2017. The antibiotic course has had its day. *BMJ* 358: j3418.

Marshall, B.M., and S.B. Levy. 2011. Food animals and antimicrobials: Impacts on human health. *Clinical Microbiology Reviews* 24 (4): 718–733.

Martinez, J.L. 2009. Environmental pollution by antibiotics and by antibiotic resistance determinants. *Environmental Pollution* 157 (11): 2893–2902.

McEwen, S.A., F.J. Angulo, P.J. Collignon, and J.M. Conly. 2018. Unintended consequences associated with national-level restrictions on antimicrobial use in food-producing animals. *The Lancet Planetary Health* 2 (7): e279–ee82.

McKeown, T. 1976. *The role of medicine: Dream, mirage, or nemesis?* Princeton: Princeton University Press.

Morgan, D.J., R.P. Wenzel, and G. Bearman. 2017. Contact precautions for endemic MRSA and VRE: Time to retire legal mandates. *JAMA* 318 (4): 329–330.

Nasrin, D., P.J. Collignon, L. Roberts, E.J. Wilson, L.S. Pilotto, and R.M. Douglas. 2002. Effect of βlactam antibiotic use in children on pneumococcal resistance to penicillin: Prospective cohort study. *BMJ* 324 (7328): 28.

Neuzil, K.M., B.G. Mellen, P.F. Wright, E.F. Mitchel Jr., and M.R. Griffin. 2000. The effect of influenza on hospitalizations, outpatient visits, and courses of antibiotics in children. *New England Journal of Medicine* 342 (4): 225–231.

Norrby, S.R., C.E. Nord, and R. Finch. 2005. Lack of development of new antimicrobial drugs: A potential serious threat to public health. *The Lancet Infectious Diseases* 5 (2): 115–119.

O'Neill, J. 2015. Antimicrobial resistance: Tackling a crisis for the health and wealth of nations. In *The review on antimicrobial resistance.* Chaired by Jim O'Neill. December, 2014.

Östholm-Balkhed, Å., M. Tärnberg, M. Nilsson, L.E. Nilsson, H. Hanberger, A. Hällgren, et al. 2013. Travel-associated faecal colonization with ESBL-producing Enterobacteriaceae: Incidence and risk factors. *Journal of Antimicrobial Chemotherapy* 68 (9): 2144–2153.

Pruden, A., D.G.J. Larsson, A. Amézquita, P. Collignon, K.K. Brandt, D.W. Graham, et al. 2013. Management options for reducing the release of antibiotics and antibiotic resistance genes to the environment. *Environmental Health Perspectives* 121 (8): 878.

Rump, B., M. De Boer, R. Reis, M. Wassenberg, and J. Van Steenbergen. 2017. Signs of stigma and poor mental health among carriers of MRSA. *Journal of Hospital Infection* 95 (3): 268–274.

Safdar, N., and E.A. Bradley. 2008. The risk of infection after nasal colonization with Staphylococcus aureus. *The American Journal of Medicine* 121 (4): 310–315.

Schwaber, M.J., B. Lev, A. Israeli, E. Solter, G. Smollan, B. Rubinovitch, et al. 2011. Containment of a country-wide outbreak of carbapenem-resistant Klebsiella pneumoniae in Israeli hospitals via a nationally implemented intervention. *Clinical Infectious Diseases* 52 (7): 848–855.

Selgelid, M.J. 2008. Improving global health: Counting reasons why. *Developing World Bioethics* 8 (2): 115–125.

Smith, D.L., J. Dushoff, E.N. Perencevich, A.D. Harris, and S.A. Levin. 2004. Persistent colonization and the spread of antibiotic resistance in nosocomial pathogens: Resistance is a regional problem. *Proceedings of the National Academy of Sciences of the United States of America* 101 (10): 3709–3714.

Tacconelli, E., F. Sifakis, S. Harbarth, R. Schrijver, M. van Mourik, A. Voss, et al. 2017. Surveillance for control of antimicrobial resistance. *The Lancet Infectious Diseases.*

Tischendorf, J., R.A. de Avila, and N. Safdar. 2016. Risk of infection following colonization with carbapenem-resistant Enterobactericeae: A systematic review. *American Journal of Infection Control* 44 (5): 539–543.

Viens, A.M., C.M. Bensimon, and R.E.G. Upshur. 2009. Your liberty or your life: Reciprocity in the use of restrictive measures in contexts of contagion. *Journal of Bioethical Inquiry* 6 (2): 207–217.

White, N.J. 2017. Does antimalarial mass drug administration increase or decrease the risk of resistance? *The Lancet Infectious Diseases* 17 (1): e15–e20.

World Health Organisation. 2015. *Antibiotic resistance: Multi-country public awareness survey.* Geneva: WHO.

———. 2017a. *Ethics guidance for the implementation of the end TB strategy.* Geneva: WHO.

———. 2017b. *WHO guidelines on ethical issues in public health surveillance.* Geneva: WHO.

———. 2017c. *WHO guidelines on use of medically important antimicrobials in food-producing animals.* Geneva: WHO.

Chapter 2
Preventive Therapy for Multidrug Resistant Latent Tuberculosis Infection: An Ethical Imperative with Ethical Barriers to Implementation?

Binh Nguyen, Greg J. Fox, Paul H. Mason, and Justin T. Denholm

Abstract Multidrug resistant tuberculosis (MDR-TB) has a substantial impact on individuals and communities globally, including lengthy, expensive and burdensome therapy with high rates of treatment failure and death. Strategies to prevent disease are well established for those who acquire latent tuberculosis infection (LTBI) after exposure to drug susceptible TB (DS-TB). However, there has been limited research or programmatic experience regarding the prevention of MDR-TB. Accordingly, while global recommendations strongly emphasize the need to deliver LTBI therapy after TB exposure, most programs do not do so where MDR LTBI is identified.

The paucity of prospective randomized trial evidence for the effectiveness of MDR LTBI therapy, and concerns regarding its adverse effects, have been used to justify a reluctance to scale up programmatic interventions to prevent MDR-TB, or to participate in research evaluating such strategies. However, such a response fails to adequately balance potential risks of therapy with the substantial harms associated with inaction. Furthermore, the cost of inaction falls disproportionately on the

B. Nguyen
Woolcock Institute of Medical Research, University of Sydney, Sydney, NSW, Australia

G. J. Fox
Woolcock Institute of Medical Research, University of Sydney, Sydney, NSW, Australia

Faculty of Medicine, University of Sydney, Sydney, NSW, Australia

P. H. Mason
Woolcock Institute of Medical Research, University of Sydney, Sydney, NSW, Australia

School of Social Sciences, Monash University, Melbourne, VIC, Australia

J. T. Denholm (✉)
Department of Microbiology and Immunology, University of Melbourne, Melbourne, VIC, Australia

Victorian Tuberculosis Program at the Doherty Institute, Melbourne, VIC, Australia
e-mail: justin.denholm@mh.org.au

© The Author(s) 2020
E. Jamrozik, M. Selgelid (eds.), *Ethics and Drug Resistance: Collective Responsibility for Global Public Health*, Public Health Ethics Analysis 5,
https://doi.org/10.1007/978-3-030-27874-8_2

most vulnerable members of society, including children. Delays in implementing proven preventive strategies may also mask hidden programmatic concerns, particularly regarding the financial cost and other burdens of treating drug resistant infection. Reticence to engage with preventative therapy for MDR-TB, even in the absence of high-level evidence, may run counter to the best interests of individuals who have been exposed to MDR-TB.

This chapter will explore ethical tensions raised by expanding access to preventative therapies for MDR-TB, and consider how ethically optimal responses to this adverse condition may be evaluated. An ethical perspective on evidentiary burden will be addressed, emphasizing how MDR LTBI research may both offer, and be shaped by, paradigmatic insights into human research ethics more generally. Emerging research and illustrations from the authors programmatic engagement in Vietnam are offered as case examples, because social and community expectations and norms may challenge, or support, implementation of therapy for drug-resistant infection. Such circumstances prompt consideration of the broader questions of social impact, such as the potential for widespread preventive therapy to accelerate the development of antimicrobial resistance.

Keywords Bioethics · Infectious diseases · Public health · Social justice · Equality and human rights · Asian culture

2.1 Background

Multidrug resistant tuberculosis (MDR-TB) is a global pandemic disease, characterised as 'Ebola with wings' by public health experts due to its airborne transmission and significant patient mortality (Voelker 1998). With more than 450,000 cases reported in 2015, MDR-TB has become an established global health emergency (World Health Organization 2013a, 2015a, 2016; Hoang et al. 2015a). While treatment for drug-susceptible (DS)-TB typically involves multiple antibiotic tablets for a period of 6 months, standard MDR-TB treatment involves up to 2 years of both oral and injected antibiotics. These toxic regimens frequently cause nausea, liver damage and irreversible hearing loss, and may also require surgery or other invasive procedures (Fox et al. 2016; Torun et al. 2005). Despite the availability of these intensive regimens, MDR-TB therapy is successful in only around half of cases globally, with a high risk of treatment failure and death (Orenstein et al. 2009). In addition, the burden of MDR-TB extends far beyond its medical impact. The prolonged illness, and its associated treatment, has major financial implications for patients and their families - incurring significant financial and housing instability (Tanimura et al. 2014). In short, MDR-TB is costly, prolonged, and complex for both individuals and health care services (Keal et al. 2013).

As TB is a contagious infectious disease, those who live in close contact with affected individuals are themselves at a high risk of developing disease - around 10% of contacts with latent (asymptomatic) TB infection (LTBI) due to *M. tuberculosis* will subsequently develop disease (Marks et al. 2000). Current strategies to prevent the spread of MDR-TB mainly focus upon interrupting the transmission of infection, by identifying and treating patients with active disease (World Health Organization 2011, 2013b; Fox et al. 2013a). Such strategies are important for individuals with MDR-TB, but have limited impact on preventing disease in those around them who are likely to have already been infected prior to the diagnosis of the treated patient. Transmission studies indicate that close contacts of MDR-TB patients have an elevated risk of infection. Not only are contacts exposed directly to the recognised patient, but they also share socio-economic determinants of disease (World Health Organization 2014; Grandjean et al. 2011; Fox et al. 2017a). This confers an increased risk of exposure to other affected individuals, and susceptibility to developing this dangerous and burdensome disease (Fox et al. 2013b).

Given the high risk of developing active disease after infection with either DS- or MDR-TB, interventions to reduce risk among close contacts are of considerable clinical and public health importance (Trauer et al. 2016). In cases of DS-TB exposure, international guidelines recommend screening of exposed contacts, and a period of antibiotic therapy to prevent progression for those at highest risk of disease progression – such as children under 5 years of age, or those with HIV infection (World Health Organization 2015b). The effectiveness of preventative chemoprophylaxis for DS-TB is well established, including through randomized controlled trials in a variety of global settings (Lobue and Menzies 2010; Sterling et al. 2011). The importance of strategies for preventing MDR-TB has recently been summarised in national and international guidelines for the programmatic management of drug-resistant TB based on the risks of transmission, high morbidity and mortality, and the further development of drug resistance (World Health Organization 2014, 2015c; Ministry of Health 2015; European Centre for Disease Prevention and Control 2012). However, most guidelines have not recommended the routine use of targeted chemopreventive therapy for contacts of MDR-TB (Fox et al. 2017b), citing insufficient evidence of effectiveness in preventing disease, instead recommending programmatic surveillance of contacts. While such surveillance may lead to earlier identification of those developing disease, it fails to reduce the risk of disease among infected contacts, and hence does not prevent the consequent social and economic hardship caused by drug-resistant disease. Thus, current approaches to MDR-TB contacts recognise the high risk of disease among contacts, but do not offer routine use of preventive therapy in most settings, including Vietnam.

2.2 Discussion

2.2.1 Ongoing and Proposed Clinical Trials to Evaluate Antibiotic Therapy to Prevent Drug-Resistant Infection

In this section, we will introduce ongoing and planned clinical trials that aim to establish the effectiveness of preventive therapy for MDR-TB in order to highlight ethical issues arising from this research and reflect upon possible solutions. The first of these is the V-QUIN MDR Trial - a randomised placebo-controlled trial among infected contacts of patients with MDR-TB. Contacts are recruited from district tuberculosis clinics (DTUs) where MDR-TB treatment is delivered, throughout the Southeast Asian nation of Vietnam. The trial is underway within the Programmatic Management of Drug Resistant TB (PMDT) program at 132 clinics across 10 Provinces of the country. The primary aim of the VQUIN MDR Trial is to evaluate the effectiveness of levofloxacin (a fluoroquinolone antibiotic) in the prevention of active TB among household contacts of patients with MDR-TB with latent tuberculosis infection. Adult contacts will receive either levofloxacin or placebo daily for 6 months. In accordance with international recommendations, contacts will then be monitored for disease progression over a further 2 years to detect incident TB disease.

Vietnam has the twelfth highest TB burden in the world, and is listed among the top 27 countries with the highest burden of MDR-TB (WHO 2015). The country applies regimens recommended by WHO for the treatment of MDR-TB in Vietnam, including at least 19 months of treatment. This comprises, a minimum of 6 months of intensive phase treatment (including a second line injectable antibiotic, kanamycin or capreomycin) and 13 months of continuation phase therapy. Antibiotic treatment is provided free of charge for patients meeting the eligibility criteria for the program.

Despite having a nation-wide TB Program network and a structured, well-organized health service, and reporting impressive rates of treatment completion (WHO 2015), Vietnam still faces many challenges in implementing effective MDR-TB screening and treatment. Ongoing difficulties include a lack of communication and consistency in implementing policy changes, a lack of integration between general district hospitals and the National TB Program network, and limited resources. These health-system factors contribute to a significant gap between the estimated number of cases and the number of patients commencing treatment (Hoang et al. 2015b).

In the hope of gaining more data for evidence of effective regimens in MDR-TB contacts, two other clinical trials are planned, TB-CHAMP and PHOENIx (ACTG A5300) (Clayden et al. 2015; ACTG and IMPAACT Networks 2015). Similar to V-QUIN MDR Trial using levofloxacin and placebo for intervention and control groups, TB-CHAMP study is aimed to test levofloxacin as chemoprevention in children recruited from four clinical sites in South Africa. In this trial, children under five who are household contacts of MDR-TB patients are randomized into two groups taking levofloxacin or placebo every day for 6 months and followed-up for up to 2 years (Tuberculosis child multidrug-resistant preventive therapy: TB CHAMP trial 2016). PHOENIx (ACTG A5300) run by AIDS Clinical Trials Group (ACTG) and International Maternal Pediatric Adolescent AIDS Clinical Trials Network (IMPAACT) presents another different approach in which it aims to assess the efficacy of delamanid which is a new TB drug treatment for MDR-TB and XDR TB (Xavier and Lakshmanan 2014) in treating LTBI for high risk groups of household contacts of MDR-TB patients by comparing its daily use in 6 months with isoniazid preventive therapy then follow up study subjects in 2 years. The trial is planned to be conducted in Africa, South America and Asia (ACTG and IMPAACT Networks 2015).

Observational studies of LTBI treatment are limited. The three randomized controlled clinical trials on LTBI treatment in MDR-TB contacts that we outline here could well be the only research projects currently designed to address this research gap (Mitnick et al. 2016). In the following section, we will present issues and challenges in more detail, informed by our work in Vietnam and discussion with other research teams and experts in the field. Programmatic and research ethics are contextual, and specific settings may provide useful insights into the range of issues which consideration of MDR LTBI generates. In this chapter, we aim to draw on Vietnamese experience to illustrate and reflect on key ethical issues, which may be more broadly applicable in other contexts.

2.2.2 Challenges in the Use of Antibiotics as a Research Intervention in LTBI Treatment

2.2.2.1 How to Balance Between Uncertainties and Risk of Harm: A Common Issue in Public Health Practice

Medical decision-making for LTBI is replete with medical uncertainty. Diagnostic tools with the capacity to identify resistance patterns in LTBI are unavailable. Similarly, diagnostic tools capable of identifying which cases of LTBI will progress to TB disease are lacking. Developing appropriate responses with this diagnostic repertoire requires reflective engagement with medical uncertainty and the ethical challenges of emerging practices (Mason 2014a; Mason 2014b). An effective treatment for LTBI may well be the best method to stop progression to active disease (World Health Organization 2011; Hill et al. 2008), but puzzling questions persist about who should receive treatment for LTBI. The problem is complex enough for LTBI cases where the index patient has drug sensitive TB, but even more complicated for MDR LTBI, where high risk of progression is recognized but uncertainty regarding potential side effects of the drugs and the impact of preventative therapy persists. For LTBI cases who have been exposed to MDR-TB, there is little consensus to guide clinicians and programs towards the risk/benefit of MDR LTBI treatment. In contexts where existing programmatic guidelines recommend against MDR LTBI treatment on the basis of such uncertainty, additional difficulties are faced for researchers seeking to establish a study protocol, which may not align to traditional clinical views and practices in a study country site even if evidentiary equipoise is present.

In the balance between benefits and harms to receivers of MDR-TB preventive treatment, a default approach of 'surveillance' is frequently assumed in which "strict clinical observation and close monitoring for the development of active TB disease for at least two years is preferred over the provision of preventive treatment for contacts with MDR-TB cases" (World Health Organization 2015c). However, such a conclusion perhaps does not follow reasonably from the available evidence and resource availability in resource-limited settings. First, strict observation and close monitoring in order to early detect active TB disease requires many resources for TB screening, diagnosis and case follow-up at a community level over a long period of time. In resource-limited settings where there is a lack of healthcare staff, diagnostic tools and competency, this approach is likely neither feasible nor sustainable. Even if such an approach to surveillance were in place, some may argue in support of providing additional chemoprophylactic agents such as isoniazid (INH) for contacts exposed to MDR-TB patients, thus raising a question of ethical acceptability for study protocols which use a placebo as a control arm.

Based on the fact that INH or RIF has been proven to reduce the risk of developing active TB by at least 60%, and widely adopted in international and national guidelines for LTBI as a standard treatment for those exposed to TB (World Health Organization 2015c; Ministry of Health 2015), the argument in favour of giving INH for MDR-TB contacts involves three rationales. The first reason is that a proportion of contacts with LTBI will have been infected previously, or infected by another index patient, and so will benefit from the therapy. The second is that giving INH is recommended as the standard of care in many countries, and that it would be inappropriate to deprive people of that option because even if it is less effective, it still will be somewhat effective. The third reason is INH is a relatively safe drug, particularly for those under 35 years. For small children, on the balance of risks and benefits, using INH is preferred where the risk of untreated infection leading to disseminated disease is high. However, considering that equipoise may be present given the lack of evidence to guide LTBI treatment for MDR-TB contacts, randomized controlled trials using placebo with periodic follow-ups could be considered ethical for the following reasons: (a) INH is unlikely to be effective, given the most proximate exposure is with MDR-TB; (b) in settings where there is a high rate of INH resistant TB—for example, in Vietnam, at least 17% of all newly diagnosed TB is INH resistant (Nhung et al. 2015)—the effectiveness will be less even if the person has been infected by non-MDR-TB; (c) the current standard of care for adults is either passive case-finding or screening for prevalent TB, (d) INH has a degree of toxicity associated with its use (Denholm et al. 2014). If a drug is toxic, ineffective, and unlikely to benefit patients, then it cannot be ethically administered. Given the diagnostic and therapeutic tools available, serial follow-up by chest Xray is a preferable form of active intervention that exceeds the current standard of care, and will detect cases early enough to reduce serious consequences and allow referral for free treatment.

Recognising that MDR-TB preventive treatment is important, more efforts should be made in finding out effective therapies when there is no standard treatment, or when no proven effective and safe treatment is known to exist. This brings us to our second theme about assessing the effectiveness of LTBI treatment in MDR-TB contacts. Systematic reviews highlight the lack of data and limits of studies conducted in assessing the effectiveness of LTBI regimens available in relevant settings (Fraser et al. 2006; van der Werf et al. 2012). Small observational studies on LTBI regimens have reported promising results. In these studies, a combination of first and second-line TB medications, including a fluoroquinolone antibiotic, is prescribed to both adult and child contacts as preventive therapy from 2 to 12 months with post treatment follow-up mean periods less than 2 years comparing with placebo or no intervention. High rates of treatment completion, low rates of incident TB disease with low rates of adverse events and good tolerability are reported from the studies (Lobue and Menzies 2010; Seddon et al. 2013; Bamrah et al. 2010). Put together, this evidence suggests that LTBI treatment may be beneficial and further research on LTBI alternative therapies is necessary in producing more comprehensive data on both effectiveness and safety on LTBI therapies.

As the risk of developing MDR-TB is acknowledged to be high in recent contacts, the risk of serious adverse effects from any preventative treatment would need to outweigh the potential benefits in order to justify withholding treatment. While sensible responses to identified MDR-TB exposure should prioritise those at highest risk, the institution of a surveillance strategy instead of provision of treatment with potential adverse effects would preferentially advantage those at lowest risk of developing disease. On this basis, despite imperfect objective estimation of the risks and benefits of different therapeutic options, we would argue that programs may default towards provision of potentially effective therapy for those with MDR LTBI, provided reasonable measures to minimize harms (such as pharmacovigilance programs to identify adverse effects) are in place. In parallel, however, it is also contingent on clinicians and programs continuing to strengthen knowledge of both risks and benefits associated with MDR LTBI therapy, so that future care may be optimally targeted and individuals appropriately informed.

2.2.2.2 Development of Acquired Drug Resistance during Preventive Therapy

Research collaborators and infectious disease clinicians participating in the VQUIN MDR-TB Trial raised valid concerns about administering drugs with unknown effectiveness and the potential harm of selecting new strains of drug resistant TB. The use of antibiotics whose effectiveness has not been demonstrated raises potential problems for individuals and the community. An ineffective regimen may fail to protect the individual, or even result in acquired drug resistance if taken infrequently. Consequently, acquired drug resistance, particularly fluoroquinolone resistance, may then lead to transmission of more advanced strains of drug resistant TB. In the case of the VQUIN MDR-TB Trial, concern hinged on two issues. First, fluoroquinolones (such as levofloxacin) are a part of the backbone of the standard regimen used to treat MDR-TB. Treatment of active TB requires effective multidrug therapy, in order to prevent resistance. If active disease is not excluded prior to commencing preventive therapy, single drug therapy could lead to acquired drug resistance. This concern underpins the reluctance of some physicians to use single-antibiotic preventive therapy in these patients. Second, in settings where LTBI is not routinely managed, clinicians and patients report reluctance to prescribe treatment in the absence of symptoms. In high-burden, low-income settings such as Vietnam, LTBI is not perceived as a disease or condition requiring treatment. Such concerns are compounded by lack of programmatic experience with the use of LTBI therapies more generally in resource-limited settings, as clinicians have few opportunities to confidently exclude active disease and develop experience with preventative therapies.

Responding to perceptions of risk in relation to amplification of drug resistance is challenging, particularly when public health messaging regarding good antimicrobial stewardship emphasizes the need to avoid unnecessary antibiotic use to preserve drug effectiveness (Doron and Davidson 2011). Some evidence suggests that isoniazid monotherapy to treat drug-susceptible LTBI is unlikely to contribute to drug resistance (Balcells et al. 2006). In fact, modeling data suggest that preventive therapy may actually reduce the overall prevalence of resistance in a population by its secondary effect of reduced propagation among cases that would otherwise have been generated (Fox et al. 2015). One priority is clear: assisting clinicians and community members to appropriately distinguish active from latent TB is an important issue if preventive therapy is to be scaled up. This requires concerted efforts to provide education, clear guidelines and updated knowledge of TB management and research.

2.2.3 Challenge in Conducting Research Using Fluoroquinolone in Children

A significant challenge in conducting research using levofloxacin to treat LTBI in contacts of MDR-TB patients is that fluoroquinolones are generally contraindicated in children and growing adolescents by drug manufacturers and cautiously prescribed by doctors, due to theoretical concerns about the toxicity of the drug class (Goldman and Kearns 2011). In Vietnam, the study proposal to use levofloxacin to treat MDR LTBI in children under 15 years old has caused some controversy among local scientists and members of national ethics committee in considering appropriate assessment of risks and benefits to conducting the research on children. Scientifically, the debate mainly revolves around possible adverse effects of levofloxacin to tendon and musculoskeletal system of children. This concern originated from an association between fluoroquinolone use and irreversible joint cartilage defects in juvenile animals (Ingham et al. 1977) and reversible musculoskeletal events in children (Schaaf et al. 2006). Considering that children have an increased risk of both developing active disease and more severe disseminated forms of disease, and only indirect evidence for a link between fluoroquinolones and musculoskeletal harm exists, there is a strong argument for the inclusion of children in studies such as VQUIN. However, while strategies to deal with potential risks have been considered, including enhanced adverse effect monitoring of younger participants, at present the study has only been approved for adult contacts of MDR-TB.

2.2.4   Poor Understanding about LTBI and the Use of Diagnostic Tests

LTBI is generally poorly understood (Colson et al. 2010). Interviews with community members in a variety of contexts consistently identify confusion regarding the difference between latent and active TB, the extent of risk associated with infection, and the availability of treatment to prevent progression to disease (Wieland et al. 2012). A lack of knowledge about LTBI and its attendant risks of progression to disease presents difficulties for the introduction of preventative therapies, particularly where they involve prolonged treatment with some potential adverse effects.

Such uncertainty exists among healthcare workers as well as in the general community. In the research of LTBI therapy given to contacts of MDR-TB patients in Provincial Hospitals and District TB Units in Vietnam, we have identified particular issues including low level of knowledge of LTBI and a general lack of knowledge of treatment options for both active MDR-TB and suspected MDR LTBI. While TB is commonly known as a transmissible airborne infection, the pathways to active disease following exposure are still poorly understood by most healthcare staff. Uncertainties have been repeatedly expressed over the use of diagnostic tests for LTBI, the type and duration of LTBI treatment, presumably reflecting a lack of familiarity with preventative therapies in this context.

2.2.5   Challenges in Obtaining Informed Consent and Following-up Study Participants

The requirement to obtain informed consent is central to the ethical conduct of research involving human subjects. As clinical research has become more global, bringing with it a requirement to obtain consent in different places where many disparate values are held, the obtaining of valid consent increasingly raises a range of challenges. Consent practices in resource-limited settings may be impacted by time constraints for researchers to provide detailed research information, the lack of familiarity with medical research, traditionally paternalistic doctor-patient power dynamics and communication styles, involvement of family members and community members in the decision making process, conflict of duty of healthcare provider acting as researcher (Cheah and Parker 2014; Nguyen 2016), not to mention the significant cultural dimensions involved in adapting (and asserting) this research practice in settings where it is a foreign concept (Mason et al. 2017). Suffice to say that obtaining consent in research on LTBI treatment presents complex issues posed by its research context and clinical practice in addition to the complexities of explaining LTBI to a naïve audience.

Based on our experience in conducting research on LTBI therapy given to contacts of MDR-TB patients in Vietnam, we outline challenges generated by different understandings about research and preventive treatment. When the term "research" is translated directly into Vietnamese, it arouses commonly-held negative perceptions within the lay Vietnamese population. Healthcare workers also often express concerns that patients will decline to participate in research studies due to these negative associations, and prefer to avoid the term. An inability to refer to a study as being "research" may obstruct appropriate consent practices, since participants may not be aware of the experimental nature of the involved procedures. This difficulty is further complicated by differing understandings about LTBI and the appropriateness of preventive treatment.

The practical experience of obtaining consent in Vietnam for this study has been explored in series of research staff interviews, highlighting misperceptions about preventive therapy and the low priority people give to it. A staff member who worked on the community studies of the V-QUIN TB screening commented:

> "Our Vietnamese common perception is that "no disease, no treatment". Preventive treatment is not a priority to our people, especially people in farming regions. Their educational level is low so they don't think about preventive treatment. They only buy drugs when they get sick" (Study staff – N01).

With limited information about LTBI treatment in national guidelines, and specific recommendations against MDR LTBI treatment, challenges to effective consent and study participation are likely. In the traditionally hierarchical Vietnamese healthcare system, the existence of guidelines recommending against MDR LTBI treatment is a strong disincentive to both clinicians and community members' participation in research. Research into MDR LTBI treatment may be perceived as being in opposition to existing recommendations, despite the limited evidence on which they are based.

2.2.6   Stigmatization

The stigmatization of TB may impact upon the management of and research into LTBI. In research practice, we experienced that some MDR-TB patients are self stigmatised, or are isolated by their family members due to their disease status. Such patients may want to hide the disease from household contacts and neighbors (therefore prefer going to a private clinic to keep their privacy) and have poor medical adherence. Those most concerned about stigma may also avoid providing accurate information about their household contacts. Study staff in collaboration with local healthcare providers have to explore the information on household contacts by gradually building the relationship between patients and their care-giver, if there is any. This can be achieved through talking with them and providing them more information about TB, LTBI and preventive treatment in clinical assessments performed at hospitals or district tuberculosis clinics. The same stigma may drive healthy household contacts with LTBI to avoid sharing their infection status with neighbors and friends. As a result, these infected individuals do not want any home visit made by NTP staff/study staff, and do not want to be contacted and asked about the patient. This can create difficulties and challenges in approaching potential study subjects who may participate in the research. It also impedes monitoring of drug adherence and post-treatment following up. As expressed by a study staff about stigmatization in TB:

> "TB is a social disease. It makes participant feel ashamed of getting infection, and thus not wanting to talk about their disease status or being followed up for a long period of time in the study to have their health checked. They also avoid their neighbors knowing about the disease" (Study staff – S01).

2.3 Solutions for Identified Ethical Problems and Challenges

So far, we have attempted to map out a range of issues arising in the prevention of MDR-TB, current approaches to MDR LTBI therapy and in the context of doing a randomized controlled trial in LTBI with our experiences in Vietnam as an exemplary case of high burden TB and limited resource countries. What we will propose as solutions in this chapter accordingly will be general solutions to be considered in providing LTBI treatment and in doing MDR LTBI research. They are put forward in light of the nature of the problems, challenges occurring in the research practice and practical conditions in the setting.

2.3.1 Developing a Comprehensive LTBI Research Agenda

Clinical studies are crucial in advancing medical care. They are needed to produce systematic information on pathogenesis, clinical course, potential interventions and response to treatment. Especially in LTBI, it is important to address a key ethical challenge which is balancing between uncertainties and risk of harm involved in screening and treatment provision, and from that to derive lessons and possible ways to scale up interventions at the community level. Any new scientific information or breakthroughs can bring about alterations to current accounts of scientific and ethical considerations about existing interventions. Concerns about how best to apply systematic screening of LTBI and its related accounts of risk and benefit with beneficiaries (Degeling et al. 2017) is an example which can be expected to be resolved via promising outcome of studies on new LTBI diagnostic methods and indicators of reactivation risk (Dodd and Schlesinger 2017). From our experience drawn in the setting as outlined above, we argue that systematic research, especially randomized clinical trials in evaluating the effectiveness of preventive therapy for LTBI in the contacts of patients with MDR-TB are imperative to better inform clinical decisions, to benefit future infected people and the public in general.

2.3.2   Collaboration

Efforts toward the elimination of tuberculosis call for wide collaboration of various stakeholders including, but not limited to, clinicians in both public and private sectors, public health practitioners (Hauck and Panchal 2009; Taylor et al. 2005), researchers of all groups of expertise, funders, research communities and the public (Sablan 2009). This would serve as a ground for developing a comprehensive and balanced research agenda to inform clinical and public health practice, and to create sustainable research and public health platforms with long term facilities and community support. This type of collaboration can be conducted through the format of national and international collaborative research networks, for example, The Tuberculosis Network European Trials group (TBNET) (Giehl et al. 2012), The Australasian Clinical Tuberculosis Network (ACTnet), The Union World Conference on Lung Health, and the involvement of research community, public and mass media. The primary aim of such collaborations is to engage relevant stakeholders from the beginning of a research initiative to translating research findings into practice in community level on the basis of mutual understanding of a shared account of vision and mission and benefit generating for afflicted population.

While collaborations are recognized as crucial to successful research schemes, effective collaborations, in our view, need to be characterized by 'openness' in which collaborative partners see each other having supplementary role rather competitors, and a shared common interest that is to contribute to the knowledge of the disease for the public good through sharing data and samples. Contributing to the framework of global health, such an open form of collaboration is aimed to protect the common interest towards the global health and stress on the duty to protect affected members of the public and the public good.

2.3.3   Provide Education and Raise Community
Awareness of LTBI

Provide necessary scientific training to healthcare workers and study staff.

In order to help ensure success in LTBI treatment program and research initiatives, given the issues related to the lack of knowledge of current LTBI diagnostic methods and therapies which we have outlined above, training on scientific knowledge of LTBI and the role of research in producing systematic information to inform practice is necessary. It is importantly required for frontline health workers who directly provide healthcare to patients and communities where are most needed. These can include pharmacists, nurses, clinicians, and public health officers who serve in hospitals and community clinics. As the first and sometimes the only link to essential healthcare services in limited resource settings, this group of experts hold great potential to make a direct impact on individual and community health through supporting research activities, delivering good healthcare services and community consultation. Studies and our experience have shown that interaction and communication between health workers/study doctors and patients/study participants brings significant effect to the level of recruitment, medication adherence and treatment outcomes (Horne 1999; Dwamena et al. 2012; Sumartojo 1993). In this case of LBTI and TB control, it is knowledge of current approaches for TB control and prevention available locally and internationally, important health implications of LTBI, the need to provide LTBI treatment, and associated adverse events and their management that needs to be provided and updated to health workers and study staff to equip them in delivering healthcare services at the best level and standard.

Public education.

When research participants are subject to possible sources of vulnerability, the consent process with consideration of some additional protections should aim to protect the safety and the rights of research participants and benefits of afflicted population. This duty should be taken on by ethics committees, physician-researchers, local and international research institutions and other entities involved in the research in designing, reviewing and implementing research projects. At the same time, the consent process should be developed in a way that will enhance "voluntary actions that are intended to help or benefit another individual or group of individuals" (Eisenberg and Mussen 1989a) of research participants, family members and community.

In addressing the key issues in consent and the awareness of LTBI and the role of MDR LTBI treatment in the prevention of TB, we propose that educating the public about the nature of research and its necessity for the improvement and advancement of science in medicine as an overarching plan to make people understand more about the meaning of research and therefore to encourage their participation. For the purpose of raising community awareness of LTBI and the importance of doing more research on this subject, along with the general public education as mentioned above, it also requires providing education programs about clinical research and LTBI to the population and disseminating information about LTBI treatment in the forms of national guidelines with references to international guidelines, health promotional materials and community consultation. This form of education should be carried out on a long term and regular basis. Specific aims of these plans would be (1) to change negative attitude towards research, e.g. 'being a Guinea Pig'; (2) to let people have a better understanding of research, scientific methods in treatment and healthcare, and the role of research in support of medicine; and (3) to maintain trust and nurture pro-social behavior that will enhance "voluntary actions that are intended to help or benefit another individual or group of individuals" in society (Eisenberg and Mussen 1989b) for research and public health agenda.

2.3.4 Strengthen Communication Between Research Ethics Committees (RECs) and Researchers

Facilitating cooperation between RECs and scientists/researchers is necessary for mutual understanding and a rapid response in LTBI research context. One way to achieve this is to establish and maintain communication between researchers and RECs throughout research scheme. On the part of RECs, effective communication strategies with researchers will help the RECs to improve transparency in their decisions, understand practical challenges in doing research in the local context, develop expertise in a particular topic area, understand researchers' perspectives and make researchers mutually understand the challenges and duties of RECs. This mutual understanding would eventually place both RECs and researchers in an engaged process whereby research participants are better protected and research can be conducted effectively without being subject to unnecessary delays, misunderstanding and uncertainties.

2.4 Conclusion

In this chapter, we have outlined key issues in preventative therapies for MDR-TB and challenges in conducting research to assess potentially effective MDR LTBI therapies. We have raised possible solutions, derived from our own work in the Vietnamese setting in juxtaposition with broader ethical considerations. We have argued that preventive therapy for MDR-TB should be a high priority. There is a need for appropriately conducted systematic research to address the spread of MDR-TB in limited-resource settings globally. Engagement with local cultural norms and priorities is critical for ensuring that such research is conducted in both ethical and effective fashion.

References

ACTG and IMPAACT Networks. 2015. *A5300B/IMPAACT2003B: Protecting Households On Exposure to Newly Diagnosed Index Multidrug Resistant Tuberculosis Patients (PHOENIx).* Available from: http://impaactnetwork.org/studies/IMPAACT2003B.asp.

Balcells, M.E., et al. 2006. Isoniazid preventive therapy and risk for resistant tuberculosis. *Emerging Infectious Diseases* 12 (5): 744–751.

Bamrah, S., et al.. 2010. *An ounce of prevention: Treating MDR-TB contacts in a resource-limited setting*, in *International Union of Tuberculosis and Lung Disease Conference* 2010: Berlin, Germany, 11–15 November, 2010. p. FA-1- 656-14:S180.

Cheah, P.Y., and M. Parker. 2014. Consent and assent in paediatric research in low-income settings. *BMC Medical Ethics* 15 (1): 22.

Clayden, P., et al. 2015. *Pipeline report: HIV, hepatitis C virus (HCV), and tuberculosis (TB). Drugs, diagnostics, vaccines, preventive technologies, research towards a cure, and immune-based and gene therapies in development.* New York: Treatment Action Group.

Colson, P.W., et al. 2010. Tuberculosis knowledge, attitudes, and beliefs in foreign-born and US-born patients with latent tuberculosis infection. *Journal of Immigrant and Minority Health* 12 (6): 859–866.

Degeling, C., et al. 2017. Eliminating latent tuberculosis in low-burden settings: Are the principal beneficiaries to be disadvantaged groups or the broader population? *Journal of Medical Ethics* 43: 632–636.

Denholm, J.T., et al. 2014. Adverse effects of isoniazid preventative therapy for latent tuberculosis infection: A prospective cohort study. *Drug, Healthcare and Patient Safety* 6: 145–149.

Dodd, C.E., and L.S. Schlesinger. 2017. New concepts in understanding latent tuberculosis. *Current Opinion in Infectious Diseases* 30 (3): 316–321.

Doron, S., and L.E. Davidson. 2011. Antimicrobial stewardship. *Mayo Clinic Proceedings* 86 (11): 1113–1123.

Dwamena, F., et al. 2012. Interventions for providers to promote a patient-centred approach in clinical consultations. *Cochrane Database of Systematic Reviews* 12: CD003267.

Eisenberg, N., and P.H. Mussen. 1989a. *The roots of prosocial behavior in children*. Cambridge: Cambridge University Press.

———. 1989b. *The roots of prosocial behavior in children (Cambridge studies in social and emotional development)*, 208. Cambridge: Cambridge University Press.

European Centre for Disease Prevention and Control. 2012. *Management of contacts of MDR TB and XDR TB patients*. Stockholm: ECDC.

Fox, G.J., et al. 2013a. Contact investigation for tuberculosis: A systematic review and meta-analysis. *European Respiratory Journal* 41: 134–150.

———. 2013b. Contact investigation for tuberculosis: A systematic review and meta-analysis. *The European Respiratory Journal* 41 (1): 140–156.

Fox, G.J., O. Oxlade, and D. Menzies. 2015. *Fluoroquinolone Therapy for the Prevention of Multidrug-Resistant Tuberculosis in Contacts. A Cost-Effectiveness Analysis.* (1535–4970 (Electronic)).

Fox, G.J., et al. 2016. Surgery as an adjunctive treatment for multidrug-resistant tuberculosis: An individual patient data Metaanalysis. *Clinical Infectious Diseases* 62 (7): 887–895.

———. 2017a. Latent tuberculous infection in household contacts of multidrug-resistant and newly diagnosed tuberculosis. *The International Journal of Tuberculosis and Lung Disease* 21 (3): 297–302.

———. 2017b. Preventing the spread of multidrug-resistant tuberculosis and protecting contacts of infectious cases. *Clinical Microbiology and Infection* 23 (3): 147–153.

Fraser, A., et al. 2006. Drugs for preventing tuberculosis in people at risk of multiple-drug-resistant pulmonary tuberculosis. *Cochrane Database of Systematic Reviews* 2: CD005435.

Giehl, C., et al. 2012. TBNET – Collaborative research on tuberculosis in Europe. *European Journal of Microbiology & Immunology* 2 (4): 264–274.

Goldman, J.A., and G.L. Kearns. 2011. *Flouroquinolone Use in Paediatrics: Focus on Safety and Place in Therapy - A commissioned work for the Guidelines Group for the Revision of the "Guidance for National Tuberculosis Programmes on the Management of Tuberculosis in Children", World Health Organization, 30-31* March 2010, Geneva, Switzerland. 2011.

Grandjean, L., et al. 2011. Tuberculosis in household contacts of multidrug-resistant tuberculosis patients. *International Journal of Tuberculosis & Lung Disease* 15 (9): 1164–1169.

Hauck, F.R., and A.S. Panchal. 2009. Latent tuberculosis infection: The physician's role in tuberculosis control. *American Family Physician* 79 (10): 845–846.

Hill, L., et al. 2008. *Multi-level barriers to LTBI treatment: a research note.* (1557–1920 (Electronic)).

Hoang, T.T.T., et al. 2015a. Challenges in detection and treatment of multidrug resistant tuberculosis patients in Vietnam. *BMC Public Health* 15 (1): 980.

———. 2015b. Challenges in detection and treatment of multidrug resistant tuberculosis patients in Vietnam. *BMC Public Health* 15: 980.

Horne, R. 1999. Patients' beliefs about treatment: The hidden determinant of treatment outcome? *Journal of Psychosomatic Research* 47 (6): 491–495.

Ingham, B., et al. 1977. Arthropathy induced by antibacterial fused n-alkyl-4-pyrodoine-3-carboxylic acids. *Toxicology Letters* 1: 21–26.

Keal, J., et al. 2013. P97 multi-drug resistant tuberculosis: The first UK guideline for treatment monitoring. *Thorax* 68 (Suppl 3): A119.

Lobue, P., and D. Menzies. 2010. Treatment of latent tuberculosis infection: An update. *Respirology* 15 (4): 603–622.

Marks, G.B., et al. 2000. Incidence of tuberculosis among a cohort of tuberculin-positive refugees in Australia: Reappraising the estimates of risk. *American Journal of Respiratory & Critical Care Medicine* 162 (5): 1851–1854.

Mason, P.H. 2014a. The liminal body: Comment on "Privacy in the Context of 'Re-emergent' Infectious Diseases" by Justin T. Denholm and Ian H. Kerridge. *Journal of Bioethical Inquiry* 11 (4): 565–566.

———. 2014b. *The liminal body: Comment on "Privacy in the context of 're-emergent' infectious diseases" by Justin T. Denholm and Ian H. Kerridge.* (1176–7529 (Print)).

Mason, P.H., I. Kerridge, and W. Lipworth. 2017. The global in Global Health is not a given. *The American Journal of Tropical Medicine and Hygiene* 96 (4): 767–769.

Ministry of Health. 2015. *Guidelines for tuberculosis diagnosis, treatment and prevention.*

Mitnick, C.D., et al. 2016. Programmatic management of drug-resistant tuberculosis: An updated research agenda. *PLoS One* 11 (5).

Nguyen, T.C.B. 2016. The ethics of research in rapidly evolving epidemics: an international perspective. In *Kellogg.* Oxford: University of Oxford.

Nhung, N.V., et al. 2015. The fourth national anti-tuberculosis drug resistance survey in Viet Nam. *The International Journal of Tuberculosis and Lung Disease* 19 (6): 670–675.

Orenstein, E.W., et al. 2009. Treatment outcomes among patients with multidrug-resistant tuberculosis: Systematic review and meta-analysis. *The Lancet Infectious Diseases* 9 (3): 153–161.

Sablan, B. 2009. An update on primary care management for tuberculosis in children. *Current Opinion in Pediatrics* 21 (6): 801–804.

Schaaf, H., et al. 2006. Childhood drug-resistant tuberculosis in the Western Cape Province of South Africa. *Acta Paediatrica* 95 (5): 523–528.

Seddon, J.A., et al. 2013. Preventive therapy for child contacts of multidrug-resistant tuberculosis: A prospective cohort study. *Clinical Infectious Diseases* 57 (12): 1676–1684.

Sterling, T.R., et al. 2011. Three months of rifapentine and isoniazid for latent tuberculosis infection. *The New England Journal of Medicine* 365 (23): 2155–2166.

Sumartojo, E. 1993. When tuberculosis treatment fails. A social behavioral account of patient adherence. *The American Review of Respiratory Disease* 147 (5): 1311–1320.

Tanimura, T., et al. 2014. Financial burden for tuberculosis patients in low- and middle-income countries: A systematic review. *The European Respiratory Journal* 43 (6): 1763–1775.

Taylor, Z., C.M. Nolan, and H.M. Blumberg. 2005. Controlling tuberculosis in the United States. Recommendations from the American Thoracic Society, CDC, and the Infectious Diseases Society of America. *MMWR Recommendations and Reports* 54 (RR-12): 1–81.

Torun, T., et al. 2005. Side effects associated with the treatment of multidrug-resistant tuberculosis. *The International Journal of Tuberculosis and Lung Disease* 9 (12): 1373–1377.

Trauer, J.M., et al. 2016. Risk of active tuberculosis in the five years following infection … 15%? *Chest* 149 (2): 516–525.

Tuberculosis child multidrug-resistant preventive therapy: TB CHAMP trial. 2016. Available from: http://www.isrctn.com/ISRCTN92634082.

van der Werf, M.J., et al. 2012. Lack of evidence to support policy development for management of contacts of multidrug-resistant tuberculosis patients: Two systematic reviews. *The International Journal of Tuberculosis and Lung Disease* 16 (3): 288–296.

Voelker, R. 1998. Ebola with wings. *JAMA* 280 (14): 1216–1216.

WHO. 2015. *World Health Organization global tuberculosis report, 2015.* Geneva: WHO.

Wieland, M.L., et al. 2012. Perceptions of tuberculosis among immigrants and refugees at an adult education center: A community-based participatory research approach. *Journal of Immigrant and Minority Health* 14 (1): 14–22.

World Health Organization. 2011. *Guidelines for the programmatic management of drug-resistant tuberculosis – 2011 update*. Geneva: WHO.

———. 2013a. *Global tuberculosis report 2013*. Geneva: World Health Organization.

———. 2013b. *Systematic screening for active tuberculosis: principles and recommendations*. Geneva: WHO.

———. 2014. *Companion handbook to the WHO guidelines for the programmatic management of drug-resistant tuberculosis*. Geneva: WHO.

———. 2015a. *Global tuberculosis report 2015*. Geneva: World Health Organization.

———. 2015b. *Guidelines on the management of latent tuberculosis infection*. Geneva: WHO.

———. 2015c. *Guidelines on the management of latent tuberculosis infection*. Geneva: WHO. WHO/HTM/TB/2015.01. 38.

———. 2016. *Global tuberculosis report 2016*. Geneva: WHO.

Xavier, A.S., and M. Lakshmanan. 2014. Delamanid: A new armor in combating drug-resistant tuberculosis. *Journal of Pharmacology and Pharmacotherapeutics* 5 (3): 222–224.

Chapter 3
Providing Universal Access While Avoiding Antiretroviral Resistance: Ethical Tensions in HIV Treatment

Bridget Haire

Abstract The provision of effective antiretroviral therapy is an ethical imperative, and global access to antiretroviral drugs is an important aspect of this. The other less recognised aspect of effective HIV management is in ensuring that HIV does not become resistant to the drugs used in treatment (and increasingly also in prevention), as multi-drug resistant HIV poses a major threat to the sustainability of current responses to HIV control. In resource-constrained environments, the rapid scale up of access to life-saving anti-HIV treatment was achieved using a public health approach that standardised antiretroviral regimens, minimised laboratory monitoring, and devolved responsibilities from clinicians where necessary. In recent years demand for antiretroviral treatment has increased due to new understandings of the clinical importance of early treatment, but global investment has declined. Exponential growth of the population using antiretrovirals without careful monitoring increases the risk of significant antiretroviral drug resistance. In this chapter, I consider the example of single-drug interventions to prevent parent-to-child HIV transmission, and how the implementation of that strategy increased health risks for mothers. I argue that while global antiretroviral scale up must continue, laboratory monitoring at individual and national levels needs to improve to maintain treatment effectiveness, and protocols for moving people from failing regimens need to be strengthened.

Keywords Ethics · Public health · Infectious diseases

B. Haire (✉)
Kirby Institute, University of NSW, Sydney, Australia
e-mail: b.haire@unsw.edu.au

© The Author(s) 2020   37
E. Jamrozik, M. Selgelid (eds.), *Ethics and Drug Resistance: Collective Responsibility for Global Public Health*, Public Health Ethics Analysis 5,
https://doi.org/10.1007/978-3-030-27874-8_3

3.1 Introduction

With an estimated 36.9 million people living with HIV worldwide (WHO 2018), the provision of effective antiretroviral therapy (ART) is an ethical imperative. Ensuring that people with HIV can access ART is an important part of this, but another less recognised aspect is ensuring that the HIV does not become resistant to the drugs used in the treatment, as multi-drug resistant HIV poses a major threat to the sustainability of current responses to HIV control. There are, however, ethical tensions about striking the right balance between maximising ART access and minimising the risk of the emergence of drug resistant HIV. In resource-constrained environments these tensions are concerned with relative investment in drugs as compared with laboratory monitoring, and sometimes 'blaming' discourses regarding non-adherent people with HIV.

In 1996 it was established that antiretroviral drugs used in combination could control HIV replication ('viral load') in individuals and thus prevent HIV from destroying the CD4 cells that protect against infection (Arts and Hazuda 2012). Over time, the control of viral replication was shown to prevent people with HIV from becoming immune suppressed and developing the opportunistic infections associated with acquired immune deficiency syndrome (AIDS) (Moore and Chaisson 1999). ART could also reverse immune damage, to some extent (Kaufmann et al. 2005). Thus, combination ART could prevent AIDS and prolong, and perhaps save, the lives of people with HIV. Initial drug regimens however were often highly complex. The drugs were associated with serious side effects and were prohibitively expensive, as were the laboratory tests used to monitor both the impact of the drugs on viral load (viral load tests) and the degree of immune functionality (measured by CD4 cell tests). Hence there were significant barriers to access to these regimens for the majority of people living with HIV.

The stark injustice of life saving medication being inaccessible to the majority of those living with HIV, concentrated predominantly in low and middle income countries (LMIC) and especially in sub-Saharan Africa, was the catalyst for a worldwide treatment access movement that put pressure on governments, drug companies and non-government organisations to find solutions to make ART access more equitable. In response to the treatment access movement the World Health Organization launched its first 'public health' program for HIV treatment, an approach that aimed for universal access, starting with the initial goal of getting 3 million people onto ART by 2005, the '3 by 5 Programme' (WHO 2003a, b). This approach was supported and facilitated by the creation of two major donor organisations – The Global Fund for HIV, TB and Malaria (the Global Fund), and the (US) President's Emergency Program for AIDS Relief (PEPFAR) – that supplied drugs to LMIC (Smart 2006). Even with these donors, it was not feasible to mimic the highly individualised approach to ART prescription in affluent countries that relied heavily on expensive and resource-intensive laboratory monitoring. Accordingly, the program standardised both ART regimens (the drugs) and monitoring (including tests that measure the effect of the drugs on viral load and those that measured the impact of

HIV on the immune system). In some LMIC, nurse-practitioners performed roles usually reserved for doctors. Thus, the 'public health' approach was designed to facilitate ART delivery in resource-poor settings with limited health infrastructure, simplifying, standardising and devolving responsibilities as required (WHO 2003a, b). (Note, while the '3 by 5 Programme' was the initial iteration of this approach, the WHO has continued to champion universal access programmatically using a public health approach. For simplicity's sake, in this chapter I will treat the WHO universal access/public health approach as a strategy that started with '3 by 5' and has continued to this day, albeit with various adjustments and shifts in nomenclature.)

This chapter will first describe the mechanism for the development of HIV drug resistance and detail elements of the 'public health' approach to ART. It will then discuss how and why the 'public health' approach to ART in lower income countries was a reasonable trade-off that enabled widespread access to HIV treatment, despite increasing the likelihood of the development of drug resistance. In the medium to long term, however, strategies to avoid, contain and diagnose drug resistance will be critical to the sustained management of the epidemic, and to reaching the global goal of virtual elimination of new HIV infections. The tensions between a public health approach that focuses on access and an increasing emphasis on effective use of resources will be explored, using the history of mother-to-child prevention programs as a case study. The increasing use of ART-based prevention strategies for sexual transmission will also be considered. Finally consideration will be given to issues likely to impact on responses to drug resistant HIV in a context where guidelines recommend universal treatment, but access to uninterrupted drug supplies and high quality monitoring standards are limited.

3.2   Drug Resistant HIV

HIV is a highly mutable virus, meaning that as it replicates it makes errors and hence changes and evolves over time within the human host. These changes are random, though the principles of natural selection mean that over time and in the absence of treatment, changes that make the virus more replication competent will crowd out those that make it less fit.

Drug resistant HIV develops when a person is taking ART that is insufficient to fully suppress viral replication. This creates the conditions of selective pressure, so that viral copies that are less sensitive to the drugs have a competitive advantage over other viral copies, and begin to push the evolution of HIV in the particular body to become increasingly resistant. If the ART regimen is sufficient to fully suppress replication, this will not occur and hence resistance will not develop. Standard ART regimens will fully suppress viral replication if adherence is good, if no biological event disrupts individual drug absorption and metabolism, and unless the person was infected with a strain of HIV that had already evolved to be resistant to some or all of the drugs in the particular ART regimen.

In terms of the potential development of ART resistance, all ART drugs are not equal – the strength of antiretroviral action can differ, and some drugs are more vulnerable to the development of resistance. ART drugs work by targeting and blocking selected enzymes or proteins of HIV that enable it to replicate and infect cells. Changes to particular HIV regions that the ART drugs target make the drugs less effective. Some regions of HIV are more prone to mutation than others, and some drugs lose their effectiveness in the face of HIV evolution faster than others. For example, a single mutation reduces susceptibility to the drug nevirapine (this is called having a low genetic barrier to resistance) (Luber 2005). Drug resistance is also not a black and while phenomenon, in that a particular mutation or set of mutations may diminish susceptibility to particular drugs, but not completely stop working in terms of controlling HIV replication. For example, a person taking ART who has a detectable but low viral load may have some drug resistance, but may also be receiving clinical benefit from the drugs.

ART drugs are classified into classes that are determined by where and how they target HIV replication. While ART drugs within the designated classes work in slightly different ways, developing resistance to one drug in a class can also confer either resistance (or at least reduced sensitivity) to other drugs in the class. Accordingly, if a person has to change from one ART regimen to another, the second regimen needs to comprise drugs to which there is no likely pre-existing cross resistance to maximise the likely success. Further, in high income countries like Australia, when a person is first diagnosed with HIV, genetic resistance testing determines whether the person has been infected with a resistant strain of HIV that is less likely to respond optimally to standard treatment (if resistance is found, different regimens are used).

In summary, there are four main causes of drug resistance: acquiring a drug resistant strain; being prescribed a regimen that is incompletely suppressive (usually single or dual therapy combinations, or a triple combination that is too weak); non-adherence (failing to take pills as prescribed and on time); and taking drugs intermittently (where fluctuating blood levels of medication allow HIV replication under selective pressure).

3.3 A Word on HIV Monitoring

There are two highly significant tests used to monitor HIV disease progression and the effectiveness of ART in an individual: Tests that count CD4 cells per cubic ml of blood, and viral load tests. CD4 cells are immune cells that are targeted by HIV. A healthy person would have a CD4 cell count in excess of 500. Once a person's CD4 cell count drops to 200 or below s/he is at serious, imminent risk of HIV-related opportunistic infections. A person with HIV who has a CD4 cell count above 200 but below 500 is showing some HIV-related damage. There is a strong relationship between declines in CD4 cell count and disease progression, and in CD4 cell recovery and effective ART treatment (Mellors et al. 1997). CD4 counts are thus a useful tool in the management of HIV disease.

Viral load tests measure viral RNA in the peripheral blood. The higher the viral RNA number is, the greater the level of viral replication, and the greater the risk of disease progression (Mellors et al. 1996). 'Undetectable viral load' means that viral replication is suppressed below the level of detection on the test used (tests have become much more sensitive since they were first available). An undetectable viral load is the individual goal of ART treatment. At the strategic level, having 90% of people who know their HIV status on ART, and 90% of those with an undetectable viral load is a current global goal (UNAIDS 2017).

3.4   Key Elements of the 'Public Health' Approach to HIV

To maximise access to life-saving ART, WHO prioritised getting ART drugs into the bodies of people with HIV, and maximising the capacity to deliver ART by processes of standardisation, simplification and, where necessary, syndromic management rather than laboratory monitoring. The core underlying principles were identified as urgency, equity and sustainability (Macklin 2004).

Standardised first line and second line ART regimens were a critical aspect of the approach. In determining the drugs used in these regimens, a balance had to be struck between user-friendliness (pill burden, dosing routine, side effect profile), effectiveness (probability of viral load suppression), and cost (WHO 2003a). Further, the second line regimen had to be effective against the ART resistance likely to develop (for whatever reason) to the drugs in the first line regimen. (Third line regimens are also detailed in guidelines.) These standardised regimens were presented as algorithms that were flexible enough to allow for the different availability of certain drugs in particular countries due to purchasing arrangements, and to allow some substitutions for people who had different health needs, such as those who were pregnant, had other infections like TB or viral hepatitis. The standard algorithms for first and second line treatment have changed since the first iteration of the WHO universal access program, as further information about drug efficacy emerged (for example, a first line regimen recommended in 2003 was found to be insufficient of suppress very high viral loads found in people initiating treatment in advanced disease), and alternative drugs became available to replace ART associated with significant toxicities (Gulick et al. 2004). Both inadequate viral suppression and high levels of side effects are significant for the development of resistance – incomplete suppression create the conditions for the emergence of resistance, and ART toxicities can reduce adherence, also potentially resulting in incomplete suppression.

The issue of how to prioritise people for treatment access was another critical element of the approach. Utility, equity and concern for the worst-off (the principle of maximin) were key considerations (Macklin 2004). Initial WHO guidance for resource constrained settings recommended that ART should be commenced in those at high and proximate risk of clinical disease progression and/or death (those diagnosed with Stage IV HIV disease regardless of CD4 cell count; those with CD4 cell counts below 200 regardless of disease staging; those with stage III HIV disease and CD4 counts below 35) (WHO 2003a). Targeting ART at those at experiencing or at imminent risk of serious HIV disease was acceptable in a moral sense as it prioritised treatment for those with greatest clinical need, and it also greatly reduced the pool of people eligible for treatment, as only a fraction of those living with HIV would require treatment under such guidance. The downside to this approach was that many people tested and found HIV positive would not be eligible for treatment, increasing the potential of significant loss to follow-up (Rosen and Fox 2011). It is important to note that while WHO guidelines play an important normative role in shaping standards, they are in no sense binding, and the expectation of their use is that countries draw from them and adapt them according to relevant contextual factors (WHOF 2006).

Laboratory monitoring was simplified in WHO guidance documents, with CD4 cell tests recommended but not mandatory both at baseline and for subsequent monitoring of treatment efficacy at 6–12 month intervals. Haemoglobin testing would occur depending on the drug regimen (it was required if AZT was included.) Other blood work such as white blood cell and liver function tests were to be ordered as determined by clinical symptoms, and viral load testing – the gold standard for measuring ART response – not even mentioned. Developing guidelines for the surveillance of drug resistance was however one of the itemised strategic actions of the initial '3 by 5 Programme' (WHO 2003a, b). The low level of laboratory monitoring was contentious, given that the longer a person stays on a failing regimen, the greater the opportunity both for the development of resistance and the loss of clinical benefit. Accordingly, a randomised control trial (RCT) investigated whether (and to what extent) ART could be delivered safely without laboratory monitoring (the DART study). This study, first reported in 2009 and published in 2010, found that differences in outcomes occurred after the second year of treatment. Investigators argued that this justified only introducing CD4 cell tests at 24 months, and prioritising spending on ART drugs rather than resource-intensive monitoring (DART Trial Team 2010). Of note, this study only looked at clinical outcomes, in people who already had advanced HIV disease, not at the drug resistance outcomes.

3.5   Changes in Eligibility for ART

At the inception of the global universal access program, there was no clear scientific evidence as to the best time to commence ART, except that it clearly prolonged life for those who commenced with low CD4 counts (below 200). From the mid 1990s, many argued on the basis of modelling from other infectious diseases that HIV replication should be suppressed as early and as completely as possible. 'Hit hard, hit early', was the catch cry (Ho 1995). Mitigating against this was the fact that serious HIV disease did not usually occur until 6–10 years after infection, and the problem that ART regimens in the mid to late 1990s were hard to take (with high pill burdens, specific food and or fasting requirements for dosing particular drugs, and high rates of side effects). Of course the rapacious cost of ART was also significant – for example in 1997, 6 months use of the drug AZT was costed at $US 800 – it is now an estimated $US 237 (Pharmacychecker 2017).

From the late 90s, a series of RCTs clarified issues pertaining to treatment commencement, indicating that better outcomes were achieved with earlier treatment initiation. These included CIPRA-HT001, a study in Haiti that showed that people who started ART with CD4 counts higher than 200 had a significantly reduced risk of death (Severe et al. 2010). This result was underscored by findings from the SMART Study, which was designed to test whether structured ART treatment interruptions were safe (people with HIV took treatment interruptions or 'holidays' for a variety of reasons, including relief from side effects). The SMART Study found that not only did treatment interruptions increase the risk of disease progression at any CD4 level, but that the greater time spent with a CD4 count below 350, the greater the risk of HIV disease progression and death (Strategies for Management of Antiretroviral Therapy Study Group 2008). These findings together were the catalyst for a revision of WHO guidelines for resource-constrained environments to recommend ART commencement at the higher CD4 level of 350 (WHO 2009). Two further studies have been critical in pushing WHO ART guidelines toward recommending treatment at higher levels – the HPTN 052 Study, which showed that early ART treatment reduced onward HIV transmission to sexual partners by 96 % (Cohen et al. 2011), and then the START Study, reported in 2015, which showed a clear clinical benefit from early treatment initiation (at CD4 levels of 500 and above), with people who initiated treatment immediately showing significant benefit compared with those who did not initiate ART until their CD4 counts had dropped to 350 or below (The Insight START Study Group 2015). This finding resulted in a revision of WHO guidelines, such that immediate ART therapy is now recommended for any person diagnosed with HIV, regardless of CD4 cell count.

This change to WHO ART guidelines is based in robust evidence about the beneficial impact of immediate ART on health, and where taken up in national guidelines, may reduce the numbers of people diagnosed with HIV who get lost to follow -up due to being deemed ineligible for immediate ART. It also has some problematic implications. Firstly, it increases exponentially the estimates of people living with HIV who require ART, which may increase the likelihood of rationing taking place where supplies are inadequate. Secondly, notwithstanding the fact that some resource-poor countries may continue to align their national guidelines to earlier iterations of the WHO guidance, the expansion of eligibility for ART puts greater demand on ART stocks, and increases the risks of stock-outs, which could have an impact on the development of drug resistance. Even if drug supplies were perfect, human adherence will not be, so the logical result of millions more people taking lifelong ART over many decades must be an increase in drug resistance. Thirdly, increasing the number of people eligible for treatment may increase pressure at country level to continue to prioritise wider access to ART drugs at the expense of improving monitoring, exponentially expanding the risk that large scale poorly monitored ART use poses with regard to development and transmission of drug resistant ART (at higher CD4 cell counts, the risk of death is far more remote that at lower one, so the risks of drug resistance are not so obviously balanced by the over-whelming benefit of prevention of illness and death). Fourthly, there is a major para-digm shift in terms of health communication in shifting ART treatment from something required for people with ill or declining health status to treating people who are well, and requiring this treatment to be maintained over a lifespan. Thus, well-targeted health promotion, innovative and effective adherence support, improved monitoring to detect signs of resistance early, and affordable, tolerable second, third and fourth line (or rescue) regimens all need to be integrated into this ambitious expansion of ART eligibility.

3.6   ART in Pregnancy in LMIC: A Case Study

"Once started, antiretroviral therapy is for life...." states the introduction to the '3 by 5 Programme' (WHO 2003a, b). The reason that ART treatment, once started, is to be continued for life is to maximise health by maintaining a supressed HIV viral load, and preventing the development of drug resistance that could occur if ART is started and stopped (as the various drugs in combination ART regimens have differ-ent half-lives, it is complex to stop a regimen in a way that ensures that there is no lingering, suboptimal antiviral activity that could provide the conditions for drug resistance to develop). A brief look at the history of interventions aimed at prevent-ing vertical infection (from mother-to-child), however, shows that the maxim of ART therapy being for life has frequently not been the case with regard to provi-sions of ART to pregnant women. The case study below, I suggest, is an example of how the prevention benefit of ART was initially prioritised above the healthcare of pregnant women with HIV.

The prevention of HIV transmission from mother to child (PMTCT) was the first instance of effective biomedical HIV prevention, achieved in 1994 by the ACTG 076 study (Connor et al. 1994). Famously, this breakthrough spawned a series of placebo-controlled trials in LMIC using less intensive, cheaper regimens (see Wade et al. 1998). This chapter will not delve into the ethical quagmire surrounding the use of placebos despite an effective preventative intervention having been estab-lished, as this has been exhaustively covered elsewhere (see Macklin 2001). Instead, this chapter will recount and analyse what occurred programmatically in PMTCT following the results of the HIVNET 012 Trial, which established that a single dose of the ART drug nevirapine to both mother and infant reduced transmission in the first 14–16 weeks of life by nearly 50% in a breastfeeding population (Guay et al. 1999). This regimen, while not as effective as the ACTG 076 regimen, was consid-ered to be a major breakthrough for LMIC. Unlike ACTG 076, it did not require Caesarian section, it did not require treatment uptake in mid pregnancy, it did not require intravenous ART during delivery, and it appeared to work in breastfeeding populations. All of these factors made it significantly more feasible to implement at scale in LMIC. (Of note, when the Food and Drug Administration looked at the data from this trial, it found serious anomalies in the conduct of the trial. Accordingly questions remain regarding the validity of this study's results (Cohen 2004; Institute of Medicine Committee 2005).

One of the issues with the early WHO guidelines that recommended ART treatment only for people with signs of immune damage and/or HIV disease was the problem it created regarding pregnant women with HIV. Despite pregnant women in high income countries achieving excellent health and prevention outcomes with early ART, it meant that pregnant women in LMIC without clinical signs of HIV disease and with CD4 counts above levels recommended for treatment were deemed not to require ART for their own health, despite knowledge that ART would reduce transmission to infants. This created an artificial distinction between the treatment use of ART and prophylactic use in pregnant women – prophylactic use only was recommended. In other words, a pregnant woman with HIV but in otherwise good health would get access to ART to protect her infant, but this would be short-term.

Despite its appeal in terms of feasibility, the nevirapine single-dose[1] regimen had a very specific problem. Although a potent ART drug, nevirapine as noted earlier has a low genetic barrier to resistance, and a long half-life. Thus the drug persists in the blood stream for a long time after a dose, but at concentrations insufficient to fully suppress viraemia. This creates a good environment for the development of drug resistance, and a single dose is enough for this to occur. Exposing women and their infants to this regimen could thus potentiate the development of drug resistance, which might reduce the efficacy of the intervention for any subsequent pregnancies, and limit treatment options for the mother and potentially for the infant, should the infant be HIV infected despite the use of nevirapine prophylaxis.

The publication of the long-term results HIVNET 012 trial was accompanied by a controversial commentary by Karen Palmore Beckerman (2003) that argued that the implementation of single-dose nevirapine would leave 'between 20–100% of women who received prophylaxis resistant to [the class of drugs to which nevirapine belongs]'. While the claim that up to 100% would develop resistance was hyperbolic (Wilfert 2003), the issues she raised were prescient. Beckerman argued that pregnant women with HIV should be treated with effective ART regimens both for their own sake and to enable their survival – citing the devastating example that three women who participated in HIVNET 012 died of AIDS within 56 days of delivery. She suggested that adding a combination ART 'tail' to the nevirapine regimen could reduce the likelihood of resistance (a suggestion later taken up in WHO guidelines in 2010), and that the primary focus be on sustained and sustainable ART treatment.

Beckerman's (2003) article built on an earlier commentary of hers (Beckerman 2002) where – again in response to the publication of short-course trial to reduce mother-to-child transmission – she argued that the risks of programs using single or

---

[1] This regimen is referred to in the literature as 'single dose' though it in fact involves two doses – one to mother and one to infant. For simplicity's sake I'm adopting the common usage.

dual ART drugs were increasing as the prospect of good treatment access in LMIC grew, given that women who had used short-course ART would be more likely to experience treatment failure due to drug resistance. With limited treatment options available, failing first line ART options would increase the risk of the mother dying, which would in turn increase risk for surviving children, Beckerman argued.

There is a considerable literature on the clinical implications of nevirapine resistance, with much focus on whether or not it 'fades' over time such that subsequent introduction of an ART regimen containing nevirapine might still have clinical utility as either treatment or prophylaxis (Johnson et al. 2005; Lockman et al. 2007, McConnell et al. 2007; Stringer et al. 2010). Regardless of whether or not resistance 'fades' to a level that would result in nevirapine-based combination being effective in women exposed to single-dose PMTCT, it is clear that, as others have noted, the prevention of transmission to infants took precedence over the health security of women with HIV, at least for a time when these single-dose nevirapine programs were widespread (Eyakuze et al. 2008). As noted in a meta-analysis, these programs ultimately resulted in a high burden of drug resistance in women and children, with significant potential to contribute to increase failure of first line ART therapy (Arrivé et al. 2007; Samuel et al. 2016).

From 2003, WHO guidance recognised that pregnant women should not be prescribed ART regimens likely to result in drug resistance, and had specifically provided a warning about the single-dose nevirapine regimen. By 2012 the policy of women cycling on and off ART with successive pregnancies was superseded by 'Option B+'[2] – a recommendation that pregnant women with HIV should be prescribed fully suppressive ART regimens, and kept on them for life, regardless of CD4 cell count or HIV disease stage. While Option B+ has not been fully implemented in all LMIC, it is the only PMTCT option that fully applies the 'rules' of effective ART treatment – that regimens should be fully suppressive, and taken for life. Ironically, there have been reports of significant loss to follow up with these programs, perhaps because women with HIV had being educated to accept intermittent therapy unless they themselves were in poor health (Tenthani et al. 2014)

Reflecting on the history of PMTCT, there are tensions between the prevention imperative, emerging knowledge about optimal timing of ART, the status of women's health, equity, and the constraints of the early stages of implementation of the morally praiseworthy but hugely ambitious goal of universal access. In the late 90s and early 2000s, the nevirapine single-dose regimen had appeal in terms of its simplicity, its low drug burden, and the fact that it could be implemented during

---

[2] Option B+ was added to two previous options, option A and Option B. Briefly, Under Option A, women received ARV prophylaxis prenatally and during delivery, along with an antiretroviral postpartum "tail" regimen to reduce risk of drug resistance, and their infants received postpartum antiretroviral prophylaxis throughout the duration of breastfeeding. With Option B, all pregnant and lactating women with HIV were offered ARV – beginning in the antenatal period and continuing throughout the duration of breastfeeding. At the end of breastfeeding those women deemed to not yet require ARV for their own health would discontinue the prophylaxis and continue to monitor their CD4 count, eventually re-starting ARV when their CD4 cell count fell below 350 cells.

delivery. In short, it had been shown to be better than nothing at a time when 'nothing' was still seen in some quarters as an acceptable comparator. While the shift away from the single dose regimen happened relatively quickly (at least in terms of normative guidance, if not in practice), its impacts are likely to live on in the form of suboptimal response to first line ART regimens and transmitted drug resistance (Kébé et al. 2014).

It took 6 years from the initial reporting of the SMART study – which showed that starting and stopping ART was detrimental – for PMTCT guidance to recommend life-long treatment for women who initiated ART due to pregnancy. On the one hand, this seems shocking. On the other hand, due to PMTCT programs even with their limitations, women in LMIC tended to access ART earlier than men, and hence have had lower risk of death from HIV disease than men (Beckham et al. 2016; Taylor-Smith et al. 2010), so further strengthening care access for women may have seemed inequitable in some contexts. It could be argued that the history of PMTCT in LMIC demonstrates a programmatic emphasis on prevention without adequate regard for the potential development of drug resistance, and subsequent drug failure in mothers and pregnant women with HIV. It is probably more reasonable, however to say that the complexity of implementing best practice PMTCT in the midst of programmatic scale up of ART access in a context where knowledge about ART changed significantly made compromise and incremental steps forward hard to avoid, if not inevitable. Nevertheless, exposure to sub-optimal therapy through PMCTC programs has left a legacy of drug resistance in some settings (Rowley et al. 2016; Cambiano et al. 2013).

Now that lifelong ART from the point of diagnosis is the WHO recommendation for everyone, exponentially more people will be accessing ART prior to any clinical indications of HIV. That number will be expected to keep increasing, as ART access should enable people with HIV to live normal lifespans. In this context the risk of the development of drug resistance must necessarily increase, given the difficulties of maintaining good-enough adherence for large populations over long periods of time. While the low-level clinical monitoring approach to HIV disease had an important role in enabling the 'public health' roll out of ART in LMIC, this approach would not be sustainable in populations accessing treatment much earlier in HIV disease.

3.7 Addressing ART Resistance

Addressing the risk of drug resistance requires strategic action at many levels. Action is required at the community and local levels, to develop and sustain programs that support individual adherence and that work to address systemic problems that can affect adherence (by improving systems of clinic appointments and transport and drug access, for example). Scaling up monitoring and investing in viral load testing will become critical so that people who are on failing regimens – for whatever reason – are identified quickly and either supported to improve adherence or switched to second line therapy. Development of viral load and CD4 testing technologies that are better adapted to use in remote and/or resource constrained settings, and which do not require highly trained laboratory staff, should obviously be prioritised to reduce the dependence of remote and under-resourced clinics on distant laboratory services. Research and development into new effective ART that are well tolerated and are effective against current common resistance patterns in HIV – and new modes of delivery, such as periodic injectables – are of course highly desirable.

While some commentators have suggested withdrawing or withholding ART from non-adherent people rather than providing practical support and transitioning them as necessary to second line ART, this perspective positions poor adherence as an individual (and moral) issue (Chawana & Bogaert 2011). A perspective of adherence that is informed by public health and human rights, on the other hand, recognises that while adherence is an individual behaviour, it is highly determined by structural factors and social context, which can and should be modified to make adherence as simple and as socially acceptable and socially desirable as possible. There is evidence that people in certain demographics (such as young people) may have more trouble with adherence (Haire 2015). Withholding or withdrawing ART on the basis of poor adherence from young people, for example, would not only remove the benefits of early treatment from the demographic with the most to gain, it would also remove the prevention benefit from a population group highly likely to be sexually active. Accordingly, it would seriously undermine global targets for universal access. Recognising, working with, and seeking to ameliorate adherence problems in the social groups that experience them, while advocating for ART that provide different delivery options such as periodic injection, makes sense from a public health and human rights perspective. Universal access goals aim for 90% of people with HIV knowing their status, 90% of those being on ART and 90% of those having undetectable viral loads by 2020 (UNAIDS 2017). Using 2017 figures, this would require more than 14 million more people to be on ART within 2 years – considerable investment in supportive and responsive programming to achieve, support and sustain adherence must be factored into such ambitious plans.

3.8 Biomedical Prevention and Drug Resistance

Biomedical prevention is the use of antiretroviral drugs to prevent HIV transmission. It includes ART use after exposure to prevent transmission (post-exposure prophylaxis or PEP), ART use prior to exposure in HIV negative people at high risk of HIV acquisition to prevent transmission (pre-exposure prophylaxis or PrEP), and ART use in HIV positive people to fully supress HIV replication, preventing onward transmission to sexual/injecting partners ('treatment-as-prevention'). The latter two strategies are highly effective and well supported by both RCT and observational data (data on PEP is less robust as it has not been tested in an RCT). PMTCT programs contain elements of two biomedical approaches: maternal viral load is reduced to reduce infectiousness, like treatment as prevention, and infants receive

ART to prevent or abort HIV acquisition, similar to PrEP and PEP. The preventative aspect of early ART treatment is one aspect (in addition to the clinical benefit for the person with HIV) that makes implementation of ART programs in people immediately after diagnosis attractive to governments and donors – it has the potential to reduce onward infection. Similarly, while investment in PrEP programs has been slow, there is a steady increase in availability (AVAC n.d.).

In adherent participants, both PrEP and treatment -as -prevention have shown close to 100% protection in trials (Grant et al. 2014; Anderson et al. 2012; Bavinton et al. 2018; Rodgers et al. 2016, 2018; Cohen et al. 2011). There has however been at least one verified example of HIV transmission occurring in an adherent PrEP-taker, due to being exposed to HIV that was resistant to both the drugs used in the PrEP combination (Knox et al. 2016). With regard to treatment -as -prevention, having a lowered viral pool in a population would be expected to translate into fewer transmissions at population level. To have confidence in treatment -as -prevention at an individual level, however, the sexual partner with HIV needs to be confident that his/her viral load is sustained at an undetectable level, and that requires access to viral load tests at regular intervals. A combination of PrEP and treatment-as-prevention is being trialled in LMIC in serodiscordant couples, where the HIV positive partner is initiated on ART and the negative partner on PrEP, until such time as the positive partner has a sustained undetectable viral load. For this to succeed programmatically, viral load monitoring is required for the ongoing protection of the negative partner (should the positive partner be non-adherent, the negative partner would run the risk of acquiring HIV, and possibly drug resistant HIV), and to maintain the efficacy of the regimen for the positive partner. In this new era of biomedical prevention and all people with HIV being eligible for ART, the need to implement viral load monitoring at scale is increasingly being recognised (WHO 2016).

3.9   Monitoring Drug Resistance Beyond the Individual

In high income countries, when people are diagnosed with HIV, a blood sample is sent to a laboratory for genotypic testing – a form of analysis that provides information of whether the person has been infected with a strain of HIV that is resistant (or less susceptible) to particular ART drugs. The person is then prescribed an appropriate regimen that will be effective against the person's HIV.[3]

In LMIC surveillance of HIV drug resistance includes country specific surveys of HIV genotypes to determine incidence of transmitted drug resistance in

---

[3] Genotyping HIV also has some risks in high income countries, such as the potential to use these data to track 'infection trees' – map pathways of infection. This is ethically problematic in terms of maintaining an enabling environment, as it uses information gleaned from people who test positive for HIV intended to maximise their health outcomes for a purpose that could potentially lead to criminalisation.

populations, through a WHO network of designated laboratories (Bertagnolio et al. 2008). In LMIC this information is directed towards making country or region-wide decisions about first line treatment, and other programs such as the drugs used in post-exposure prophylaxis. In addition, early warning indicators provide information about critical aspects of whether/how people are accessing ART. These include on-time pill pickup; retention in care at 12 months following ART initiation; whether pharmacies experience stock-outs; prescribing practices (whether regimens prescribed meet national or international guidance regarding adequate viral suppression); and, in settings where viral load testing is implemented, viral load suppression at 12 months following ART initiation. These early warning indicators are well designed to signal structural problems, like interruptions in drug supply and poor prescriber compliance with guidelines, and access issues such as whether drug collection systems are 'good enough' (reliable and convenient) for people to be able to pick up pills on time. In 2017, WHO reported 'brisk' implementation of these systems, with 26 countries having completed or currently completing surveys, and 14 having reported data (WHO 2017). Of note, in countries or regions where drug stock outs are frequent, the benefit of putting people with high CD4 counts onto ART would need to be weighed against the individual and public health concern of the development of resistance, should ART be repeatedly started and stopped (for people with CD4 counts lowered by HIV, the clinical benefit of the ART is likely to outweigh the risk). Ideally, of course, the stock out issues should be addressed effectively.

3.10   Conclusion

Although hope and hype about vaccines and cures for HIV circulate, ART is currently the most effective tool for the elimination of HIV as a public health threat, both in terms of treatment and prevention (in combination with or as an adjunct to condoms). Preventing the emergence of drug resistance at levels that compromises treatment efficacy and ART based prevention is crucial.

With PMTCT programs, the emphasis on the simplest and cheapest regimens, maintained well after the implications for drug resistance were apparent, has already compromised optimal response to ART for too many women and children. This cannot be repeated in the expansion of ART access. Standardised first, second and third line ART has facilitated the rapid scale up of ART in LMIC, and saved many lives. To build on this, a highly strategic approach to monitoring drug resistance in populations and individuals needs to be taken, with viral load monitoring supporting more rapid change to second and third line therapies as required, rather than running the risk of keeping people on failing regimens. Supporting adherence at health service, community, family and individual levels also requires investment – investment in research, to find context-appropriate solutions that work, in communities to provide labour, and in systems to streamline the medical and dispensing processes and practices that can facilitate or hinder access and adherence. While refinements in drug development may make adherence simpler in years to come, the challenge is to ensure that programming now is responsive to problem of resistance, and that existing drugs and monitoring are carefully deployed to optimise longevity of tools currently available.

References

Anderson, P.L., D.V. Glidden, A. Liu, S. Buchbinder, J.R. Lama, J.V. Guanira, et al. 2012. Emtricitabine-tenofovir concentrations and pre-exposure prophylaxis efficacy in men who have sex with men. *Science Translational Medicine* 4 (151): 151ra125.

Arrivé, E., M.-L. Newell, D.K. Ekouevi, M.-L. Chaix, R. Thiebaut, B. Masquelier, et al. 2007. Prevalence of resistance to nevirapine in mothers and children after single-dose exposure to prevent vertical transmission of HIV-1: a meta-analysis†. *International Journal of Epidemiology* 36 (5): 1009–1021.

Arts, E.J., and D.J. Hazuda. 2012. HIV-1 antiretroviral drug therapy. *Cold Spring Harbor Perspectives in Medicine* 2 (4): a007161. https://doi.org/10.1101/cshperspect.a007161.

AVAC. PrEP Watch. Scaling up country updates. http://www.prepwatch.org/scaling-up/country-updates

Bavinton, B.R., A.N. Pinto, N. Phanuphak, et al. 2018. Viral suppression and HIV transmission in serodiscordant male couples: An international, prospective, observational, cohort study. *Lancet*. Published online July 16.

Beckerman, K.P. 2002. Mothers, orphans, and prevention of paediatric AIDS. *The Lancet* 359 (9313): 1168–1169.

———. 2003. Long-term findings of HIVNET 012: The next steps. *The Lancet* 362 (9387): 842–843.

Beckham, S.W., C. Beyrer, P. Luckow, M. Doherty, E.K. Negussie, and S.D. Baral. 2016. Marked sex differences in all-cause mortality on antiretroviral therapy in low- and middle-income countries: A systematic review and meta-analysis. *Journal of the International AIDS Society* 19 (1): 21106.

Bertagnolio, S., I. Derdelinckx, M. Parker, J. Fitzgibbon, H. Fleury, M. Peeters, et al. 2008. World Health Organization/HIVResNet drug resistance laboratory strategy. *Antiviral Therapy* 13 (Suppl 2): 49–57.

Cambiano, V., S. Bertagnolio, M.R. Jordan, J.D. Lundgren, and A. Phillips. 2013. Transmission of drug resistant HIV and its potential impact on mortality and treatment outcomes in resource-limited settings. *The Journal of Infectious Diseases* 207 (Suppl 2): S57–S62.

Chawana, R., and D.K. van Bogaert. 2011. Risk management in HIV/AIDS: Ethical and economic issues associated with restricting HAART access only to adherent patients. *African Journal of AIDS Research* 10 (sup1): 369–380.

Cohen, Jon. 2004. Allegations raise fears of backlash against AIDS prevention strategy. *Science* 306 (5705): 2168–2169. https://doi.org/10.1126/science.306.5705.2168.

Cohen, M.S., Y.Q. Chen, M. McCauley, T. Gamble, M.C. Hosseinipour, N. Kumarasamy, et al. 2011. Prevention of HIV-1 infection with early antiretroviral therapy. *The New England Journal of Medicine* 365 (6): 493–505.

Connor, E.M., R.S. Sperling, R. Gelber, P. Kiselev, G. Scott, M.J. O'Sullivan, et al. 1994. Reduction of maternal-infant transmission of human immunodeficiency virus type 1 with Zidovudine treatment. *The New England Journal of Medicine* 331 (18): 1173–1180.

DART Trial Team. 2010. Routine versus clinically driven laboratory monitoring of HIV antiretroviral therapy in Africa (DART): A randomised non-inferiority trial. *The Lancet* 375 (9709): 123–131. https://doi.org/10.1016/S0140-6736(09)62067-5.

Eyakuze, C., D.A. Jones, A.M. Starrs, and N. Sorkin. 2008. From PMTCT to a more comprehensive aids response for women: A much-needed shift. *Developing World Bioethics* 8 (1): 33–42.

Grant, R.M., P.L. Anderson, V. McMahan, A. Liu, K.R. Amico, M. Mehrotra, et al. 2014. Uptake of pre-exposure prophylaxis, sexual practices, and HIV incidence in men and transgender women who have sex with men: A cohort study. *The Lancet Infectious Diseases* 14 (9): 820–829.

Guay, L.A., P. Musoke, T. Fleming, D. Bagenda, M. Allen, C. Nakabiito, et al. 1999. Intrapartum and neonatal single-dose nevirapine compared with zidovudine for prevention of mother-to-child transmission of HIV-1 in Kampala, Uganda: HIVNET 012 randomised trial. *The Lancet* 354 (9181): 795–802.

Gulick, Roy M., Heather J. Ribaudo, Cecilia M. Shikuma, Stephanie Lustgarten, Kathleen E. Squires, William A. Meyer III, Edward P. Acosta, Bruce R. Schackman, Christopher D. Pilcher, Robert L. Murphy, William E. Maher, Mallory D. Witt, Richard C. Reichman, Sally Snyder, Karin L. Klingman, and Daniel R. Kuritzkes. 2004. Triple-Nucleoside regimens versus Efavirenz-containing regimens for the initial treatment of HIV-1 infection. *New England Journal of Medicine* 350 (18): 1850–1861. https://doi.org/10.1056/NEJMoa031772.

Haire, B. 2015. Preexposure prophylaxis-related stigma: Strategies to improve uptake and adherence – A narrative review. *HIV/AIDS – Research and Palliative Care* 7: 241–249. https://doi.org/10.2147/HIV.S72419.

Ho, David D. 1995. Time to hit HIV, early and hard. *New England Journal of Medicine* 333 (7): 450–451. https://doi.org/10.1056/nejm199508173330710.

Institute of Medicine (US) Committee on Reviewing the HIVNET 012 Perinatal HIV Prevention Study. 2005. *Review of the HIVNET 012 perinatal HIV prevention study*. National Academy of Sciences. Washington, DC: National Academies Press.

Johnson, J.A., Li J-f, L. Morris, N. Martinson, G. Gray, J. McIntyre, et al. 2005. Emergence of drug-resistant HIV-1 after Intrapartum administration of single-dose Nevirapine is substantially underestimated. *The Journal of Infectious Diseases* 192 (1): 16–23.

Kaufmann, G.R., H. Furrer, B. Ledergerber, L. Perrin, M. Opravil, P. Vernazza, M. Cavassini, E. Bernasconi, M. Rickenbach, B. Hirschel, and M. Battegay. 2005. Characteristics, determinants, and clinical relevance of CD4 T cell recovery <500 Cells/µL in HIV type 1 – Infected individuals receiving potent antiretroviral therapy. *Clinical Infectious Diseases* 41 (3): 361–372. https://doi.org/10.1086/431484.

Kébé, K., L. Bélec, H.D. Ndiaye, S.B. Gueye, A.A.M. Diouara, S. Ngom, et al. 2014. The case for addressing primary resistance mutations to non-nucleoside reverse transcriptase inhibitors to treat children born from mothers living with HIV in sub-Saharan Africa. *Journal of the International AIDS Society* 17 (1): 18526.

Knox, D.C., D.H. Tan, P.R. Harrigan, and P.L. Anderson. 2016. *HIV-1 infection with multiclass resistance despite Preexposure Prophylaxis (PrEP)*. 169aLB. In Conference on retro viruses and opportunistic infections, 22–25 February 2016, Boston, Massachusetts.

Lockman, S., R.L. Shapiro, L.M. Smeaton, C. Wester, I. Thior, L. Stevens, F. Chand, J. Makhema, C. Moffat, A. Asmelash, P. Ndase, P. Arimi, E. van Widenfelt, L. Mazhani, V. Novitsky, S. Lagakos, and M. Essex. 2007. Response to antiretroviral therapy after a single, peripartum dose of nevirapine. *New England Journal of Medicine* 356 (2): 135–147. https://doi.org/10.1056/NEJMoa062876.

Luber, A.D. 2005. Genetic barriers to resistance and impact on clinical response. *Medscape General Medicine* 7 (3): 69.

Macklin, R. 2001. After Helsinki: Unresolved issues in international research. *Kennedy Institute of Ethics Journal* 11 (1): 17–36. Project MUSE. https://doi.org/10.1353/ken.2001.0005.

———. 2004. *Ethics and equity in access the HIV treatment 3 by 5 initiative: Background paper consultation on ethics and equitable access to treatment and care for HIV/AIDS*. http://www.who.int/ethics/background-macklin2.pdf

McConnell, M.S., J.S.A. Stringer, A.P. Kourtis, P.J. Weidle, and S.H. Eshleman. 2007. Use of single-dose nevirapine for the prevention of mother-to-child transmission of HIV-1: Does development of resistance matter? *American Journal of Obstetrics & Gynecology* 197 (3): S56–S63.

Mellors, J.W., C.R. Rinaldo, P. Gupta, R.M. White, J.A. Todd, and L.A. Kingsley. 1996. Prognosis in HIV-1 infection predicted by the quantity of virus in plasma. *Science* 272 (5265): 1167–1170.

Mellors, J.W., A. Munoz, J.V. Giorgi, et al. 1997. Plasma viral load and CD4+ lymphocytes as prognostic markers of HIV-1 infection. *Annals of Internal Medicine* 126 (12): 946–954.

Moore, R.D., and R.E. Chaisson. 1999. Natural history of HIV infection in the era of combination antiretroviral therapy. *AIDS* 13 (14): 1933–1942.. PMID 10513653. https://doi.org/10.1097/00002030-199910010-00017.

Pharmacychecker. 2017. https://www.pharmacychecker.com/generic/price-comparison/zidovudine/300+mg. Accessed 29 July 2017.

Rodger AJ et al. for the PARTNER Study Group. 2016. Sexual activity without condoms and risk of HIV transmission in serodifferent couples when the HIV-positive partner is using suppressive antiretroviral therapy. *JAMA* 316 (2): 1–11. https://doi.org/10.1001/jama.2016.5148. http://jama.jamanetwork.com/article.aspx?doi=10.1001/jama.2016.5148. Accessed 12 July 2016.

Rodger AJ, V. Cambiano, T. Bruun, et al. 2018. *Risk of HIV transmission through condomless sex in MSM couples with suppressive ART: The PARTNER2 Study extended results in gay men.* In 22nd International AIDS conference (AIDS 2018), Amsterdam, the Netherlands. Oral Abstract WEAX0104LB.

Rosen, S., and M.P. Fox. 2011. Retention in HIV care between testing and treatment in Sub-Saharan Africa: A systematic review. *PLOS Medicine* 8 (7): e1001056.

Rowley, C.F., I.J. MacLeod, D. Maruapula, B. Lekoko, S. Gaseitsiwe, M. Mine, et al. 2016. Sharp increase in rates of HIV transmitted drug resistance at antenatal clinics in Botswana demonstrates the need for routine surveillance. *Journal of Antimicrobial Chemotherapy* 71 (5): 1361–1366.

Samuel, R., M.N. Julian, R. Paredes, R. Parboosing, P. Moodley, L. Singh, et al. 2016. HIV-1 drug resistance by ultra-deep sequencing following short course zidovudine, single-dose nevirapine, and single-dose tenofovir with emtricitabine for prevention of mother-to-child transmission. *JAIDS Journal of Acquired Immune Deficiency Syndromes* 73 (4): 384–389.

Severe, Patrice, Marc Antoine Jean Juste, Alex Ambroise, Ludger Eliacin, Claudel Marchand, Sandra Apollon, Alison Edwards, Heejung Bang, Janet Nicotera, Catherine Godfrey, Roy M. Gulick, Warren D. Johnson Jr., Jean William Pape, and Daniel W. Fitzgerald. 2010. Early versus standard antiretroviral therapy for HIV-infected adults in Haiti. *New England Journal of Medicine* 363 (3): 257–265. https://doi.org/10.1056/NEJMoa0910370.

Smart, T. 2006. *PEPFAR and Global Fund both highly effective, but is the funding sustainable?* AIDSMAP. http://www.aidsmap.com/PEPFAR-and-Global-Fund-both-highly-effective-but-is-the-funding-sustainable/page/1424121. Accessed 23 June 2006.

Strategies for Management of Antiretroviral Therapy Study Group, Sean Emery, Jacqueline A. Neuhaus, Andrew N. Phillips, Abdel Babiker, Calvin J. Cohen, Jose M. Gatell, Pierre-Marie Girard, Birgit Grund, Matthew Law, Marcelo H. Losso, Adrian Palfreeman, and Robin Wood. 2008. Major clinical outcomes in antiretroviral therapy (ART)-naive participants and in those not receiving ART at baseline in the SMART study. *The Journal of Infectious Diseases* 197 (8): 1133–1144. https://doi.org/10.1086/586713.

Stringer, J.S.A., M.S. McConnell, J. Kiarie, O. Bolu, T. Anekthananon, T. Jariyasethpong, et al. 2010. Effectiveness of non-nucleoside reverse-transcriptase inhibitor-based antiretroviral therapy in women previously exposed to a single intrapartum dose of nevirapine: A multi-country, prospective cohort study. *PLOS Medicine* 7 (2): e1000233.

Taylor-Smith, K., H. Tweya, A. Harries, E. Schoutene, and A. Jahn. 2010. Gender differences in retention and survival on antiretroviral therapy of HIV-1 infected adults in Malawi. *Malawi Medical Journal: The Journal of Medical Association of Malawi* 22 (2): 49–56.

Tenthani, Lyson, Andreas D. Haas, Hannock Tweya, Andreas Jahn, Joep J. van Oosterhout, Frank Chimbwandira, Zengani Chirwa, Wingston Ng'Ambi, Alan Bakali, Sam Phiri, Landon Myer, Fabio Valeri, Marcel Zwahlen, Gilles Wandeler, and Olivia Keiser. 2014. Retention in care under universal antiretroviral therapy for HIV infected pregnant and breastfeeding women ("Option B+") in Malawi. *AIDS* 28 (4): 589–598. https://doi.org/10.1097/QAD.0000000000000143.

The INSIGHT START Study Group. 2015. Initiation of antiretroviral therapy in early asymptomatic HIV infection. *New England Journal of Medicine* 373 (9): 795–807. https://doi.org/10.1056/NEJMoa1506816.

UNAIDS. 2017. *90–90–90 – An ambitious treatment target to help end the aids epidemic.* http://www.unaids.org/en/resources/documents/2017/90-90-90

Wade, N.A., G.S. Birkhead, B.L. Warren, T.T. Charbonneau, P.T. French, L. Wang, et al. 1998. Abbreviated regimens of zidovudine prophylaxis and perinatal transmission of the human immunodeficiency virus. *New England Journal of Medicine* 339 (20): 1409–1414.

Wilfert, C.M. 2003. HIVNET 012 and Petra. *The Lancet* 363 (9404): 244–245.

World Health Organization. 2003a. *3 by 5 Strategy: Making it happen.* http://www.who.int/3by5/publications/documents/en/3by5StrategyMakingItHappen.pdf

———. 2003b. *Scaling up antiretroviral therapy in resource-limited settings: Treatment guidelines for a public health approach.* http://www.who.int/hiv/pub/prev_care/en/ARTrevision2003en.pdf?ua=1

———. 2006. *Antiretroviral therapy for HIV infection in adults and adolescents: Recommendations for a public health approach (2006 revision).* http://www.who.int/hiv/pub/ART/adult/en

———. 2009. *Rapid advice: antiretroviral therapy for HIV infection in adults and adolescents.* http://www.who.int/hiv/pub/arv/rapid_advice_art.pdf

——— 2016. *Global Action Plan on HIV Drug Resistance 2017–2021.* https://www.who.int/hiv/drugresistance/hivdr_darft_gap.pdf

———. 2017. *HIV drug resistance report 2017.* Geneva: World Health Organization. http://apps.who.int/iris/bitstream/handle/10665/255896/9789241512831-eng.pdf;jsessionid=9AFD8EB2096DEF42908467A854A90A9E?sequence=1

———. 2018. *WHO HIV data and statistics.* http://www.who.int/hiv/data/en

Chapter 4
Ethics and Antimalarial Drug Resistance

Phaik Yeong Cheah, Michael Parker, and Nicholas P. J. Day

Abstract There has been impressive progress in malaria control and treatment over the past two decades. One of the most important factors in the decline of malaria-related mortality has been the development and deployment of highly effective treatment in the form of artemisinin-based combination therapies (ACTs). However, recent reports suggest that these gains stand the risk of being reversed due to the emergence of ACT resistance in the Greater Mekong Subregion and the threat of this resistance spreading to Africa, where the majority of the world's malaria cases occur, with catastrophic consequences. This chapter provides an overview of strategies proposed by malaria experts to tackle artemisinin-resistant malaria, and some of the most important practical ethical issues presented by each of these interventions. The proposed strategies include mass antimalarial drug administrations in selected populations, and mandatory screening of possibly infected individuals prior to entering an area free of artemisinin-resistant malaria. We discuss ethical issues such as tensions between the wishes of individuals versus the broader goal of

P. Y. Cheah (✉)
Mahidol Oxford Tropical Medicine Research Unit (MORU), Faculty of Tropical Medicine, Mahidol University, Bangkok, Thailand

Centre for Tropical Medicine and Global Health, Nuffield Department of Clinical Medicine, University of Oxford, Oxford, UK

The Ethox Centre, Nuffield Department of Population Health, University of Oxford, Oxford, UK
e-mail: Phaikyeong@tropmedres.ac

M. Parker
The Ethox Centre, Nuffield Department of Population Health, University of Oxford, Oxford, UK

Wellcome Centre for Ethics and Humanities, Nuffield Department of Population Health, University of Oxford, Oxford, UK
e-mail: michael.parker@ethox.ox.ac.uk

N. P. J. Day
Mahidol Oxford Tropical Medicine Research Unit (MORU), Faculty of Tropical Medicine, Mahidol University, Bangkok, Thailand

Centre for Tropical Medicine and Global Health, Nuffield Department of Clinical Medicine, University of Oxford, Oxford, UK

© The Author(s) 2020
E. Jamrozik, M. Selgelid (eds.), *Ethics and Drug Resistance: Collective Responsibility for Global Public Health*, Public Health Ethics Analysis 5,
https://doi.org/10.1007/978-3-030-27874-8_4

malaria elimination, and the risks of harm to interventional populations, and conclude by proposing a set of recommendations.

4.1 The Problem, Context and Background

Malaria is the most important parasitic disease of man. It remains a major cause of death in tropical countries, and an important cause of illness, particularly in childhood. There were an estimated 435,000 deaths from malaria in 2017, of which over 90% were in Africa (Global Malaria Programme, World Health Organization 2018). Although there has been impressive progress in malaria control and treatment over the past two decades in recent years progress has stalled and there has been a resurgence of malaria in Southeast Asia, where antimalarial drug resistance is increasingly prevalent (Global Malaria Programme, World Health Organization 2018). One of the most important factors in the decline of malaria related mortality since the 1990s has been the development and deployment of highly effective treatment in the form of artemisinin-based combination therapies (ACTs) (Bhatt et al. 2015). ACTs are currently the mainstay of antimalarial treatment throughout the world, recommended by the World Health Organization as the first line treatment globally for falciparum malaria (World Health Organization 2016a). Their widespread deployment, along with the expanded use of insecticide treated bed nets, accounts for a large part of the reduction in malaria deaths in Africa over the past decade (White et al. 2014).

The effectiveness of current and future interventions are, however, at risk from the emergence of new forms of drug resistance. In the early 2000s malaria parasites that were partially resistant to the artemisinins emerged in Western Cambodia. The problem was identified and characterised in 2008 (Dondorp et al. 2009; Noedl et al. 2008). The hallmark of infection by these resistant parasites was slow parasite clearance rather than outright treatment failure (Dondorp et al. 2009). These slow clearing infections were associated with mutations in the *PfKelch13* gene, with multiple *PfKelch13* mutations described (though each parasite only carried one) (Ariey et al. 2014). By 2014 slow clearing malaria infections caused by *PfKelch13* mutation-carrying artemisinin resistant parasites could be found in Myanmar, Laos, Thailand, Vietnam and Yunnan province in China, and by 2016 *Pfkelch13* mutants were identified in Arunachal Pradesh state in northeastern India (Ashley et al. 2014; Tun et al. 2016; Mishra et al. 2016).

The initial response to the emergence of artemisinin resistance had two aspects. The major concerns were that: i. artemisinin resistance might lead to or combine with partner drug resistance resulting in resistance to ACTs. That is, to the loss of the drug combinations that are the mainstay of malaria treatment; and ii. that were this to be the case, artemisinin resistance and then ACT resistance then has the potential to spread from Southeast Asia to Africa, where the majority of the world's malaria cases occur, with catastrophic consequences. There are precedents for this latter concern; in the last 60 years first chloroquine resistance and then sulfadoxine-pyrimethamine resistance arose in Western Cambodia and subsequently spread to Africa, leading to millions of deaths (Verdrager 1986; Roper et al. 2004; Trape et al. 1998).

Recognizing the risks and consequences of spread to Africa, the World Health Organization initially developed a plan to contain rather than eliminate the problem (World Health Organization 2011). However, this approach was criticized by some at the time who called for initiatives aimed at eliminating all malaria in the Greater Mekong Subregion, on the grounds that the resistant parasites were in fact already at that stage contained (Dondorp et al. 2017). They argued that the view that elimination would be an appropriate strategy was supported by mathematical modelling that showed that as malaria was controlled and transmission fell, the proportion of infections that were resistant would increase – the 'last man standing' would be resistant (Maude et al. 2012). This modelling suggested that resistance could not be eliminated without eliminating all malaria in the affected regions.

Bolstered by advocacy for global malaria eradication by the Bill and Melinda Gates Foundation, the WHO did eventually change its policy from containment to one of elimination (Global Malaria Programme WHO 2015), but the discovery through molecular studies that *PfKelch13* mutations had arisen spontaneously and independently multiple times within the Greater Mekong Subregion (GMS) led to what some saw as a reduction in the urgency to eliminate malaria to prevent the 'spread' of resistant parasites (Takala-Harrison et al. 2015). Surveillance for resistance and an aim to eliminate malaria *everywhere* was now the policy of WHO (World Health Organization 2017; Global Malaria Programme WHO 2016). This necessarily spread resources more widely and reduced the focus on drug resistance. Despite this there has been considerable investment in malaria control and elimination efforts in the GMS, with at least initially substantial reductions in malaria transmission (Dondorp et al. 2017). These successes have, however, to some extent masked the continued threat of increasingly drug resistant malaria parasites emerging and spreading in the region – as the mathematical modelling had predicted (Maude et al. 2009) – with outbreaks of artemisinin-resistant malaria occurring in areas previously considered malaria free (Imwong et al. 2015).

In the 10 years since the first description of artemisinin resistance the concern that ACT resistance and failure would develop has come to pass, with rising mefloquine resistance on the Thai-Myanmar border and piperaquine resistance in Cambodia, Thailand, Southern Laos and Viet Nam (Phyo et al. 2016a; Amaratunga et al. 2016; Imwong et al. 2017). Recent evidence that most of the resistance to dihydroartemisinin-piperaquine is due to the geographical *spread* of a particularly fit artemisinin-resistant parasite clone which has picked up piperaquine resistance has rekindled the debate about whether urgent focus should be placed on an accelerated effort to eliminate artemisinin-resistant malaria in Southeast Asia, with the aim of preventing the spread of ACT resistance to Africa (World Health Organization 2016b).

Although the global pipeline for new malaria drugs in development is healthier than it has been for decades, all the most promising candidates (schizonticidals, that kill the asexual blood stage of the parasite which causes the clinical manifestations of malaria) are at least five years away from being available in the market (Phyo et al. 2016b). The RTS,S/AS01 malaria vaccine which recently received a favourable scientific opinion from the European Medicines Agency (EMA) is only partially protective (Vandoolaeghe and Schuerman 2018). This means that protecting the efficacy of the currently available antimalarial drugs is of global importance. The spread of ACT resistance to Africa would threaten the loss of millions of lives, especially those of young African children. This will be a global health issue – untreatable malaria worldwide arising from resistance in Southeast Asia. A problem arising in a specific location that has the potential to threaten global health is not unique to malaria. Many 'global health problems' arise in low-income settings, for example outbreaks of Ebola in West Africa and Zika in South America, and are recognized as having worldwide implications. In the case of Ebola and Zika the World Health Organization formally declared a 'Public Health Emergency of International Concern' (PHEIC), but for the emergence of artemisinin resistant malaria in Southeast Asia WHO has as of 2018 declined to do this despite some experts calling for it to do so (World Health Organization 2016b; Talisuna et al. 2012).

4.1.1 How Should the Problem of Artemisinin Resistant Malaria be Tackled?

There is now broad agreement amongst experts that to prevent the spread of artemisinin-resistant *P. falciparum* it is necessary to completely interrupt *P. falciparum* transmission (Maude et al. 2009), and that a programme of accelerated malaria elimination is warranted in the GMS and surrounding areas. The scientific consensus is that a combination of strategies is required to achieve this (Dondorp et al. 2017; World Health Organization 2017). These include:

1. *Ongoing surveillance* with a network of village malaria workers (VMWs) in endemic areas trained and equipped to provide early detection and treatment of malaria cases.
2. *Targeting of the asymptomatic malaria reservoir* in so called malaria 'hotspots', identified through surveys of healthy individuals employing highly sensitive methods of parasite detection such as large volume quantitative PCR and highly sensitive rapid diagnostic tests (RDTs). This may take two forms:

   a. *Mass drug administration* (MDA) -WHO agrees that targeted mass antimalarial drug administration may play an important role in malaria elimination (World Health Organization, Global Malaria Programme 2015a).
   b. *Mass screening and treatment* (MSAT) using novel highly sensitive diagnostics.

3. *Vector control* with insecticide treated bed nets, despite these being less effective in the GMS than in Africa because of the biting habits of many of the vector species (biting in the forest rather than in houses, and in the early evening or morning).
4. *Targeting 'source' populations* such as forest workers and migrants, rather than 'sink' populations secondarily affected. This requires an understanding of transmission dynamics and population movements – important in the GMS where cross border movement/migration is common.
5. *Mandatory screening* may be necessary of possibly infected individuals entering an area free of artemisinin-resistant malaria (Houston and Houston 2015; Tatarsky et al. 2011).
6. Use of *effective antimalarial treatments*. Most currently approved ACTs consist of only two drugs (a fast acting short half-life artemisinin and a longer half-life partner drug) and are vulnerable to the development of resistance. The testing and deployment of new triple artemisinin-based combination therapies (TACTs) has been recommended, and several of these are currently being tested in the GMS and beyond (Dondorp et al. 2017).

In the face of the global threat posed by increasing ACT resistance, there is therefore now an emerging expert consensus that the combination of strategies outlined above is the most effective way of halting or slowing its international spread provide strong ethical arguments for their rapid adoption. Each of the interventions listed above present a wide range of interconnected challenges – including, scientific, technological, governmental, economic and ethical – all of which will need to be overcome if the elimination of malaria in the region is to be achieved. In addition to these practical scientific, and political challenges, the success of each of the interventions also depends upon the addressing of a number of important practical ethical questions, which need to be taken into account in their design and implementation. In the next section, we outline some of the most important practical ethical issues presented by each of the interventions proposed above.

4.2 Practical Ethical Issues Arising in These Interventions

4.2.1 Ongoing Surveillance

In many countries in the Greater Mekong Subregion, networks of VMWs are the cornerstone of malaria surveillance and the delivery of malaria-related interventions. These networks are usually run either directly by the national malaria control programmes (NMCPs) or by NGOs working with the NMCPs, but may also be put in place by private providers such as companies running palm oil plantations. The coverage of such networks has increased impressively in many areas in SEA, particularly in border areas, conflict zones and areas underserved by government health programmes. VMWs are consulted by villagers suffering from fever, and are equipped with RDTs and antimalarial treatments. Where a substantial proportion of febrile illnesses are indeed caused by malaria VMWs are a valuable resource for the local population. However, the success of malaria control and elimination efforts increasingly means that a diminishing proportion of the febrile illnesses they encounter are caused by malaria. Unfortunately, VMWs are not usually equipped to deal with these alternative causes of illness. This means that as malaria rates decline there is a risk that villagers will cease consulting VMWs as more of them are told that because their fever is not clinical malaria infection, no diagnosis or treatment of the cause of their illness is available or offered. Unless these village workers are retrained for a wider role as 'village *health* workers' able to manage other febrile illnesses or simple primary health problems, they will become increasingly irrelevant and demotivated. The consequences of this would be diminishing effectiveness of the malaria surveillance network itself, at the point in the elimination process when it is most needed and, in the absence of a wider role, for the goal of malaria elimination becoming a disincentive for the VMWs (and NMCPs), for many of whom being a VMW is a source of their livelihood. This suggests that, even if the elimination of malaria is the primary goal, there are strong arguments in favour of the provision of resources for access to health care beyond malaria in the region.

4.2.2   Mass Drug Administration (MDA)

MDA in the context of malaria elimination consists of mass treatment with a schizonticide to kill the asexual blood stage of the parasite which causes the clinical manifestations of malaria combined with a transmission blocking drug to kill gametocytes. Giving such treatment to all members of a community should eliminate the asymptomatic parasite reservoir and speed up the interruption of malaria transmission. Where and when it is warranted has been the subject of much debate, but the consensus is that MDA should be targeted at communities with high transmission and a large asymptomatic reservoir (World Health Organization 2011; von Seidlein and Dondorp 2015). This requires a functioning surveillance system to identify such communities, with surveys of healthy individuals with highly sensitive tests to estimate accurately the scale of the asymptomatic reservoir.

There are a number of important ethical considerations when determining when and where to use MDA. The first of these arises out of the fact that MDA by its nature involves administering drugs to individuals who will not benefit *directly* from the treatment, i.e. to healthy people in the interests of the wider community and the broader goal of elimination. In the case of transmission blocking drugs this is the entire community, and for the schizonticidal drugs this is the substantial proportion of the community who are not infected with malaria parasites. However, if the MDA is effective and malaria is eliminated from the area, all individuals will benefit *indirectly* by living in a malaria free community. In the GMS the schizonticide currently used in MDA is dihydroartemisinin-piperaquine (DP), which in the treatment of malaria is considered a safe drug. Studies of the safety of DP in this context have shown that DP is safe (Tripura et al. 2018), but widespread deployment of DP exposes much larger numbers of individuals such that its rare but serious side effects may occur (Cheah and White 2016). The transmission blocking drug currently used in MDA is primaquine, which targets the transmissible sexual gametocytes not killed by the schizonticide but has little or no impact on the asexual parasites which cause disease. Primaquine is an oxidative drug which causes haemolysis (the rupture of red blood cells) in G6PD-deficient individuals (median 8% of the population in malaria endemic areas) when given in the large doses needed to radically cure vivax malaria (killing the hypnozoites in the liver) (Howes et al. 2012). However to kill gametocytes (rather than cure malaria) a single much lower dose is required, one considered safe to be administered to all individuals without prior G6PD testing (World Health Organization, Global Malaria Programme 2015b; Bancone et al. 2016).

A second ethically significant consideration is a worry that there is a risk that with the widespread deployment of antimalarial drugs in MDA the resulting increased drug pressure may itself lead to drug resistance, particularly in the case where elimination is not achieved. It has been argued on theoretical grounds that this is unlikely, but the risk however low highlights the importance of achieving elimination in areas where MDA is deployed (White 2017). This suggests that the initiation of MDA is only ethically justified where there is a genuine commitment to complete the elimination task. Once the process of MDA has begun, important ethical issues are presented relating to the question of when such an initiative should be ended.

Thirdly, the effectiveness of MDA is predicated upon high population coverage (World Health Organization 2017; Newby et al. 2015). Achieving this is a challenge for several reasons: explaining the rationale for taking antimalarials when asymptomatic can be difficult in the absence of an understanding of the underlying scientific concepts, target communities are often remote with poor access and populations can be highly mobile. For these reasons, effective community engagement efforts are essential, so that individuals are informed of the risks and benefits of malaria elimination efforts in general and MDA in particular (Adhikari et al. 2016; Peto et al. 2018). For effective community engagement those implementing MDA need to understand and adapt the information they provide and the form of the engagement they adopt, to the cultural and practical requirements of each community. Engagement with community leaders is essential, and coverage can be promoted by offering healthcare alongside MDA (Sahan et al. 2017; Pell et al. 2017). Effective community engagement may also minimize risks of coercion or counterproductive misunderstanding of the aims of the public health authorities (Parker and Allen 2013). The ethical issues around these concerns are similar in some respects to those encountered in the context of vaccination campaigns.

4.2.3   Mass Screening and Treatment (MSAT)

Mass screening and treatment of village populations has been suggested as an alternative strategy to MDA for speeding up malaria elimination. Its advantage over MDA is that only those individuals with proven asymptomatic malaria infection will be exposed to antimalarial drugs and their attendant risks, negating many of the ethical concerns described in the MDA section above. However, the likely success of this is limited by the sensitivity of the tests available for detecting low levels of parasites in the blood. Because the current tests are laboratory based there is inevitably considerable delay between sampling and result, which appears to limit the effectiveness of MSAT (von Seidlein 2014). Highly sensitive rapid diagnostic tests have now been developed, but these are only now being tested in the field (Slater et al. 2015). Such tests have the potential to be much quicker but they are not as sensitive as the laboratory based tests, and it possible that up to half of asymptomatic carriers will be 'missed'. However, the contribution to malaria transmission of individuals with very low parasitaemias at the time of testing is uncertain, and the results of studies of the field effectiveness of MSAT with highly sensitive RDTs are awaited with interest. If MSAT with highly sensitive RDTs does turn out to be effective, effective community engagement will be as important as it is with MDA.

4.2.4   Vector Control

The distribution of insecticide treated nets (ITNs) in Africa has had a major impact on malaria there, and as a result it has become almost an article of faith in the global malaria community that ITNs should be considered the most important single intervention in the battle against malaria (Bhatt et al. 2015). In a WHO-sponsored meeting on tackling artemisinin resistant malaria in the GMS the chairman suggested that *all* the additional resources being made available to counter resistance in the region should be spent on ITNs (*NPJD personal communication*). Unfortunately, Southeast Asian malaria vectors and populations do not behave like African vectors and populations, with most transmission occurring in the forest rather than in dwellings (Gryseels et al. 2015; Smithuis et al. 2013a). Several studies have now confirmed the limited efficacy of bed nets in malaria elimination effort in this region (Smithuis et al. 2013b; Satitvipawee et al. 2012). An important ethical issue here is around resource allocation, and overcoming established (but not evidence-based) pro-ITN sentiment amongst international and national policy makers. ITNs do have an important role to play and are relatively cheap to deploy, but given the evidence of differences in vector behaviours in Africa and the GMS, the relative allocation of limited resources should be driven by evidence-based health economics studies (Drake et al. 2015).

4.2.5 Targeting 'Source' Populations

In the GMS, malaria transmission is concentrated in poor, hard-to-reach, highly mobile populations. Transmission is mainly occupationally related, highest among men who travel into forested areas to work. Many of the most at-risk populations are disenfranchised minority groups, often living in border regions, with little or no health infrastructure. Understanding the drivers of transmission in these populations entails acquiring better knowledge of population movement/migration, much of which is 'illegal'. Working with these populations requires sensitivity not only to the cultural contexts but also to the uncertain legal status of many of the individuals. Several of the more endemic areas are mired in armed conflicts, and many populations are vulnerable as refugees or economic migrants without papers. The area currently with the highest endemicity in the GMS, for example, is Rakhine State in Myanmar, currently undergoing considerable civil strife and large-scale movement of populations. NGOs, government workers and researchers have to work within their own externally determined constraints, limiting their ability to engage 'source' populations. Even if the not inconsiderable task of eliminating malaria from many of these areas were to be achieved, political difficulties, the mobile nature of these populations, and changes in the ability to access them would leave them vulnerable to the reintroduction of malaria. Furthermore, as immunity will have waned because of the intervention the public health consequences of this reintroduction could potentially be worse than if malaria had not been eliminated in the first place.

4.2.6 Mandatory Screening

There are a number of situations in which mandatory screening for asymptomatic malaria may be indicated to prevent individuals unwittingly spreading drug resistant malaria parasites. Following the disastrous importation of cholera into Haiti by African UN peacekeepers (leading to 8300 deaths), for example, a call has been made for Southeast Asian Peacekeepers to be screened for malaria before they travel to missions in Africa (Houston and Houston 2015). This would prevent peacekeepers from importing drug resistant malaria to a drug sensitive region, and could be implemented by the UN. Another situation where mandatory screening could potentially be introduced in the context of eliminating artemisinin-resistant malaria within the GMS, would be screening local people moving between areas where malaria has and has not been eliminated. Although practically difficult to implement, this has the potential to be of real importance in geographical locations with highly mobile populations – such as along the Thai-Myanmar Border. There are a number of practical barriers to implementing such a policy, which make it unlikely to be introduced at present. However, its possibility raises important ethical questions about the legitimacy of overriding personal autonomy in the global public interest and its limits.

4.2.7   Triple Artemisinin Combination Therapies (TACTs)

The rationale for TACTs is similar to that of triple or quadruple therapy in HIV, tuberculosis, leprosy and other infectious diseases – to prevent the development of resistance. Two TACTs are currently being studied in 16 sites in Asia and 1 site in Africa (web identifier: NCT02453308): dihydroartemisinin-piperaquine + mefloquine; and artemether-lumefantrine + amodiaquine. The combination of a short acting artemisinin with two long acting partner drugs ensures that parasites are less likely to encounter only one long acting partner drug at any one time, minimizing the chance of resistance developing. In addition, it is hypothesized that these triple therapies could exploit potential inverse relationships between the parasite molecular resistance mechanisms to the paired long-acting partner drugs. It is thought that the wide implementation of triple therapy in malaria will slow the spread of multidrug-resistant malaria in areas where artemisinin and partner drug resistance is well established, and slow down or prevent the emergence of drug resistance in areas where resistance has not yet emerged. It is in the latter case where TACTs should be most effective.

There are several ethical issues to be considered here. TACTs differ from other examples of combination therapy in that the objective is to prevent antimalarial drug resistance at the population rather than at the individual level. Unlike in chronic infections such as TB and HIV development of ACT resistance within an individual patient during treatment is rare. Hence individuals are potentially exposed to the additional side effects of three rather than two drugs for little or no benefit to themselves; it is against the interest of the individual patient (usually a child) to take three rather than two drugs. If the strategy works the benefit will be to the population, which will only indirectly benefit the individual. In addition, TACTs are expected to be most effective at countering resistance in areas where resistance has not yet developed to any of the components, so that the long acting partner drugs will protect each other from the development of resistance and both will protect the short acting artemisinin component. Hence the areas where they will be most effective will be the ones where currently ACTs remain highly effective at the individual level.

Preliminary evidence of triple therapy is promising but safety and efficacy data are not widely available yet. Even if evidence is available, populations where ACTs still work such as in Africa where the majority of malaria cases are in children under five, may not readily change their prescribing behaviours. Other practical problems might be access to the triple therapy, availability of co-formulated drugs and the problem of substandard and falsified drugs especially in the private and informal sectors (Liverani et al., Chap. 5, this volume).

4.3 Summary of Ethical Considerations

Above we have attempted to illustrate the ethical complexity of the implementation of the strategies widely agreed by experts to be necessary for the control of resistance to antimalarial drugs. It is clear that there are strong ethical arguments in favour of the implementation of such strategies in the global public interest. However, the considerations outlined above suggest that such interventions raise important ethical questions both about the nature and scope of implementation itself and about the obligations of countries both outside and within the region to those who are to bear its costs. The success of an elimination strategy based on these elements will depend upon these problems being addressed. In the remainder of this chapter, we summarise what we consider to be some of the most important ethical tensions and outline some preliminary thoughts about ways these might be addressed.

4.3.1 Autonomy and Consent Versus the Global Benefit

There are a number of different ways in which the implementation of the intervention strategies above raises important questions about respect for autonomy. In some cases, these interventions may lead to tensions between the interests and wishes of individuals and the global benefit. During implementation of each strategy, individuals in selected communities are subjected to interventions – 'treatment' or surveillance – not for their own good but for the common good of current and future populations both locally and internationally (see discussion on common goods in Chaps. 8 and 9) (Jamrozik and Selgelid, Chap. 1, this volume-a; Smith and Coast, Chap. 17, this volume). Under current circumstances, such interventions are voluntary: making MDA compulsory or imposing travel restrictions on people who have come from areas with artemisinin resistant malaria is not considered achievable or justified at present. However, it is possible that as in other global health contexts this judgement might change and that individuals may lose their right to opt out of, for example, MDA. Draconian measures have been taken to contain dangerous contagions such as H5N1 influenza, SARS, MERS-CoV, Ebola virus, Lassa fever, and multidrug-resistant tuberculosis, which involved restriction of liberty in order to protect the public. Were compulsory approaches to be considered in the context of malaria, this would present important questions about the legitimacy of restrictions of liberty per se but also questions about how this was in fact undertaken, and about the nature of the obligations of the wider global community – particularly wealthy countries – to those who are subjected to such interventions in the global health interest.

Questions about autonomy also arise in the context of voluntary approaches. Where individuals and their communities are being asked to decide about participation in the strategies outlined above, it is vital that best efforts are made to ensure that any consent they give is grounded in a good understanding of the implications. However, the evidence is that valid consent is likely to be difficult to achieve in such contexts. This places particular importance on the roles of wider communication, community participation, political involvement and other forms of public engagement preceding and during the intervention. It has to be acknowledged that even in the context of well-resourced, evidence-based approaches to consent and community engagement, understanding is likely to be partial given the complexities of malaria transmission and how these inventions work. This need not mean that the choice to participate is invalid but it does mean that the moral basis for the intervention cannot rest on consent alone, even when the choice to participate is voluntary. This suggests that those who are responsible for the conduct of such strategies have obligations to ensure that they are conducted to high ethical standards, and that appropriate protections, and possibly compensation, are in place.

4.3.2   Risk Benefit

The potential benefits of malaria elimination are substantial, including the direct burden averted and economic growth through improved educational attainment and productivity; these gains were estimated recently to far outstrip the costs required to achieve them (Purdy et al. 2013). That said it is clear from the discussion above that those who bear the consequences of malaria elimination efforts are not those who will benefit directly. The majority of the populations with the highest prevalence of resistant malaria and of submicroscopic malaria in Southeast Asia are poor and mobile forest workers (Phommasone et al. 2016; Tripura et al. 2017). This is unsurprising as vulnerability to malaria – as is also true of many other infectious diseases – is largely a consequence of social determinants of health such as poverty, malnutrition and insufficient access to healthcare. Malaria burden is both a consequence and an illustration of global inequities. These populations are already burdened by their circumstances and environments. Yet they are the very individuals who will likely to be shouldering the burdens of any global intervention to curb resistant malaria. In the case of MDA, entire communities are treated whether or not they are unwell with malaria. That means that many individuals who are neither ill nor carriers of the parasite will be asked (or required) to take drugs and therefore be at some risk of potential adverse drug reactions. This uneven distribution of individual risks and inconveniences – that is individuals in SEA shouldering the burden for the benefit of good health outcomes primarily in the interests of populations elsewhere – is a key moral challenge. Whilst at the macroeconomic level the costs of malaria elimination are outweighed by the benefits, this may well not be true at the level of the individuals involved. Important ethical questions concern the question of when, if at all, the imposition of risks of harm on (often vulnerable) individuals is legitimate in the interests of others and the limits of this. The ethical questions here concern not only those related to whether the imposition of such burdens is justified but also, where this is the case, both the approach adopted to such implementation and the nature and scope of our obligations to those upon whom it is imposed. Is there, for example, an obligation to compensate such populations?

4.3.3 Data and Sample Sharing

Ethical issues also arise with regard to the international collaboration required to ensure high scientific and public health standards in the interventions. This is important because it is the achievement of such high standards (and hence the potential for success) that justify the imposition of risks and restrictions of liberty on vulnerable populations. In order to make the most informed decisions about planning interventions to eliminate malaria, there is a need to ensure that there is access to as much good quality data as possible. That is, it depends crucially upon data sharing. However, there is generally a lack of transparency and confidence in the quality of available malaria data. This can be due to poor quality data and underreporting of cases, which can in turn be due to variable availability of diagnostic tools such as rapid diagnostic tests and blood slides, unsurprising given that data are frequently collected under resource-starved conditions. This contributes to the lack of trust in the data on antimalarial resistance, and in the data used in mathematical modeling and the resulting predictions. An additional key problem is that many national malaria control programmes do not readily share their malaria data for political, economic and national security reasons. Data related to population movement and migration that could aid interventions such as MDA and engagement with "source populations" are, for example, particularly difficult to access. Although there is advantage in data sharing, it is also acknowledged that it can pose a number of ethical challenges around issues of privacy, potential stigma and economic harms (Mishra et al. 2016; Bull et al. 2015; Cheah et al. 2015). This suggests that questions about the ethical tensions between the interests of individuals and the global health interest also arise at the level of institutions, health ministries, and countries.

4.3.4 Scientific Disagreement About the Best Way Forward

We outlined a number of proposed interventions above, and given the current state of the evidence there are valid debates in the scientific community about what action or actions are appropriate where. Interventions such as TACTs and MDA are still under study. An important, as yet unresolved, scientific debate is about the way resistance spreads or emerges. There are data supporting both sides of the argument – geographical spread of sporadic vs. spontaneous distributed emergence (Lu et al. 2017). However, there is also a limited window of opportunity to act. There are strong moral reasons for acting to prevent the spread of resistance both within and beyond Southeast Asia. Inaction will almost inevitably lead to a repeat of history – the loss of safe, inexpensive and highly effective treatments and an increase in cases of severe malaria and related deaths. It may also mean that a once in a generation opportunity, capitalizing on the combination of the availability of political will and effective tools to take a big step towards the eradication of malaria, will have been missed. There remains, however, a degree of scientific uncertainty. This raises ethical questions about the level of scientific consensus required for action. Understanding is likely to remain imperfect. Is it legitimate to initiate strategies such as those outlined above on the basis of good but imperfect understanding? The answer to this question cannot be 'never' because there is widespread agreement about the urgency of the situation and a residual degree of scientific uncertainty will always remain.

4.4   Conclusions

In this chapter, we have described practical ethical issues arising in currently proposed interventions (and the lack of them) to reduce the risk of the movement of resistance to the current best antimalarial drugs from Southeast Asia to Africa, as well as to prevent resistance emerging in Africa. Whilst strong ethical reasons for such interventions are provided by the seriousness and scale of the threat and the existence of a degree of scientific consensus on this strategy, its implementation is ethically complex. We have outlined some of the most important practical ethical problems presented by each of possible components of the proposed strategy, and have also argued that even if the various interventions were to be ethically justified this would not be the end of the ethical debate. We have argued that important ethical questions about the mode of implementation of the interventions and about the obligations of the wider community to those they affect would remain.

Some of the most important of such considerations are those relating to fairness in the selection of interventional populations. All populations which meet a set of criteria for an intervention, be it MDA, TACT or travel restrictions, should where practical have the same intervention. The intervention should be evidence-based and justified, and all relevant stakeholders should be involved in the decision-making process and have meaningful input into the deliberations. The manner and context in which decisions are made should be reasonable, fair and transparent.

In addition to ethical considerations relating to the selection of such populations, we have argued that important obligations exist for those countries and governments that can afford it to assist and perhaps to compensate those individuals who are subject to such interventions and may experience harms as a result of their participation (Upshur 2002). Meeting these obligations may call for the provision of compensation, for example where businesses suffer due to lack of mobility or where people suffer from side effects of MDA. Or perhaps, in the form of community level benefits such as improved healthcare facilities. An important aspect of the obligations of the wider world to those who live in the region is that any intervention is well-planned and adequately resourced. It is clear that curbing antimalarial resistance, similar to resistance of other antimicrobials, is both a global priority and a global responsibility (Jamrozik and Selgelid, Chap. 1, this volume-b). Both scientifically and in terms of effective public health interventions, solutions to this problem are inevitably going to be collaborative. In addition to the provision of adequate resources, communities, countries, researchers and funders must be encouraged to work together. It is our view that four key requirements for a successful and appropriate collaborative approach to addressing emerging ACT resistance, and hence ethically important requirements of those who propose such interventions, are as follows:

i. *Encouraging and funding more research.* Research should be conducted to address the gaps needed for each of the interventions proposed by the scientific community such as determining the safety of DP for MDA, the efficacy and safety of triple therapy, and determining the way resistance spreads. More evidence would help channel resources to the correct people and places, and facilitate a scientific consensus.
ii. *Retraining and supporting village malaria workers* so that they are able to manage other febrile diseases and hence remain relevant and retain community support. This could be provision of education and strengthening support from provincial health departments.
iii. *Encourage collection of quality malaria data, and sharing and pooling of these data.* A data sharing initiative, the WorldWide Antimalarial Resistance Network (www.WWARN.org) was established by malaria researchers in 2009 to facilitate collaborative study groups working to answer specific research questions using pooled analyses. WWARN has had considerable success in pooling individual-participant data from multiple clinical trials from academic groups and pharmaceutical companies, but has been less successful with NMCPs. Individual research groups have also established data sharing mechanisms via a managed access route (Cheah and Day 2017).
iv. *Engaging with affected communities in creative and sensitive ways.* Some work has already been conducted to engage forest workers, minority groups and mobile populations, and much more is needed (Lim et al. 2017). This will improve understanding of the science behind malaria and malaria elimination and will facilitate interventions such as MDAs and MSATs.

References

Adhikari, B., N. James, G. Newby, L. von Seidlein, N.J. White, N.P. Day, et al. 2016. Community engagement and population coverage in mass anti-malarial administrations: A systematic literature review. *Malaria Journal* 15 (1): 523.

Amaratunga, C., P. Lim, S. Suon, S. Sreng, S. Mao, C. Sopha, et al. 2016. Dihydroartemisinin-piperaquine resistance in Plasmodium falciparum malaria in Cambodia: A multisite prospective cohort study. *The Lancet Infectious Diseases* 16 (3): 357–365.

Ariey, F., B. Witkowski, C. Amaratunga, J. Beghain, A.C. Langlois, N. Khim, et al. 2014. A molecular marker of artemisinin-resistant Plasmodium falciparum malaria. *Nature* 505 (7481): 50–55.

Ashley, E.A., M. Dhorda, R.M. Fairhurst, C. Amaratunga, P. Lim, S. Suon, et al. 2014. Spread of artemisinin resistance in Plasmodium falciparum malaria. *The New England Journal of Medicine* 371 (5): 411–423.

Bancone, G., N. Chowwiwat, R. Somsakchaicharoen, L. Poodpanya, P.K. Moo, G. Gornsawun, et al. 2016. Single low dose primaquine (0.25 mg/kg) does not cause clinically significant haemolysis in G6PD deficient subjects. *PLoS One* 11 (3): e0151898.

Bhatt, S., D.J. Weiss, E. Cameron, D. Bisanzio, B. Mappin, U. Dalrymple, et al. 2015. The effect of malaria control on Plasmodium falciparum in Africa between 2000 and 2015. *Nature* 526 (7572): 207–211.

Bull, S., P.Y. Cheah, S. Denny, I. Jao, V. Marsh, L. Merson, et al. 2015. Best practices for ethical sharing of individual-level health research data from low- and middle-income settings. *Journal of Empirical Research on Human Research Ethics* 10 (3): 302–313.

Cheah, P.Y., and N.P.J. Day. 2017. Data sharing: Experience from a tropical medicine research unit. *Lancet* 390 (10103): 1642.

Cheah, P.Y., and N.J. White. 2016. Antimalarial mass drug administration: Ethical considerations. *International Health* 8 (4): 235–238.

Cheah, P.Y., D. Tangseefa, A. Somsaman, T. Chunsuttiwat, F. Nosten, N.P. Day, et al. 2015. Perceived benefits, harms, and views about how to share data responsibly: A qualitative study of experiences with and attitudes toward data sharing among research staff and community representatives in Thailand. *Journal of Empirical Research on Human Research Ethics* 10 (3): 278–289.

Dondorp, A.M., F. Nosten, P. Yi, D. Das, A.P. Phyo, J. Tarning, et al. 2009. Artemisinin resistance in Plasmodium falciparum malaria. *The New England Journal of Medicine* 361 (5): 455–467.

Dondorp, A.M., F.M. Smithuis, C. Woodrow, and L.V. Seidlein. 2017. How to contain artemisinin- and multidrug-resistant falciparum malaria. *Trends in Parasitology* 33 (5): 353–363.

Drake, T.L., S.S. Kyaw, M.P. Kyaw, F.M. Smithuis, N.P. Day, L.J. White, et al. 2015. Cost effectiveness and resource allocation of Plasmodium falciparum malaria control in Myanmar: A modelling analysis of bed nets and community health workers. *Malaria Journal* 14: 376.

Global Malaria Programme WHO. 2015. *World Malaria Report 2015*. Available from: http://apps.who.int/iris/bitstream/10665/200018/1/9789241565158_eng.pdf.

———. 2016. *World Malaria Report 2016*. Available from: http://apps.who.int/iris/bitstream/10665/254912/1/WHO-HTM-GMP-2017.4-eng.pdf.

Global Malaria Programme, World Health Organization. 2018. *World Malaria Report 2018*. Available from: https://www.who.int/malaria/publications/world-malaria-report-2018/report/en/.

Gryseels, C., L. Durnez, R. Gerrets, S. Uk, S. Suon, S. Set, et al. 2015. Re-imagining malaria: Heterogeneity of human and mosquito behaviour in relation to residual malaria transmission in Cambodia. *Malaria Journal* 14: 165.

Houston, S., and A. Houston. 2015. Screening and treating UN Peacekeepers to prevent the introduction of artemisinin-resistant malaria into Africa. *PLoS Medicine* 12 (5): e1001822.

Howes, R.E., F.B. Piel, A.P. Patil, O.A. Nyangiri, P.W. Gething, M. Dewi, et al. 2012. G6PD deficiency prevalence and estimates of affected populations in malaria endemic countries: A geostatistical model-based map. *PLoS Medicine* 9 (11): e1001339.

Imwong, M., T. Jindakhad, C. Kunasol, K. Sutawong, P. Vejakama, and A.M. Dondorp. 2015. An outbreak of artemisinin resistant falciparum malaria in Eastern Thailand. *Scientific Reports* 5: 17412.

Imwong, M., K. Suwannasin, C. Kunasol, K. Sutawong, M. Mayxay, H. Rekol, et al. 2017. The spread of artemisinin-resistant Plasmodium falciparum in the Greater Mekong subregion: A molecular epidemiology observational study. *The Lancet Infectious Diseases* 17 (5): 491–497.

Jamrozik, E., and M. Selgelid. 2020a. The ethics and politics of antimicrobial resistance: Moral responsibility and the justifications of policies to preserve antimicrobial effectiveness. In *Ethics and drug resistance: Collective responsibility for Global Public Health*. Cham: Springer.

———. 2020b. Drug-resistant infection: Causes, consequences, and responses. In *Ethics and drug resistance: Collective responsibility for Global Public Health*. Cham: Springer.

Lim, R., R. Tripura, T.J. Peto, M. Sareth, N. Sanann, C. Davoeung, et al. 2017. Drama as a community engagement strategy for malaria in rural Cambodia. *Wellcome Open Research* 2: 95.

Liverani, M.H.L., M. Khan, and R. Coker. this volume. Antimicrobial resistance and the private sector in Southeast Asia. In *Ethics and drug resistance: Collective responsibility for Global Public Health*. Cham: Springer.

Lu, F., R. Culleton, M. Zhang, A. Ramaprasad, L. von Seidlein, H. Zhou, et al. 2017. Emergence of indigenous artemisinin-resistant plasmodium falciparum in Africa. *The New England Journal of Medicine* 376 (10): 991–993.

Maude, R.J., W. Pontavornpinyo, S. Saralamba, R. Aguas, S. Yeung, A.M. Dondorp, et al. 2009. The last man standing is the most resistant: Eliminating artemisinin-resistant malaria in Cambodia. *Malaria Journal* 8: 31.

Maude, R.J., D. Socheat, C. Nguon, P. Saroth, P. Dara, G. Li, et al. 2012. Optimising strategies for Plasmodium falciparum malaria elimination in Cambodia: Primaquine, mass drug administration and artemisinin resistance. *PLoS One* 7 (5): e37166.

Mishra, N., R.S. Bharti, P. Mallick, O.P. Singh, B. Srivastava, R. Rana, et al. 2016. Emerging polymorphisms in falciparum Kelch 13 gene in Northeastern region of India. *Malaria Journal* 15 (1): 583.

Newby, G., J. Hwang, K. Koita, I. Chen, B. Greenwood, L. von Seidlein, et al. 2015. Review of mass drug administration for malaria and its operational challenges. *The American Journal of Tropical Medicine and Hygiene* 93 (1): 125–134.

Noedl, H., Y. Se, K. Schaecher, B.L. Smith, D. Socheat, M.M. Fukuda, et al. 2008. Evidence of artemisinin-resistant malaria in western Cambodia. *The New England Journal of Medicine* 359 (24): 2619–2620.

Parker, M., and T. Allen. 2013. Questioning ethics in global health. In *Ethics in the field: Contemporary challenges*, Studies of the Biosocial Society, ed. J. MacClancy and A. Fuentes, vol. 7, 24–41. New York: Berghahn Books.

Pell, C., R. Tripura, C. Nguon, P. Cheah, C. Davoeung, C. Heng, et al. 2017. Mass anti-malarial administration in western Cambodia: A qualitative study of factors affecting coverage. *Malaria Journal* 16 (1): 206.

Peto, T.J., R. Tripura, C. Davoeung, C. Nguon, S. Nou, C. Heng, et al. 2018. Reflections on a community engagement strategy for mass antimalarial drug administration in Cambodia. *The American Journal of Tropical Medicine and Hygiene* 98 (1): 100–104.

Phommasone, K., B. Adhikari, G. Henriques, T. Pongvongsa, P. Phongmany, L. von Seidlein, et al. 2016. Asymptomatic Plasmodium infections in 18 villages of southern Savannakhet Province, Lao PDR (Laos). *Malaria Journal* 15 (1): 296.

Phyo, A.P., E.A. Ashley, T.J.C. Anderson, Z. Bozdech, V.I. Carrara, K. Sriprawat, et al. 2016a. Declining efficacy of artemisinin combination therapy against P. Falciparum malaria on the Thai-Myanmar Border (2003-2013): The role of parasite genetic factors. *Clinical Infectious Diseases* 63 (6): 784–791.

Phyo, A.P., P. Jittamala, F.H. Nosten, S. Pukrittayakamee, M. Imwong, N.J. White, et al. 2016b. Antimalarial activity of artefenomel (OZ439), a novel synthetic antimalarial endoperoxide, in patients with Plasmodium falciparum and Plasmodium vivax malaria: An open-label phase 2 trial. *The Lancet Infectious Diseases* 16 (1): 61–69.

Purdy, M., M. Robinson, K. Wei, and D. Rublin. 2013. The economic case for combating malaria. *The American Journal of Tropical Medicine and Hygiene* 89 (5): 819–823.

Roper, C., R. Pearce, S. Nair, B. Sharp, F. Nosten, and T. Anderson. 2004. Intercontinental spread of pyrimethamine-resistant malaria. *Science* 305 (5687): 1124.

Sahan, K., C. Pell, F. Smithuis, A.K. Phyo, S.M. Maung, C. Indrasuta, et al. 2017. Community engagement and the social context of targeted malaria treatment: A qualitative study in Kayin (Karen) State, Myanmar. *Malaria Journal* 16 (1): 75.

Satitvipawee, P., W. Wongkhang, S. Pattanasin, P. Hoithong, and A. Bhumiratana. 2012. Predictors of malaria-association with rubber plantations in Thailand. *BMC Public Health* 12: 1115.

Slater, H.C., A. Ross, A.L. Ouedraogo, L.J. White, C. Nguon, P.G. Walker, et al. 2015. Assessing the impact of next-generation rapid diagnostic tests on Plasmodium falciparum malaria elimination strategies. *Nature* 528 (7580): S94–S101.

Smith, R.D., and J. Coast. this volume. The economics of resistance through an ethica lens. In *Ethics and drug resistance: Collective responsibility for Global Public Health*. Cham: Springer.

Smithuis, F.M., M.K. Kyaw, U.O. Phe, I. van der Broek, N. Katterman, C. Rogers, et al. 2013a. Entomological determinants of insecticide-treated bed net effectiveness in Western Myanmar. *Malaria Journal* 12: 364.

———. 2013b. The effect of insecticide-treated bed nets on the incidence and prevalence of malaria in children in an area of unstable seasonal transmission in western Myanmar. *Malaria Journal* 12: 363.

Takala-Harrison, S., C.G. Jacob, C. Arze, M.P. Cummings, J.C. Silva, A.M. Dondorp, et al. 2015. Independent emergence of artemisinin resistance mutations among Plasmodium falciparum in Southeast Asia. *The Journal of Infectious Diseases* 211 (5): 670–679.

Talisuna, A.O., C. Karema, B. Ogutu, E. Juma, J. Logedi, A. Nyandigisi, et al. 2012. Mitigating the threat of artemisinin resistance in Africa: Improvement of drug-resistance surveillance and response systems. *The Lancet Infectious Diseases* 12 (11): 888–896.

Tatarsky, A., S. Aboobakar, J.M. Cohen, N. Gopee, A. Bheecarry, D. Moonasar, et al. 2011. Preventing the reintroduction of malaria in Mauritius: A programmatic and financial assessment. *PLoS One* 6 (9): e23832.

Trape, J.F., G. Pison, M.P. Preziosi, C. Enel, A. Desgrees du Lou, V. Delaunay, et al. 1998. Impact of chloroquine resistance on malaria mortality. *Comptes Rendus de l'Académie des Sciences Série III* 321 (8): 689–697.

Tripura, R., T.J. Peto, C.C. Veugen, C. Nguon, C. Davoeung, N. James, et al. 2017. Submicroscopic Plasmodium prevalence in relation to malaria incidence in 20 villages in western Cambodia. *Malaria Journal* 16 (1): 56.

Tripura, R., T.J. Peto, N. Chea, D. Chan, M. Mukaka, P. Sirithiranont, et al. 2018. A controlled trial of mass drug administration to interrupt transmission of multidrug-resistant falciparum malaria in Cambodian villages. *Clinical Infectious Diseases* 67 (6): 817–826.

Tun, K.M., A. Jeeyapant, M. Imwong, M. Thein, S.S. Aung, T.M. Hlaing, et al. 2016. Parasite clearance rates in upper Myanmar indicate a distinctive artemisinin resistance phenotype: A therapeutic efficacy study. *Malaria Journal* 15: 185.

Upshur, R.E. 2002. Principles for the justification of public health intervention. *Canadian Journal of Public Health* 93 (2): 101–103.

Vandoolaeghe, P., and L. Schuerman. 2018. The RTS, S/AS01 malaria vaccine in children aged 5-17 months at first vaccination. *The Pan African Medical Journal* 30: 142.

Verdrager, J. 1986. Epidemiology of the emergence and spread of drug-resistant falciparum malaria in South-East Asia and Australasia. *The Journal of Tropical Medicine and Hygiene* 89 (6): 277–289.

von Seidlein, L. 2014. The failure of screening and treating as a malaria elimination strategy. *PLoS Medicine* 11 (1): e1001595.

von Seidlein, L., and A. Dondorp. 2015. Fighting fire with fire: Mass antimalarial drug administrations in an era of antimalarial resistance. *Expert Review of Anti-Infective Therapy* 13 (6):
715–730.
White, N.J. 2017. Does antimalarial mass drug administration increase or decrease the risk of
resistance? *The Lancet Infectious Diseases* 17 (1): e15–e20.
White, N.J., S. Pukrittayakamee, T.T. Hien, M.A. Faiz, O.A. Mokuolu, and A.M. Dondorp. 2014.
Malaria. *Lancet* 383 (9918): 723–735.
World Health Organization. 2011. *A global plan for artemisinin resistance containment.* Available
from: http://apps.who.int/iris/bitstream/10665/44482/1/9789241500838_eng.pdf.
———. 2016a. *Guidelines for the treatment of malaria*, 3rd ed. Available from: http://apps.who.
int/iris/bitstream/10665/162441/1/9789241549127_eng.pdf.
———. 2016b. *Minutes of the Evidence Review Group meeting on the emergence and spread of
multidrug-resistant Plasmodium falciparum lineages in the Greater Mekong subregion 2016.*
Available from: http://www.who.int/malaria/mpac/mpac-mar2017-erg-multidrug-resistance-
session6.pdf?ua=1.
———. 2017. *A framework for malaria elimination 2017.* Available from: http://apps.who.int/iris/
bitstream/10665/254761/1/9789241511988-eng.pdf.
World Health Organization, Global Malaria Programme. 2015a. *The role of mass drug administration, mass screening and treatment, and focal screening and treatment for malaria 2015.*
Available   from: http://www.who.int/malaria/publications/atoz/role-of-mda-for-malaria.
pdf?ua=1.
———. 2015b. *Policy brief on single-dose primaquine as a gametocytocide in Plasmodium
falciparum malaria 2015.* Available from: http://www.who.int/malaria/publications/atoz/
who_htm_gmp_2015.1.pdf.

Chapter 5
Antimicrobial Resistance and the Private Sector in Southeast Asia

Marco Liverani, Lauren Oliveira Hashiguchi, Mishal Khan, and Richard Coker

Abstract Southeast Asia is considered a regional hotspot for the emergence and spread of antimicrobial resistance (AMR). A commonality across countries in the region, particularly those with lower incomes such as Cambodia, Myanmar, Lao PDR and Vietnam, is the high utilisation of private healthcare providers, often unregulated, which may play a role in driving AMR. In this chapter we discuss challenges to the control of AMR in Southeast Asia, with a focus on the role of the private sector. After providing an overview of the problem and current policy responses, we consider ethical issues of equity and fairness that may arise from the implementation of established and proposed interventions.

Keywords Drug resistance · Ethics · Medicine & Pubic Health · Private sector · Southeast Asia.

M. Liverani (✉)
Department of Global Health and Development, London School of Hygiene and Tropical Medicine, London, UK

Faculty of Public Health, Mahidol University, Bangkok, Thailand

School of Tropical Medicine and Global Health, Nagasaki University, Nagasaki, Japan
e-mail: Marco.Liverani@lshtm.ac.uk

L. Oliveira Hashiguchi
Department of Global Health and Development, London School of Hygiene and Tropical Medicine, London, UK

School of Tropical Medicine and Global Health, Nagasaki University, Nagasaki, Japan
e-mail: Lauren.Hashiguchi@lshtm.ac.uk

M. Khan · R. Coker
Department of Global Health and Development, London School of Hygiene and Tropical Medicine, London, UK

Faculty of Public Health, Mahidol University, Bangkok, Thailand
e-mail: Mishal.Khan@lshtm.ac.uk; richard.coker@lshtm.ac.uk

© The Author(s) 2020
E. Jamrozik, M. Selgelid (eds.), *Ethics and Drug Resistance: Collective Responsibility for Global Public Health*, Public Health Ethics Analysis 5,
https://doi.org/10.1007/978-3-030-27874-8_5

5.1 Diversity, Epidemiology and Surveillance Capacity

Southeast Asia is a loosely defined geographic region, whose configuration is variable and depends on different political, institutional, and cultural perspectives. For the purpose of this chapter, we refer to Southeast Asia as the ten member countries of the Association of Southeast Asian Nations (ASEAN), that is, Brunei, Cambodia, Indonesia, Lao PDR, Malaysia, Myanmar, the Philippines, Singapore, Thailand, and Vietnam. As such, this region includes small, wealthy countries such as Singapore (5.6 million population, US$ 52,960 per capita) and populous, lower middle-income countries such as Indonesia (261.2 million population, 3570 US per capita) (The World Bank 2017). It is also characterised by socio-cultural differences and vast diversity in ecosystems including farming, natural habitats, and urbanisation. Although gaps remain in our understanding of AMR in Southeast Asia, available evidence indicates that this is an important and growing challenge (Cherau et al. 2017). For example, studies have identified a high prevalence of antimicrobial resistant infections in hospitalized paediatric populations in several regional countries (Al-Taiar et al. 2013; Stoesser et al. 2013; Turner et al. 2016). The Asian Network for Surveillance of Resistant Pathogens (ANSORP) reported rates of pneumococcal resistance to penicillin exceeding 50% in some contexts (Song et al. 1999) and that resistance had spread across the region. Resistance to enteric pathogens is becoming increasingly prominent (Coker et al. 2017), with studies in Thailand and Cambodia identifying high *Campylobacter coli* and *Campylobacter jejuni* rates of resistance to ciprofloxacin among isolate samples from children hospitalized with acute diarrhoea (Bodhidatta et al. 2002; Meng et al. 2011). In terms of respiratory infections, the World Health Organization (WHO) 2016 Global Tuberculosis Report lists Myanmar, Indonesia, Thailand, Philippines, and Vietnam among the 27 countries bearing the highest burden of multidrug resistant tuberculosis in the world (WHO 2015). Drug-resistant malaria is also prevalent across the region, including resistance of the malaria parasite *P. falciparum* to chloroquine, sulfadoxine–pyrimethamine, and most recently, and of considerable concern, artemisinin (WHO 2017). It is well documented that Western Cambodia, near the border with Thailand, has been an epicentre of antimalarial resistance since the 1950s (Dondorp et al. 2010; Alam et al. 2011; Vinayak et al. 2010). Despite much documentation of resistance throughout the region (Ashley et al. 2014; Imwong et al. 2017), the full epidemiological profile of resistance to antimalarials in Southeast Asia, as with AMR more broadly, is not known.

Fragmented surveillance systems and weak laboratory capacity remain major barriers to acquiring quality surveillance data and information for AMR in the region, especially in LMICs (Lee and Wakabayashi 2013). These barriers were echoed in the findings of the Joint External Evaluation (JEE) (WHO 2016), a WHO-led process to assess country capacity to prevent, detect, and respond to public health threats. In Southeast Asia, JEEs have been conducted in Lao PDR, Vietnam, and Cambodia. Vietnam and Cambodia, the only countries with public JEE reports at present (Joint External Evaluation of the IHR Core Capacities of the Kingdom of

Table 5.1 Joint external evaluation scoring for AMR Technical Area Indicator, Cambodia and Vietnam (Joint External Evaluation Tool and Process Overview 2016; Joint External Evaluation of the IHR Core Capacities of the Kingdom of Cambodia 2017; Joint External Evaluation of IHR Core Capacities of Viet Nam 2017)

	AMR detection	Surveillance of infections caused by AMR pathogens	Healthcare associated infection prevention and control programs	Antimicrobial stewardship activities
Cambodia	3	2	2	2
	Developed capacity	Limited capacity	Limited capacity	Limited capacity
Vietnam	2	2	3	2
	Limited capacity	Limited capacity	Developed capacity	Limited capacity

Cambodia 2017; Joint External Evaluation of IHR Core Capacities of Viet Nam 2017), reported limited capacity for surveillance of infections caused by AMR pathogens and advocated improved stewardship to prevent inappropriate use of antimicrobials (Table 5.1) (see also Chap. 23 in this book).

5.2 Private Health Services and AMR

The private sector is the dominant health care provider in many countries in the region. For example, about 70% of Cambodian patients first seek treatment from the private sector and private drug sellers are the preferred healthcare providers for the majority of those who are ill (NIS Cambodia 2014). Across the region, out-of-pocket payment is a common method to finance health care, despite good progress toward universal health coverage in Thailand, Malaysia, Singapore, and Indonesia (Van Minh et al. 2014). Out-of-pocket health expenditure as a percentage of total health expenditure is more than 50% in Cambodia, Singapore, the Philippines, and Myanmar (Table 5.2) (The World Bank 2017). A study including data from 47 countries indicated that out-of-pocket health expenditure was strongly correlated with AMR in low-income countries, with the authors concluding that high demand for the private sector may be related to higher AMR owing to heightened incentives among private providers to overprescribe and less standardized quality assurance of antimicrobials (Alsan et al. 2015).

Prior to discussing the role of private providers in AMR, it is important to briefly describe the diversity of the private health sector in Southeast Asia and reasons for its popularity. Broadly, private providers include persons operating outside of the government-financed system, alone or in groups, to provide diagnosis, treatment or advice to individuals for health-related concerns. In Southeast Asia, as in other regions, the private sector includes a variety of providers, ranging from large private hospitals, small clinics and pharmacies to road-side informal drug vendors and traditional healers (Khan 2016). The level of training varies greatly. Some private

Table 5.2 Health sector expenditure indicators in ASEAN (The World Bank 2017)

	Population (thousands), 2016	GDP (billions) 2016	Health expenditure as proportion of the GDP, 2014	Public health expenditure as % of total health expenditure, 2014	Out-of-pocket health expenditure as % of total expenditure on health, 2014
Brunei	423,196	11.4	2.7%	93.9%	6.0%
Cambodia	15,762,370	20.0	5.6%	22.0%	95.2%
Indonesia	261,115,456	932.3	2.9%	37.8%	46.9%
Lao PDR	6,758,353	15.9	1.9%	50.5%	38.9%
Malaysia	31,187,265	296.4	4.1%	55.2%	35.3%
Myanmar	52,885,223	67.4	2.3%	45.9%	50.7%
Philippines	103,320,222	304.9	4.7%	32.3%	53.7%
Singapore	5,607,283	297.0	4.9%	41.7%	54.8%
Thailand	68,863,514	406.8	4.1%	77.8%	11.9%
Vietnam	92,701,100	202.6	7.1%	54.1%	36.8%

practitioners have no training or claim to have qualifications that they do not have, while others have several years of specialist training. In addition to allopathic healthcare providers, there are alternative therapeutic approaches, which include homeopathy and traditional healing. Private providers also vary in terms of the fees charged. Some are highly priced and accessible only to a fraction of the population while others are more accessible and may offer flexible payment arrangements.

Why do patients use the private sector? A recent systematic review comparing the performance of private and public health-care systems in LMICs found that patients' preferences for private providers were related to shorter waiting times, better hospitality, increased time spent with doctors, cleanliness of facilities, longer and flexible opening times, personal attitude, and better availability of staff (Basu et al. 2012). However, quality of care in terms of competence and adherence to guidelines is often low (Morgan et al. 2017).

In terms of AMR specifically, over-prescribing or over-selling of antimicrobial drugs by for-profit healthcare providers appears to be fairly common, although inappropriate prescribing practices have also been reported in public hospitals and health centres (Apisarnthanarak et al. 2008; Om et al. 2017; Yeung et al. 2011). A large study of over 400 healthcare providers including drug shops, private clinics and hospitals in Vietnam found that 79% would dispense antibiotics for common colds with fever and only 19% had knowledge of antibiotic prescribing according to national guidelines (Hoa et al. 2009). Drug sellers were more likely to dispense antibiotics inappropriately than other types of healthcare providers in this study. Other studies in the region have shown that, among drugs sold, antimicrobials are very common as they are reported to be the most profit-generating (Gollogly 2002; Chuc and Tomson 1999).

Drug quality is also a problem in the region. There is evidence from Cambodia, particularly on antimalarial drugs (Novotny et al. 2016), and from neighbouring countries such as Vietnam, that unregulated drug shops often sell poor-quality medications. According to the WHO, substandard drugs are products whose composition and ingredients do not meet the correct scientific specifications (WHO 2003). This could be due to inappropriate storage at high temperature and humidity or poor quality assurance during the manufacturing process. Both of these conditions are more likely to occur in less-developed countries. Counterfeit drugs are considered a subset of substandard drugs that are deliberately and fraudulently mislabelled with respect to identity and/or source. Antibiotics and antimalarials are at particular risk of targeting by counterfeiters and drug manufacturers that use poor practices owing to large volumes of sales, their relatively low production cost, and the challenges met by regulatory mechanisms and their enforcement. In 2013, a multi-governmental investigation across Cambodia, Indonesia, Lao PDR, Myanmar, Thailand, and Vietnam found that nearly one third of both antimalarials and antibiotics were of poor quality and potentially counterfeited (Weraphong et al. 2013). Similarly, a 2004 cross-sectional survey of pharmacies and drug shops in Myanmar found an "alarming high proportion" of counterfeit artesunate (Dondorp et al. 2004), although a recent survey provides encouraging evidence that the quality of artemisinin-based combination therapies has improved (Yeung et al. 2015).

As noted earlier, this chapter focuses on the use of antimicrobials in human health. But the use in animals is also important, and is a major contributor to AMR globally (Nhung et al. 2015), as further discussed in Chaps. 7 and 18. Subtherapeutic use of antibiotics in livestock for growth promotion or prophylaxis is of particular concern, as the low doses used can lead to the emergence of drug resistance (Van Boeckel et al. 2015). While there are regulations around the use of growth promoters in Europe and America, the use of antibiotics by unregulated veterinarians and drug sellers in Southeast Asia is poorly understood. Multiple studies in the region do, however, indicate a high prevalence of confirmed resistance to ciprofloxacin, and gentamicin, and of particular concern, colistin among *E. coli* isolates in pig and chicken farms (Nhung et al. 2015, 2016; Nguyen et al. 2015, 2016). Colistin is one of the last line of drugs available for treating multidrug resistant Gram-negative pathogens in humans.

5.3 Policy Challenges in Tackling AMR

Most governments in Southeast Asia have recognised the dangers of AMR and taken action to control the use of antimicrobials through policy making, legal provisions, and program implementation. Legal frameworks to regulate the pharmaceutical supply chain have been improved in many countries, providing stronger legal bases to counter the problem of substandard medicines and poor management practices (Lamy and Liverani 2015). Operational capacities of drug regulatory authorities have also increased, leading to more effective quality control, closure of unlicensed businesses and a crackdown of the trade in counterfeit medicines, with major enforcement operations being conducted throughout the Mekong region (Interpol 2015). Further, new laws have been introduced to regulate drug sellers, particularly for antimalarial medications, although provisions are variable across regional countries. In Thailand, for example, the sale of antimalarial medications in the private sector is banned since 1995, but in other countries regulations are less stringent. In Myanmar, there is no prohibition, while in Lao PDR, the ban applies only to grocery stores, general retailers and itinerant vendors (Akulayi et al. 2017).

Despite increasing commitment, the control of AMR remains a major policy challenge in Southeast Asia, particularly in relation to the sale and use of antibiotics. In Thailand, for example, pharmacists can dispense antibiotics without prescription, leaving more room to profit-motivated sale (Apisarnthanarak et al. 2008; Saengcharoen and Lerkiatbundit 2010). In 2007, the "Antibiotic Smart Use" programme was piloted in community health centres and hospitals to reduce unnecessary prescription of antibiotics for respiratory infections, diarrhoea, and simple wounds. Based on promotional material and performance-based incentives, this programme was subsequently scaled up nationwide and described as a workable model to improve the rational use of antibiotics in Thailand (Akulayi et al. 2017). However, efforts to engage private pharmacies have been less effective (Chalker et al. 2005). In other countries, such as Vietnam and Indonesia, regulations on prescribing and dispensing of antibiotics are more restrictive, but enforcement has been difficult to achieve due to either lack of resources, weak sanctions, or challenges in monitoring compliance in highly diversified markets (Nga et al. 2014; Widayati et al. 2011; Mao et al. 2015). Antimicrobial use is particularly difficult to control in remote rural areas, where access to public health services is more limited, patients tend to self-medicate, and medicines of dubious origin are more likely to be available in road stalls and other informal outlets (Lon et al. 2006; Hadi et al. 2010; Om et al. 2017).

In general, AMR is a complex, multi-dimensional health issue which requires a comprehensive multi-sector policy approach, able to tackle structural drivers and determinants across human and animal health sectors. However, policy development and program implementation have often targeted particular diseases, especially in LMICs where donors have prioritised vertical health programmes. In Cambodia, for example, innovative social marketing schemes have been introduced to improve access to good-quality antimalarials in the private sector and control of pharmaceutical drivers of drug resistance, such as the use of artemisinin monotherapies (Yeung et al. 2011; Yamey et al. 2012). As in other countries, however, no comprehensive policy response to promote the responsible use of medicines has been developed to date. Professional development, supervision, and behaviour change activities are driven by vertical disease programmes as well as the systems for the collection, analysis, and dissemination of surveillance data on antimicrobial resistance. Engagement with the agricultural sector has also been relatively weak in many countries, with gaps in legal provisions and operational capacities to regulate and monitor antibiotic use for animal health prophylaxis and growth promotion (Archawakulathep et al. 2014). Further challenges result from the diversity of the livestock sector in Southeast Asia. While supervision is more feasible in large production units, there is less or no control of antibiotic use in smallholders and contracted farmers (Om and McLaws 2016).

In recognition of these challenges and the need for a more comprehensive policy approach, action plans on AMR have been adopted in Cambodia, Indonesia, the Philippines, Thailand, and Vietnam, while the Lao PDR, Malaysia and Singapore are currently developing their national plans (Cheng 2016). In line with WHO global guidelines (WHO 2015), these plans have been informed by One Health approaches and a commitment to strengthening cooperative arrangements across sectors and national authorities. To this end, inter-ministerial and inter-agency committees have been established to guide policy implementation and monitor progress. Since these plans have been adopted only recently, it is too early to assess outcomes and their value in promoting more effective responses to AMR. However, it is expected that the achievement of the stated policy and governance goals will require a significant increase in resource allocation, which may be challenging to achieve in LMICs. Further, as in other contexts, the divisive forces of institutional mandates and professional interests are likely to counteract the imperative of multi-sectoral cooperation, requiring political will at the highest level of policymaking. Finally, as we will see below, achieving a good balance between *access* to antimicrobials, as well as other vital resources for human and animal health, and curbing inappropriate use and *excess* is arguably one the most important challenges ahead (Das and Horton 2017), posing key ethical questions about policy choices in the short and the long term.

5.4 Ethical Issues

The challenges to controlling the emergence of AMR in Southeast Asia are multisectoral and complex, requiring policies that can address the diverse range of contributing factors, including behaviour and practices, gaps in health service provision and regulations, and macro- and micro-economic drivers in the antimicrobial supply chain and market at the national, regional, and global level. In the process, particular attention should be given to the private health sector and food production systems, as these are arguably the largest channels for the distribution, sale, and use of a wide range of antimicrobials in many regional countries. Yet policy and planning in this area is problematic. In addition to governance challenges noted above, policy options to control AMR in the private sector raise ethical issues of equity and fairness, which have been recognised only recently in research and policy communities (Heyman et al. 2014; Littmann and Viens 2015). While a reduction of antibiotics use is necessary to avert the potentially disastrous impact of future drug-resistant pandemics, a single focus on this policy goal may result in undesirable consequences in the short term, particularly in relation to equitable access to essential goods and services, as we discuss below.

Access to Health Services Informal providers account for a significant fraction of diagnosis and treatment services in parts of Southeast Asia, raising concerns about the quality of care and their role in driving AMR. Efforts are being made in some countries to ban unlicensed practitioners. However, "village doctors" and other informal providers may be the only accessible source of primary care in some areas, particularly in remote rural communities where public health facilities are hard to reach and other options are not available. In such contexts, regulatory enforcement may be necessary to control potentially harmful practices and key drivers of AMR, but may also reduce access to care where it is needed most. Engagement with the informal sector, including training and supervision programmes, might be a solution to this problem. Yet informal providers have weak legitimacy in relation to the formal health system. The implicit recognition of unqualified providers as legitimate health workers might be challenging if conducted or sanctioned by government agencies and is likely to face strong opposition from established professional categories. On the other hand, informal providers may be reluctant to participate in projects with public health authorities for fear of being exposed (Khan et al. 2015).

Access to Medicines Policy and regulations on prescription and dispensing of antimicrobials raise similar concerns (see also Chap. 24 in this book). Since 1998, the WHO has urged Member States to "prohibit the dispensing of antimicrobials without the prescription of a qualified health care professional" and to strengthen legislation "to prevent the sale of antibiotics on the informal market" (WHO 1998). In many countries, however, the government health system or other authorised suppliers might not be able to reach all population groups or gaps may exist in their ability to meet the demand for antimicrobials. In Cambodia and Lao PDR, for example, health volunteers have been appointed in remote communities to diagnose suspected malaria cases by using rapid diagnostic tests, administer subsidised artemisinin-based combination therapy (ACT), and refer severe patients to the nearest public health facility (Liverani et al. 2017; Alum et al. 2017). These programmes have been successful in improving adherence to malaria policy and treatment where other health services are either lacking or inadequate. However, informal drug outlets and grocery stores remain the only accessible and affordable suppliers of medicines in some areas or for some hard-to-reach groups such as mobile and migrant workers. In addition, stronger quality control and higher standards are likely to increase the cost of medicines, making them less affordable to the poor. While subsidization of particular categories of antimicrobials, such as anti-malarial and anti-tuberculosis treatments, has been implemented with good results (Novotny et al. 2016; Hill and Mao 2007), this approach might not be feasible for mass-market drugs such as antibiotics.

Access to Food As noted earlier, evidence indicates that the use of antibiotics in the livestock sector is an important driver of AMR (Witte 1998). Thus, it is suggested that this practice should be reduced or banned outright. However, it is also known that regulatory restrictions may have significant economic effects on farmers, as the use of antibiotics allows larger numbers of healthy animals to be maintained with lower cost to the producer (National Research Council 1999). Further, the effects of restrictions or bans are likely to be unequally distributed across livestock production systems. While large companies and production units may replace antibiotics with other disease prevention practices and veterinary services, smallholders may not be able to do so, bearing higher cost of regulatory change. Finally, more stringent regulations raise concerns of fairness and equity in access to food. Low and sub-therapeutic use of antibiotics contributes to growth promotion and improved feed efficiency, resulting in lower costs of meat and eggs (National Research Council 1999). While the effect of reduced antibiotics use on the price of meat products is difficult to gauge with precision, organic foods are generally more expensive and therefore less affordable to lower income groups (Chander et al. 2011). As Littman (Littmann and Viens 2015) pointed out, "these considerations do not appear to weigh heavily enough to justify the continuation of current practices"; however, we should recognise, "who will be disadvantaged by proposed policy changes, and discuss what kind of subsidy or compensation may be warranted".

5.5 Conclusion

In conclusion, AMR is a complex problem that extends far beyond the technical challenge of reducing or improving the use of antimicrobials, as interventions are likely to have externalities on other important issues for human well-being and livelihood. Thus, in Southeast Asia as in other contexts, policy and planning in this area, and modelling and pilot exercises that provide evidence, will need to incorporate a wide range of issues and a calculation of costs and benefits to individuals, enterprises, and across the whole of society. This will likely require us to find a difficult balance between the urgent need to address the rising challenge of AMR while addressing the moral obligations to broad public health benefits and limiting economic hardship. And all of this within complex social and political environments. Simple solutions are likely to induce undesirable, and perhaps unforeseen, consequences.

References

Alum, A., A. Andrada, J. Archer, E. Auko, K. Bates, P. Bouanchaud, et al. 2017. The malaria testing and treatment landscape in the southern Lao People's Democratic Republic (PDR). *Malaria Journal* 16 (1): 169.

Akulayi, L., A. Alum, A. Andrada, J. Archer, E. D. Arogundade, E. Auko, et al. 2017. Private sector opportunities and threats to achieving malaria elimination in the Greater Mekong Subregion: Results from malaria outlet surveys in Cambodia, the Lao PDR, Myanmar, and Thailand. *Malaria Journal* 16: 180.

Alam, M.T., S. Vinayak, K. Congpuong, C. Wongsrichanalai, W. Satimai, L. Slutsker, et al. 2011. Tracking origins and spread of sulfadoxine-resistant Plasmodium falciparum dhps alleles in Thailand. *Antimicrobial Agents and Chemotherapy* 55 (1): 155–164.

Alsan, M., L. Schoemaker, K. Eggleston, N. Kammili, P. Kolli, and J. Bhattacharya. 2015. Out-of-pocket health expenditures and antimicrobial resistance in low- and middle-income countries. *The Lancet Infectious Diseases* 15 (10): 1203–1210.

Al-Taiar A, M.S. Hammoud, L. Cuiqing, J.K.F. Lee, K.-M. Lui, N. Nakwan, et al. 2013. Neonatal infections in China, Malaysia, Hong Kong and Thailand. *Archives of Disease in Childhood – Fetal and Neonatal Edition* 98 (3): F249–F255.

Anomaly, Jonathan. this volume. Antibiotics and animal agriculture: The need for global collective action. In *Ethics and drug resistance: Collective responsibility for Global Public Health*, 299–310. Cham: Springer.

Apisarnthanarak, A., J. Tunpornchai, K. Tanawitt, and L.M. Mundy. 2008. Nonjudicious dispensing of antibiotics by drug stores in Pratumthani, Thailand. *Infection Control and Hospital Epidemiology* 29 (6): 572–575.

Archawakulathep, A., C.T. Thi Kim, D. Meunsene, D. Handijatno, H.A. Hassim, H.R.G. Rovira, et al. 2014. Perspectives on antimicrobial resistance in livestock and livestock products in ASEAN countries. *Thai Journal of Veterinary Medicine* 44: 5–13.

Ashley, E.A., M. Dhorda, R.M. Fairhurst, C. Amaratunga, P. Lim, S. Suon, et al. 2014. Spread of artemisinin resistance in Plasmodium falciparum malaria. *The New England Journal of Medicine* 371 (5): 411–423.

Basu, S., J. Andrews, S. Kishore, R. Panjabi, and D. Stuckler. 2012. Comparative performance of private and public healthcare systems in low- and middle-income countries: A systematic review. *PLoS Medicine* 9 (6): e1001244.

Boden, Lisa and Dominic Mellor. this volume. Epidemiology and ethics of antimicrobial resistance in animals. In Ethics and drug resistance: *Collective responsibility for Global Public Health*, 109–121. Cham: Springer.

Bodhidatta, L., N. Vithayasai, B. Eimpokalarp, C. Pitarangsi, O. Serichantalergs, and D.W. Isenbarger. 2002. Bacterial enteric pathogens in children with acute dysentery in Thailand: Increasing importance of quinolone-resistant Campylobacter. *The Southeast Asian Journal of Tropical Medicine and Public Health* 33 (4): 752–757.

Chalker, J., S. Ratanawijitrasin, N.T.K. Chuc, M. Petzold, and G. Tomson. 2005. Effectiveness of a multi-component intervention on dispensing practices at private pharmacies in Vietnam and Thailand – A randomized controlled trial. *Social Science and Medicine* 60: 131–141.

Chander, M., B. Subrahmanyeswari, R. Mukherjee, and S. Kumar. 2011. Organic livestock production: An emerging opportunity with new challenges for producers in tropical countries. *Revue Scientifique et Technique* (International Office of Epizootics) 30 (3): 969–983.

Cheng, L. 2016. *State of play of antimicrobial resistance research and surveillance in Southeast Asia*. Bonn, Germany: SEA-EU-NET II. Available from: https://sea-eu.net/object/document/274/attach/20161201_State_of_Play_on_AMR_Research__Surveillance_-_FINAL.pdf[Accessed May 2020]

Chereau, F., L. Opatowski, M. Tourdjman, and S. Vong. 2017. Risk assessment for antibiotic resistance in South East Asia. *BMJ* 358: j3393. https://doi.org/10.1136/bmj.j3393

Chuc, N.T., and G. Tomson. 1999. "Doi moi" and private pharmacies: A case study on dispensing and financial issues in Hanoi, Vietnam. *European Journal of Clinical Pharmacology* 55 (4): 325–332.

Coker, R.J., B.M. Hunter, J.W. Rudge, M. Liverani, and P. Hanvoravongchai. 2017. Emerging infectious diseases in southeast Asia: Regional challenges to control. *Lancet* 377 (9765): 599–609.

Das, P., and R. Horton. 2017. Antibiotics: Achieving the balance between access and excess. *Lancet* 387 (10014): 102–104.

Dondorp, A.M., P.N. Newton, M. Mayxay, W. Van Damme, F.M. Smithuis, S. Yeung, et al. 2004. Fake antimalarials in Southeast Asia are a major impediment to malaria control: Multinational cross-sectional survey on the prevalence of fake antimalarials. *Tropical Medicine & International Health* 9 (12): 1241–1246.

Dondorp, A.M., S. Yeung, L. White, C. Nguon, N.P.J. Day, D. Socheat, et al. 2010. Artemisinin resistance: Current status and scenarios for containment. *Nature Reviews in Microbiology* 8 (4): 272–280.

Gollogly, L. 2002. The dilemmas of aid: Cambodia 1992–2002. *Lancet* 360 (9335): 793–798.

Hadi, U., P. van den Broek, E.P. Kolopaking, N. Zairina, W. Gardjito, I.C. Gyssens, et al. 2010. Cross-sectional study of availability and pharmaceutical quality of antibiotics requested with or without prescription (over the counter) in Surabaya, Indonesia. *BMC Infectious Diseases* 10: 203.

Heyman, G., O. Cars, M.-T. Bejarano, and S. Peterson. 2014. Access, excess, and ethics – Towards a sustainable distribution model for antibiotics. *Upsala Journal of Medical Sciences* 119 (2): 134–141.

Hill, P.S., and T.E. Mao. 2007. Resistance and renewal: Health sector reform and Cambodia's national tuberculosis programme. *Bulletin of the World Health Organization* 85: 631–636.

Hoa, N.Q., M. Larson, N.T. Kim Chuc, B. Eriksson, N.V. Trung, and C.L. Stalsby. 2009. Antibiotics and paediatric acute respiratory infections in rural Vietnam: Health-care providers' knowledge, practical competence and reported practice. *Tropical Medicine & International Health* 14 (5): 546–555.

Imwong, M., K. Suwannasin, C. Kunasol, K. Sutawong, M. Mayxay, H. Rekol, et al. 2017. The spread of artemisinin-resistant Plasmodium falciparum in the Greater Mekong subregion: A molecular epidemiology observational study. *The Lancet Infectious Diseases* 17 (5): 491–497.

Interpol. 2015. Falsified and illicit medicines worth USD 7 million seized across Asia in INTERPOL-led operation. Available from: https://www.interpol.int/en/News-and-Events/News/2015/Falsified-and-illicit-medicines-worth-USD-7-million-seized-across-Asia-in-INTERPOL-led-operation [Accessed May 2020].

WHO. 2017a. Joint External Evaluation of IHR Core Capacities of Viet Nam. Geneva: World Health Organization.

WHO. 2017b. Joint External Evaluation of the IHR Core Capacities of the Kingdom of Cambodia. Geneva: World Health Organization.

WHO. 2016. Joint External Evaluation Tool and Process Overview. Geneva: World Health Organization.

Khan, M. 2016. *Health markets and antibiotics: Unlikely places you can buy them.* Health Policy and Planning Debated Blog, London School of Hygiene and Tropical Medicine. Available from: http://blogs.lshtm.ac.uk/hppdebated/2016/11/29/health-markets-and-antibiotics/#respond [Accessed May 2020].

Khan, M.S., S. Salve, and J.D.H. Porter. 2015. Engaging for-profit providers in TB control: Lessons learnt from initiatives in South Asia. *Health Policy and Planning* 30 (10): 1289–1295.

Lamy, M., and M. Liverani. 2015. Tackling substandard and falsified medicines in the Mekong: National responses and regional prospects. *Asia & the Pacific Policy Studies* 2 (2): 245–254.

Lee, Y., and M. Wakabayashi. 2013. Key informant interview on antimicrobial resistance (AMR) in some countries in the western pacific region. *Global Health* 9: 34.

Littmann, J., and A.M. Viens. 2015. The ethical significance of antimicrobial resistance. *Public Health Ethics* 8 (3): 209–224.

Liverani, M., C. Nguon, R. Sok, D. Kim, P. Nou, S. Nguon, et al. 2017. Improving access to health care amongst vulnerable populations: A qualitative study of village malaria workers in Kampot, Cambodia. *BMC Health Services Research* 17 (1): 335.

Lon, C.T., R. Tsuyuoka, S. Phanouvong, N. Nivanna, D. Socheat, C. Sokhan, et al. 2006. Counterfeit and substandard antimalarial drugs in Cambodia. *Transactions of the Royal Society of Tropical Medicine and Hygiene* 100 (11): 1019–1024.

Mao, W., H. Vu, Z. Xie, W. Chen, and S. Tang. 2015. Systematic review on irrational use of medicines in China and Vietnam. *PLoS ONE* 10: e0117710.

Meng, C.Y., B.L. Smith, L. Bodhidatta, S.A. Richard, K. Vansith, B. Thy, et al. 2011. Etiology of diarrhea in young children and patterns of antibiotic resistance in Cambodia. *The Pediatric Infectious Disease Journal* 30 (4): 331-335.

Morgan, R., T. Ensor, and H. Waters. 2017. Performance of private sector health care: Implications for universal health coverage. *Lancet* 388 (10044): 606–612.

NIS Cambodia. 2014. Cambodia Demographic and Health Survey 2014. Phnom Penh, Cambodia: National Institute of Statistics.

National Research Council. 1999. *The use of drugs in food animals: Benefits and risks.* Washington, DC: The National Academies Press.

Nga, D.T.T., N.T.K. Chuc, N.P. Hoa, N.Q. Hoa, N.T.T. Nguyen, H.T. Loan, et al. 2014. Antibiotic sales in rural and urban pharmacies in northern Vietnam: an observational study. *BMC Pharmacology and Toxicology* 15 (1): 6.

Nguyen, V.T., J.J. Carrique-Mas, T.H. Ngo, H.M. Ho, T.T. Ha, J.I. Campbell, et al. 2015. Prevalence and risk factors for carriage of antimicrobial-resistant Escherichia coli on household and small-scale chicken farms in the Mekong Delta of Vietnam. *The Journal of Antimicrobial Chemotherapy* 70 (7): 2144–2152.

Nguyen, N.T., H.M. Nguyen, C.V. Nguyen, T.V. Nguyen, M.T. Nguyen, H.Q. Thai, et al. 2016. Use of colistin and other critical antimicrobials on pig and chicken farms in Southern Vietnam and its association with resistance in commensal Escherichia coli bacteria. *Applied and Environmental Microbiology* 82 (13): 3727–3735. http://www.ncbi.nlm.nih.gov/pmc/articles/PMC4907207/

Nhung, N.T., N.V. Cuong, J. Campbell, N.T. Hoa, J.E. Bryant, V.N.T. Truc, et al. 2015. High levels of antimicrobial resistance among Escherichia coli isolates from livestock farms and synanthropic rats and shrews in the Mekong Delta of Vietnam. *Applied and Environmental Microbiology* [Internet] 81 (3): 812–820.

Nhung, N.T., N.V. Cuong, G. Thwaites, J. Carrique-Mas. 2016. Antimicrobial usage and antimicrobial resistance in animal production in Southeast Asia: A review. *Antibiotics* 5 (4). pii: E37.

Novotny, J., A. Singh, L. Dysoley, S. Sovannaroth, and H. Rekol. 2016. Evidence of successful malaria case management policy implementation in Cambodia: Results from national ACTwatch outlet surveys. *Malaria Journal* 15 (1): 194.

Om, C., and M.-L. McLaws. 2016. Antibiotics: Practice and opinions of Cambodian commercial farmers, animal feed retailers and veterinarians. *Antimicrobial Resistance and Infection Control* 5: 42.

Om, C., F. Daily, E. Vlieghe, J.C. Mclaughlin, M.-L. Mclaws. 2017. Pervasive antibiotic misuse in the Cambodian community: antibiotic-seeking behaviour with unrestricted access. *Antimicrobial Resistance and Infection Control* [Internet] 6: 30. Available from: https://doi.org/10.1186/s13756-017-0187-y

Saengcharoen, W., and S. Lerkiatbundit. 2010. Practice and attitudes regarding the management of childhood diarrhoea among pharmacies in Thailand. *The International Journal of Pharmacy Practice* 18 (6): 323–331.

Song, J.-H., N.Y. Lee, S. Ichiyama, R. Yoshida, Y. Hirakata, W. Fu, et al. 1999. Spread of drug-resistant Streptococcus pneumoniae in Asian Countries: Asian Network for Surveillance of Resistant Pathogens (ANSORP) study. *Clinical Infectious Diseases* 28 (6): 1206–1211.

Stoesser, N., C.E. Moore, J.M. Pocock, K.P. An, K. Emary, M. Carter, et al. 2013. Pediatric bloodstream infections in Cambodia, 2007 to 2011. *The Pediatric Infectious Disease Journal* 32 (7): e272–e276.

The World Bank. 2017. *World Bank Data Bank* Available from: http://data.worldbank.org.

Turner, P., S. Pol, S. Soeng, P. Sar, L. Neou, P. Chea, et al. 2016 Aug. High prevalence of antimicrobial-resistant gram-negative colonization in hospitalized Cambodian infants. *The Pediatric Infectious Disease Journal* 35 (8): 856–861.

Van Boeckel, T.P., C. Brower, M. Gilbert, B.T. Grenfell, S.A. Levin, T.P. Robinson, et al. 2015. Global trends in antimicrobial use in food animals. *Proceedings of the National Academy of Sciences of the United States of America* 16: 1–6.

Van Minh, H., N.S. Pocock, N. Chaiyakunapruk, C. Chhorvann, H.A. Duc, P. Hanvoravongchai, et al. 2014. Progress toward universal health coverage in ASEAN. *Global Health Action* 7. https://doi.org/10.3402/gha.v7.25856.

Vinayak, S., M.T. Alam, T. Mixson-Hayden, A.M. McCollum, R. Sem, N.K. Shah, et al. 2010. Origin and evolution of sulfadoxine resistant Plasmodium falciparum. *PLOS Pathogens* 6 (3):e 1000830.

Weraphong, J., S. Pannarunothai, T. Luxananun, N. Junsri, and S. Deesawatsripetch. 2013. Catastrophic health expenditure in an urban city: Seven years after universal coverage policy in Thailand. *The Southeast Asian Journal of Tropical Medicine and Public Health* 44 (1): 124–136.

WHO. 2015. *Use of high burden countries for TB by WHO in the post-2015 era*. Geneva, Switzerland: World Health Organization.

Widayati, A., S. Suryawati, C. de Crespigny, and J.E. Hiller. 2011. Knowledge and beliefs about antibiotics among people in Yogyakarta City Indonesia: A cross sectional population-based survey. *BMC Research Notes* 4 (1): 491.

Witte, W. 1998. Medical consequences of antibiotic use in agriculture. *Science* 279: 996–997.

WHO 1998. *Emerging and other communicable diseases: Antimicrobial resistance*. World Health Assembly (WHA5117). Geneva, Switzerland: World Health Organization.

———. 2003. *Fact sheet 275: Substandard and counterfeit medicines*. Geneva: World Health Organization.

———. 2015. *Global action plan on antimicrobial resistance*. WHO Press, 1–28. Geneva, Switzerland: World Health Organization. Available from: http://www.who.int/drugresistance/global_action_plan/en/ [Accessed May 2020].

———. 2017. *Global database on antimalarial drug efficacy and resistance*. Geneva, Switzerland: World Health Organization.

Yamey, G., D. Montagu, and M. Schäferhoff. 2012. Piloting the affordable medicines facility-malaria: What will success look like? *Bulletin of the World Health Organization* 90 (6): 452–460.

Yeung, S., E. Patouillard, H. Allen, and D. Socheat. 2011. Socially-marketed rapid diagnostic tests and ACT in the private sector: Ten years of experience in Cambodia. *Malaria Journal* 10 (1): 243.

Yeung, S., H.L.S. Lawford, P. Tabernero, C. Nguon, A. van Wyk, N. Malik, et al. 2015. Quality of antimalarials at the epicenter of antimalarial drug resistance: Results from an overt and mystery client survey in Cambodia. *The American Journal of Tropical Medicine and Hygiene* 92 (Suppl 6): 39–50.

Chapter 6
Hospital Infection Prevention and Control (IPC) and Antimicrobial Stewardship (AMS): Dual Strategies to Reduce Antibiotic Resistance (ABR) in Hospitals

Gwendolyn L. Gilbert and Ian Kerridge

Abstract In this chapter we review the development of hospital infection prevention and control (IPC) since the nineteenth century and its increasingly important role in reducing the spread of antibiotic resistance (ABR). Excessive rates of hospital-acquired infection (HAI) fell dramatically, towards the end of the nineteenth century, because of improved hygiene and surgical antisepsis, but treatment remained rudimentary until effective antibiotics became widely available in the mid-twentieth century. While antibiotics had profound clinical benefits, their widespread appropriate and inappropriate use in humans and animals inevitably led to the emergence of antibiotic resistance (ABR). Within 50 years, this could no longer be offset by a reliable supply of new drugs, which slowed to a trickle in the 1980s. In hospitals, particularly, high rates of (often unnecessary) antibiotic use and ABR are exacerbated by person-to-person transmission of multi-drug resistant organisms (MDRO), which have, so far, largely resisted the introduction of antimicrobial stewardship (AMS) programs and repeated campaigns to improve infection prevention and control (IPC). Despite clear evidence of efficacy in research settings, both AMS and IPC programs are often ineffective, in practice, because of, *inter alia,* insufficient resourcing, poor implementation, lack of ongoing evaluation and failure to consult frontline staff. In this chapter we review reasons for the relatively low priority given to preventive programs despite the ethical obligation of healthcare organisations to protect current and future patients from preventable harm. The imminent

G. L. Gilbert (✉)
Sydney Health Ethics, University of Sydney, Sydney, NSW, Australia

Marie Bashir Institute for Emerging Infections and Biosecurity, University of Sydney, Sydney, NSW, Australia
e-mail: lyn.gilbert@sydney.edu.au

I. Kerridge
Sydney Health Ethics, University of Sydney, Sydney, NSW, Australia

Department of Haematology, Royal North Shore Hospital, Sydney, NSW, Australia

E. Jamrozik, M. Selgelid (eds.), *Ethics and Drug Resistance: Collective Responsibility for Global Public Health*, Public Health Ethics Analysis 5,
https://doi.org/10.1007/978-3-030-27874-8_6

threat of untreatable infections may provide an impetus for a shared organisational and professional commitment to promoting the cultural and behavioural changes needed to successfully reduce the burdens of ABR and drug-resistant HAIs.

Keywords Medicine and public health · Infectious diseases · History of medicine · Infection prevention and control · Antimicrobial resistance · Antimicrobial stewardship

6.1 Introduction

Antibiotic resistance (ABR[1]) has been described as a "slowly emerging disaster" (Viens and Littmann 2015). The risk of acquired ABR was recognised before the first antibiotics were widely used and its dire consequences have been understood by experts for many years, but the magnitude of the threat and the urgent need for radical solutions to limit its impact have been widely acknowledged only recently. In this chapter, we argue that it is not too late to mitigate the disaster and check its progress, at least in hospital settings, if certain contributory factors are acknowledged and addressed without delay. These factors, we suggest, include naïve optimism, ignorance, hubris and nihilism on the part of pharmaceutical companies, healthcare professionals (mainly prescribing doctors) and health administrators, among others.

By the first half of the twentieth century, improvements in living conditions in industrialised countries contributed to rapidly falling infectious disease mortality (Armstrong et al. 1999). At the same time, surgical antisepsis, clean wards, hand washing by clinicians and skilled nursing care (Larson 1989; Gill and Gill 2005), had diminished the risk of death in hospitals from puerperal or postoperative sepsis (Gawande 2012). But the remedies for serious infections, such as bleeding, purging or toxic infusions of arsenic, mercury or opiates, probably hastened, more than they postponed, death (Funk et al. 2009).

Antibiotics changed all that. From the beginning they were hailed as miracle drugs. Doctors embraced their use, not only to cure once life-threatening diseases, but also, because they seemed so free of adverse effects, to treat minor infections or even a perceived risk of infection. But, as early as 1945, Alexander Fleming, who shared a Nobel Prize for the discovery of penicillin, warned: "…the public will demand [penicillin]…then will begin an era…of abuses. The microbes are educated to resist penicillin and a host of penicillin-fast organisms is bred out which can be

---

[1] Most of what follows applies to antimicrobial resistance (AMR), generally, but antibiotics are by far the most commonly prescribed antimicrobials, in human medicine, which is the focus of this chapter; therefore, our discussion will mainly focus on ABR.

passed to other individuals ......In such a case the thoughtless person playing with penicillin treatment is morally responsible for the death of the man who finally succumbs to infection with the penicillin-resistant organism. " (A. Fleming, 1945,[2] quoted in (Bartlett et al. 2013)) And, indeed, within a few years most hospital isolates of previously susceptible *Staphylococcus aureus* were penicillin resistant (Barber and Rozwadowska-Dowzenko 1948).

Although they recognised that overuse would promote resistance, pharmaceutical companies aggressively promoted the benefits and safety of antibiotics to doctors and directly to the public. And, in the 1950s they recognised an even larger market when Thomas Jukes, at Lederle laboratories, demonstrated the growth-promoting effect of antibiotics in food-producing animals: "Animals receiving 10 ppm of chlortetracycline in the diet developed resistance in their intestinal bacteria in less than two days...[but] their growth rate increased. ....we concluded that if [resistant pathogens] appeared ... usefulness of the antibiotic supplement would vanish and farmers would stop feeding it. ...The [company] decision was strongly opposed by the veterinarians at Lederle, but their wishes were abruptly denied by Dr. Malcolm [Lederle President], who made an individual decision to go ahead. Competition was right on our heels..." (Jukes 1985). And it soon caught up: "The power of the calliope in the antibiotic bandwagon increased spectacularly during the [1950s] while poultry, pigs, and patients danced to its tune." (T. D. Luckey, 1959,[3] quoted in (Podolsky 2017)).

By the mid-twentieth century, it was predicted that antibiotics (and vaccines) would put an end to infectious diseases. "[T]he belief that infectious diseases had been successfully overcome was pervasive in biomedical circles - including ... a Nobel Laureate, medical Dean, and other thought leaders - from as early as 1948........" (Spellberg and Taylor-Blake 2013). It was widely assumed that if bacteria developed resistance to one drug, as they often did, there would soon be better ones to replace it; and, for many years, there were. Indeed there was such confidence that infections could be easily cured, that preventing them became a lower priority. But by the 1970s there was mounting disquiet about the emergence of ABR. Methicillin resistant *S. aureus* (MRSA) (Jevons et al. 1963) and transmissible resistance in Gram-negative bacteria (Datta and Pridie 1960) had emerged in the 1960s and their prevalence was increasing, especially in hospitals (Chambers 2001; Aminov 2010; Ventola 2015). The first International Conference on Nosocomial Infections, in 1970, was followed by the Study on the Efficacy of Nosocomial Infection Control (SENIC), in the USA (Forder 2007); in Australia, the handbook of "Antibiotic Guidelines" was published for the first time, in 1978 (Harvey et al. 2003). Anxiety that antibiotics would progressively lose efficacy, became more acute in the 1980s, when the seemingly unlimited flow of new antibiotics slowed to

---

[2] Alexander Fleming. Penicillin"s finder assays its future. New York Times 26 June 1945: 21

[3] Luckey TD. Antibiotics in nutrition. In: Goldberg HS, ed. Antibiotics: Their Chemistry and Non-Medical Uses. Princeton: D. Van Nostrand; 1959:174–321.

a trickle, and pharmaceutical companies turned their attention to more profitable projects.

Antibiotic use is no longer regarded as an unquestioned good. Antimicrobial resistance (AMR) is now seen as a threat to global health security and the "end of the antibiotic era" predicted; it is broadly accepted that urgent measures are needed to salvage what we can of the "antibiotic miracle": more prudent use of existing antimicrobial agents; novel strategies to promote development of new ones; and greater attention to preventing the spread of drug resistance organisms (WHO 2012; CDC 2013).

6.2 Hospital Infection in the "Pre-Antibiotic Era"

In nineteenth century Europe, medical science advanced rapidly; there was increasing demand for new hospitals, where university-trained doctors could develop and experiment with new remedies and advance their knowledge. Anaesthetics increased the scope of surgery (Gawande 2012) and pregnant women were more likely to be admitted to 'lying-in' hospitals, where advances in obstetrics offered relief from excessively long, difficult labour (Loudon 1986). But hospitals were overcrowded, dirty and poorly ventilated; infectious disease outbreaks were common and maternal mortality from childbed fever much higher in hospitals than in the community.

Alexander Gordon, an Aberdeen physician, had recognised puerperal fever as a "specific contagion or infection" that could be carried between parturient women on the hands of her attendants, in 1795 (Gordon 1795), but his work was ignored until the 1840s, when James Young Simpson, in Edinburgh (Selwyn 1965), Oliver Wendell Holmes, in Boston (Holmes 2001), and Ignaz Semmelweis (Nuland 1979; Carter 1981), in Vienna, independently came to similar conclusions. Simpson also recognised that puerperal and surgical fevers were "intercommunicable" and coined the term "hospitalism" to describe outbreaks of surgical infection, which he believed were so serious that "...every patient placed upon an operating table ... is in ... greater danger than a soldier entering one of the bloodiest and most fatal battlefields" (J. Y. Simpson, 1859 quoted in (Selwyn 1965)).

In Vienna, Semmelweis was troubled by the much higher maternal mortality, in a clinic staffed by doctors and medical students, than in an otherwise similar clinic staffed by midwives. After months of investigation, he realised that the only significant difference between the clinics was that, unlike the midwives, the students and doctors frequented the mortuary, dissecting the bodies of women who had died from childbed fever. When they returned to the clinic, they carried on their hands "cadaver particles which are not entirely removed by the ordinary method of washing the hands with soap....[Therefore] the hands of the examiner must be cleansed with chlorine, not only after handling cadavers, but likewise after examining patients" (Semmelweis 1983). After he enforced this strict hand washing regime, the maternal mortality in the doctors' clinic rapidly fell to a level similar to that in the midwives' clinic (Nuland 1979). Despite the evidence, Semmelweis' findings lacked a conceptual basis, 20 years in advance of Pasteur's germ theory of disease, and were largely rejected by his peers. His accusation that anyone who did not follow his recommendations would be 'murderers', undoubtedly contributed to their antagonism (Pittet 2004; Saint et al. 2010a).

During the Crimean war, in 1854, the British public was scandalised by a newspaper report of deplorable conditions in the British Army hospital at Scutari. When Florence Nightingale was sent to investigate, she found vermin- and lice-infested wards, excreta on walls and floors, injured soldiers in dirty, bloodstained clothes and frequent infectious disease outbreaks. Nightingale believed that disease was caused by filth and foul air (miasmas); she overcame bitter opposition from the military surgeons and formidable logistic barriers to implement a strict regime - immediate wound care; clean dressings, clothes and bedding; nutritious food; and regular cleaning and ventilation of wards. Her meticulous records showed that soldiers were far more likely to die from preventable infection than war wounds. In the January–March quarter of 1855, the mortality at Scutari was 33%, by the July–September quarter it was 2%. While critics have belittled her achievements, her methods remain the basis of good nursing care and hospital infection control (Larson 1989; Gill and Gill 2005).

In the 1860s, Joseph Lister's knowledge of Pasteur's germ theory informed his belief that the almost inevitable (and often fatal) suppuration that complicated compound fractures and amputations was due to "minute organisms suspended in [the atmosphere], which owed their energy to their vitality" (Lister 1867). By liberal use of carbolic acid to soak wound dressings and disinfect his hands, instruments and the operative site, he was able to manage most compound fractures without amputation and the post-amputation mortality fell from 46% (16/35) in 1864–6 to 15% (6/40) in 1867–9 (Newsom 2003). Many of his compatriots ridiculed his methods, but they were gradually accepted, particularly in Europe. His acknowledged place as the "father of antiseptic surgery" owes much to its basis in the germ theory, which gave it an authority that Semmelweis' earlier work lacked.

As antisepsis and later asepsis, environmental hygiene and skilled nursing care were gradually incorporated into hospital practice, hospital infection rates fell and hospitals became places of healing rather than dying.

6.3   The Antibiotic Era

Acquired antibiotic resistance (ABR), of bacterial pathogens that affect humans, is mainly due to nearly 75 years' of both appropriate and inappropriate antibiotic use in human medicine and dissemination of resistant organisms. Antibiotic use in agriculture and veterinary practice and environmental contamination are also implicated, but their contributions are contested and vary from region to region (Landers et al. 2012; Marshall and Levy 2011; Chang et al. 2015). The dynamics are complex and incompletely understood (Turnidge and Christiansen 2005) but, in general, exposure of bacteria to antibiotics exerts powerful selection pressure; the greater the total amount and the broader the antibacterial spectra of antibiotics used in any setting, such as a hospital (Willemsen et al. 2009; Xu et al. 2013) or community (Goossens et al. 2005; Bell et al. 2014), the higher the prevalence of ABR. More antibiotics (by weight) are prescribed for non-human animals than people and environmental contamination is a major contributor to ABR. Nevertheless, although inappropriate prescribing is probably the main contributor to drug resistant infections humans, it is now accepted that control of AMR/ABR requires a One Health approach (Robinson et al. 2016). However, the focus of this chapter is on ABR in hospitals, where multidrug resistant organisms (MDROs)[4] are most obvious and prevalent[5] and preventive measures most likely to be effective.

Most life-threatening infections are treated in hospitals, where the greatest variety of antibiotics is used, in repeated courses or for prolonged periods. In hospitals, busy healthcare professionals often carry MDROs on their hands, exposing patients to the risk of healthcare-associated infection (HAI) or colonisation with an MDRO. Hospital laboratory reports increase prescribers' awareness of ABR and, perhaps, promote a (mistaken) perception that it is ubiquitous; this may encourage defensive prescribing – e.g. of multiple or broad-spectrum agents - to avert treatment failure. Paradoxically, increased awareness of ABR is not necessarily reflected in increased adherence to measures designed to prevent it. Now that it is recognised that profligate antibiotic use promotes ABR and inadequate infection prevention and control (IPC) facilitates transmission, the challenge is to break these intersecting vicious cycles, particularly in hospitals, where they are most apparent.

6.4 Antibiotic Use and Stewardship in Hospitals

Antibiotics are prescribed very frequently in hospitals; studies in USA and Australia have shown that more than 50% of hospital patients receive at least one antibiotic and up to 50% of prescriptions are unnecessary or inappropriate, according to prescribing guidelines (Turnidge et al. 2016; Baggs et al. 2016; Reddy et al. 2015). Making the right antibiotic prescribing decision is difficult, even for an experienced practitioner, particularly when the diagnosis is uncertain. For patients with sepsis - especially those most at risk of life-threatening infections, e.g. who are neutropenic or immune-compromised - delay can lead to serious complications or death from septic shock (Kumar et al. 2006). But antibiotics prescribed empirically are often continued, even after an alternative diagnosis is made, or not reviewed, despite laboratory results that indicate the empirical choice was ineffective or unnecessarily

---

[4] MDROs are resistant to more than one - usually several – classes of antibiotic; they include methicillin resistant *Staphylococcus aureus* (MRSA), vancomycin-resistant enterococci (VRE) and extended spectrum β lactamase- and carbapenemase-producing (ESBL and CPE, respectively) Enterobacteriaceae.

[5] In countries where antibiotics can be used in humans or animals without restriction, high rates of ABR occur in the community and in hospitals.

broad-spectrum (Braykov et al. 2014). Antibiotics are also often given in combination, in the wrong dose, by the wrong route or for too short or too long a period (Dryden et al. 2011; Gilbert 2015). Any of these errors can contribute to ABR, without concomitant benefit to the patient herself and, potentially, with significant harm including an increased risk of *Clostridium difficile* infection, MDRO acquisition or prolonged disruption of the gut microbiome, with potentially serious adverse effects (Dingle et al. 2017; Becattini et al. 2016).

Diagnostic uncertainty is not the only, or even the most common, reason for inappropriate prescribing. The prescriber's rational concern can transform into excessive risk aversion, without regard for antibiotic conservation or potential adverse effects on the patient. A junior hospital doctor may consider ABR "morally and professionally important..." but "of limited concern at the bedside" (Broom et al. 2014). Fear of missing something or being criticised, by peers or superiors, outweighs consideration of long-term risks to future patients or the environment. She may prescribe an antibiotic, even if she thinks it unnecessary or futile, because of inexperience, her consultant's routine practice or a duty of benevolence towards her patient that makes her want to do something. Junior doctors are required to make complicated prescribing decisions, often in the face of conflicting, inconsistent (or no) advice or feedback (Mattick et al. 2014). Broom et al. (Broom et al. 2014) concluded that "..social risks, including the peer-based and hierarchical reputational consequences associated with 'not doing enough'" are far more potent than the risk of ABR (Broom et al. 2014).

Over the past 10 years, programs have been introduced into hospital practice in many countries with the aim of minimising inappropriate antibiotic therapy. Antimicrobial stewardship (AMS) programs aim to ensure that patients are given antibiotics when they need them – "the right drug, at the right time, in the right dose and for the right duration" (Dryden et al. 2011; Doernberg and Chambers 2017) – with the least possible selection pressure. They consist of 'bundles' of interventions, including: restrictions on the use of certain key antibiotics, except with specific authorisation; prescriber education and academic detailing; audit of prescribing patterns, with feedback to prescribers; optimisation of laboratory testing, including rapid diagnostics; and technological support such as electronic access to microbiology results and computerised decision support systems (Davey et al. 2017).

AMS programs depend on multidisciplinary teams - including infectious disease physicians, clinical microbiologists, specialist antimicrobial pharmacists and/or IPC professionals - whose complementary expertise and spheres of influence provide mutual support and greater authority than each has, individually. The specialist pharmacist's expertise in drug dosing, interactions and administration and her role in implementing regulations, such as automatic stop orders or restricted drug authorisation, and auditing prescribing records, are critical to an AMS program. Nevertheless, even the most experienced pharmacist or AMS team can encounter resistance from a senior specialist who may regard their advice as a threat to her autonomy and clinical experience (Broom et al. 2016).

How effective an AMS program will be depends on what it includes and how it is implemented. A recent systematic review (Davey et al. 2017) showed that providing advice and feedback to clinicians improved prescribing and reduced over-all antibiotic use more than simply imposing rules and restrictions, suggesting that AMS programs that support prescribers help to mitigate the fear of censure or litiga-tion that often drives inappropriate prescribing. Overall, studies of AMS show that it can reduce inappropriate prescribing, pharmacy costs and avoidable drug reac-tions; improve therapeutic drug monitoring; shorten hospital length of stay by an average of 2 days; and may reduce rates of *C. difficile,* fungal and MDRO infections (Davey et al. 2013; Baur et al. 2017). Although some studies have confirmed the cost-effectiveness of AMS, more high-quality research is needed (Coulter et al. 2015).

Clearly, eliminating inappropriate antibiotic use is necessary, but not sufficient, to reduce the impact of ABR, which is exacerbated by hospital spread of MDROs.

6.5 Hospital Infection Prevention and Control (IPC) and ABR

6.5.1 Healthcare-Associated Infections and Their Consequences

According to WHO ".....HAI is the most frequent adverse event in health care [but] its true global burden remains unknown because of the difficulty in gathering reli-able data" (WHO 2011). A systematic review of HAIs in high and middle/low-income countries showed that 3.5%–12% of hospitalised patients develop at least one HAI (WHO 2011), of which 50%, or more, are potentially preventable (Harbarth et al. 2003; Umscheid et al. 2011). The estimated number of people, globally, who die from drug-resistant infections each year – currently at least 700,000 - is pre-dicted to increase to ten million by 2050 (O'Neill 2016). HAIs caused by MDROs are more difficult to treat, more likely to be fatal and more costly than comparable HAIs due to antibiotic-susceptible pathogens (Cosgrove 2006; de Kraker et al. 2011). HAI risks are associated with patient factors: severity of illness and co-morbidities such as chronic organ failure, malnutrition, immune-suppression, seri-ous trauma or contaminated surgery; and organisational factors: bed occupancy rate; staff workload; hospital environment and infrastructure; prevalence of endemic or introduced MRDO pathogens; adherence of healthcare workers to basic hygiene principles (Clements et al. 2008; Daud-Gallotti et al. 2012; Scheithauer et al. 2017). Hospital IPC policies are designed to minimise these risks and the burden of HAIs and ABR. Unlike other adverse hospital events, MDRO colonisation and HAIs are not confined to individuals; they are communal threats that affect other patients, hospital staff and the wider community and add to the burden of AMR.

Even without clinically significant infection, MDRO colonisation has significant impacts. Patients colonised with certain high-impact MDROs[6] are identified by active admission screening and isolated in single rooms, with contact precautions.[7] These are expensive measures and they can adversely affect patient care and wellbeing. Patients in contact isolation are, on average, visited by healthcare workers less often and for shorter periods; less likely to be examined by a doctor; more likely to suffer non-infectious adverse effects, such as falls, pressure sores and fluid and electrolyte imbalance; and more likely to express dissatisfaction with their hospital care, than other patients (Saint et al. 2003; Stelfox et al. 2003; Morgan et al. 2014). They may feel stigmatised and anxious about risks to family members; they and their visitors are often inadequately informed or given conflicting information about the implications of MDRO colonisation and how to protect themselves and others (Wyer et al. 2015; Seibert et al. 2017). Contact isolation is also burdensome to healthcare workers. Compliance with hand hygiene and use of personal protective equipment (and, thus, the effectiveness of contact isolation) is often relatively poor and likely to deteriorate as the number of isolated patients increases (Clock et al. 2010; Dhar et al. 2014).

Moreover, there is conflicting evidence (Cohen et al. 2015; Morgan et al. 2017), as to whether active screening and contact isolation prevent MDRO transmission more effectively than less expensive and burdensome measures such as strict adherence to standard precautions[8] (Huskins et al. 2011) and/or targeted contact isolation of patients with other risk factors (e.g. diarrhoea, open wounds) (Djibre et al. 2017). This raises the question as to whether these practices are ethically justified, based on the precautionary principle - i.e. that they might prevent harm to others - or cost-effective. Patients are rarely asked for their consent to be screened or informed of the consequences of a positive result, although contact isolation will restrict their liberty for others', but not their own, benefit. It is arguable that these measures *are* ethically justifiable if: the specific MRDO for which they are implemented is particularly dangerous; patients are fully informed, before they are screened, of the reasons, implications and benefits of MDRO colonisation; and there is adequate staffing to ensure they are implemented with optimal effectiveness and minimal risk. An alternative approach would be to promote strict adherence to standard precautions, by all staff, behind a "veil of ignorance" by assuming that any patient might be MDRO-colonised and engaging patients - when their condition permits - and their visitors as active participants in IPC (Ahmad et al. 2016). If given an opportunity and adequate information, patients can make positive contributions to IPC, including how to minimise MDRO transmission and the adverse effects of contact isolation (Wyer et al. 2015).

---

[6] MDROs for which special control measures are used are chosen according to criteria such as: extent of resistance, transmissibility, virulence and local prevalence: e.g. they often include MRSA, VRE and CPE.

[7] Contact precautions include routine use of gown and gloves when caring for patients in isolation, in addition to standard precautions, which include hand hygiene as the main component, and use of personal protective equipment when exposure to patient's blood or body fluid is anticipated

[8] Mainly strict compliance with hand hygiene.

6.5.2 Hospital IPC Programs

Given the adverse individual and communal effects and excess costs of HAIs and MDRO colonisation, healthcare organisations and professionals have an unequivocal duty-of-care and ethical responsibility to take appropriate measures to prevent them. Hospital IPC programs include, *inter alia*: hand hygiene; appropriate use of personal protective equipment; aseptic technique for invasive procedures; environmental hygiene; air flow; bundles of measures to prevent certain HAI syndromes, such as device-related blood stream infections; surveillance of selected HAIs and feed-back of data to clinicians; and isolation of infected and MDRO-colonised patients, with the caveats outlined above (Sydnor and Perl 2011).

It is usually impossible to trace an HAI or MRDO transmission event to a specific action, omission or individual healthcare worker, because there are inevitable time gaps and multiple factors and people involved (McLaws 2015). HAIs "do not carry fingerprints … to identify the offending healers who failed the patient." (Palmore and Henderson 2012).

The effectiveness of an IPC program depends on organisational commitment, adequate resources and participation of everyone in the hospital community. Despite the compelling ethical imperative to "do no harm", the economic burden of HAIs (Stone 2009) and proven cost-effectiveness (Dick et al. 2015) of IPC programs, they often struggle to attract the necessary support and resources. Their beneficiaries, like those of any preventive program, are unknown "statistical lives" whose demands are far less compelling than those of known "identified lives" who need immediate, often expensive, interventions (Cookson et al. 2008; Beauchamp and Childress 2009). The typically low priority of IPC is reflected in a vicious cycle of inadequate resources, poor compliance - and, hence, limited effectiveness - which can then seem to justify further cost cutting.

The basic principles of hospital IPC were recognised in the nineteenth century and incorporated into routine hospital policy in the latter part of the twentieth, when it became clear that antibiotics, alone, could not prevent morbidity and mortality from HAIs. Nevertheless, implementation and maintenance of IPC programs remain a major challenge. Variation in HAI rates, between otherwise comparable hospitals, presumably reflects differences in organizational cultures, policies, working conditions (Daud-Gallotti et al. 2012; Krein et al. 2010) and professional attitudes, behaviours and leadership (Saint et al. 2010a), suggesting that improvement is possible.

6.5.3 The Central Role of Hand Hygiene in IPC

"In the absence of the possibility to directly link individual infectious outcomes to individual hand hygiene failures… hand hygiene performance remains the only measure to judge the degree of system safety…." (Stewardson et al. 2011). Despite the proven effectiveness of hand hygiene in preventing MDRO transmission (Pittet et al. 2000; Johnson et al. 2005), compliance is often poor. The discovery, about

20 years ago, that it could be improved by use of alcohol-based hand rub (ABHR), was a major breakthrough. ABHR has significant advantages over traditional hand washing with soap and water: it takes less time, is less irritant to hands, accessible at the point-of-care and in settings without access to clean water and has more rapid and potent antibacterial action (Pittet et al. 2000; Widmer 2000). In 2007, "My Five Moments of Hand Hygiene" was introduced to improve healthcare worker training and standardise monitoring and reporting of hand hygiene compliance (Sax et al. 2007); in 2009, the "Five Moments" were incorporated into WHO hand hygiene guidelines. Since then, there have been numerous studies and reviews of factors affecting compliance and interventions to improve it (Erasmus et al. 2009; Huis et al. 2012; Neo et al. 2016; Kingston et al. 2016).

One review reported an overall average compliance of 40%; it was lower in ICUs (30–40%) than other settings (50–60%); among doctors (32%) than nurses (48%) and for moment one (before patient contact; 21%) than moment four (after patient contact; 47%). Performing dirty tasks, availability of ABHR, and performance feedback were associated with better compliance (Erasmus et al. 2009). A review of interventions reported an average pre-intervention compliance of 34% with variable, but modest improvements (8–31%) and mean post-intervention compliance of 57%. Multimodal interventions included various combinations of staff education, facility design and planning, HAI surveillance and/or compliance monitoring with feedback, financial incentives and active support by clinical champions and hospital administrations (Kingston et al. 2016).

These studies illustrate the enormous effort entailed in achieving even modest, often short-term, improvements. They contrast with the, apparently, much better compliance achieved by the Australian National Hand Hygiene Initiative, which was established in 2009 as a "standardised hand hygiene culture-change program throughout all Australian hospitals to improve … compliance among Australian health care workers; establish a validated system of …auditing to allow local, national and international benchmarking…" (Grayson et al. 2011). Between 2009 and June 2017, overall compliance increased, steadily, from 63% to 85%, which clearly represents major improvement, to above the national benchmark (70%) (http://www.hha.org.au/). But it obscures significant variation (e.g. between hospitals, professional groups and moments) and sampling biases, suggesting aggregated national data can be misleading. Moreover, the estimated auditing cost is AU$2.2 million per annum for an annual improvement (adjusted for sampling) of 1% compliance (Azim and McLaws 2014).

How should these data be interpreted? Routine audits, by direct observation, necessarily involve short periods of observation (20–30 minutes) at times of convenience and so are not representative of 24 hour/whole-of-hospital activity; it was estimated that <2% of total daily hand hygiene opportunities are sampled during a 60-minute audit (Fries et al. 2012). Auditing by direct observation is subject to the Hawthorne effect[9] (Srigley et al. 2014) and observer bias; it does not assess whether hand hygiene is performed correctly and, when compared with continuous

---

[9] Hawthorne effect: individuals modify their behaviour in response to awareness of being observed.

automated monitoring, overestimates compliance (Kwok et al. 2016). Furthermore, there is no consensus as to what target compliance rate is necessary or realistic (Mahida 2016). Video or electronic monitoring systems would reduce workload, measure compliance more consistently and could improve it, by providing rapid feedback and prompts (Srigley et al. 2015), but experience with their use is limited and there are many logistic, industrial, and ethical challenges (Palmore and Henderson 2012; Conway 2016). There is clearly a need to establish realistic compliance targets, more accurate, less labour intensive auditing methods and a more holistic approach to IPC monitoring.

6.5.4   Doctors and IPC

There is extensive evidence than doctors' hand hygiene compliance is consistently less than that of nurses, overall, albeit highly variable (Pittet et al. 2004; Cantrell et al. 2009). In one hospital it was more than 80% among physicians and paediatricians but only 30% among surgeons and anaesthetists (Pittet et al. 2004). Compliance has been associated with an emotional, outcome-oriented, rather than a 'rational', thinking style – typically associated with nurses and doctors, respectively (Sladek et al. 2008) - and inversely correlated with professional education level i.e. senior doctors were less compliant than junior doctors or nurses (Duggan et al. 2008). These differences matter: senior doctors have status and power within hospital communities and their attitudes and behaviours disproportionately influence those of other staff (Lankford et al. 2003). Poorly compliant, peripatetic junior doctors can act as "super-spreaders", with many opportunities to transmit pathogens between the numerous individual patients they encounter each day (Temime et al. 2009; Hornbeck et al. 2012). Doctors are more likely to perform hand hygiene *after* patient contact, to avoid a perceived personal risk, than *before* contact, to protect patients (Scheithauer et al. 2011). Many believe these are equivalent, but overlook the many opportunities for contamination of their hands, from bed curtains, patient notes, door handles, computer keyboards etc., between patients. In a focus-group study of hospital staff, most non-physician participants said they noticed the hand hygiene practices of other staff and rated doctors least compliant. Doctors were confident of their hand hygiene knowledge but discounted its importance before patient contact. They rarely noticed the practices of others, apart from their senior colleagues; medical students said that senior doctors' hand hygiene practices influenced their own (Jang et al. 2010a, 2010b).

The reasons for some doctors' apparent lack of commitment to IPC may lie in the medical practice model, which focuses on individual patient's clinical problems, which require investigation, decision-making, intervention, often with tangible results. IPC policies fit poorly with this model; they lack obvious utility, since they do not, meaningfully, influence clinical practice. The common perception that HAIs and MDRO acquisition are rare, but unavoidable, can promote a sense of nihilism – that they are inevitable features of contemporary healthcare. Often this is attributable to ignorance of the impacts of HAIs and benefits of IPC, which is partly because of inadequate surveillance and feedback to clinicians.

Some doctors' apparent indifference or even hostility towards IPC may also reflect aspects of professional and organisational cultures. Traditionally, IPC has been a nursing responsibility; the role of infection control practitioners' (usually nurses) is to implement IPC policies on behalf of hospital management, who have to report, against mandatory IPC performance indicators, to a central authority (e.g. Ministry of Health). Responsibility for monitoring these indicators, such as hand hygiene compliance, is often devolved to nurse managers, who are held to account if targets are not met. But they have little influence over doctors, who choose to ignore rules they see as unnecessary or excessive or who object, on principle, to any regulation, imposed by nurses or managers. "Senior doctors consider themselves exempt from following policy and practice within a culture of perceived autonomous decision-making that relies more on personal knowledge and experience than formal policy" (Charani et al. 2013). This professional antipathy to IPC, may also reflect the historical - but gradually changing - gender distribution between nursing, which has been a largely female profession, and medicine, traditionally dominated by men, particularly in senior positions. Doctors' attitudes to IPC are consistent with a more general failure - there are many exceptions - to engage in quality improvement activities or comply with organisational policies, which have been linked to an entrenched medical culture (Jorm and Kam 2004) and/or to the perceived loss of professional autonomy and dominance associated with managerialism (Davies and Harrison 2003).

How widespread these attitudes are and how the hospital management handles them will determine the success or failure of hospital-wide quality improvement programs such as IPC or AMS. If they are tolerated or seen as "too hard" to address, the morale of other staff and the success of the program will be compromised and recalcitrant doctors' skepticism about its relevance, reinforced. On the other hand, some IPC rules *are* unnecessarily rigid and officiously enforced. They may seem straightforward to their authors, but poor compliance is sometimes due to clinicians' finding them confusing, incompatible with the realities of frontline practice or inappropriate in some settings (Hor et al. 2017). There are faults on both sides when, ideally, all "sides" should be working collaboratively to promote patient safety. The issues need to be addressed if healthcare management and staff are to fulfil their moral responsibility to support and participate in programs that promote patient safety. Individuals "are not somehow 'outside' and separate from 'systems': they create, modify and are subject to the social forces that are an inescapable feature of any organizational system; each element acts on the other" (Aveling et al. 2016). The success of any program is likely to hang on the extent to which the values and goals of all of its members – particularly the most influential - align with, and contribute to, those of the organisation.

6.5.5   The Organization's Role in IPC/AMS Programs

Government and private healthcare funding bodies generally mandate that each hospital has established IPC and AMS programs and reports regularly, sometimes publicly, against mandatory performance targets. This does not necessarily guarantee the programs' success; there are wide variations in practices and outcomes between apparently similar hospitals. Most studies of successful IPC/AMS interventions have focused on "what" works, rather than "why" it works. The components of organisational culture likely to determine the success or failure of program implementation are leadership, shared vision and values, inter-professional relationships, resources and service priorities (Krein et al. 2010). Successful implementation of IPC/AMS, requires commitment by hospital management, strong clinical leadership (Saint et al. 2010b), highly motivated champions (Damschroder et al. 2009) and interdisciplinary departmental teams. The goals, benefits and measures of success of the programs need to be clearly defined, but flexible enough to allow local modification, based on the knowledge and experience of frontline staff. Imposing cultural change from without is less likely to be sustainable than allowing frontline staff to discover how to change it from within (Iedema et al. 2015; Zimmerman et al. 2013). Measures of success should be defined, monitored and rewarded, at least by timely feedback, if not more tangible recognition.

6.6   Conclusions

AMR is an acknowledged threat to global health security and will not be adequately addressed by development of new antibiotics. The most urgent priority is to curtail the inappropriate use of antibiotics and spread of MDROs, which are most prevalent in hospitals where they are also most amenable to control. Despite evidence that properly implemented hospital AMS/IPC programs can reduce the burden of ABR, the increasing prevalence of preventable HAIs, show that many healthcare organisations have either not recognised, or failed to meet, the challenge. In this chapter, we have identified some of the barriers to successful implementation of AMS/IPC programs; although they are usually mandatory, in hospitals, their quality and outcomes vary. The organisational characteristics most likely to assure successful implementation include: commitment to a shared vision and values; strong leadership, governance and systems; respectful, collaborative inter-professional relationships and fair, cost-effective resource allocation.

Healthcare organisations and hospital managers have ethical and legal responsibilities to protect existing patients, employees and the public – and future patients whose treatment will be compromised by a lack of effective antimicrobial therapy - from preventable harm originating in hospital facilities or activities. Unfortunately, preventive programs are often a low priority because their beneficiaries are unknown future persons whose claims are eclipsed by known, present persons and powerful professional or commercial interests. Preventive programs also generally lack the solid, cost-effectiveness data that administrators demand before committing resources, especially if it is at the expense of therapies. A case for adequate resources to sustain AMS/IPC programs would, ideally, include *local,* as well as published, statistics on current rates, costs and adverse consequences of HAIs and ABR and personal histories of known patients who have suffered adverse effects from an HAI, contact isolation or inappropriate antibiotic administration.

Successful AMS/IPC policies will be adaptable to unit/department-specific requirements rather than rigidly imposed rules and restrictions, which fail to account for variable, unpredictable clinical exigencies and so are liable to be ignored or circumvented. Effective policy implementation needs frontline ownership of "practice-based guidelines", based on local knowledge, including potential patient participation.

Policies and implementation plans often fail to clearly define their goals or how success will be measured. Evidence of success that can be rapidly fed back to staff is a strong motivator of adherence, but many hospital managers fail to invest in HAI surveillance and feedback to clinicians that is relevant, accessible and timely enough to motivate improvement. Despite the importance of hand hygiene compliance, its prominence as a single (often inaccurate) measure of IPC practice risks neglecting other important cultural and behavioural factors – teamwork, interdisciplinary co-operation and motivation – that determine the effectiveness of a hospital's AMS/IPC programs.

Securing the commitment, of an often small, but powerful, minority of senior medical staff, who regard AMS/IPC programs as a threat to professional autonomy and status, remains a challenge for many hospitals. It requires, as a minimum, respectful consultation during program planning, recruitment of clinical leaders and champions and, once a decision is made to adopt it, clarity that all staff are expected to support and participate in programs to which the organisation is committed.

References

Ahmad, R., et al. 2016. Defining the user role in infection control. *The Journal of Hospital Infection* 92 (4): 321–327.

Aminov, R.I. 2010. A brief history of the antibiotic era: Lessons learned and challenges for the future. *Frontiers in Microbiology* 1: 134.

Armstrong, G.L., L.A. Conn, and R.W. Pinner. 1999. Trends in infectious disease mortality in the United States during the 20th century. *JAMA* 281 (1): 61–66.

Aveling, E.L., M. Parker, and M. Dixon-Woods. 2016. What is the role of individual accountability in patient safety? A multi-site ethnographic study. *Sociology of Health & Illness* 38 (2): 216–232.

Azim, S., and M.L. McLaws. 2014. Doctor, do you have a moment? National Hand Hygiene Initiative compliance in Australian hospitals. *The Medical Journal of Australia* 200 (9): 534–537.

Baggs, J., et al. 2016. Estimating National Trends in inpatient antibiotic use among US hospitals from 2006 to 2012. *JAMA Internal Medicine* 176 (11): 1639–1648.

Barber, M., and M. Rozwadowska-Dowzenko. 1948. Infection by penicillin-resistant staphylococci. *Lancet* 2 (6530): 641–644.

Bartlett, J.G., D.N. Gilbert, and B. Spellberg. 2013. Seven ways to preserve the miracle of antibiotics. *Clinical Infectious Diseases* 56 (10): 1445–1450.

Baur, D., et al. 2017. Effect of antibiotic stewardship on the incidence of infection and colonisation with antibiotic-resistant bacteria and Clostridium difficile infection: A systematic review and meta-analysis. *The Lancet Infectious Diseases* 17 (9): 990–1001.

Beauchamp, T.L., and James F. Childress. 2009. *Principles of biomedical ethics*. 7th ed. New York: Oxford University Press.

Becattini, S., Y. Taur, and E.G. Pamer. 2016. Antibiotic-induced changes in the intestinal microbiota and disease. *Trends in Molecular Medicine* 22 (6): 458–478.

Bell, B.G., et al. 2014. A systematic review and meta-analysis of the effects of antibiotic consumption on antibiotic resistance. *BMC Infectious Diseases* 14: 13.

Braykov, N.P., et al. 2014. Assessment of empirical antibiotic therapy optimisation in six hospitals: An observational cohort study. *The Lancet Infectious Diseases* 14 (12): 1220–1227.

Broom, A., J. Broom, and E. Kirby. 2014. Cultures of resistance? A Bourdieusian analysis of doctors' antibiotic prescribing. *Social Science & Medicine* 110: 81–88.

Broom, A., et al. 2016. A qualitative study of hospital pharmacists and antibiotic governance: Negotiating interprofessional responsibilities, expertise and resource constraints. *BMC Health Services Research* 16: 43.

Cantrell, D., et al. 2009. Hand hygiene compliance by physicians: Marked heterogeneity due to local culture? *American Journal of Infection Control* 37 (4): 301–305.

Carter, K.C. 1981. SemmelWeis and his predecessors. *Medical History* 25 (1): 57–72.

CDC. 2013. *Antibiotic threats in the USA.*.

Chambers, H.F. 2001. The changing epidemiology of Staphylococcus aureus? *Emerging Infectious Diseases* 7 (2): 178–182.

Chang, Q., et al. 2015. Antibiotics in agriculture and the risk to human health: How worried should we be? *Evolutionary Applications* 8 (3): 240–247.

Charani, E., et al. 2013. Understanding the determinants of antimicrobial prescribing within hospitals: The role of "prescribing etiquette". *Clinical Infectious Diseases* 57 (2): 188–196.

Clements, A., et al. 2008. Overcrowding and understaffing in modern health-care systems: Key determinants in meticillin-resistant Staphylococcus aureus transmission. *The Lancet Infectious Diseases* 8 (7): 427–434.

Clock, S.A., et al. 2010. Contact precautions for multidrug-resistant organisms: Current recommendations and actual practice. *American Journal of Infection Control* 38 (2): 105–111.

Cohen, C.C., B. Cohen, and J. Shang. 2015. Effectiveness of contact precautions against multidrug-resistant organism transmission in acute care: A systematic review of the literature. *The Journal of Hospital Infection* 90 (4): 275–284.

Conway, L.J. 2016. Challenges in implementing electronic hand hygiene monitoring systems. *American Journal of Infection Control* 44 (5 Suppl): e7–e12.

Cookson, R., C. McCabe, and A. Tsuchiya. 2008. Public healthcare resource allocation and the rule of rescue. *Journal of Medical Ethics* 34 (7): 540–544.

Cosgrove, S.E. 2006. The relationship between antimicrobial resistance and patient outcomes: Mortality, length of hospital stay, and health care costs. *Clinical Infectious Diseases* 42 (Suppl 2): S82–S89.

Coulter, S., et al. 2015. The need for cost-effectiveness analyses of antimicrobial stewardship programmes: A structured review. *International Journal of Antimicrobial Agents* 46 (2): 140–149.

Damschroder, L.J., et al. 2009. The role of the champion in infection prevention: Results from a multisite qualitative study. *Quality & Safety in Health Care* 18 (6): 434–440.

Datta, N., and R.B. Pridie. 1960. An outbreak of infection with Salmonella typhimurium in a general hospital. *The Journal of Hygiene (London)* 58: 229–240.

Daud-Gallotti, R.M., et al. 2012. Nursing workload as a risk factor for healthcare associated infections in ICU: A prospective study. *PLoS One* 7 (12): e52342.

Davey, P., et al. 2013. Interventions to improve antibiotic prescribing practices for hospital inpatients. *Cochrane Database of Systematic Reviews* 4: CD003543.

————. 2017. Interventions to improve antibiotic prescribing practices for hospital inpatients. *Cochrane Database of Systematic Reviews* 2: CD003543.

Davies, H.T., and S. Harrison. 2003. Trends in doctor-manager relationships. *BMJ* 326 (7390): 646–649.

de Kraker, M.E., et al. 2011. Mortality and hospital stay associated with resistant Staphylococcus aureus and Escherichia coli bacteremia: Estimating the burden of antibiotic resistance in Europe. *PLoS Medicine* 8 (10): e1001104.

Dhar, S., et al. 2014. Contact precautions: More is not necessarily better. *Infection Control and Hospital Epidemiology* 35 (3): 213–221.

Dick, A.W., et al. 2015. A decade of investment in infection prevention: A cost-effectiveness analysis. *American Journal of Infection Control* 43 (1): 4–9.

Dingle, K.E., et al. 2017. Effects of control interventions on Clostridium difficile infection in England: An observational study. *The Lancet Infectious Diseases* 17 (4): 411–421.

Djibre, M., et al. 2017. Universal versus targeted additional contact precautions for multidrug-resistant organism carriage for patients admitted to an intensive care unit. *American Journal of Infection Control* 45 (7): 728–734.

Doernberg, S.B., and H.F. Chambers. 2017. Antimicrobial stewardship approaches in the intensive care unit. *Infectious Disease Clinics of North America* 31 (3): 513–534.

Dryden, M., et al. 2011. Using antibiotics responsibly: Right drug, right time, right dose, right duration. *The Journal of Antimicrobial Chemotherapy* 66 (11): 2441–2443.

Duggan, J.M., et al. 2008. Inverse correlation between level of professional education and rate of handwashing compliance in a teaching hospital. *Infection Control and Hospital Epidemiology* 29 (6): 534–538.

Erasmus, V., et al. 2009. A qualitative exploration of reasons for poor hand hygiene among hospital workers: Lack of positive role models and of convincing evidence that hand hygiene prevents cross-infection. *Infection Control and Hospital Epidemiology* 30 (5): 415–419.

Forder, A.A. 2007. A brief history of infection control - past and present. *South African Medical Journal* 97 (11 Pt 3): 1161–1164.

Fries, J., et al. 2012. Monitoring hand hygiene via human observers: How should we be sampling? *Infection Control and Hospital Epidemiology* 33 (7): 689–695.

Funk, D.J., J.E. Parrillo, and A. Kumar. 2009. Sepsis and septic shock: A history. *Critical Care Clinics* 25 (1): 83–101. viii.

Gawande, A. 2012. Two hundred years of surgery. *The New England Journal of Medicine* 366 (18): 1716–1723.

Gilbert, G.L. 2015. Knowing when to stop antibiotic therapy. *The Medical Journal of Australia* 202 (11): 571.

Gill, C.J., and G.C. Gill. 2005. Nightingale in Scutari: Her legacy reexamined. *Clinical Infectious Diseases* 40 (12): 1799–1805.

Goossens, H., et al. 2005. Outpatient antibiotic use in Europe and association with resistance: A cross-national database study. *Lancet* 365 (9459): 579–587.

Gordon, A.. 1795. *Treatise on the Epidemic Puerperal Fever of Aberdeen*. Internet archive open knowledge commons and Harvard Medical School ed. 1795, London: G.G. & J. Robinson.

Grayson, M.L., et al. 2011. Outcomes from the first 2 years of the Australian National Hand Hygiene Initiative. *The Medical Journal of Australia* 195 (10): 615–619.

Harbarth, S., H. Sax, and P. Gastmeier. 2003. The preventable proportion of nosocomial infections: An overview of published reports. *The Journal of Hospital Infection* 54 (4): 258–266. quiz 321.

Harvey, K., J. Dartnell, and M. Hemming. 2003. Improving antibiotic use: 25 years of antibiotic guidelines and related initiatives. *Communicable Diseases Intelligence Quarterly Report* 27 (Suppl): S9–S12.

Holmes, O.W. 2001. *The contagiousness of puerperal fever*. New York: P.F. Collier & Son, 1909–14; Bartleby.com, 2001. http://www.bartleby.com/38/5/. [Date of Printout]. The Harvard Classics ed. The Harvard Classics. Vol. XXXVIII, part 5. New York: P.F. Collier & Son.

Hor, S.Y., et al. 2017. Beyond hand hygiene: A qualitative study of the everyday work of preventing cross-contamination on hospital wards. *BMJ Quality and Safety* 26 (7): 552–558.

Hornbeck, T., et al. 2012. Using sensor networks to study the effect of peripatetic healthcare workers on the spread of hospital-associated infections. *The Journal of Infectious Diseases* 206 (10): 1549–1557.

Huis, A., et al. 2012. A systematic review of hand hygiene improvement strategies: A behavioural approach. *Implementation Science* 7: 92.

Huskins, W.C., et al. 2011. Intervention to reduce transmission of resistant bacteria in intensive care. *The New England Journal of Medicine* 364 (15): 1407–1418.

Iedema, R., S.-Y. Hor, M. Wyer, G.L. Gilbert, C. Jorm, C. Hooker, and M.V.N. O'Sullivan. 2015. An innovative approach to strengthening health professionals' infection control and limiting hospital-acquired infection: video-reflexive ethnography. *BMJ Innovations* 1: 157–162.

Jang, J.H., et al. 2010a. Focus group study of hand hygiene practice among healthcare workers in a teaching hospital in Toronto, Canada. *Infection Control & Hospital Epidemiology* 31 (2): 144–150.

———. 2010b. Physicians and hand hygiene practice: A focus group study. *The Journal of Hospital Infection* 76 (1): 87–89.

Jevons, M.P., A.W. Coe, and M.T. Parker. 1963. Methicillin resistance in staphylococci. *Lancet* 1 (7287): 904–907.

Johnson, P.D., et al. 2005. Efficacy of an alcohol/chlorhexidine hand hygiene program in a hospital with high rates of nosocomial methicillin-resistant Staphylococcus aureus (MRSA) infection. *The Medical Journal of Australia* 183 (10): 509–514.

Jorm, C., and P. Kam. 2004. Does medical culture limit doctors' adoption of quality improvement? Lessons from Camelot. *Journal of Health Services Research & Policy* 9 (4): 248–251.

Jukes, T.H. 1985. Some historical notes on chlortetracycline. *Reviews of Infectious Diseases* 7 (5): 702–707.

Kingston, L., N.H. O'Connell, and C.P. Dunne. 2016. Hand hygiene-related clinical trials reported since 2010: A systematic review. *The Journal of Hospital Infection* 92 (4): 309–320.

Krein, S.L., et al. 2010. The influence of organizational context on quality improvement and patient safety efforts in infection prevention: A multi-center qualitative study. *Social Science & Medicine* 71 (9): 1692–1701.

Kumar, A., et al. 2006. Duration of hypotension before initiation of effective antimicrobial therapy is the critical determinant of survival in human septic shock. *Critical Care Medicine* 34 (6): 1589–1596.

Kwok, Y.L., C.P. Juergens, and M.L. McLaws. 2016. Automated hand hygiene auditing with and without an intervention. *American Journal of Infection Control* 44 (12): 1475–1480.

Landers, T.F., et al. 2012. A review of antibiotic use in food animals: Perspective, policy, and potential. *Public Health Reports* 127 (1): 4–22.

Lankford, M.G., et al. 2003. Influence of role models and hospital design on hand hygiene of healthcare workers. *Emerging Infectious Diseases* 9 (2): 217–223.

Larson, E. 1989. Innovations in health care: Antisepsis as a case study. *American Journal of Public Health* 79 (1): 92–99.

Lister, J. 1867. Antiseptic principles in the practice of surgery. *British Medical Journal* (Sept, 21): 246–248.

Loudon, I. 1986. Deaths in childbed from the eighteenth century to 1935. *Medical History* 30 (1): 1–41.

Mahida, N. 2016. Hand hygiene compliance: Are we kidding ourselves? *The Journal of Hospital Infection* 92 (4): 307–308.

Marshall, B.M., and S.B. Levy. 2011. Food animals and antimicrobials: Impacts on human health. *Clinical Microbiology Reviews* 24 (4): 718–733.

Mattick, K., N. Kelly, and C. Rees. 2014. A window into the lives of junior doctors: Narrative interviews exploring antimicrobial prescribing experiences. *The Journal of Antimicrobial Chemotherapy* 69 (8): 2274–2283.

McLaws, M.L. 2015. The relationship between hand hygiene and health care-associated infection: it's complicated. *Infection and Drug Resistance* 8: 7–18.

Morgan, D.J., K.S. Kaye, and D.J. Diekema. 2014. Reconsidering isolation precautions for endemic methicillin-resistant Staphylococcus aureus and vancomycin-resistant enterococcus. *JAMA* 312 (14): 1395–1396.

Morgan, D.J., R.P. Wenzel, and G. Bearman. 2017. Contact precautions for endemic MRSA and VRE: Time to retire legal mandates. *JAMA* 318 (4): 329–330.

Neo, J.R., et al. 2016. Evidence-based practices to increase hand hygiene compliance in health care facilities: An integrated review. *American Journal of Infection Control* 44 (6): 691–704.

Newsom, S.W. 2003. Pioneers in infection control-Joseph Lister. *The Journal of Hospital Infection* 55 (4): 246–253.

Nuland, S.B. 1979. The enigma of Semmelweis--an interpretation. *Journal of the History of Medicine and Allied Sciences* 34 (3): 255–272.

O'Neill, J. 2016. Tackling drug-resistant infections globally: final report and recommendations. In *The review on antimicrobial resistance*, ed. J. O'Neill. London, UK.

Palmore, T.N., and D.K. Henderson. 2012. Big brother is washing...Video surveillance for hand hygiene adherence, through the lenses of efficacy and privacy. *Clinical Infectious Diseases* 54 (1): 8–9.

Pittet, D. 2004. The Lowbury lecture: Behaviour in infection control. *The Journal of Hospital Infection* 58 (1): 1–13.

Pittet, D., et al. 2000. Effectiveness of a hospital-wide programme to improve compliance with hand hygiene. Infection Control Programme. *Lancet* 356 (9238): 1307–1312.

———. 2004. Hand hygiene among physicians: Performance, beliefs, and perceptions. *Annals of Internal Medicine* 141 (1): 1–8.

Podolsky, S.H. 2017. Historical perspective on the rise and fall and rise of antibiotics and human weight gain. *Annals of Internal Medicine* 166 (2): 133–138.

Reddy, S.C., et al. 2015. Antibiotic use in US hospitals: Quantification, quality measures and stewardship. *Expert Review of Anti-Infective Therapy* 13 (7): 843–854.

Robinson, T.P., et al. 2016. Antibiotic resistance is the quintessential one health issue. *Transactions of the Royal Society of Tropical Medicine and Hygiene* 110 (7): 377–380.

Saint, S., et al. 2003. Do physicians examine patients in contact isolation less frequently? A brief report. *American Journal of Infection Control* 31 (6): 354–356.

Saint, S., J.D. Howell, and S.L. Krein. 2010a. Implementation science: How to jump-start infection prevention. *Infection Control and Hospital Epidemiology* 31 (Suppl 1): S14–S17.

Saint, S., et al. 2010b. The importance of leadership in preventing healthcare-associated infection: Results of a multisite qualitative study. *Infection Control and Hospital Epidemiology* 31 (9): 901–907.

Sax, H., et al. 2007. 'My five moments for hand hygiene': A user-centred design approach to understand, train, monitor and report hand hygiene. *The Journal of Hospital Infection* 67 (1): 9–21.

Scheithauer, S., et al. 2011. Suspicion of viral gastroenteritis does improve compliance with hand hygiene. *Infection* 39 (4): 359–362.

———. 2017. Workload even affects hand hygiene in a highly trained and well-staffed setting: A prospective 365/7/24 observational study. *The Journal of Hospital Infection* 97 (1): 11–16.

Seibert, G., et al. 2017. What do visitors know and how do they feel about contact precautions? *American Journal of Infection Control*.

Selwyn, S. 1965. Sir James Simpson and Hospital cross infection. *Medical History* 9: 241–248.

Semmelweis, I. 1983. *The etiology, concept and prophylaxis of childbed fever*, Wisconsin Publications in the History of Science and Medicine. London: University of Wisconsin Press.

Sladek, R.M., M.J. Bond, and P.A. Phillips. 2008. Why don't doctors wash their hands? A correlational study of thinking styles and hand hygiene. *American Journal of Infection Control* 36 (6): 399–406.

Spellberg, B., and B. Taylor-Blake. 2013. On the exoneration of Dr. William H. Stewart: debunking an urban legend. *Infectious Diseases of Poverty* 2 (1): 3.

Srigley, J.A., et al. 2014. Quantification of the Hawthorne effect in hand hygiene compliance monitoring using an electronic monitoring system: A retrospective cohort study. *BMJ Quality and Safety* 23 (12): 974–980.

———. 2015. Hand hygiene monitoring technology: A systematic review of efficacy. *The Journal of Hospital Infection* 89 (1): 51–60.

Stelfox, H.T., D.W. Bates, and D.A. Redelmeier. 2003. Safety of patients isolated for infection control. *JAMA* 290 (14): 1899–1905.

Stewardson, A., et al. 2011. Back to the future: Rising to the Semmelweis challenge in hand hygiene. *Future Microbiology* 6 (8): 855–876.

Stone, P.W. 2009. Economic burden of healthcare-associated infections: An American perspective. *Expert Review of Pharmacoeconomics & Outcomes Research* 9 (5): 417–422.

Sydnor, E.R., and T.M. Perl. 2011. Hospital epidemiology and infection control in acute-care settings. *Clinical Microbiology Reviews* 24 (1): 141–173.

Temime, L., et al. 2009. Peripatetic health-care workers as potential superspreaders. *Proceedings of the National Academy of Sciences of the United States of America* 106 (43): 18420–18425.

Turnidge, J., and K. Christiansen. 2005. Antibiotic use and resistance--proving the obvious. *Lancet* 365 (9459): 548–549.

Turnidge, J.D., et al. 2016. Antimicrobial use in Australian hospitals: How much and how appropriate? *The Medical Journal of Australia* 205 (10): S16–S20.

Umscheid, C.A., et al. 2011. Estimating the proportion of healthcare-associated infections that are reasonably preventable and the related mortality and costs. *Infection Control and Hospital Epidemiology* 32 (2): 101–114.

Ventola, C.L. 2015. The antibiotic resistance crisis: Part 1: Causes and threats. *P T* 40 (4): 277–283.

Viens, A.M., and J. Littmann. 2015. Is antimicrobial resistance a slowly emerging disaster? *Public Health Ethics* 8 (3): 255–265.

WHO. 2011. *Report on the burden of endemic health care-Associated infection worldwide.* Geneva: WHO.

———, The evolving threat of antimicrobial resistance. Options for action. 2012 WHO: Geneva.

Widmer, A.F. 2000. Replace hand washing with use of a waterless alcohol hand rub? *Clinical Infectious Diseases* 31 (1): 136–143.

Willemsen, I., et al. 2009. Correlation between antibiotic use and resistance in a hospital: Temporary and ward-specific observations. *Infection* 37 (5): 432–437.

Wyer, M., et al. 2015. Involving patients in understanding hospital infection control using visual methods. *Journal of Clinical Nursing* 24 (11–12): 1718–1729.

Xu, J., et al. 2013. Surveillance and correlation of antimicrobial usage and resistance of Pseudomonas aeruginosa: A hospital population-based study. *PLoS One* 8 (11): e78604.

Zimmerman, B., et al. 2013. Front-line ownership: Generating a cure mindset for patient safety. *Healthcare Papers* 13 (1): 6–22.

Chapter 7
Epidemiology and Ethics of Antimicrobial Resistance in Animals

Lisa Boden and Dominic Mellor

Abstract Despite a large and rapidly growing volume of research activity and output, primarily on the biological bases of antimicrobial resistance (AMR), epidemiological understanding of the causal mechanisms at play behind the apparent recent global rise in prevalence of AMR has, arguably, progressed very little. Despite this inconvenient fact, political imperative and expedience, among other drivers, have given substantial impetus to an interventionist approach against what are considered to be the culprits for the apparent growing prevalence of AMR and its impacts. Concern about the rise in prevalence of microbial infections that are resistant to therapeutic agents designed to kill them has arisen almost exclusively in relation to human health. (Public awareness and concern about antihelmintic resistance, for which the impacts are much more substantial for animal health, at least in developed temperate countries, are trivial by comparison). Nevertheless, antimicrobial drugs have been, and are, widely used in animal health and production throughout the world, and the contribution of this diverse usage to the 'global AMR problem' has historically been controversial. There is growing acceptance, notwithstanding the limitations in causal understanding noted previously, of AMR as an ecological problem of competing populations of microorganisms experiencing both natural and anthropogenic selection pressures in compartments that transcend species and other boundaries. Typifying what is described as a 'One Health' problem, AMR is therefore considered to be most amenable to conjoint mitigation efforts in all compartments: i.e. interventions in human health, animal health, food and the environment in a coherent manner.

In animals, this calls into question the motivations and practices for antimicrobial drug usage, the majority of which are justified on the basis of promoting animal

L. Boden
Global Academy of Agriculture and Food Security, The Royal (Dick) School of Veterinary Studies and The Roslin Institute, University of Edinburgh, Edinburgh, UK
e-mail: lisa.boden@ed.ac.uk

D. Mellor (✉)
School of Veterinary Medicine, College of Medical, Veterinary and Life Sciences, University of Glasgow, Glasgow, UK
e-mail: dominic.mellor@glasgow.ac.uk

© The Author(s) 2020   109
E. Jamrozik, M. Selgelid (eds.), *Ethics and Drug Resistance: Collective Responsibility for Global Public Health*, Public Health Ethics Analysis 5,
https://doi.org/10.1007/978-3-030-27874-8_7

health and welfare and securing a food supply for a growing human population. Not surprisingly, there are great differences in animal husbandry and food demand, and in availability, access and regulation of antimicrobial usage in animals, and in surveillance of AMR, which are likely to be starkest between developed and developing countries. Thus, it is unlikely that the impacts of AMR, and the impacts of efforts to mitigate AMR that are directed to the 'animal compartment' of the ecosystem, will be felt equally across the world.

Keywords AMR · Ethics · One Health · Veterinary · Animal · Causality

7.1   Introduction: Evolutionary History of Antimicrobial Resistance as a Natural Phenomenon

Antimicrobial resistance has occurred as a natural phenomenon for millennia, as a response to inhibitory substances produced by microbial populations competing for resources in different ecosystems (Hall and Barlow 2004). Human discovery of the existence of these substances was exploited in the early twentieth century leading to the development of the first antibacterial drugs for therapeutic use. The rapid discovery and development of both natural and synthetic antimicrobial drugs (active against bacteria, fungi, viruses, and protozoa), and their widespread use to treat and prevent human and animal infectious diseases, took place throughout the latter half of the twentieth century. As predicted by Alexander Fleming himself, as therapeutic use of antibiotics grew there closely followed the emergence of untreatable strains of organisms that had hitherto responded to treatment.

Any microorganism that isn't inhibited or killed by appropriate, effective antimicrobials is classified as resistant (Ridge et al. 2011). This phenomenon is now widely explained in terms of a 'selection pressure' being exerted on the populations of microorganisms which are exposed to antimicrobial agents. Such microbial populations are usually comprised of an almost unimaginable number and diversity of individual organisms, amongst which the target 'pathogen' population for antimicrobial therapy may constitute only a fraction of those exposed to the agent. Under these circumstances, those organisms susceptible to the agent are inactivated, and cease to compete for resources, and those that are equipped with mechanism(s) to resist the effects of the agent thrive through access to the resources no longer consumed by the inactivated organisms (Levin et al. 2000). In the case of therapeutic use of antimicrobial agents, if the organisms equipped with mechanism(s) to resist the effects of the agent are members of the 'pathogen' population, the result is likely to be treatment failure and prolonged clinical disease for the patient. If the organisms equipped with mechanism(s) to resist the effects of the agent are members of the non-pathogen population (usually referred to as commensals), the effect is to select for populations that carry resistance mechanisms, but is unlikely to result in treatment failure at that time. Nevertheless, and especially for bacteria, many of the mechanisms to resist the effects of antimicrobial agents are coded for by genes carried on transmissible genetic elements which can be passed between organisms of the same or different species or genera. Thus, selection for resistance in members of the non-pathogen population could subsequently lead to transfer to 'pathogen' populations and increased likelihood of subsequent treatment failure.

7.2 Drug Resistance as an Animal or Public Health Concern

Treatment failure and prolonged and/or more serious clinical disease for the patient are probably much more widely recognized consequences of drug resistance in human than in veterinary medicine. In efforts to mitigate these principally public health effects of antimicrobial resistance in the developed world, much attention has been focussed on human healthcare settings, particularly hospitals and care homes (Edwards et al. 2012). Efforts to improve infection prevention and control (IPC) in these settings are paramount, because they are needed not only to reduce the transmission between people of infectious agents likely to require drug therapy, but also the transmission of antimicrobial resistance genes. Principles of IPC form the mainstays of current strategies to combat antimicrobial resistance and are coupled with measures to promote 'better' prescribing of antimicrobial agents through guidelines and audit, and initiatives to educate the healthcare profession and the public. Judicial antimicrobial use (i.e. "better prescribing") is based on evidence which supports using a particular agent against a particular organism in a particular patient for a particular reason and period of time (e.g. British Veterinary Association 2015).

The contribution and nature of veterinary use of antimicrobial agents to the problem of antimicrobial resistance on a global scale has been a controversial issue for many years, with conflicting, polarized views espoused by different respected research groups (Aarestrup et al. 2000; Collignon 2013; Mather et al. 2013; van Bunnik and Woolhouse 2017). Much has been made of the widespread and varied use of antimicrobial agents in veterinary medicine and the potential for this to contribute to the (largely) public health problem posed by antimicrobial resistance. In this, it is noteworthy that, with few exceptions, relatively little is made of treatment failure and prolonged and/or more serious clinical disease for animal patients. The use of the broad term 'antimicrobial' has been unhelpful in this regard as it is inclusive of agents such as anti-coccidials, which are important for animal health and food security, but have no effect on other organisms (such as bacteria or fungi) which drive resistance in humans (Mendelson et al. 2017). Mendelson et al. (2017) argue for more precision in the language around drug-resistant infections, and for more specific terms such as "antibiotic" or "antifungal" to be used in preference to "'antimicrobials" when referring to medicines against a specific type of organism.

There is a growing acceptance that drug-resistant infections are a 'One Health' problem that transcends species (and other) barriers (Karesh et al. 2012). Expanding on this, it is obvious that agents that carry resistance determinants exist very well and evolve outside animal or human hosts. There is increasing recognition of the need to consider environmental reservoirs, and inanimate vehicles of transmission, not least among which are foods of animal origin – home produced and imported – and integrate these into thinking about the ecology and epidemiology of resistance (Wellington et al. 2013). Resistance genes may spread directly from people to animals and from animals to people through food-borne and environmental contamination (via wastewater, soil and the spread of contaminated manure from livestock and wild animals) (Casey et al. 2013; Kim et al. 2013; Davis and Rutkow 2012; Johnson and Becker 2010; Cantas et al. 2013; Roe and Pillai 2003; Soonthornchaikul and Garelick 2009; Levy et al. 1976; Thanner et al. 2016; Literak et al. 2011). However, the relative importance of these routes of transmission is uncertain; exposure is complex and not unequivocally in one direction (Carlet et al. 2011; Mather et al. 2013; Zhu et al. 2013). Additionally, there are other epidemiological pathways (such as transmission of resistance via irrigation and waste-water, plant production and the disposal or presence of disinfectants and heavy metals in the environment) that have been shown to be associated with the emergence of drug resistance. These are not well researched due to the lack of analytical methods to monitor contaminants in waste, surface and drinking water and soil (Thanner et al. 2016). Thus, the epidemiological drivers associated with the selection for and against resistant organisms in animals are unlikely to be different to the causal mechanisms believed to exist in humans or the environment. There are important caveats to this assertion: (1) many parts of the relevant ecosystem are ignored by surveillance approaches adopted to date, notably the environment and food (home produced and imported), (2) it is difficult to partition antibiotic resistance into that which arises naturally from bacterial competition in various ecological niches and that which is selected for anthropogenically through therapeutic or other use of drugs and/or biocides. Of course, the relative intensity of the various selection pressures for the emergence of drug-resistant infections is itself driven in part by broader socio-economic issues (e.g. poverty, sanitation, hygiene and public health resources) which are harder to quantify and even more difficult for governments to address. These caveats arise due to the limitations of the evidence provided by surveillance for drug-resistant infections, which has largely focused on trying to compare observations of antibiotic resistance in bacterial isolates from human patients to those of antibiotic resistance in animal populations (often using different methods of antimicrobial resistance determination) in developed countries.

High rates of drug-resistant infections are found in densely populated, developing countries where there is corruption and unreliable enforcement of laws and regulations pertaining to the practice of human and veterinary medicine (Collignon et al. 2015). Individuals who are exposed to resistant bacteria or fungi in these hotspots through international travel and medical tourism can subsequently import resistance into other countries, resulting in rapid global spread (World Health Assembly Resolution 51.9 1998).

There is still a need to convince some sectors that this is a shared problem with shared accountability and shared responsibility, but that principle is pretty much implicit in nationally and internationally agreed accords (e.g. Department of Health 2013; O'Neill 2016). Based on experience from the UK, plans to implement the recommendations of such accords, whilst all claiming to be taking a 'One Health' approach, can vary even among devolved administrations. In this sense, creation of a bespoke 'One Health' agency, with appropriately balanced multidisciplinary representation (and buy-in and trust), acting in a collegiate manner to coordinate a collective and coherent response seems intuitively more likely to succeed than simply hoping that separate agencies will be able to work in parallel towards a common aim without 'funding the arrows' (Campbell 2006).

In conclusion, considering the epidemiology of drug-resistance in animals in isolation is missing the point.

7.3 Antimicrobial Use in Animals

Similar to antimicrobial use in humans, there is substantial variation throughout the world in the availability, regulation, control and administration of antimicrobials to animals. Whilst these factors themselves are not expected to alter the postulated causal mechanisms by which drug use affects drug resistance in microorganisms, they are likely to modify the extent to which these mechanisms have the opportunity to act. Simply put, it appears that greater usage of particular antimicrobial drugs, in a relevant time and place, is positively correlated with greater proportions of isolates tested for susceptibility in that time and place being designated as resistant. Thus, greater control over the quality and supply of antimicrobial drugs and greater regulation and professionalism over their administration should correlate with reduced proportions of resistant isolates.

It is worthy of note that this general observation appears also to be true of anthelmintic drugs, although the ecology of macroparasites is of course different to microbes. However, in the developed world, the problem of anthelmintic resistance is more of an immediate issue for animals than humans, which means that it has received much less attention from a public health point of view. Nevertheless, sophisticated guidance on control of parasitic infestations, based on understanding their ecology, and seeking to preserve the efficacy of anthelmintic drugs, are seen as critical to future food security (e.g. see Sustainable Control of Parasites in Sheep (http://www.scops.org.uk/)).

A range of antimicrobial drugs is used widely to treat disease in domestic animals through a large and internationally varying number of preparations and through a number of different routes of administration. In most developed countries of the world, the classes of antimicrobial drug available for animal administration, and the preparations and routes by which they can be administered, have to be specifically licensed for animal use and are highly regulated, especially for animal species likely to enter the human food chain. In these countries, there are also strictly enforced 'withdrawal periods' which define for how long an animal must be 'off' treatment before products form that animal can be used for human consumption. In some parts of the world, so-called 'off-label' use (i.e. an unlicensed product and/or an unlicensed route of administration) is sometimes permitted under derogation for the treatment of companion animals. Lists of critically important antimicrobials (CIAs) in humans and animals have been agreed, and are periodically updated, by the OIE and WHO and others, and the use of these drugs should be restricted to treating infections that have been demonstrated to be susceptible (and resistant to less important, so-called first-line, drugs) (OIE list of antimicrobials of veterinary importance 2007; WHO list of Critically Important Antimicrobials (CIA) 2017). In developing countries, due to limitations in infrastructure, such regulations, where they exist, are much more difficult to make effective. Coupled with a very high demand due to a high disease burden and often high population densities, and issues such as uncontrolled, unauthorized markets and counterfeit drugs, the conditions appear likely to favour intense selection for antimicrobial resistance and subsequent dissemination, though there are few reliable data to provide firm evidence of this.

Much of the controversy around use of antimicrobials in animals has concentrated on their use as growth promoters. It has been considered that routine addition of some antimicrobial compounds (usually antibacterial drugs and often at sub-therapeutic concentrations) to livestock feed increases food conversion efficiency by more than enough to outweigh the cost of adding the drugs, though this is disputed (Graham et al. 2007). However, other agents, such as anti-coccidials used by the poultry sector, are also included in this broad classification (Mendelson et al. 2017). Use of antibiotic growth promoters is considered by many to be particularly undesirable due to the selection pressure being applied in an almost unrelenting way to the populations of microorganisms colonizing these animals, especially in instances in which drugs are used at sub-therapeutic concentrations, as this is thought to select more strongly for antimicrobial resistance. Much research has sought to investigate the impact of the use of antimicrobials for growth promotion and has been interpreted by most as demonstrating a positive association with increased prevalence of resistant organisms in exposed microbial populations (O'Neill (2015)). On the basis of this evidence, their use as growth promoters has been banned in many parts of the world, notably Europe, yet still persists in others.

In the EU, "the use of agents from classes which are or may be used in human or veterinary medicine (i.e. where there is a risk of selecting for cross-resistance to drugs used to treat bacterial infections)" as growth promoters has not been permitted since 2006 (Regulation (EC) No 1831/2003/EC) and withdrawal periods for antimicrobial use prior to animal slaughter are designed to ensure that there are no antimicrobial residues in food. In the USA, non-therapeutic use is still widespread in industrial farming. The USA FDA has historically been slow to respond to calls to reduce antimicrobial use and unwilling to exert its authority over the antimicrobial approval process (see Natural Resources Defense Council, et al. v. United States Food and Drug Administration, et al.). Until recently, the FDA implemented a voluntary approach to antimicrobial conservation that encouraged drug companies to withdraw approvals for antimicrobials for non-therapeutic use and replace them with approvals for other uses such as chemoprophylaxis (USDA 2012a, b). However, this policy has had little real effect on antimicrobial usage because, in many cases, the doses and durations of drug use for chemoprophylaxis and growth promotion are the same (Outterson 2014).

In emerging economies, such as the BRICS and MINT countries, it is anticipated that non-therapeutic use of antibiotics in animals may exponentially increase because of increasing intensification of agriculture/aquaculture, high prevalences of production and endemic diseases (which are likely to be better controlled in other countries) and lack of resources to ensure appropriate governance over antimicrobial use (Carlet et al. 2011; Van Boeckel 2015). In some of these countries, AMR and antimicrobial conservation aren't on the "political agenda" at all (Grace 2015 at p 11–12) because addressing other issues such as poverty, starvation, malnutrition through (un)sustainable livestock and farmed fish production are more urgent for the current population. The impact of antimicrobial use on accessibility of animal food sources hasn't been quantified for most countries, partly due to the variability and uncertainty regarding the quantities of antimicrobials used even in similarly intensive systems elsewhere (Rushton et al. 2014), and there is no agreement on the desired levels of antimicrobial consumption. Even if an enforced ban on the non-therapeutic use of antimicrobials were to be introduced, in some places, the absence of national R&D investment means that there are few alternative mechanisms (such as vaccines) to concurrently improve animal husbandry and avoid production losses which could paradoxically increase AMR prevalence through off-label or unprescribed use of poor quality or counterfeit antimicrobials. The relative costs of not using antimicrobials on the security of the global food supply and the success of the Sustainable Development Goals (in eradicating poverty and hunger in the current generation) have not been compared to AMR treatment failure in people (in future generations). Although developed countries arguably have sufficient means to assist developing countries address some of these issues, they have so far focussed on their own national priorities in order to achieve wider international societal benefits (Clift 2013).

Distinction is made in veterinary medicine (at least in the developed world) between prophylactic (administering antimicrobial drugs to prevent anticipated infection) and metaphylactic use of antimicrobials (administering antimicrobial drugs to clinically well members of a population in contact with an index case of infectious disease to prevent anticipated infection), which is not recognized in human medicine. Prophylaxis in veterinary medicine is criticized by many as an excuse for poor infection prevention and control (often referred to as 'biosecurity' in animal production) in a particular animal husbandry system, but presents ethical dilemmas to veterinarians who struggle with the notion of withholding treatment in the face of what is considered to be almost inevitable disease and associated welfare compromise. In many instances, the clinical use of antimicrobials in animals is empirical and sensitivity testing of an isolate of the putative causative agent of infection is not carried out. The reasons for this are largely to do with cost and expediency in starting treatment to improve the clinical condition of the patient. Thus, data that characterize the resistance profiles of clinical isolates from animals represent a very small proportion of the putative infections treated by veterinarians, and their use for epidemiological purposes often appears to overlook this fact. However, this is also true of human medicine.

The notion that antimicrobial resistant organisms somehow 'arise' in animals, driven by selection pressures applied by veterinary use of antimicrobials, and pass to humans, principally through the food chain, has been a popular model of antimicrobial resistance 'acquisition' for some time. However, some recent research shows that transmission is likely to occur in both directions between animals and people directly and indirectly through the environment and various fomites (Mather et al. 2013). As discussed in the introduction, it is probably more reasonable to think of animals and people as inextricably linked 'samplers' of a shared environmental pool of organisms subject to different selection pressures in different compartments.

7.4   Surveillance for Antimicrobial Resistance in Animals

'Better' surveillance for antimicrobial resistance in animals is a more or less universal feature of international concordats, calls to action and other declarations on antimicrobial resistance (Department of Health 2013; O'Neill 2016). As the bedrock of epidemiology, surveillance activity, and the intelligence generated by it, offers the greatest potential to understand fully the causal relationships at play within the complex landscape ecology underlying the antimicrobial resistance 'phenomenon' (Singer et al. 2006). Nevertheless, there are a great many limitations, often overlooked, that can apply to data derived from surveillance of infections in animals (and people) for the purposes of exploring causal relationships of relevance to the emergence and spread of antimicrobial resistance. Chief among these is the question of how well the isolate(s) derived from the sample(s) collected from the individual(s) and/or their environment(s) under study represent the actual nature and dynamics of the interactions between host and microbial populations and the selection pressures experienced by them. Confounding this, in many instances, is a lack of standardization of the microbiological methods applied and inconsistent definitions of how antimicrobial sensitivity or lack of it is defined for individual drug/bug combinations derived from different species (animals and humans).

The World Health Organization (WHO) has led international efforts to develop action plans to monitor and reduce drug-resistant infections (WHO Strategy 2001). However, without a legally-binding mandate, it has been difficult implement this strategy within Member States. Indeed, between 1998 and 2015 (when the WHO first published its global action plan), there have been at least seven World Health Assembly (WHA) Resolutions promoting surveillance of drug-resistant infections, but still no internationally coordinated or standardized AMR surveillance strategy within either the human or veterinary sectors. Although collaboration between the WHO, Food and Agricultural Organization of the United Nations (FAO) and World Organization of Animal Health (OIE) is improving coordination between human and animal surveillance, within individual Member States the two sectors remain distinct and regulated independently of one another. Surveillance, if undertaken at all, has hitherto been implemented separately by each sector, without harmonization or standardization of approach.

Existing national action plans within Member States describe broad strategic aims to mitigate, and reduce current rates of drug-resistant infections and resultant treatment failures. These include: improvements in infection prevention and control, education and training initiatives, optimization of prescribing practices (particularly in animals with respect to critically important antimicrobials) and incentives to innovate new, effective therapies. For the most part, these aims have not been translated yet into specific actions with explicit timescales for delivery or agreed outcome metrics. In some countries, technical and financial constraints, such as lack of existing public health infrastructure, access to diagnostic technologies and changing public attitudes towards public health mean that surveillance is poor (STAG-AMR 2013; WHO 2011). Accurate inferences (and/or between-country comparisons) about antibiotic or antifungal consumption rates aren't always possible because of the scarcity and variability of available information on consumption, species treated, routes of administration, dose rate, and pharmaceutical costs (Rushton et al. 2014 at p 11). Other important differences, such as husbandry practices, diversity of available drugs and prescribing habits of veterinarians, mean that comparisons of antimicrobial use based on animal demographics are potentially misleading. For example, Chile is the second largest producer of farmed salmon and the only important developing country producer. However, it uses around 300 times more antibiotics than the largest salmon producer, Norway because Norway has the resources to access and implement vaccination and alternative husbandry measures instead of antimicrobials to control diseases (Grace 2015 at page 11). Thus, it is currently not possible to accurately chart progress towards AMR reduction or identify early-warning performance indicators for actions that aren't effective in achieving these aims.

Despite these limitations, surveillance and research data generated, principally in the developed world, are subject to ever more sophisticated epidemiological analyses to infer risk factors and causal relationships for resistance emergence, persistence and spread. Entrenched in this aim is the notion that population patterns of drug-resistant infections can be explained by a complex web of multiple interconnected factors, which if identified, can be used to inform and target interventions to improve public health (Krieger 1994). Increasingly, these analyses seek to be comparative and inferential about the impacts of antimicrobial use in one species (animals or man) on antimicrobial resistance in the other, as the 'One Health' construct of drug resistance becomes more widely accepted. However, caution is required in the interpretation of these findings. 'Causal webs' are not unbiased; they are necessarily hierarchical, focusing "attention on risk factors closest to the outcome under inspection", inevitably prioritizing biological factors (amenable to medical or veterinary intervention) over other broader social or environmental determinants which could be addressed through social action (Krieger 1994; Thanner et al. 2016). For example, addressing a lack of access to education, sanitation and adequate healthcare (including infection prevention and control, scarcity of new antibiotics, poor prescribing practices and absence of concurrent diagnostic tools to ensure appropriate treatment) will affect rates of disease, which will influence subsequent amounts of antimicrobial used (Buckland Merrett 2013). Reduction of antimicrobial use may reduce selection pressure for resistance, but at the same reduces the numbers of antimicrobial 'customers' and thus incentives to innovate new drugs. Equally, economic strategies that are driven by sales rewards will inevitably conflict with principles of conservation and IPC. Understanding of the epidemiology of AMR (and this broader socio-economic and ecological 'web of causation') and an appreciation for the gaps in communication and coordination between stakeholders and regulators involved across multiple sectors and disciplines at each of these foci is therefore necessary to ensure policy decisions are robust and ethical.

7.5   Summary and Conclusion

The epidemiology of antimicrobial resistance in animals seems very unlikely to be extricable or distinct from that in people. A powerful epidemiological model of the causation and dissemination of antimicrobial resistance as a feature of microbiological ecology among a very complex web of interconnected host and environmental compartments has emerged and gained widespread acceptance under the construct of 'One Health'. However, the limitations in the quality of the data that have been used to build this model leave room for uncertainty about its validity. In addition, there needs to be 'thinking time and space' to consider and account for cognitive biases in such models and to incorporate socio-economic modifiers of such biological models in order to inform efficient and effective measured interventions. As rapidly developing scientific advances offer the potential to improve the quality (e.g. WGS) and representativeness (e.g. through properly and purposively designed comparative surveillance programmes) of data, it is to be hoped that epidemiological inference will be able to keep pace and offer better, and what yet may be surprising, insights into the problem of antimicrobial resistance and ways in which it might be tackled.

References

Aarestrup, F.M., Y. Agerso, P. Gerner-Schmidt, M. Madsen, and L.B. Jense. 2000. Comparison of antimicrobial resistance phenotypes and resistance genes in *Enterococcus faecalis* and *Enterococcus faecium* from humans in the community, broilers and pigs in Denmark. *Diagnostic Microbiology and Infectious Disease* 27: 127–137.

British Veterinary Association. 2015. *Responsible use of antimicrobials in veterinary practice.* Available from: https://www.bva.co.uk/uploadedFiles/Content/News,_campaigns_and_policies/Policies/Medicines/responsible-use-of-antimicrobials-in-veterinary-practice.pdf.

Buckland Merrett, G.L. 2013. *Tackling antibiotic resistance for greater global health security.* Centre on Global Health Security. Available from: http://www.chathamhouse.org/sites/files/chathamhouse/public/Research/Global%20Health/1013bp_antibioticresistance.pdf.

Campbell, A. 2006. *The Australian natural resource management knowledge system.* Available from: http://lwa.gov.au/products/pr061081.

Cantas, L., S.Q.A. Shah, L.M. Cavaco, C.M. Manaia, F. Walsh, M. Popowska, H. Garelick, H. Bürgmann, and H. Sørum. 2013. A brief multi-disciplinary review on antimicrobial resistance in medicine and its linkage to the global environmental microbiota. *Frontiers in Microbiology* 4: 96. https://doi.org/10.3389/fmicb.2013.00096.

Carlet, J., P. Collignon, D. Goldmann, H. Goossens, Gyssens ICJ, S. Harbarth, V. Jarlier, S.B. Levy, B. N'Doye, D. Pittet, R. Richtmann, W.H. Seto, J.W.M. van der Meer, and A. Voss. 2011. Society's failure to protect A precious resource: Antibiotics. *The Lancet* 378 (9788): 369–371.

Casey, J.A., F.C. Curriero, F.C. Cosgrove, K.E. Nachman, and B.S. Schwartz. 2013. High-density livestock operations, crop field application of manure, and risk of community-associated methicillin-resistant Staphylococcus Aureus infection. *Journal of the American Medical Association of Internal Medicine* 173 (21): 1980–1990. https://doi.org/10.1001/jamainternmed.2013.10408.

Clift, C. 2013. The UK's new antimicrobial resistance strategy. Available at https://www.chathamhouse.org/media/comment/clift/amr (last accessed 31 July 2015).

Collignon, P. 2013. Superbugs in food: A severe public health concern. *The Lancet Infectious Disease* 13 (8): 641–643.

Collignon, P., A. Prema-chandra, S. Senanayake, and F. Khan. 2015. Antimicrobial resistance: The major contribution of poor governance and corruption to this growing problem. *PLoS One* 10 (3): e0116746. https://doi.org/10.1371/journal.pone.0116746.

Davis, M.F., and L. Rutkow. 2012. Regulatory strategies to combat antimicrobial resistance of animal origin: Recommendations for a science-based US approach. *Tulane Environmental Law Journal* 25: 327–387.

Department of Health. 2013. *UK five year antimicrobial resistance strategy 2013 to 2018.* Available from: https://www.gov.uk/government/uploads/system/uploads/attachment_data/file/244058/20130902_UK_5_year_AMR_strategy.pdf.

Edwards, R., E. Charani, N. Sevdalis, B. Alexandrou, E. Sibley, D. Mullett, H.P. Loveday, L.N. Drumright, and A. Holmes. 2012. Optimisation of infection prevention and control in acute health care by use of behaviour change: A systematic review. *Lancet Infectious Diseases* 12: 318–329. https://doi.org/10.1016/S1473-3099(11)70283-3.

Grace, D. 2015. *Review of evidence on antimicrobial resistance and animal agriculture in developing countries.* ILRI Available from: http://www.evidenceondemand.info/review-of-evidence-on-antimicrobial-resistance-and-animal-agriculture-in-developing-countries.

Graham, J.P., J.J. Boland, and E. Silbergeld. 2007. Growth promoting antibiotics in food animal production: An economic analysis. *Public Health Reports* 122: 79–87.

Hall, B.G., and M. Barlow. 2004. Evolution of the serine beta-lactamases: Past, present and future. *Drug Resistance Updates* 7: 111–123. https://doi.org/10.1016/j.drup.2004.02.003.

Johnson, R., and G. Becker. 2010. *US-Russia meat and poultry trade issues.* Washington, DC: Congressional Research Service for Congress. Available from: http://nationalaglawcenter.org/wp-content/uploads/assets/crs/RS22948.pdf.

Karesh, W.B., A. Dobson, J.O. Lloyd-Smith, J. Lubroth, M.A. Dixon, M. Bennett, S. Aldrich, T. Harrington, P. Formenty, E.H. Loh, C.C. Machalaba, M.J. Thomas, and D.L. Heymann. 2012. Ecology of zoonoses: Natural and unnatural histories. *The Lancet* 380: 1936–1945.

Kim, B.F., L.I. Laestadius, R.S. Lawrence, R.P. Martin, S.E. Mckenzie, K.E. Nachman, T.J.S. Smith, and P. Truant. (2013). *Industrial food animal production in America: Examining the impact of the Pew Commission's Priority Recommendations.* John's Hopkins Centre for a Liveable Future. Available at http://www.jhsph.edu/research/centers-and-institutes/johns-hopkins-center-for-a-livable-future/_pdf/research/clf_reports/CLF-PEW-for%20Web.pdf.

Krieger, N. 1994. Epidemiology and the web of causation: Has anyone seen the spider? *Social Science Medicine* 39 (7): 887–903.

Levin, B.R., V. Perrot, and N. Walker. 2000. Compensatory mutations, antibiotic resistance and population genetics of adaptive evolution in bacteria. *Genetics* 154: 985–997.

Levy, S.B., G.B. FitzGerald, and A.B. Macone. 1976. Changes in intestinal flora of farm personnel after introduction of a tetracycline-supplemented feed on a farm. *New England Journal of Medicine* 295: 583–588. https://doi.org/10.1056/NEJM197609092951103.

Literak, I., R. Petro, M. Dolejska, E. Gruberova, H. Dobiasova, J. Petr, et al. 2011. Antimicrobial resistance in fecal *Escherichia coli* isolates from healthy urban children of two age groups in relation to their antibiotic therapy. *Antimicrobial Agents and Chemotherapy* 55: 3005–3007. https://doi.org/10.1128/AAC.01724-10.

Mather, A.E., S.W.J. Reid, D.J. Maskell, J. Parkhill, M.C. Fookes, S.R. Harris, D.J. Brown, J.E. Coia, M.R. Mulvey, M.W. Gilmour, L. Petrovska, E. de Pinna, M. Kuroda, M. Akiba, H. Izumiya, T.R. Connor, M.A. Suchard, P. Lemey, D.J. Mellor, D.T. Haydon, and N.R. Thomson. 2013. Distinguishable epidemics of multidrug-resistant Salmonella Typhimurium DT104 in different hosts. *Science* 341: 1514–1517. https://doi.org/10.1126/science.1240578.

Mendelson, M., M. Balasegaram, T. Jinks, C. Pulcini, and T. Sharland. 2017. Antibiotic resistance has a language problem. *Nature* 545 (7652): 23–25.

Natural Resources Defense Council. et al. 2012a. v. United States Food and Drug Administration, et al. 884 F.Supp.2d 127 (S.D.N.Y.).

Natural Resources Defense Council. et al. 2012b. v. United States Food and Drug Administration, 872 F.Supp.2d 318 (S.D.N.Y.).

O'Neill, J. 2015. *Antimicrobials in agriculture and the environment: Reducing unnecessary use and waste.* Available from: https://amr-review.org/sites/default/files/160525_Final%20paper_with%20cover.pdf.

———. 2016. *Tackling drug-resistant infections globally: Final report and recommendations.* Available from: https://amr-review.org/sites/default/files/160525_Final%20paper_with%20cover.pdf.

OIE list of antimicrobials of veterinary importance. 2007. Available from: https://www.oie.int/fileadmin/Home/eng/Our_scientific_expertise/docs/pdf/AMR/A_OIE_List_antimicrobials_July2019.pdf

Outterson, K. 2014. *New business models for sustainable antibiotics.* Centre on global health security working group papers 1–31. Available from: http://www.chathamhouse.org/sites/files/chathamhouse/public/Research/Global%20Health/0214SustainableAntibiotics.pdf.

Regulation (EC) No 1831/2003/EC Of The European Parliament and of the Council of 22 September 2003 On additives for use in animal nutrition, replacing directive 70/524/EEC on additives in feeding-stuffs. 2003. Official Journal of the European Union L 268: 0029–0043.

Ridge, K.W., K. Hand, M. Sharland, I. Abubakar, and D.M. Livermore. 2011. Antimicrobial resistance. In *The annual report of the chief medical officer volume 2, Infections and the rise of antimicrobial resistance.* Available at https://www.gov.uk/government/uploads/system/uploads/attachment_data/file/138331/CMO_Annual_Report_Volume_2_2011.pdf

Roe, M.T., and S.D. Pillai. 2003. Monitoring and identifying antibiotic resistance mechanisms in bacteria. *Poultry Science* 82: 622–626.

Rushton, J., J. Pinto Ferreira, and K.D. Stärk. 2014. *Antimicrobial resistance: The use of antimicrobials in the livestock sector,* OECD Food, Agriculture and Fisheries Papers No. 68. Paris: OECD Publishing. https://doi.org/10.1787/5jxvl3dwk3f0-en.

Singer, R.S., M.P. Ward, and G. Maldonado. 2006. Can landscape ecology untangle the complexity of antimicrobial resistance? *Nature Reviews Microbiology* 4: 943–952.

Soonthornchaikul, N., and H. Garelick. 2009. Antimicrobial resistance of campylobacter species isolated from edible bivalve molluscs purchased from Bangkok markets, Thailand. *Foodborne Pathogens and Disease* 6: 947–951. https://doi.org/10.1089/fpd.2008.0236.

Strategic and Technical Advisory Group on Antimicrobial Resistance (STAG-AMR). Report of the first meeting Geneva, 19–20 September 2013. Available from: http://www.who.int/drugresistance/stag/amr_stag_meetingreport0913.pdf.

Sustainable Control of Parasites in Sheep. Available from: http://www.scops.org.uk/.

Thanner, S., D. Drissner, and F. Walsh. 2016. Antimicrobial resistance in agriculture. *MBio* 7 (2): e02227–e02215. https://doi.org/10.1128/mBio.02227-15.

Van Boeckel, T.P. 2015. Global Trends In Antimicrobial Use In Food Animals. *Proceedings of the National Academy of Science* 112 (18): 649–5654. https://doi.org/10.1073/pnas.1503141112.

van Bunnik, B.A.D., and M.E.J. Woolhouse. 2017. Modelling the impact of curtailing antibiotic usage in food animals on antibiotic resistance in humans. *Royal Society Open Science* 4: 161067. https://doi.org/10.1098/rsos.161067.
Wellington, E.M.H., A.B.A. Boxall, P. Cross, E.J. Feil, W.H. Gaze, P.M. Hawkey, A.S. Johnson-Rollings, D.L. Jones, N.M. Lee, W. Otten, C.M. Thomas, and A.P. Williams. 2013. The role of the natural environment in the emergence of antibiotic resistance in Gram-negative bacteria. *Lancet Infectious Diseases* 13: 155–165.
World Health Assembly Resolution 51.9. 1998. Emerging and other communicable diseases: Antimicrobial resistance. In *51st world health assembly*. Available from: http://apps.who.int/medicinedocs/index/assoc/s16335e/s16335e.pdf?ua=1.
World Health Organisation. 2001. *WHO global strategy for containment of antimicrobial resistance*. Available from: http://www.who.int/drugresistance/WHO_Global_Strategy_English.pdf.
———. 2011. *Tackling antibiotic resistance from a food safety perspective in Europe*. Available from: http://www.euro.who.int/__data/assets/pdf_file/0005/136454/e94889.pdf at page 9.
World Health Organisation List of Critically Important Antimicrobials (CIA). 2017. Available from: https://www.who.int/foodsafety/publications/antimicrobials-sixth/en/
Zhu, Y.-G., T.A. Johnson, J.-Q. Su, M. Qiao, G.-X. Guo, R.D. Stedtfield, S.A. Hashsham, and J.M. Tiedje. 2013. Diverse and abundant antibiotic resistance genes in Chinese swine farms. *Proceedings of the National Academy of Science* 110 (9): 3435–3440.

Part II
Theoretical Approaches to Ethics and Drug Resistance

Chapter 8
The Virtuous Physician and Antimicrobial Prescribing Policy and Practice

Justin Oakley

Abstract In this chapter, I outline some key patient-centred medical virtues and several community-centred medical virtues, and I consider what sorts of antimicrobial prescribing decisions such virtues would lead physicians to make. I argue that practically-intelligent virtuous physicians should also have an awareness of the sorts of cognitive biases that are especially likely to distort their antimicrobial prescribing decisions, and I urge physicians to develop ways of avoiding or counteracting such biases. Further, I argue that effectively addressing the impact of these biases and other countervailing factors that inhibit virtuous prescribing practices is the responsibility not only of individual physicians, but also of institutions and regulators. I outline some strategies that individual physicians, institutions, and healthcare policymakers could develop to help physicians hit the targets of those patient-centred and community-centred medical virtues, and to thereby play their part in redressing the problems of antimicrobial resistance.

Keywords Medical virtue · Practical intelligence · Justice · Antibiotic overprescribing · Cognitive bias

The detrimental health impact of antimicrobial resistance raises significant questions about physicians' antimicrobial prescribing decisions, so it is important to investigate what sorts of prescribing decisions about antimicrobials would be made by a virtuous physician in various contexts. Antimicrobial resistance has become a

An earlier version of this chapter was presented at the 2018 Effective Altruism Global Conference, Melbourne, where I received some helpful comments on this chapter. I also received some useful feedback on these ideas in medical ethics seminars at Monash University, and I thank students at these seminars for this. I am very grateful to Zeb Jamrozik for his resourcefulness and his excellent suggestions about relevant sources, and I wish to thank Michael Selgelid and Zeb for their invaluable advice with this chapter.

J. Oakley (✉)
Monash Bioethics Centre, Monash University, Melbourne, VIC, Australia
e-mail: justin.oakley@monash.edu

© The Author(s) 2020   125
E. Jamrozik, M. Selgelid (eds.), *Ethics and Drug Resistance: Collective Responsibility for Global Public Health*, Public Health Ethics Analysis 5,
https://doi.org/10.1007/978-3-030-27874-8_8

major global health concern, as the resistance of disease-causing microorganisms to antimicrobials can completely undermine the effectiveness of many antibiotics and other antimicrobials that are commonly used to treat a variety of diseases, and to lower the risks of post-surgical infections. The dwindling effectiveness of many antimicrobials has been implicated in a growing number of patient deaths world-wide, often from conditions such as respiratory and wound infections which had previously been responsive to antimicrobial treatments. Addressing what sorts of antimicrobial prescribing decisions a virtuous physician would make seems especially challenging for virtue ethics approaches to medical practice. For such approaches typically evaluate physician decision-making by reference to patient-centred role virtues like medical beneficence, medical courage, and trustworthiness, but tend to say little about broader medical virtues that help physicians act in the best health interests of the community, which is a crucial consideration in ethically justifiable clinician decision-making regarding antimicrobials.

In this chapter, I discuss some key patient-centred medical virtues and several candidate community-centred medical virtues, and I consider what sorts of antimicrobial prescribing decisions such virtues would lead physicians to make. I draw out the moral significance of these community-centred medical virtues by examining certain cases where prescribing an antimicrobial seems likely to be in the best interests of the patient in question, but seems unlikely to be in the best health interests of the community overall. In doing so, I consider some analogies with other sorts of cases involving seemingly divergent virtue directives, such as those involving psychiatrists breaching patient confidentiality to protect third parties from harm. Also, I argue that practically-intelligent virtuous physicians should have an awareness of the sorts of cognitive biases that are especially likely to distort their antimicrobial prescribing decisions, and I urge physicians to develop ways of avoiding or counteracting such biases. Further, effectively addressing the impact of these biases and other countervailing factors that inhibit virtuous prescribing practices is, I argue, the responsibility not only of individual physicians, but also of institutions and regulators. Therefore, I also consider the prospects of certain institutional and regulatory initiatives which aim to reduce antimicrobial over-prescribing by highlighting and undermining such biases in clinical practice.

8.1 Antimicrobial Resistance and Virtue Ethics

Rising global concerns about antimicrobial resistance has prompted many national and international agencies, professional medical associations, and policymakers to investigate what sorts of factors are contributing to this problem, and to the resulting significant increases in morbidity and mortality rates of certain diseases in many countries around the world. The accelerating resistance of microorganisms to antibiotics and other antimicrobial agents commonly used in medical treatment has substantially reduced – and at times entirely removed – effective microbial treatment options for some patients with serious conditions such as pneumonia, tuberculosis, or septicaemia, and a number of those patients have subsequently died. Further, antibiotic-resistant 'superbugs' like methicillin-resistant staphylococcus aureus (MRSA) have been found in significant quantities in some of the world's major hospitals, and many patients worldwide have died from hospital-acquired MRSA infections.[1] Antimicrobial resistance is a major problem in high-income countries and also in low- and middle-income countries. For example, China has the world's highest prevalence of MRSA, and this bacterium is associated with higher mortality rates for hospitalised patients there.[2] Also, the prevalence of multi-drug resistant tuberculosis is relatively high in countries such as India and Russia – it has recently been predicted that 12.4% of all cases of TB will soon be multi-drug resistant in India, and that 32.5% of all cases of TB will soon be multi-drug resistant in Russia.[3] It has also been reported that over 50% of microorganisms involved in hospital-acquired infections in Greece seem to be resistant to *all* available antibiotics.[4] Effective antimicrobial treatments have therefore become a somewhat scarce resource in clinical practice in a number of contexts, and many countries have now introduced new government regulations and clinical guidelines aimed at encouraging more responsible use and stewardship of antimicrobials. While antibiotics are one among several antimicrobial agents, I will focus here on the prescription of antibiotics, as inappropriate antibiotic prescribing appears to be a major contributor to antimicrobial resistance, which is why many of the efforts to improve antimicrobial prescribing practices are directed at antibiotic prescribing in particular.

This accelerating growth in antimicrobial resistance is due to a range of factors, including widespread farming practices, patient misuse of antimicrobials, and inappropriate antimicrobial prescribing practices by physicians. These contributing factors are now being targeted by a range of initiatives from government agencies, national and international medical and health care organisations, and at the level of hospital management. Thus, the WHO recently modified its longstanding *Essential Medicines List* by introducing recommendations that prioritise the use of some antibiotics over others in certain contexts – for example, the list recommends that amoxicillin be made widely available to treat common infections, such as pneumonia, but that other antibiotics should be reserved for use as a last resort when a life-threatening infection has failed to respond to other antibiotics.[5] And the clinical practice guidelines about antibiotic prescribing in Australia's *Therapeutic Guidelines: Antibiotic* also encourage the responsible use by clinicians of antibiotics in treating a wide variety of conditions, and thus discourage the over-prescribing of antibiotics.[6] However, questions have arisen about the effectiveness of clinical

---

[1] Staphylococcus aureus already exist in (e.g.) our respiratory tract, and some of those bacteria will already be antimicrobial-resistant, but antibiotic treatment allows those antimicrobial resistant S. aureus to potentially multiply.

[2] See Dan Cui et al. (2017), and Zhenjiang Yao et al. (2015).

[3] Aditya Sharma et al. (2017).

[4] Stavros Saripanidas (2016).

[5] See http://www.who.int/mediacentre/news/releases/2017/essential-medicines-list/en/.

[6] Antibiotic Expert Groups (2019).

guidelines aimed at improving antibiotic prescribing. For example, a recent Australian study concluded that "Antibiotics are prescribed for ARIs [acute respiratory infections] at rates 4–9 times as high as those recommended by clinical guidelines. The potential for reducing rates of antibiotic prescription and to thereby reduce rates of antibiotic-related harms, particularly bacterial resistance, is therefore substantial."[7] Physician decision-making in relation to antibiotics has thus been identified as a major contributor to the problem of antimicrobial resistance,[8] and so this remains a promising context on which to focus in developing effective ways of redressing this problem. In this chapter, therefore, I concentrate on the antimicrobial prescribing behaviour of physicians, and on what would plausibly be required of a virtuous physician in regard to antimicrobial prescribing.

The emphasis in virtue ethics on the character and moral psychology of virtuous agents have led some to regard this approach as excessively individualistic, and so it might be wondered whether virtue ethics is capable of providing useful guidance on addressing antimicrobial resistance, which is a complex multi-dimensional problem on an international scale. However, as noted above, physician prescribing behaviour is a major contributing factor to antimicrobial resistance, and so to this extent, focusing on what decisions a virtuous physician would make in relation to the prescription of antimicrobials would seem to offer a promising line of response to this problem. Further, recent empirically-informed accounts of the Aristotelian idea of practical intelligence (*phronesis*) emphasise the importance of agents being aware of, and having strategies to counteract, common decision-making biases that can divert their virtuous dispositions from hitting the target of the relevant virtue.[9] Physicians are evidently not immune from such biases,[10] and these biases may well contribute to physicians' inappropriate antimicrobial prescribing in certain sorts of cases.[11] So applying recent empirically-informed accounts of virtues and practical intelligence to physicians' roles and to their antimicrobial prescribing behaviour would seem to be a useful way of approaching the problems of antimicrobial resistance.

Generally speaking, virtue ethics evaluates actions by asking, 'what sort of person would do a thing like that?' For example, we can ask whether an action was generous or mean-spirited, courageous or cowardly, and we can examine whether this is the sort of thing which a kind person or a just person would do in the circumstances. Virtue ethics can therefore draw upon such considerations to provide the elements of a criterion of right action. A virtue ethics criterion of right action can be

---

[7] Amanda R. McCullough et al. (2017).

[8] See e.g. the US Centers for Disease Control and Prevention statement, 'About Antimicrobial Resistance': "The use of antibiotics is the single most important factor leading to antibiotic resistance around the world. Simply using antibiotics creates resistance. These drugs should only be used to manage infections." https://www.cdc.gov/drugresistance/about.html. [accessed 1 November 2018].

[9] See Justin Oakley (2018a, b).

[10] See Blumenthal-Barby and Krieger (2015) and Saposnik et al. (2016).

[11] See e.g. Jeffrey et al. (2014).

stated initially in broad terms as holding that: An action is right if, and only if, it is what an agent with a virtuous character would do in the circumstances.[12] Virtue ethicists have also applied such an appeal to exemplars in developing accounts of right action in the context of various professional roles, such as medical and legal practice. An influential version of these accounts highlights links between the proper goal/s of the profession in question – such as health and justice for medicine and law, respectively – and an Aristotelian conception of human flourishing, or *eudaimonia*. The proper goal/s of a profession can, in turn, provide the basis for an account of the role virtues for that profession. Thus in the case of medical practice, the role virtues for doctors can be understood as those character traits that enable doctors to serve the goal of health. These traits have been thought to include patient-centred dispositions such as medical beneficence, medical courage, and trustworthiness, along with community-centred dispositions such as justice.[13]

8.2 Medical Virtues and Antimicrobial Prescribing

We saw in the previous section that inappropriate prescribing decisions by physicians in relation to antibiotics is highlighted as one of the most important contributing factors to antimicrobial resistance and its resulting problems. Physicians in many countries are evidently over-prescribing antibiotics, and for a variety of conditions. For example, a recent study by Katherine Fleming-Dutra and colleagues found that during 2010–2011 physicians in US clinics prescribed antibiotics for acute respiratory infections at double the rates that were clinically appropriate for such infections – in other words, around half of those antibiotic prescriptions were clinically inappropriate.[14] Also, the 2018 national survey of antimicrobial prescribing practices in Australian hospitals found that 21.4% of antimicrobials prescribed during that year were classified as inappropriate prescriptions.[15] And a recent study using survey data from 2010–2015 on antibiotic prescribing rates in general practice found that Australian GPs are prescribing antibiotics for acute respiratory infections at rates up to nine times higher than those recommended by the national *Therapeutic Guidelines: Antibiotic*. These researchers found, for example, that "Despite the recommendation in these guidelines against prescribing antibiotics for bronchitis, general practitioners were estimated to have prescribed antibiotics for patients with bronchitis in 85% of cases during 2010–2015".[16] A subsequent interview with study co-author Chris Del Mar reported that he felt "doctors were acting out of a misplaced sense of caution, as many conditions requiring treatment with antibiotics

---

[12] Rosalind Hursthouse (1999).

[13] See e.g. Justin Oakley and Dean Cocking (2001).

[14] Katherine E. Fleming-Dutra et al. (2016).

[15] Australian Commission on Safety and Quality in Health Care (2020).

[16] McCullough et al. (2017), op. cit., p. 68.

share similar symptoms to those that do not.... doctors needed to stop considering a prescription for antibiotics as their first port of call. 'The idea that we have as GPs, that writing a prescription for antibiotics just in case, is probably not so effective..., That in fact not using antibiotics at all is actually safe, you don't end up with missed cases of meningitis and Lemierre's disease – all sorts of really nasty things.'"[17]

Given that patients frequently request antibiotics for acute respiratory infections, and are usually keen to get better sooner, it is perhaps not surprising that there is a tendency among physicians in many countries to over-prescribe antibiotics for such infections in responding to such requests. But while yielding to patient pressure results in physicians over-prescribing antibiotics, there is also evidence that antibiotics are prescribed inappropriately because physicians overestimate patients' expectations of receiving these drugs. Thus, in a 1997 study of antibiotic prescribing decisions by UK General Practitioners, Macfarlane and colleagues found that "Non-clinical factors influence the decision to prescribe antibiotics for nearly a half of those receiving one. Patient pressure was cited most frequently...and identified by the Audit Commission as an important reason for the excess use of antibiotics in the community. Pressure from patients to prescribe antibiotics, particularly for respiratory symptoms, has been identified as the commonest reason for doctors' discomfort with prescribing decisions. General practitioners can, however, overestimate patients' expectations. A quarter of our patients received antibiotics when they stated that before the consultation they had not wanted antibiotics".[18] A survey of 544 patients of inner London general practitioners found further evidence that physicians are overly influenced by their perceptions of patients' hopes and expectations of receiving a medication prescription, and those perceptions frequently seemed to be inaccurate: "A third [31%] of the prescriptions [which were sometimes for antibiotics] written in this study were either not indicated or not hoped for, with 3% being neither indicated nor hoped for.... The writing of nonindicated prescriptions was primarily associated with the doctor's sense of feeling pressurised [by the patient to write a prescription]".[19] This survey also found that "In a fifth [22%] of consultations in which a prescription was written, the prescription was not strictly indicated on purely medical grounds".[20] Overall, the strongest influence on a physician's decision to prescribe a medication was found in this study to be the physician's perceptions of their patients' hopes and expectations about receiving that medication.

---

[17] Angus Randall (2017).

[18] John Macfarlane et al. (1997), p. 1213. There is also evidence that UK General Practitioners have overestimated patients' expectations of receiving a prescription, not only in cases involving antibiotics, but also with other sorts of medication: see Nicky Britten (1995), p. 1084: "When general practitioners are surveyed they describe high levels of demand, but objective evidence consistently suggests that doctors overestimate patients' expectations. Reanalysis of published data shows that about a fifth of patients leave general practice consultations with prescriptions they did not expect."

[19] Nicky Britten and Obioha Ukoumunne (1997), p. 1509.

[20] Ibid, p. 1510.

In some of the above types of cases where physicians prescribe antibiotics inappropriately, they may well be acting from a beneficent motivation to serve the best interests of their patient in some way – particularly where, as mentioned above, physicians prescribe an antibiotic "just in case" the patient has a serious, albeit rare, type of infection where antibiotic treatment at that stage would likely benefit the patient in the longer-term. However, physicians are frequently in a position to accurately diagnose that a patient has the less serious condition of acute bronchitis, and to recognise that it is in the long-term best interests of such a patient not to be prescribed an antibiotic now.[21] Where physicians prescribing antibiotics are acting in misguided ways from motives of beneficence towards their patients, these cases can be usefully characterised as involving a failure of such motives to hit the target of the patient-centred virtue of medical beneficence.[22] Cognitive biases seem to be an important factor in explaining why physicians' beneficent motives can fail to hit the target of the virtue of medical beneficence in their prescribing decisions. An illuminating recent study of Australian GP trainees (registrars) describes their decision-making processes in deciding whether or not to prescribe an antibiotic for the patient. A number of these registrars explained that, while they realised prescribing an antibiotic was often not in the patient's best interests, and that it was important to avoid antibiotic over-prescribing, the registrars indicated how they rationalised prescribing an antibiotic in certain cases – for instance, telling themselves that it was in the patient's best interests to receive an antibiotic. The study found that: "Many registrars were concerned about patient safety and avoiding subsequent hospital presentations [for example, one registrar said]: *'I tend to probably cover things a bit more because...if something goes wrong I want to make sure that the patient is going to be safe.' (Registrar 9)*".[23] In certain cases this seems to be an overly cautious approach, which may indicate that a registrar is unduly influenced by the well-known 'framing effect', whereby possible losses loom larger to the agent than do possible gains.[24] The researchers also mentioned that: "Barriers to evidence-based prescribing included role-modelling outdated practices, or setting a precedent of

---

[21] See McCullough et al., op. cit. For a useful systematic review of studies on how individuals treated with an antibiotic for respiratory and urinary infections can subsequently develop a higher level of bacterial resistance to that antibiotic for several months, see Céire Costelloe et al. (2010).

[22] Christine Swanton emphasizes the importance of virtues 'hitting the target' of the contextually-relevant virtue, and she discusses various examples of candidate virtues that fail to hit their target (see Christine Swanton 2003). Similarly, Dean Cocking and I have argued that each virtue involves being guided by a 'regulative ideal': "To say that an agent has a regulative ideal is to say that they have internalised a certain conception of correctness or excellence, in such a way that they are able to adjust their motivation and conduct so that it conforms – or at least does not conflict – with that standard" (Oakley and Cocking 2001, op. cit., p. 25).

[23] Anthea Dallas et al. (2014), p. e564. While antibiotic over-prescribing seemed due in part to certain cognitive biases in the registrars, "many registrars described a dissonance between their attitudes to guidelines and their prescribing behaviours, producing dissatisfaction with their own prescribing...: *'There's probably been times where I've given them the script and kind of felt a bit disappointed in myself afterwards.'(Registrar 14)"* p. e563.

[24] See Amos Tversky and Daniel Kahneman (1981).

prescribing that created patient expectations and pressure on the registrar: '*I do know one supervisor in particular will give his patients antibiotics even for something that sounds very viral, and therefore when I see his patients, I feel I'm expected to do that as well, because his patients have been seeing him for many years. So they expect it too, so I'm definitely more likely to give his patients antibiotics even when I don't think it's justified.'"(Registrar 10)*[25] This sort of approach may manifest a form of 'authority bias', whereby the registrar shows undue deference towards the antibiotic prescribing practices of his or her clinical supervisor.[26]

According to some influential recent empirically-informed accounts of virtue and virtue ethics, acting virtuously requires (among other things) agents to employ certain deliberative strategies to counteract common decision-making biases and other countervailing factors which can impede virtuous actions. These accounts develop a comprehensive conception of virtuous character-traits, which include an awareness of situational factors that conduce to or inhibit virtuous behaviour.[27] The misguided beneficent efforts of those registrars who were (or could reasonably have been) aware that prescribing an antibiotic was not in their patient's best interests can be helpfully understood as failures to exercise the central Aristotelian virtue of *phronesis*, or practical intelligence, in this context.[28] Broadly speaking, *phronesis* is an overarching normative disposition that regulates the more specific dispositions involved in particular virtues, to enable their more context-sensitive expression. Developing such practical intelligence can help well-motivated agents to avoid two potential deficiencies that can undermine their attempts to hit the target of the relevant virtue – i.e. moral ineptitude, and failures of meticulousness. Moral ineptitude is shown by agents who are well-intentioned but lack sufficient practical know-how or emotional intelligence to succeed in bringing about the good which they intend to bring about. Unmeticulous agents have the relevant practical know-how and emotional intelligence, but their efforts fall short as they do not apply strategies to circumvent decision-making biases (and similar countervailing factors) that are prevalent in the context.[29] Thus, the registrars who prescribed antibiotics against their patients' longer-term best interests failed to exercise the virtue of practical intelligence here, because the registrars were not sufficiently meticulous in preventing cognitive biases and other countervailing factors from undermining their efforts to make prescribing decisions in their patients' best interests.[30]

---

[25] Dallas et al., p. e565.

[26] The phenomenon of 'authority bias' was famously observed in Stanley Milgram's experiments on obedience to authority. See Stanley Milgram (2010).

[27] See e.g. Daniel C. Russell (2009), e.g. p. 140; Nancy E. Snow (2009), p. 563; and Nancy E. Snow (2010), especially her pp. 34–7 discussion of how people can effectively confront and combat any prejudices they might hold. These conceptions of virtue and virtue ethics have been developed in response to 'situationist' critiques of earlier accounts of virtue ethics.

[28] Aristotle (1980); and Russell (2009), op. cit.

[29] For further details about these executive failings, and their relevance to practical intelligence and medical virtues, see Oakley (2018b), op. cit. See also Oakley (2018a), op. cit.

[30] See also Daniel Russell's helpful discussion of how agents may ascertain the specific ends of virtuous actions, in: Daniel C. Russell (2015).

Indeed, there is evidence of other cognitive biases diverting physicians' prescribing decisions from patients' best interests. For example, a recent US study found that as the number of prescribing decisions made by each primary care physician increased through the morning and into the afternoon, each physician became more likely to issue an antibiotic prescription that was not in the patient's best interests. This study suggests that good prescribing behaviour can be undermined by a form of 'decision fatigue' (a phenomenon seen in many other contexts), which seemed to lead the physicians to become less able to resist providing inappropriate antibiotic prescriptions.[31] So, developing ways of effectively addressing these (and potentially other) cognitive biases that appear to be contributing to physicians prescribing antibiotics in cases where this is likely to be contrary to the patient's long-term best interests, would seem a promising strategy supported by an empirically-informed virtue ethics approach (and perhaps also by certain other ethical approaches) to the problems of antimicrobial resistance.

8.3   Community-Centred Medical Virtues and Antimicrobial Prescribing Practice and Policy

Virtue ethics approaches to medical practice have tended to focus predominantly on patient-centred virtues, such as medical beneficence and medical courage, which help an individual doctor serve well the best interests of their patients. But while it is important that these virtues correct any physician tendencies to make unreflective assumptions about what prescribing decisions are best for their patients, patient-centred virtues are not the only role virtues which are relevant to ethically justifiable prescribing behaviour and the problem of antimicrobial resistance. For the antimicrobial prescribing decisions by physicians have also contributed to broader community harms, such as the diminishing effectiveness of antibiotic treatments for other patients, and the scarcity of more expensive, last resort, antibiotics due to their increasing use as first-line treatments (and which have become unaffordable to patients in some countries). So, ethically responsible antibiotic prescribing practices by physicians must also take into account the broader effects that their prescribing decisions are likely to have on the community. The virtuous antimicrobial prescriber thus needs to take account of a wider moral universe, beyond that of the best interests of their own patients. They would therefore be guided in their antimicrobial prescribing decisions by community-centred medical virtues, such as justice and a readiness to serve the broader community, along with patient-centred virtues such as medical beneficence. The virtue of justice requires physicians to allocate fairly the medical resources under their control.[32] In Aristotelian terms, allocating

---

[31] Linder et al. (2014), op. cit.

[32] For a useful discussion of how medical benevolence should be constrained by the virtue of justice, see Roger Crisp (2015).

medical resources fairly (particularly when medical resources are scarce) can be understood as making allocation decisions in such a way as to provide each person who is affected by the decision an equal chance of developing and exercising their capabilities to live a flourishing human life. These capabilities include those elaborated in the *Nicomachean Ethics*, such as being able to understand the world, to engage in practical reasoning about our lives, and to form personal relationships with others.[33] The readiness to serve others is another community-centred medical virtue, which requires physicians to make their services broadly available to the community, and so (for example) to avoid picking and choosing their patients according to the physician's personal preferences. This virtue is plausibly understood as applying to physicians *qua* professionals, who ought to provide this readiness to serve the community in return for being granted a monopoly of expertise over the provision of key goods – i.e. those that serve patients' health.[34]

In the context of antimicrobial prescribing, the virtue of justice requires physicians to consider whether prescribing antibiotics for patients in certain circumstances is likely to detrimentally affect the availability of effective antibiotics for other patients, even if there are grounds for believing that an antibiotic prescription is likely to be of some benefit to the former patients. While it is plausible to believe that physicians often serve the health of the community best by making antibiotic prescribing decisions that are in the best interests of their own patients, there will be cases where these two goals come into conflict – because the community's interest in constraining antimicrobial resistance can sometimes be served best by the physician refraining from providing an antibiotic which may be of some benefit to their patient. Where the likely benefit to the patient in such conflict cases is only marginal, the virtue of justice will require a physician not to prescribe antibiotic treatment to their own patient. Of course, it is possible that justice could also sometimes require withholding from a patient an antibiotic that is likely to be of greater than marginal benefit – for instance, where there is only a single last-resort antibiotic available, and one patient is likely to derive much greater benefit from receiving this antibiotic than is another patient, then justice could arguably require that the first patient be provided with this antibiotic. (This is analogous to situations where justice can plausibly require the allocation of the only available ICU bed to a patient who is likely to benefit more from this resource than will another patient, even if this second patient may suffer considerable harm as a consequence.) In what follows, I will concentrate on the first kind of conflict cases.

Suppose a child presents to their physician with acute otitis media. Prescribing an antibiotic in such cases would appear likely to confer marginal benefits for the child – but as Collignon explains, "with an absolute reduction in pain in only 5%. Most cases resolve spontaneously. Seventeen children must be treated to prevent

---

[33] See the capabilities approach developed by Martha C. Nussbaum (2006). Nussbaum draws on Aristotle's view that "the form of government is best in which every man, whoever he is, can act best and live happily" (*Politics* 1324a23–5). See also Justin Oakley (1994); and Millar, M. (2020).

[34] This community-centred virtue becomes especially relevant in discussions of ethically justifiable limits on conscientious objection by physicians.

one child having some pain after two days. Antibiotics have no effect on hearing problems or other complications".[35] In these sorts of cases, despite the possible minor benefits of the child receiving an antibiotic, the detrimental impact which such patterns of antibiotic prescribing have on antimicrobial resistance in the broader community suggests that the virtue of justice would require physicians not to prescribe antibiotics in such cases, and to provide the child with other medication, to relieve their symptoms. The dictates of justice in these sorts of cases can be compared with what justice would plausibly require of psychiatrists deliberating about whether to breach patient confidentiality to protect third parties from harm. Where a patient confides to his or her psychiatrist a credible threat to significantly harm a third party, and the psychiatrist is able to take steps to see that the third party is warned about this threat, the virtue of justice plausibly requires the psychiatrist to take such steps in the interests of this third party.[36] While maintaining absolute patient confidentiality might sometimes be in the best interests of the patient involved, and so might be thought consistent with the virtue of medical beneficence, the overarching virtue of justice requires that confidentiality be breached here. (Indeed, in many such cases this course of action will not be contrary to the *virtue* of medical beneficence – for instance, where confiding such a threat is actually a 'cry for help' from the patient – even if beneficence provides some grounds for maintaining confidentiality here).

But while the virtue of justice and its concerns for the broader interests of the community can in certain circumstances require physicians to refrain from providing a patient with an antibiotic that is likely to benefit that patient, there is evidence that physicians relegate these broader interests to the periphery of their antibiotic prescribing decision-making. A 2002 US survey of 400 generalist physicians and 429 infectious diseases specialists found that: "The risk of contributing to the problem of antibiotic resistance was ranked lowest, overall and by generalists, and second lowest by ID specialists"[37] The researchers concluded that "…neither generalists nor infectious disease specialists emphasize the relative societal risks of antimicrobial drug selection in their treatment decisions for patients with community-acquired pneumonia. Instead, they emphasize providing the newest and best treatments for each individual patient even though this approach may not be supported by current guidelines or public health policy".[38] One explanation of the insufficient weight given by these doctors to the risk of generating antimicrobial resistance was an overconfidence that their own antibiotic prescribing decisions were unlikely to negatively impact on antimicrobial resistance. This can be seen as an example of the cognitive bias known as 'the overconfidence effect', whereby agents have greater confidence in their judgements than is warranted by the evidence. There is much

---

[35] Peter J. Collignon (2002), p. 328.

[36] See, for example, the much-discussed 1969 case of Tatiana Tarasoff, as described in Marilyn McMahon (1992), pp. 12–16.

[37] Joshua P. Metlay et al. (2002), p. 90.

[38] Ibid., p. 93.

evidence that physician decision-making can be distorted by an overconfidence bias, which is one of the most frequently studied biases in medical decision-making.[39] Further evidence that cognitive biases can lead physicians to give insufficient weight to their own contributions to antimicrobial resistance is provided in an illuminating recent survey of 889 US physicians, which indicated that they often lacked insight into the broader harms of their own antibiotic prescribing decisions. Most of the respondents expressed concern about the problem of antimicrobial resistance. However, the researchers found that: "While 62% of respondents agreed that other doctors overprescribe antibiotics, only 13% agreed that they themselves overprescribe antibiotics".[40] The researchers concluded that "While most respondents agreed that other doctors overprescribe antibiotics, a much smaller proportion (especially of faculty) felt that they themselves overprescribe".[41] This significant underestimation by physicians of the contribution that their own antibiotic prescribing decisions are likely having on the broader problems of antimicrobial resistance can be characterised as an example of confirmation bias, where an agent interprets information in a way that confirms a view that they already hold, regardless of whether this information actually supports or undermines that view.

Thus, cognitive biases also seem to be diverting physicians' community-centred dispositions to act in accordance with the virtue of justice from hitting its target, in the context of their antibiotic prescribing decisions. A promising way of addressing this problem is for physicians to develop various 'debiasing' strategies, which can help them carry out what justice requires of them here in more practically intelligent ways. For example, Ian Scott and colleagues suggest that physicians' awareness of their cognitive biases, and their ability to counteract the detrimental effects such biases can have on medical decision-making, can be enhanced by providing them with powerful narratives of patients who have been harmed (e.g. by antimicrobial resistance), and by reflective practice and role modelling.[42] But while the prescribing decisions of individual physicians are clearly a significant contributor to the problem of antimicrobial resistance, effectively addressing this problem goes well beyond the responsibility of each individual physician to 'smarten up' their own decision-making about antibiotic prescribing by strengthening their medical virtues on their own initiative. For these individual efforts must be complemented by the responsibilities of governments, international organisations, policymakers, and healthcare institutions to create institutional and regulatory environments which are conducive to physicians hitting the targets of the role virtues of medical beneficence and justice. For it can sometimes be difficult for physicians in their antibiotic prescribing behaviour to succeed in hitting the targets of the virtues of medical beneficence and justice, when they are working in contexts where they are frequently

---

[39] See Saposnik et al. (2016), op. cit. Overconfidence and availability bias are the two most frequently researched biases in medical decision making, and they appear to be highly prevalent in this context (though their precise prevalence levels here have not been determined).

[40] Lilian Abbo et al. (2011), p. 715.

[41] Ibid., p. 716.

[42] Ian A Scott et al. (2017). See also Ateev Mehrotra and Jeffrey A. Linder (2017).

confronted with patients' requests for antibiotics, in circumstances where prescribing an antibiotic would be clinically inappropriate.

A good example of such government efforts is the Australian *'Choosing Wisely'* initiative, which aims to reduce antibiotic overprescribing (and other forms of poor clinical practice) by providing patients and physicians with guidelines and specific examples of when medications such as antibiotics would be inappropriate. For instance, one such guideline advises against prescribing antibiotics for patients with uncomplicated acute bronchitis.[43] These guidelines and examples on the *Choosing Wisely* website encourage discussions between physicians and patients about the appropriateness or otherwise of such medications in the circumstances, and make the boundaries of good medical practice more transparent to patients and doctors. Similarly, the UK Behavioural Insights Team successfully reduced the overprescription of antibiotics by sending letters to GPs in practices with relatively high rates of antibiotic prescription, stating that "80% of practices in your local area prescribe fewer antibiotics per head than yours".[44] Providing these doctors with such benchmarking information helped nudge them to recognise and counteract their biases towards prescribing antibiotics, and thereby helped enable these doctors' medical dispositions to hit their proper targets.[45]

Also, a worthwhile institutional initiative here could be to address authority bias through the creation of 'safe spaces' for junior doctors to be able to anonymously report established practices of poor antibiotic prescribing, without necessarily jeopardising their professional relationships with consultants and senior physicians.

8.4 Conclusion

The antibiotic prescribing decisions made by physicians have clearly played a significant role in increasing antimicrobial resistance. Due to their monopoly of expertise in their professional roles, physicians are especially well-placed to be aware of, and to combat, the broader problems which certain sorts of antimicrobial prescribing decisions can result in. Virtuous physicians owe the community a readiness to serve it by taking steps to reduce – and hopefully eliminate altogether – antibiotic prescribing decisions that are contrary to the virtues of medical beneficence and justice. In this chapter, I have outlined some strategies that individual physicians, institutions, and healthcare policymakers can develop to help physicians to hit the targets of those virtues, and to thereby play their part in redressing the problems of antimicrobial resistance.

---

[43] See *Choosing Wisely Australia, https://www.choosingwisely.org.au/.*
[44] Michael Hallsworth et al. (2016). See also David Halpern (2015).
[45] See also Kiran Iyer (2017), and Fay Niker (2018).

References

Abbo, Lilian, et al. 2011. Faculty and resident physicians' attitudes, perceptions, and knowledge about antimicrobial use and resistance. *Infection Control and Hospital Epidemiology* 32 (7): 714–718.

Antibiotic Expert Groups. 2019. *Therapeutic guidelines: Antibiotic, version 16*. Melbourne, Therapeutic Guidelines. https://tgldcdp.tg.org.au/guideLine?guidelinePage=Antibiotic&from page=etgcomplete

Aristotle. 1980. *The Nicomachean Ethics*. Trans. W.D. Ross. Oxford: Oxford University Press.

Australian Commission on Safety and Quality in Health Care. 2020. *Antimicrobial prescribing practice in Australian hospitals: Results of the 2018 Hospital National Antimicrobial Prescribing Survey*, January 2020. https://www.ncas-australia.org/ncas-publications.

Blumenthal-Barby, J.S., and H. Krieger. 2015. Cognitive biases and heuristics in medical decision-making: A critical review. *Medical Decision Making* 35 (4): 539–557.

Britten, Nicky. 1995. Patients' demands for prescriptions in primary care. *British Medical Journal* 310 (29): 1084–1085.

Britten, Nicky, and Obioha Ukoumunne. 1997. The influence of patients' hopes of receiving a prescription on doctors' perceptions and the decision to prescribe: A questionnaire survey. *British Medical Journal* 315 (7121): 1506–1510.

Collignon, Peter J. 2002. Antibiotic resistance. *Medical Journal of Australia* 177 (6): 325–329.

Costelloe, Céire, et al. 2010. Effect of antibiotic prescribing in primary care on antimicrobial resistance in individual patients: Systematic review and meta-analysis. *British Medical Journal* 340 (18): c2096.

Crisp, Roger. 2015. The duty to do the best for one's patient. *Journal of Medical Ethics* 41: 220–223.

Cui, Dan, et al. 2017. Use of and microbial resistance to antibiotics in China: A path to reducing antimicrobial resistance. *Journal of International Medical Research* 45 (6): 1768–1778.

Dallas, Anthea, et al. 2014. Antibiotic prescribing for the future: Exploring the attitudes of trainees in general practice. *British Journal of General Practice* 64 (626): e561–e567.

Fleming-Dutra, Katherine E., et al. 2016. Prevalence of inappropriate antibiotic prescriptions among US ambulatory care visits, 2010–2011. *Journal of the American Medical Association* 315 (17): 1864–1873.

Hallsworth, Michael, et al. 2016. Provision of social norm feedback to high prescribers of antibiotics in general practice: A pragmatic national randomised controlled trial. *The Lancet* 387 (10029): 1743–1752.

Halpern, David. 2015. *Inside the nudge unit*. London: WH Allen.

Hursthouse, Rosalind. 1999. *On virtue ethics*, Oxford: Oxford University Press.

Iyer, Kiran. 2017. Nudging virtue. *Southern California Interdisciplinary Law Journal* 26: 469–492.

Linder, Jeffrey A., et al. 2014. Time of day and decision to prescribe antibiotics. *JAMA Internal Medicine* 174 (12): 2029–2031.

Macfarlane, John, et al. 1997. Influence of patients' expectations on antibiotic management of acute lower respiratory tract illness in general practice: Questionnaire study. *British Medical Journal* 315 (7117): 1211–1214.

McCullough, Amanda R., et al. 2017. Antibiotics for acute respiratory infections in general practice: Comparison of prescribing rates with guideline recommendations. *Medical Journal of Australia* 207 (2): 65–69.

McMahon, Marilyn. 1992. Dangerousness, confidentiality, and the duty to protect. *Australian Psychologist* 27 (1): 12–16.

Mehrotra, Ateev, and Jeffrey A. Linder. 2017. Tipping the balance towards fewer antibiotics. *JAMA Internal Medicine* 176 (11): 1649–1650.

Metlay, Joshua P., et al. 2002. Tensions in antibiotic prescribing: Pitting social concerns against the interests of individual patients. *Journal of General Internal Medicine* 17: 87–94.

Milgram, Stanley. 2010. *Obedience to authority: An experimental view*. London: Pinter and Martin.

Millar, M. 2020. A capability perspective on antibiotic resistance, inequality, and child development. In *Ethics and drug resistance: Collective responsibility for global public health*. Cham: Springer.

Niker, Fay. 2018. Policy-led virtue cultivation: Can we 'nudge' citizens towards developing virtues? In *The theory and practice of virtue education*, ed. T. Harrison and D. Walker, 153–168. London: Routledge.

Nussbaum, Martha C. 2006. *Frontiers of justice: Disability, nationality, species membership*. Cambridge, MA: Harvard University Press.

Oakley, Justin. 1994. Sketch of a virtue ethics approach to health care resource allocation. *Monash Bioethics Review* 13 (4): 27–33.

———. 2018a. Toward an empirically informed approach to medical virtues. In *The Oxford handbook of virtue*, ed. Nancy E. Snow, 571–590. Oxford: Oxford University Press.

———. 2018b. Creating regulatory environments for practical wisdom and role virtues in medical practice. In *Cultivating moral character and virtue in professional practice*, ed. David Carr, 83–95. London: Routledge.

Oakley, Justin, and Dean Cocking. 2001. *Virtue ethics and professional roles*, Cambridge: Cambridge University Press.

Angus Randall. 2017. Antibiotics: GPs prescribing to patients at up to nine times higher than current guidelines. ABC Radio, *The World Today*, 11 July 2017. http://www.abc.net.au/news/2017-07-10/australians-overloaded-with-antibiotics-research/8693600.

Russell, Daniel C. 2009. *Practical intelligence and the virtues*. New York: Oxford University Press.

———. 2015. What virtue ethics can learn from utilitarianism. In *The Cambridge companion to utilitarianism*, ed. Ben Eggleston and Dale E. Miller, 258–279. Cambridge: Cambridge University Press.

Saposnik, G., D. Redelmeier, C.C. Ruff, and P.N. Tobler. 2016. Cognitive biases associated with medical decisions: A systematic review. *BMC Medical Informatics and Decision Making* 16 (138): 1–14.

Saripanidas, Stavros. 2016. Antibiotic abuse in Greece. *British Medical Journal* 355: i6328.

Scott, Ian A., et al. 2017. Countering cognitive biases in minimising low value care. *Medical Journal of Australia* 206 (9): 407–411.

Sharma, Aditya, et al. 2017. Estimating the future burden of multidrug-resistant and extensively drug-resistant tuberculosis in India, the Philippines, Russia, and South Africa: A mathematical modelling study. *The Lancet: Infectious Diseases* 17 (7): 707–715.

Snow, Nancy E. 2009. How ethical theory can improve practice: Lessons from Abu Ghraib. *Ethical Theory and Moral Practice* 12 (5): 555–568.

———. 2010. *Virtue as social intelligence: An empirically grounded theory*. New York: Routledge.

Swanton, Christine. 2003. *Virtue ethics: A pluralistic view*. Oxford: Oxford University Press.

Tversky, Amos, and Daniel Kahneman. 1981. The framing of decisions and the psychology of choice. *Science* 211 (4481): 453–458.

Yao, Zhenjiang, et al. 2015. Healthcare associated infections of Methicillin-resistant *Staphylococcus aureus*: A case-control-control study. *PLoS One* 10 (10): e0140604.

Chapter 9
Moral Responsibility and the Justification of Policies to Preserve Antimicrobial Effectiveness

Alberto Giubilini and J. Savulescu

Abstract Restrictive policies that limit antimicrobial consumption, including therapeutically justified use, might be necessary to tackle the problem of antimicrobial resistance. We argue that such policies would be ethically justified when forgoing antimicrobials constitutes a form of easy rescue for an individual. These are cases of mild and self-limiting infections in otherwise healthy patients whose overall health is not significantly compromised by the infection. In such cases, restrictive policies would be ethically justified because they would coerce individuals into fulfilling a moral obligation they independently have. However, to ensure that such justification is the strongest possible, states also have the responsibility to ensure that forgoing antimicrobials is as easy as possible for patients by implementing adequate compensation measures.

Keywords Bioethics · Public health ethics · Antimicrobial resistance · Collective responsibility · Easy rescue

A. Giubilini (✉)
Wellcome Centre for Ethics and Humanities, University of Oxford, Oxford, UK

Oxford Uehiro Centre for Practical Ethics, University of Oxford, Oxford, UK
e-mail: alberto.giubilini@philosophy.ox.ac.uk

J. Savulescu
Oxford Uehiro Centre for Practical Ethics, University of Oxford, Oxford, UK

Visiting Professorial Fellow in Biomedical Ethics, Murdoch Children's Research Institute, Melbourne, Australia

Distinguished Visiting Professor on Law, Melbourne Law School, University of Melbourne, Melbourne, Australia

© The Author(s) 2020
E. Jamrozik, M. Selgelid (eds.), *Ethics and Drug Resistance: Collective Responsibility for Global Public Health*, Public Health Ethics Analysis 5,
https://doi.org/10.1007/978-3-030-27874-8_9

9.1   The Problem of Antimicrobial Resistance

When Alexander Fleming was awarded the Nobel Prize for Medicine in 1945, he warned in his acceptance speech of the risk that some bacteria could develop resistance to penicillin, which he had discovered 17 years earlier. It cannot be overstated how longsighted Fleming was. Today, "bacteria are resistant to nearly all antibiotics that were earlier active against them" (Herrmann and Laxminarayan 2010, p. 4.2), and "700,000 people die of resistant infections every year" (O'Neill 2016, p. 4). As estimated by the Review on Antimicrobial Resistance commissioned by the UK government, "by 2050 10 million lives a year are at risk due to drug resistant infections, as are 100 trillion USD of economic output" (O'Neill 2016). Examples of diseases associated with antimicrobial resistance (AMR) include tuberculosis, gonorrhoea, typhoid fever, and group A streptococcus (Van der Velden et al. 2013). The problem of AMR has been framed by some in terms of a "slowly emerging disaster" (Littmann and Viens 2015) and of a "global health security issue" (Balasegaram et al. 2015).

AMR is a naturally occurring phenomenon: microbes naturally tend to adapt to antimicrobials and develop resistance. However, AMR is accelerated by human behaviour (Jamrozik and Selgelid 2019), and particularly the use and abuse of antimicrobials, both in livestock (Anomaly 2019; Giubilini et al. 2017; O'Neill 2015; Anomaly 2009) and in humans (O'Neill 2016). This paper is focussed on AMR caused by *human* consumption of antimicrobials, and therefore on the human factor in the development of AMR. There is a positive correlation between antimicrobial resistance rates and antimicrobial consumption in humans (Van der Velden et al. 2013, pp. 318–19). In fact, in the case of antibiotics, it is now widely accepted that "the use of antibiotics is the single most important factor leading to antibiotic resistance around the world: simply using antibiotics creates resistance" (CDC, About antimicrobial resistance), even when antibiotic use is medically indicated to treat infections (which is what makes AMR a particularly difficult ethical issue). Thus, the paradox of antibiotic and certain other kind of antimicrobial consumption is that while certain antimicrobials can be beneficial to individuals, their use also poses a threat to public health, to the lives of millions of people, and to the world economy. If they are not used more wisely than they currently are, i.e. if their consumption is not reduced, we might face a "post antibiotic" era (Alanis 2005; WHO 2014) characterized by two undesirable outcomes. First, we might no longer have effective means of treating severe infections. Second, medicine's achievements that require effective antibiotics, such as organ transplantation, cancer chemotherapy, and major surgery, might no longer be available (WHO, Antimicrobial resistance).

To be sure, the problem of reducing antibiotic and certain type of antimicrobial consumption mainly concerns those countries with easy access to and massive consumption of antimicrobials. These are mainly high-income countries (HICs). The consumption rate of antimicrobials in most low- and middle-income countries (LMICs), although it increased dramatically in recent years, remains much lower than that in HICs, and in many LMICs the burden of infectious disease still outweighs the burden of resistance; the increase in consumption in such countries ought to be closely monitored, but drastic measures to reduce antibiotic consumption need to be implemented *primarily* in HICs (Klein et al. 2018). Reduction of therapeutically justified antimicrobial consumption in LMICs for conservation purposes might have undesirable outcomes, especially in those countries where the level of public health is already poor. There might be exceptions, of course; for instance, India is a lower-income country and one of the major contributors to resistance through very high consumption rates (Kumar et al. 2013; WHO India 2018), so perhaps *significant* restrictions on antibiotic access – that is, restrictions that apply also to therapeutically justified use – should be implemented in such context as well. However, to make our arguments the least controversial possible, we take them to apply only to HICs where a good level of public health might allow leaving certain mild and self-limiting infections untreated without posing significant costs on patients.

This paper is focussed on two often neglected aspects of the problem of AMR, namely on the individual moral obligations with regard to antimicrobial consumption and on the type of justification that health authorities could and should offer for restricting access to antimicrobials in order to preserve antimicrobial effectiveness. As we will argue in Sect. 9.3, these two aspects are closely related: the state has an obligation to provide the strongest justification possible for restrictive interventions aimed at limiting antimicrobial consumption, and the strongest justification is one based on the existence of an individual moral obligation to forgo antimicrobials, perhaps (as we will suggest in Sect. 9.4) even in case of mild self-limiting infections that do not significantly risk worsening the individual's general health. The fact that even justified use of antimicrobials, i.e. the use that is medically necessary to treat infections (as opposed to unjustified use, e.g. when antibiotics are prescribed for viral infections, a practice that is not uncommon unfortunately), contributes to antimicrobial resistance determines a moral conflict between individual interest and collective interest that might require individuals to make sacrifices for the sake of the common good.

Such moral conflict gives rise to a "tragedy of the commons" scenario (Hollis and Maybarduk 2015), which will be presented in Sect. 9.2: we will argue, following Garrett Hardin, that the tragedy of the commons has an ethical solution, which will be presented in Sect. 9.4, where we will argue that there is a *moral* obligation to contribute to the preservation of the common good of antimicrobial effectiveness which might entail a moral obligation not to use antimicrobials in some cases of mild, self-limiting infections, provided that certain conditions obtain and that the state takes all the measures that are necessary to ensure that forgoing antimicrobials in those cases approximates a form of "easy rescue".

Thus, in Sect. 9.4 we will draw the policy implications of this ethical solution to the tragedy of the commons: the moral obligation to sometimes forgo antimicrobials, grounded in a duty of easy rescue, strengthens the justification for state interventions that prohibit or discourage the use of antimicrobials in certain cases. Since states should be able to provide the strongest justification possible for implementing restrictive policies that discourage or prohibit antimicrobial consumption, appeal to individual moral obligations to sometimes forgo antimicrobials has a political weight in terms of justification of state-imposed restrictions on antimicrobial use.

However, as we will suggest in conclusion of Sect. 9.4, the moral obligation of the state to provide the strongest justification possible for restricting access to antimicrobials implies an obligation on the state to ensure that forgoing antimicrobials does represent a form of easy rescue, i.e. that individuals bear as small a cost as possible for leaving certain infections untreated with antimicrobials (assuming this measure is necessary). This means, in practice, that the state has moral obligations towards individuals who are requested to forgo antimicrobial for the sake of the collective good, such as the obligation to provide enhanced medical follow up and to adequately compensate, financially or in other ways (e.g. higher priorities on waiting lists for other medical treatments), these individuals.

9.2 Public Goods, Tragedy of the Commons, and Policies to Address Antimicrobial Resistance

It has been claimed that "if effective antibiotics are seen as a public good, their overuse may be likened to the tragedy of the commons scenario" (Littmann and Viens 2015, p. 214). This statement is not entirely correct in the way the notion of 'public good' is applied. As we are going to see in this section, antibiotic and more generally antimicrobial effectiveness shares an essential feature with common goods or common pool resources, rather than with public goods. Let's see more in detail what this terminology means when the concepts of "public goods" and of "common goods" are applied to the case of antimicrobial effectiveness.

Some benefits associated with antimicrobial effectiveness can certainly be considered public goods. These include freedom from infectious diseases (Selgelid 2007, p. 226), the containment of infectious diseases (Woodward and Smith 2003, p. 10), or the reduced risk of infection by a resistant disease (Smith and Coast 2003, p. 78). These benefits are public goods in the technical sense of the term: they are non-excludable, in the sense that people cannot be excluded from benefitting from them; and they are non-rivalrous, in the sense that the fact that a person benefits from them does not affect the way and the extent to which others benefit as well (Cowen 2008). More precisely, freedom from infectious diseases, infectious disease containment, and the reduced risk of infection can be conceived as *global public goods*, i.e. goods "exhibiting a significant degree of publicness (i.e. non-excludability and non-rivalry) across national boundaries" (Woodward and Smith 2003, p, 8).

Effective antimicrobials are the means through which such public goods are preserved. But antimicrobial effectiveness is a common good rather than a public good because antimicrobial effectiveness is rivalrous in consumption, in virtue of what Anne Schwenkenbecher in a chapter of this volume calls "the antimicrobial footprint" of antimicrobial consumptions: "simply using antibiotics [or other antimicrobials] creates resistance" (CDC, About Antimicrobial resistance) and "more consumption of antibiotics directly leads to more resistance" (O'Neill 2016, p. 17), regardless of whether antimicrobial use is therapeutically justified or not. The use of the resource of antimicrobial effectiveness through antimicrobial consumption erodes the resource and therefore diminishes its availability. As put by Jonny Anomaly, "individually rational choices produce substantial social costs by creating reservoirs of antibiotic-resistant bacteria in human hosts and more generally in our shared microbial environment" (Anomaly 2013, p. 753). It is important to point out that, while antimicrobial use might benefit individuals in the short term, this is not necessarily the case in the long term: individuals can incur a portion of the costs associated with antimicrobial resistance (Anomaly 2013, p. 752). For example, the individual who takes antibiotics can become a carrier of resistant bacteria, which means that the individual is at greater risk in case of subsequent infections (Cars et al. 2008).

It has been observed that the effectiveness of antimicrobials can "be modelled as a natural resource in much the same way as are fish, tree, oil, and other resources" (Herrmann and Laxminarayan 2010, p. 4.3). According to John Conly, for example,

> antimicrobial resistance may be likened to [the] overfishing scenario, to cattle overgrazing the grass in the commons or to deforestation on Easter Island, which led to population dying out. Antimicrobial resistance is a consequence of continued overuse of antibiotics combined with the constant growth of resistance overtime. (WHO 2010)

Typically, consumption of natural resources determines a 'tragedy of the commons' scenario (Ostrom et al. 1999,). Garrett Hardin first described the "tragedy of the commons" in an article he published in *Science* (Hardin 1968). Hardin illustrated the problem through the example of a commons to which some herdsmen have access. The tragedy occurs when many herdsmen, acting merely out of self-interest, have their cattle overgraze the commons, thus eroding the resource (Hardin 1968, p. 1244). Particularly in a context of growing population, all herdsmen acting in the same way deplete the common good. In the context of antimicrobial effectiveness, the same problem arises (Hollis and Maybarduk 2015). As Jonny Anomaly put it, "the benefits of [antibiotic] use are borne by the individual, the costs are socialised, and the consequent harm is the product of many independent actions" (Anomaly 2013, p. 752).

According to Hardin, one characteristic of a tragedy of the commons is the fact that there is no "technical" solution to the problem, meaning that there is no solution "that requires a change only in the techniques of the natural sciences, demanding little or nothing in the way of change in human values or ideas of morality" (Hardin 1968, p. 1243). On the contrary, Hardin argued, the solution requires a "fundamental extension" of our morality (Hardin 1968, p. 1243), i.e. acting in view of protecting the public interest rather than in a merely self-interested way.

However, Hardin seemed to be sceptical about the possibility that such an extension of morality could occur, at least without some external coercive intervention (which on most accounts of morality would arguably undermine the authentically "moral" nature of the change invoked). He believed that the best way to solve a tragedy of the commons is through some degree of "mutually agreed upon" coercion (Hardin 1968, p. 1247), for example in the form of taxation that would allow to internalize the negative externalities of individual consumption.[1]

---

[1] Another solution Hardin proposed is the privatization of the commons (Hardin 1968, p. 1245), which in the case of antimicrobials might consist in "extending the period of exclusivity, possibly

And indeed, some have suggested that the negative externalities of antimicrobial use could be internalised through the introduction of user fees or consumption taxes (Littmann and Viens 2015; Anomaly 2013; Herrmann and Laxminarayan 2010). Internalization is typically achieved by "taxing negative externalities (…) at a rate that would offset the social cost of the activities that generate the externalities, and then (ideally) using the revenues from the tax to fund socially useful projects" (Anomaly 2009, p. 433). The idea behind the concept of internalization is that it is those who are responsible for a negative externality that should bear the cost for it.

However, it might not be possible to fully internalise the collective cost of antimicrobial consumption and to disincentivize consumers through a tax: selection for AMR will probably continue to occur in spite of the tax and in spite of the investment of the revenue in strategies to contain AMR such as research on new antimicrobials. For this reason, it has been suggested that we might need at some point to introduce policies to restrict antimicrobial use only to the most serious infections, i.e. that it might be necessary to prevent, if necessary through legal prohibition, their use in the case of milder self-limiting infections (Foster and Grundmann 2006, p. 179). We will return to this point in Sect. 9.4.

Policies might therefore aim to disincentivize individuals from using antimicrobials, in the case of taxation; or they might coercively impose restrictions on antimicrobial use, in the case of outright prohibition of antimicrobial use in certain circumstances. As said before, we are interested here not so much in determining which policies would be more effective and ethically acceptable. Thus, we do not intend to provide arguments for or against taxation of antibiotics or restrictions on their use. Rather, we are interested in what justification an authority might offer for implementing restrictive policies, whatever form they take. While such policies might be justified by considerations of public interest, it is important that the type of justification that a state can provide for interventions that limit or discourage antimicrobial use be the strongest justification possible, given that such interventions might require individuals to sacrifice some significant personal interest, such as the interest in accessing effective health care, for the sake of the common good. As we are going to argue in the next section, considerations about the morality of individual choices regarding antimicrobial consumption can strengthen the justification for such policies.

9.3   Morality and Antimicrobial Consumption

Restrictive and coercive interventions might be efficacious at protecting the common goods associated with AMR containment and in preserving the common good of antimicrobial effectiveness. However, the "tragedy of the commons" scenario

---

indefinitely" of the patents, thus giving "the patentee the ability to charge high prices and thus indirectly restrain overuse by some users" (Hollis and Maybarduk 2015, p. 33).

which preservation of antimicrobial effectiveness gives rise to suggests that there is an independent ethical dimension to the problem of AMR. In this section we will explain why discussing such an independent ethical dimension is important from a political point of view, i.e. in view of justifying coercive policies aimed at restricting antimicrobial use.

Establishing an independent *moral* responsibility to make a more appropriate use of antimicrobials would make the justification that a state could provide for policies that restrict or discourage antimicrobial use the strongest possible. And, arguably, the state does have a duty to provide individuals with the strongest justification possible for implementing policies that restrict or discourage a treatment that is in the individuals' best interest, as is often the case with antimicrobials.

Restrictive policies might be justified in light of a state's duty to protect certain common goods and public goods, such as the public goods associated with AMR containment. These possible interventions range from those that are minimally intrusive to those that more substantially infringe rights or important interests of individuals. Thus, at one end of the spectrum we find policies such as information campaigns, nudging, or incentivisation of certain pro-social behaviours; and at the other end we find more restrictive policies such as taxation and compulsion. In all such cases, the authority enforcing such policies should be able to provide the strongest justification possible for its interventions, but the more restrictive the policies become, the more difficult it is for the authority to meet such requirement. A state's duty to protect public goods and common goods by itself does not provide the strongest justification possible for interventions that sacrifice important individual interests. The justification would be stronger if, in addition to considerations of public interest, there were independent individual moral obligations to make those individual sacrifices that are required by the restrictive policy (Giubilini et al. 2018) Thus, one way to strengthen the justification for state interventions such as taxation of antimicrobials or prohibition of certain antimicrobial uses is to identify a pre-existing individual *moral* obligation not to use antimicrobials in certain cases. Such independent moral obligation to prioritize other-regarding choices over self-interested choices would make the case for introducing restrictive policies that limit or discourage antimicrobial use as compelling as possible: such policies would simply impose or encourage choices that individuals have an independent moral obligation to make anyway.

9.4 Individual Responsibility and Duty of Easy Rescue: The Ethical Solution to the Tragedy of the Commons and the Responsibilities of the State

We have noted above that, as Hardin himself acknowledged, the tragedy of the commons is first and foremost an *ethical* problem. Therefore, it has an ethical solution that is independent from the justification for legislative coercive solutions. Now, "ethical solution" means, according to Hardin's phrasing, a change in "human values" informing human behaviour. Thus, in the case of antimicrobial consumption, the ethical solution to the tragedy of the commons consists in justifying the existence of an individual *moral* obligation to act contrary to one's (short term) self interest in order to contribute to the preservation of the common good of antimicrobial effectiveness. In other words, the ethical solution consists in finding a justification for a moral duty not to use antimicrobials, if necessary even at the cost of leaving some infections untreated, in order to protect the common good of antimicrobial effectiveness. In this section we are going to provide this justification.

It is plausible to assume that individuals have what might be called a "duty of easy rescue" (Savulescu 2007). That is, if doing X (or refraining from doing Y) entails a small cost to an individual and a large benefit (or prevention of a large harm) to others, the individual has a clear prima facie moral obligation to do X (or to refrain from doing Y). Morality is essentially different from prudence, and requires the sacrifice of one's own interests for others. It is debatable what kind of individual sacrifices morality requires, but if morality requires anything, almost everyone would agree that it certainly requires at least small sacrifices for the sake of preventing great harm. This does not mean that an individual does not have a moral obligation when the cost to her is not small. However, in easy rescue case, such moral obligation seems uncontroversial. A formulation of the duty of easy rescue was famously provided by Peter Singer, according to whom "if it is in our power to prevent something bad from happening, without thereby sacrificing anything of *comparable* moral importance, we ought, morally, to do it" (Singer 1972, p. 230, emphasis added). A roughly equivalent, though way less demanding, formulation has been provided by Tim Scanlon, according to whom "[i]f we can prevent something very bad from happening to someone by making a slight or even moderate sacrifice, it would be wrong not to do so" (Scanlon 1998, p. 224). The fact that some version of a duty of easy rescue can be defended both from a utilitarian perspective- in the case of Singer – and a contractualist perspective- in the case of Scanlon – supports the idea that it is a fundamental requirement of morality on which reasonable people could agree.

The existence of a duty of easy rescue implies an individual moral obligation to forgo antimicrobials for the sake of the common good of antimicrobial effectiveness, *when forgoing antimicrobials comes at a sufficiently small cost to individuals*. This would be the case, for example, when antimicrobials are not necessary to treat an infection (for example in the case of viral infections), in which case there would be no cost at all in forgoing antimicrobials. But forgoing antimicrobials would also come at a relatively small cost to individuals when individuals have minor self-limiting infections or low risk mild infections (for example, skin infections which could be treated topically with antiseptics) that do not significantly worsen the general state of health of the individual, where the risks of complications is adequately monitored, and when the individual is adequately compensated for any financial and non-financial cost she might incur as a consequence of leaving that infection untreated. As some have suggested (Foster and Grundmann 2006) it *might* at some point be necessary to leave such self-limiting infections untreated in order to preserve antimicrobial effectiveness.

The moral duty of easy rescue represents the ethical solution to the tragedy of the commons in the context of antimicrobial consumption that Hardin advocated: individuals have a moral obligation to prioritize the public interest in the preservation of the commons of antimicrobial effectiveness over their own interest in treating with antimicrobials any type of infection (including mild and self-limiting ones), because doing so comes at a small cost to individuals, at least when all that is required of individuals is to leave self-limiting mild or minor infections untreated in circumstances in which this would not significantly worsen individual health and where adequate compensatory measures are in place.

Now, since it is necessary that a large number of individuals fulfil a duty of easy rescue in order for a public benefit to obtain, we can say that there is not only an individual, but also a *collective* responsibility to forgo antimicrobials when doing so is a way of fulfilling a duty of easy rescue. Now, in some cases of collective action, any individual contribution to a collective good, or to the prevention of a collective harm, is insignificant; for example, the contribution each individual could make to the realisation of herd immunity through individual vaccination, or the contribution each individual could make to the prevention of global warming by avoiding driving just for fun, are both negligible. However, in the case of containment of antimicrobial resistance every single individual forgoing antimicrobials could make a difference, because every single individual could become the carrier of resistant microbes that are then passed onto other people. For example, it has been shown that individuals who take antibiotics for respiratory and urinary infections might develop bacterial resistance that could last up to 12 months (Costelloe et al. 2010). Thus, the need to preserve the common good of antimicrobial effectiveness implies not only a collective responsibility, but also an individual responsibility not to use antimicrobials so as to benefit others by contributing to the preservation of antimicrobial effectiveness, at least as long as forgoing antimicrobials comes at a small enough cost to individuals.

When forgoing antimicrobials is a moral duty of easy rescue, we can claim that the state is in the position to fulfil its obligation to provide the strongest justification possible for prohibiting or at least discouraging the use of antimicrobials: as we said in Sect. 9.3, the state would be discouraging or preventing individuals from doing only what they have a moral obligation not to do anyway. Therefore, being the state in the position to fulfil its moral obligations in implementing restrictive measures, such restrictive measures regarding antimicrobial use are ethically justified. As one of us put it, "when the cost to us of forgoing some activity is small (…) and the harm to others which thereby does not occur is great (prevention of serious disease), then liberals might require that the state prevent this harm" (Savulescu 2007, p. 10).

"Duty to rescue" laws exist in many European countries; for instance, in Germany and France it is illegal not to assist a person in danger when providing assistance entails a small or no risk to the potential rescuer. From what we have said so far, laws restricting antimicrobial use could be ethically justified as special cases of "duty to rescue" laws of this kind.

The individual moral responsibility to forgo antimicrobials in certain circumstances implies that there are two other actors with specific moral responsibilities, besides patients with minor or mild self-limiting infections.

First, doctors have the responsibility to assess whether a certain infection is self-limiting and, more in general, whether leaving a certain infection untreated is compatible with a good enough level of individual health. In one important sense, the fact that the state has the strongest justification possible for imposing bans on anti-microbial use or for discouraging through taxation antimicrobial consumption takes the responsibility for decisions about antimicrobial prescription out of the hands of doctors: if a certain infection is mild and self limiting and leaving it untreated does not significantly worsen the general state of health of the patient, then the state, rather than the doctor, is justified in preventing the use of antimicrobials for the sake of the common good. However, the doctors would still have the important responsibility of determining whether the conditions specified in such laws would apply, and thus they would still be the ultimate gate-keepers of antimicrobials.

Second, but equally important, healthcare systems and states have important moral responsibilities too. We have said that having an uncontroversial individual moral obligation based on a duty of easy rescue to sometimes forgo antimicrobials strengthens the justification for state interventions that forbid or discourage antimicrobial use. Therefore, the state or the healthcare system have the responsibility to ensure that forgoing antimicrobials does represent a form of easy rescue, i.e. that it is not too burdensome for patients, so that the justification for state intervention is the strongest possible. This means, at the very least, that patients who are denied antimicrobials should be carefully monitored and provided with adequate and enhanced medical follow up to make sure that forgoing antimicrobial treatment does not significantly worsen the general state of health of the individual. It is the responsibility of states and of healthcare systems to ensure that adequate measures are in place in order to guarantee monitoring and medical supervisions of those patients who sacrifice their self-interest for the sake of the collective good, so that their sacrifice would represent a form of easy rescue.

However, we claim that the responsibility of states extends beyond the provision of medical supervision and follow up. We propose that a further measure that might be implemented in order to ensure that the rescue is an easy rescue is that of *compensating* individuals who make a sacrifice for the sake of the collective good. Compensation – financial or of other kind – would be an appropriate measure because it would provide individuals with an additional reason for forgoing antimicrobials in certain circumstances, it would be fair in consideration of the sacrifice individuals are making for the sake of the collective good, and, more importantly, if the right type of compensation is offered, it could make their condition easier to bear, thus approximating a form of easy rescue and providing a further reason for the existence of an individual ethical obligation. For example, those who forgo antimicrobials and leave infections untreated could be offered first call on future medical treatments, even for conditions that are not related to the current infection; or they might be offered discounts when they buy other medicines, such as painkillers, or discounts on future medical treatments; alternatively, they might be offered outright financial compensation, for example in the form of tax relief or by directly paying them, which would account for any financial loss that might result from enduring a prolonged infections. In this way, they would derive at least some benefit from the sacrifice they are making, which would render the sacrifice easier to bear, and at the same time society would be "making up" for the sacrifice it is imposing on these individuals.

Granted, preserving antimicrobial effectiveness by financially compensating those who forgo antimicrobials might involve a significant cost for the state. However, we saw in the introduction that there will be significant costs associated with AMR if we don't intervene now; therefore, investing money now to compensate and to guarantee adequate medical follow up to those who make sacrifices in order to contain AMR might be a good strategy also from the point of view of cost-effectiveness.

9.5 Conclusions

When is it necessary to leave infections untreated in order to preserve the common good of antimicrobial effectiveness? This is an empirical issue and the answer depends on whether and to what extent the societal optimum of antimicrobial consumption differs from the individual optimum. In other words, it concerns the issue whether and to what extent the level of individual consumption that is consistent with satisfactory AMR containment (the societal optimum) differs from the level of consumption that would effectively treat infections in any individual (the individual optimum). As Kevin Foster and Hajo Grundmann (2006, pp. 178–9) have explained, if individual and societal optima were similar, it would be sufficient to avoid inappropriate antimicrobial use, such as using antibiotics in the case of viral infections, an abuse which often occurs (Van der Velden et al. 2013). Only inappropriate use of antimicrobials would then be morally impermissible. However, if individuals and societal optimum differed significantly, in order to protect the societal interest in AMR containment individuals might be required to leave minor and mild self-limiting infections untreated in order to reserve antimicrobials for serious major infections. In such cases, coercive policies that restrict access to *some* therapeutically justified use of antimicrobials are ethically permissible, or so we have argued. However, when implementing such restrictive policies, state and health authorities have responsibilities not only towards the collective, but also towards individuals who make (small) sacrifices for the sake of the collective interested, and who should be guaranteed adequate compensation in exchange for leaving certain infections untreated.

Acknowledgment AG and JS were funded by the Wellcome Centre for Ethics and Humanities, University of Oxford, which is supported by a Wellcome Centre Grant (203132/Z/16/Z), and by the Wellcome Trust grant 104848/Z/14/Z. Through his involvement with the Murdoch Children's Research Institute, JS was supported by the Victorian Government's Operational Infrastructure Support Program. AG was also supported by the Oxford Martin School through the Oxford Martin Program on 'Collective responsibility for infectious disease'.

References

Alanis, Alfonso J. 2005. Resistance to antibiotics: Are we in the post-antibiotics era? *Archives of Medical Research* 36 (6): 597–705.

Anomaly, Jonny. 2009. Harm to others: The social cost of antibiotics in agriculture. *Journal of Agricultural and Environmental Ethics* 22: 423–435.

———. 2013. Collective action and individual choice: Rethinking how we regulate narcotics and antibiotics. *Journal of Medical Ethics* 39: 752–756.

———. 2019. Antibiotics and animal agriculture the need for global collective action. In *Ethics and drug resistance: Collective responsibility for global public health*. Cham: Springer.

Balasegaram, Manica, et al. 2015. The global innovation model for antibiotics needs reinvention. *The Journal of Law, Ethics, and Medicine* 43 (S3): 22–26.

Baquero, Fernando, and Jose Campos. 2003. The tragedy of the commons in antimicrobial chemotherapy. *Revista espanola de quimioterapia: publicacion oficial de la Sociedad Espanola de Quimioterapia* 16 (1): 11–13.

Barlam, Tamar, and Kalpana Gutpa. 2015. Antibiotic resistance spreads internationally across borders. *The Journal of Law, Medicine, and Ethics* 43 (S3): 12–16.

Cars, Otto, et al. 2008. Meeting the challenge of antibiotic resistance. *British Medical Journal* 337: 726–728.

CDC (Center for Disease Control and Prevention). About antimicrobial resistance, updated 2015. Available at: https://www.cdc.gov/drugresistance/about.html. Last access 17 Aug 2016.

CDDEP (Center for Disease Dynamics, Economics, and Policy). 2015. *The state of the world's antibiotics*. Available at: https://cddep.org/sites/default/files/swa_2015_final.pdf. Last access 15 Nov 2016.

Coast, J., R. Smith, and M.R. Millar. 1996. Superbugs: Should antimicrobial resistance be included as a cost in economic evaluation? *Health Economics* 5 (3): 217–226.

Copp, David. 2007. The collective moral autonomy thesis. *Journal of Social Philosophy* 38: 3.

Costelloe, Ceire, et al. 2010. Effect of antibiotic prescribing in primary care on antibiotic resistance in individual patients: Systematic review and meta-analysis. *British Medical Journal* 340: c2096.

Cowen, Tyler. 2008. Public goods. In *The concise encyclopedia of economics*, ed. R. Henderson. Indianapolis: Library of Economics and Liberty. Available at: http://www.econlib.org/library/Enc/PublicGoods.html. Last access 5 Aug 2016.

Daulaire, Nils, et al. 2015. Universal access to effective antibiotics is essential for tackling antibiotic resistance. *The Journal of Law, Medicine, and Ethics* 43 (S3): 17–21.

Foster, Kevin, and Hajo Grundmann. 2006. Do we need to put society first? The potential for tragedy in antimicrobial resistance. *PLoS Medicine* 3 (2): 177–180.

Giubilini, A., et al. 2017. Taxing meat: Taking responsibility for one's contribution to antibiotic resistance. *Journal of Agricultural and Environmental Ethics* 30 (2): 179–198.

Giubilini, A., T. Douglas, J. Savulescu, and Hannah Maslen. 2018. Quarantine, isolation, and the duty of easy rescue in public health. *Developing World Bioethics* 18 (2): 182–189.

Hardin, Garrett. 1968. The tragedy of the commons. *Science* 162: 1243–1248.

Held, Virginia. 1970. Can a random collection of individuals be morally responsible? *The Journal of Philosophy* 57 (14): 471–481.

Herrmann, Markus, and Ramanan Laxminarayan. 2010. Antibiotic effectiveness: New challenges in natural resource management. *The Annual Review of Resource Economics* 2 (4): 1–14.

Hoffman, Steven, and Trygve Ottersen. 2015. Addressing antibiotic resistance requires robust international accountability mechanisms. *The Journal of Law, Medicine, and Ethics* 43 (S3): 53–64.

Hoffman, Steven, et al. 2015. International law has a role to play in addressing antibiotic resistance. *The Journal of Law, Medicine, and Ethics* 43 (S3): 65–67.

Hollis, Aidan, and Peter Maybarduk. 2015. Antibiotic resistance is a tragedy of the commons that necessitates global cooperation. *The Journal of Law, Medicine, and Ethics* 43 (S3): 33–37.

Jamrozik, E., and M.J. Selgelid. Drug-resistant infection: Causes, consequences, and responses. In *Ethics and drug resistance: Collective responsibility for global public health*. Cham: Springer.

Klein, Eili, et al. 2018. *Global increase and geographic convergence in antibiotic consumption between 2000 and 2015*. In Proceedings of the National Academy of Sciences of the United States of America.

Kumar, S.G., C. Adithan, B.N. Harish, S. Sujatha, G. Roy, and A. Malini. 2013. Antimicrobial resistance in India: A review. *Journal of Natural Science, Biology, and Medicine* 4 (2): 286–291.

Lewis, H.D. 1948. Collective responsibility. *Philosophy* 23 (84): 3–18.

Littmann, J., and A.M. Viens. 2015. The ethical significance of antimicrobial resistance. *Public Health Ethics* 8 (3): 209–224.

Littmann, J., A. Buyx, and O. Cars. 2015. Antibiotic resistance: An ethical challenge. *International Journal of Antimicrobial Agents* 46: 359–361.

Miller, Seamus. 2006. Collective responsibility: An individualist account. *Midwest Studies in Philosophy* 30: 176–193.

Narveson, J. 2002. Collective responsibility. *The Journal of Ethics* 6: 179–198.

O'Neill, Jim. 2015. Antimicrobials in agriculture and the environment: Reducing unnecessary use and waste. *The Review on Antimicrobials Resistance*. Available at: https://amr-review.org/sites/default/files/Antimicrobials%20in%20agriculture%20and%20the%20environment%20-%20Reducing%20unnecessary%20use%20and%20waste.pdf. Last access 8 Oct 2016.

O'Neill. 2016. *Tackling drug resistant infections globally: Final report and recommendations*. The Review on Antimicrobial Resistance. Available at: http://amr-review.org/sites/default/files/160518_Final%20paper_with%20cover.pdf. Last access 16 Aug 2016.

Oczkowski, Simon. 2017. Antimicrobial stewardship programmes: Bedside rationing by another name? *Journal of Medical Ethics* 43 (10): 684–687.

Ostrom, E., J. Burger, C.B. Field, R.B. Norgaard, and D. Policansky. 1999. Sustainability – Revisiting the commons: Local lessons, global challenges. *Science* 284: 278–282.

Podolski, Scott, et al. 2015. History teaches us that confronting antibiotic resistance requires stronger global collective action. *The Journal of Law, Medicine, and Ethics* 43 (S3): 27–32.

Savulescu, Julian. 2007. Future people, involuntary medical treatment in pregnancy, and the duty of easy rescue. *Utilitas* 19 (1): 1–20.

Scanlon, T. 1998. *What We Owe to Each Other*. Cambridge, MA: Harvard University Press.

Schwenkenbecher, Anne. forthcoming. Antimicrobial footprints, fairness, and collective harm. In *Ethics and drug resistance: Collective responsibility for global public health*. Cham: Springer.

Selgelid, M. 2007. Ethics and drug resistance. *Bioethics* 21 (4): 218–229.

Singer, P. 1972. Famine, affluence, and morality. *Philosophy and Public Affairs* 1 (3): 229–243.

Smith, Richard, and Joanna Coast. 2002. Antimicrobial resistance: A global response. *Bulletin of the World Health Organization* 80 (2): 126–133.

———. 2003. Antimicrobial drug resistance. In *Global public goods for health*, ed. Smith et al., 73–93. New York: Oxford University Press.

Van der Velden, A., et al. 2013. Prescriber and patient responsibilities in treatment of acute respiratory tract infection – Essential for conservation of antibiotics. *Antibiotics* 2: 316–327.

WHO. 2010. Antimicrobial resistance: Revisiting the "tragedy of the commons". *Bulletin of the World Health Organization* 88: 11. Available at http://www.who.int/bulletin/volumes/88/11/10-031110/en/.

———. 2014. *Antimicrobial resistance: Global report on surveillance 2014*. Geneva: World Health Organization.

———. Antimicrobial resistance, Fact sheet n 194, updated 2015. Available at: http://www.who.int/mediacentre/factsheets/fs194/en/. Last access 17 Aug 2016.

WHO India. 2018. *Combating antimicrobial resistance in India*. Available at: http://www.searo.who.int/india/topics/antimicrobial_resistance/Combating_Antimicrobial_Resistance_in_India/en/. Last access 19 Nov 18.

Woodward, David, and Richard D. Smith. 2003. Global public goods and health: Concepts and issues. In *Global public goods for health*, ed. Smith et al., 3–29. New York: Oxford University Press.

Chapter 10
Access to Effective Diagnosis and Treatment for Drug-Resistant Tuberculosis: Deepening the Human Rights-Based Approach

Remmy Shawa, Fons Coomans, Helen Cox, and Leslie London

Abstract The lack of access to effective diagnosis and treatment for drug-resistant tuberculosis (DR-TB) remains a persistent ethical, human rights and public health challenge globally. In addressing this challenge, arguments based on a Human Rights-Based Approach (HRBA) to health have most often been focused on the Right to Health. However, a key challenge in multidrug-resistant (MDR-) and extensively drug-resistant (XDR-) TB is the glaring absence of scientific research; ranging from basic science and drug discovery through to implementation science once new tools have been developed. Although the Right to Enjoy the Benefits of Scientific Progress and its Applications (REBSP) is a little theorised human right, it has the potential to enrich our understanding and use of the Rights-Based Approach to health. In this chapter, we argue that States' duties to *respect, protect* and *fulfil* the REBSP within and outside their borders is an important vehicle that can be drawn on to redress the lack of research into new drug development and appropriate use of existing drugs for DR-TB in high burden settings. We call for urgent attention to minimum core obligations for the REBSP and the need for a General Comment by a UN human rights monitoring body to provide for its interpretation. We also note that conceptualization of the REBSP has the potential to complement Right to Health claims intended to enhance access to treatment for DR-TB on a global scale.

R. Shawa (✉)
School of Public Health and Family Medicine, University of Cape Town,
Cape Town, South Africa

F. Coomans
Department of International and European Law, Maastricht University,
Maastricht, The Netherlands

H. Cox
Division of Medical Microbiology and Institute of Infectious Disease
and Molecular Medicine, University of Cape Town, Cape Town, South Africa

L. London
Division of Public Health Medicine, University of Cape Town, Cape Town, South Africa

155

E. Jamrozik, M. Selgelid (eds.), *Ethics and Drug Resistance: Collective Responsibility for Global Public Health*, Public Health Ethics Analysis 5,
https://doi.org/10.1007/978-3-030-27874-8_10

Keywords Drug resistance · Human rights · Tuberculosis · Scientific progress · Rights-based approach

10.1   Introduction

In this chapter we explore how the Right to Enjoy the Benefits of Scientific Progress and its Applications (REBSP), a little theorized human right found in both the Universal Declaration of Human Rights (UDHR) and the International Covenant on Economic, Social and Cultural Rights (ICESCR), can deepen our understanding of a Human Rights Based Approach (HRBA) to health, taking access to effective treatment for drug-resistant tuberculosis (DR-TB) as an example. We bring attention to the slow progress in research, development and implementation of new and repurposing of existing drugs for treating DR-TB. Further, we attempt to frame poor access to effective diagnosis and treatment as a human rights problem, not only with respect to the right to health, but also with respect to the REBSP. In locating DR-TB within this right, we articulate what we mean by scientific progress, or lack of, in DR-TB, and discuss the broad context in which scientific progress must occur. Finally, we highlight some of the challenges in the conceptualization and realization of the REBSP, and make recommendations calling for urgent attention to minimum core obligations for the right and the need for a General Comment by a UN human rights monitoring body to provide for its interpretation.

Tuberculosis remains the world's deadliest communicable disease, responsible for more than 1.6 million deaths in 2017 alone (WHO 2018). More than a century after the bacterium causing TB was first identified, it continues to kill millions of people because diagnostic tools remain poor and current life-saving, essential medicines require a minimum of 6 months to effect cure. TB is particularly difficult to both diagnose and treat when TB bacteria have become resistant to available drugs. While standard care for drug-sensitive TB requires 6 months of treatment, DR-TB treatment may take up to 2 years. Access to diagnostic tools for DR-TB remains limited with currently available technology, despite some progress such as the GeneXpert MTB/RIF (a relatively rapid test that can, in some cases, quickly diagnose TB and some types of resistance) (Evans 2011), for which accessibility is limited due to the slow pace of implementation and high costs. In 2016, only 25% of the estimated number of multidrug-resistant (MDR-) or rifampicin-resistant (RR-) TB patients emerging that year were diagnosed, and even fewer started on treatment(WHO 2017). Of those who received treatment, only 54% of MDR/RR-TB patients are successfully treated and only 26% of those treated for extensively drug-resistant (XDR-) TB were successfully treated(WHO 2017). Current treatments for RR-TB (which includes MDR- and XDR-TB) are long, and have debilitating and often severe side effects, including irreversible hearing loss in more than a third of patients (Seddon et al. 2012). Other drugs included in the currently recommended

MDR/RR-TB regimen have been known to cause renal failure, cardiac arrhythmias, and psychiatric disturbances (Skrahina et al. 2016).

As much as DR-TB is a public health problem, it is also a human rights problem because it compromises the rights and dignity of the individuals who get infected. DR-TB highlights the glaring divide that exists between high-income countries (HICs) and low-middle income countries (LMICs), as well as the divide between the rich and the poor within countries. The majority of those who get TB and DR-TB reside in LMICs (WHO 2016). In HICs, TB and DR-TB are predominantly among the vulnerable and marginalized such as migrants and refugees (Figueroa-Munoz and Ramon-Pardo 2008). This skewed burden of TB disease effectively makes it a disease of the poor, who have little capacity to pay for medical care. Eliminating TB will require special attention to these marginalized populations as part of the moral duty of HICs and LMICs alike. Therefore, framing poor access to effective treatment for DR-TB as both a public health and human rights issue calls for solutions beyond public health, into the sphere of human rights. Moreover, human rights and public health are irrevocably inter-related; the promotion of one significantly contributes to the realization of the other, while, conversely, the infringement of human rights has negative effects on public health (Mann et al. 1994).

10.2   Access to Effective DR-TB Diagnosis and Treatment

Access to effective diagnosis and treatment for DR-TB is constrained by multiple problems; three of which are discussed in this chapter. The first problem is slow, or lack of progress and innovation in TB with regard to the development of new drugs or diagnostic technology. For example, since the introduction of the inexpensive and effective four-drug (isoniazid, rifampicin, pyrazinamide and ethambutol) treatment regimen, in the 1970s (Zumla et al. 2013), there were no novel drugs developed until the appearance of bedaquiline in 2012 (Cox et al. 2015); this is in spite of TB being one of the oldest diseases, spanning centuries (Daniel 2006). The current vaccine for TB, BCG, is almost 100 years old, and is considered effective in reducing severe and disseminated TB in young children but is not effective in adults (Kernodle 2010).

In most high TB burden settings the mainstay of TB diagnosis remains sputum smear microscopy, a test essentially unchanged in a century (Steingart et al. 2006). The slow or absent innovation in TB diagnosis and treatment arises from major challenges at different stages of the research continuum from basic science through to product availability. TB drug research is hugely underfunded, and key players in the pharmaceutical industry have been withdrawing from or cutting down on their investment in TB research and drug development, predominantly due to the real or perceived lack of a profitable market (Frick 2016). While treatment of drug sensitive TB still relies on the four drug combination developed more than 40 years ago, treatment of DR-TB has utilized both older drugs previously replaced in TB treatment due to lower efficacy and/or high side effects, along with drugs that were originally meant to treat other illnesses, repurposed for TB (Zumla et al. 2013). Currently, these repurposed drugs, together with some of the new drugs on the market are being used to improve treatment success rates for MDR-TB (Ndjeka et al. 2015).

The second problem is inadequate evidence-based guidance on effective use of new or repurposed drugs. New drugs need to be registered through clinical trials, which are often lengthy and costly; and guidance on use of new drugs is restricted until further clinical trials are conducted. For example, bedaquiline is not registered for use by the FDA and other agencies in pregnant women, children or people living with HIV who are co-infected with MDR-TB (Mase et al. 2013) because there were no clinical trials on its use in these populations. Yet, in many high burden countries, co-infection with HIV and TB is very common, and in reality bedaquiline is being used in these populations (Ndjeka et al. 2015). Similarly, for repurposed medicines, the evidence base to inform guidelines for their use is inadequate (London et al. 2016; Mafukidze et al. 2016). The third problem, which is often a result of the first two, is the high cost and complexity of diagnostics and treatment, and the lack of feasible models of care for scale-up in high burden settings.

10.3   Rights Based Approach to DR-TB

To appreciate a rights-based approach to DR-TB, one needs to have an appreciation for human rights in general. Human rights are entitlements and freedoms that people have simply because they are human. They refer to moral principles or norms, which describe certain standards and moral beliefs that people have. For example, the right to food was a result of the moral belief that people should not die of starvation. Similarly, moral beliefs have been significant in defining human rights, "from resisting torture and arbitrary incarceration to demanding the end of hunger and of medical neglect" (Sen 2005). This view of human rights as morally justifiable claims entails that, if a moral claim is that no person should be enslaved to another, then the claim not to be enslaved is, by a matter of law, a human right (McFarland 2015). Likewise, the moral belief that everyone deserves equal opportunities to be in a state of mental, physical and emotional wellbeing, gives credence to the right to health. But the right to health alone is not adequate to fulfill this moral belief. Without scientific progress in prevention, treatment, or in addressing determinants of ill-health, the aspiration of good health cannot be realized. Similarly, without good trade policies, proper housing adequate food or adequate infrastructure, securing good health becomes an unrealistic aspiration.

A human rights-based approach (HRBA) to health is a way of analysing health through the framework of human rights. It entails examining the impact of health policies and programmes on human rights, and likewise, how promotion or violation of human rights impact public health. A rights-based approach to health focuses on equity of health outcomes by analyzing and addressing the inequalities, discriminatory practices (*de jure* and *de facto*) and unjust power relations which are often at the center of health problems. Under a rights-based approach, health is anchored in a system of the recognition of individuals as right-holders and states as duty-bearers. A State has obligations to realize the human rights of its people, governed by international law. Perhaps, even more importantly, a rights-based approach to health means that citizens have the power and means to hold their governments accountable to their duties and responsibilities, and in turn, governments account to their citizens in a just and transparent manner.

The right to health is included in the ICESCR, which is a legally binding treaty. Article 12(2d) provides for an obligation for States to take steps for the creation of conditions which would assure medical services and medical attention to all in the event of sickness. The United Nations Committee on Economic, Social and Cultural Rights summed up the obligations of the State as the obligation to respect, protect and fulfill. Each treaty body adopts its interpretation of the provisions of its respective treaty in the form of "general comments" or "general recommendations". The Committee's guidance on the right to health in General Comment no. 14 provides an authoritative interpretation of the right to health), and unpacks state obligations and what this right looks like in real terms.

The emergence of the rights-based approach to health in recent years brought with it a paradigm shift in the fields of public health and human rights, owing, in part, to the HIV epidemic, but also to stronger civil society movements working on both health and human rights. But what does a rights-based approach to health mean for access to effective treatment of drug-resistant tuberculosis? And what human rights, other than the right to health can this approach be anchored in? Previously, public health concerns focused on the epidemiology of diseases, analysis of risk factors and interventions to control morbidity and prevent mortality. Particularly for TB, a communicable disease, the concern was around infection control and protecting the health of the public rather than those affected (London 2008). In this "public health view" of the TB disease, the State embarks on efforts to respond to TB because it is a public health threat, and not because the State is obligated to provide for the right to health of its citizens.

10.4   The Right to Enjoy the Benefits of Scientific Progress

The REBSP is proclaimed in article 15 paragraph 1(b) of the ICESCR; that [everyone has the right] to enjoy the benefits of scientific progress and its applications. This right is closely related to other rights contained in article 15; namely, the right to take part in cultural life (article 15, paragraph 1(a); the right of everyone to benefit from the protection of moral and material interests resulting from any scientific, literary or artistic production of which they are the author (article 15, paragraph 1(c)); and the right to freedom indispensable for scientific research and creative activity (article 15, paragraph 3).

Although it is proclaimed in these two important human rights documents, the REBSP is one of the least theorised human rights, and consequently, one of the least realized. Despite its textual existence dating back to the 1940s, in the UDHR, this right is relatively new in terms of its conceptualisation (Shaver 2015). Even human rights activists and lawyers are all too often unaware of the existence of the REBSP, much less of its meaning (Chapman 2009). As such, the REBSP is one right whose conceptual content needs to be interrogated and further developed (Shaver 2015). Despite its lack of clarity, the REBSP cannot be isolated from other human rights; because of the interdependent nature of human rights (United Nations 2005), but also because science is a vehicle that is used in almost every sphere of human development (UNDP 2012). This interrelation is best exemplified in debates about unequal access to the benefits of scientific progress – not only in relation to access to essential medicines (the right to health) (FM't Hoen 2002) but also in access to seed technology (the right to food) (Dommen 2002) to scientific discoveries that can improve environmental protection (the right to a safe environment) (Maskus 2002) and to information and communication technologies (right to privacy and access to information).

Three types of obligations result from the right to enjoy the benefits of scientific progress.

10.4.1 The Obligation to Respect

For the REBSP, the State's obligation to respect this right means that the State must desist from curtailing, or interfering with people's ability to access the benefits of scientific progress. In relation to access to treatmemt for DR-TB, the state must therefore not interfere with the production of medicines and technology necessary for treatment and prevention of DR-TB. Such interference can take many forms, from having overly bureaucratic procedures for acquiring ethical clearance for research in drug development to unnecessary delay in approval of new drugs that have already been approved in other countries. Another interference might be in the form of patent protection laws and policies that prioritise protection of patents resulting from research and development ahead of access to treatment. The World Trade Organization adopted rules on intellectual property rights, which impose obligations on states to protect patents on new and existing medicines. Such rules often act as obstacles to making medicines available and accessible to all.

10.4.2 The Obligation to Protect

While a State may not engage in deliberate efforts to violate the REBSP, a failure to protect people from third parties constitutes a potential infringement on the right, and a violation of its obligation to protect. High pricing of life-saving medicines by private corporations through, for example, anti-competitive protectionism, requires the State to act in defence of people's REBSP. For example, where a pharmaceutical company holds a patent of essential TB drugs, the State can incorporate into its laws, some of the TRIPS[1] flexibilities, such as compulsory licensing. In both South Africa and Kenya, the States compelled pharmaceutical companies holding patents to Antiretroviral drugs (ARVs), to enter into "voluntary" licencing agreements with local producers (Musungu et al. 2006).

10.4.3 Obligation to Fulfill

The obligation to fulfill is the third obligation. It speaks to the State's duty to adopt positive measures, and create an enabling environment for human rights to be realised. These measures may include "legislative, administrative, judicial, promotional and other measures" (CESCR 2006). Some of the challenges in drug development, or in research and development in general, are lack of infrastructure and unfriendly regulatory policies. For the state to fulfill the REBSP for TB patients, therefore, it has to take deliberate steps, within its means, to create an environment for science to thrive. These steps do not necessarily require financial expenditure, but would entail mobilising political will for agenda setting and intentional policies to encourage the sharing of scientific knowledge.

For DR-TB, the State's duty to respect, protect and fulfil, does not solely apply to the state with the burden of the disease; countries with a low DR-TB burden, but whose multinational corporations manufacture drugs for use in high burden countries, have an ethical obligation to protect citizens of in other countries from exploitation by their multinational corporations. This speaks to what others have termed 'extraterritorial obligations'- human rights obligations that a state has beyond its borders (Coomans and Kamminga 2004). And in the context of extraterritorial obligations, states that have the skills and resources to develop better and effective medicines, need to also put in place deliberate measures that would make access to such drugs possible outside their jurisdiction.

10.4.4 Realising the REBSP

Crucial to the realisation of the REBSP is the understanding of who the right holders and duty bearers of the right are. Primary duty bearers of any given right are the States, but also the International community (other States and international agencies); they have the duty to respect, protect and fulfil rights. The people, as individuals and as a collective, are therefore rights holders, in that they have individual and collective rights, and hold claims to entitlements provided for in a particular right. One of the most important issues in human rights is to understand different actors and the relationship between rights holders and duty bearers. A distinction should

---

[1] Trade-Related Aspects of Intellectual Property Rights.

be made between duties and responsibilities. For example, under the right to health, the state has the duty to provide health care to the people, and may do so through private and public health providers; in that case, health providers assume the responsibility to provide health care, but the State still holds the duty under international law. Similarly, under the REBSP, the state has a series of obligations such as creating an enabling environment for scientific progress to thrive, ensuring that people benefit from such scientific progress; non-state actors, such us the private sector (corporations), and academic institutions, have the responsibility to meaningfully contribute to scientific progress through their work. Under international law non-state actors, such as companies or private health providers, do not have legal obligations, only responsibilities. The legal nature of the latter is much weaker.

10.4.5 Minimum Core Obligations

Core obligations are important in realising human rights as they provide a frame of reference in determining what the State needs to do, at the bare minimum (and not subject to progressive realisation[2]). Minimum core obligations require the State to demonstrate that it has made every effort to use all available resources to satisfy, as a matter of priority, those core obligations. In the General Comment on the interpretation of the nature of ICECR obligations, General Comment No. 3 (1990), the Committee on Economic, Social and Cultural Rights stated that "States parties have a core obligation to ensure the satisfaction of minimum essential levels of each of the rights enunciated in the Covenant." Core obligations should be framed as both positive steps to be taken by the State, and actions that the state will need to refrain from.

Unfortunately, for the REBSP, there are, as yet, no defined core obligations, making the application of the right difficult. As a result, there is no standard or benchmark against which people and Civil Society can compel the State's immediate efforts to realize the right. Secondly, the State lacks guidance on what it has to prioritise within its minimal resources. There is therefore an urgent need for agreed minimum core obligations under the REBSP, which would also apply to scientific progress in DR-TB.

For example, minimum core obligations in REBSP could be used to prevent harmful effects of science, to promote access to benefits and to encourage international cooperation:

(a) To monitor the potential harmful effects of science and technology, to effectively react to the findings and inform the public in a transparent way; for

---

[2] Given the resource and knowledge restraints faced by many countries, the CESCR recognizes that the fulfillment of economic and social rights can only be achieved over time, and calls for the *progressive realization* of ESCR.

instance, where drug development for TB poses threats to the people, the state does not need to wait before it intervenes and prevents such harm.

(b) To promote access to the benefits of science and its applications on a non-discriminatory basis including measures necessary to address the needs of disadvantaged and marginalized groups. This includes ensuring that TB patients, especially the poor and marginalized, benefit from scientific progress that informs the development of new drugs, the use of repurposed drugs as well as the application of such drugs in high burden settings;

(c) To take measures to encourage and strengthen international cooperation and assistance in science and technology to the benefit of all people and to comply in this regard with the States' obligations under international law. This includes, inter-alia, the state entering into international agreements that fosters exchange of knowledge and products of such knowledge. It also implies that states push for better patent laws, which promote access to scientific knowledge even for those in LMICs.

10.5 Lack of Scientific Progress in DR-TB

A critical element of scientific progress is research and development (R&D), which is key to ensuring access to effective medicines, especially when coupled with deliberate policies to address access-related challenges for the poor and the marginalized. In the current environment, research can either be for-profit or not, regardless of whether it is meant to add value to people's health. Unfortunately, investments in not-for-profit research tend to be significantly lower than investments in for-profit research. Modern research is largely driven by funding external to the researchers/scientists and that how research is financed will determine how the knowledge arising from it will be used (Yamey 2008).

While not being the only element of the right, access to the benefits of scientific progress is one that is most controversial as it involves navigating the political landscapes in the production of science itself (Besson 2015a, b) and foregrounds the need for universal agreement on what defines scientific progress (Donders 2015). Moreover, the context in which the production of science takes place, often stretches beyond national jurisdictions (Besson 2015a, b). For example, for pharmaceutical companies based in HICs to test the effectiveness of new and repurposed drugs for TB, they need to conduct clinical trials in high burden settings like South Africa. Therefore, defining scientific progress is not the responsibility of one state, but requires a shared understanding by both those on the giving and receiving end of scientific development. Progress in TB is not just about the development of new drugs. It is also about the discovery and sharing of knowledge; and ensuring that such knowledge is accessible to those who need it the most.

Nevertheless, one of the main barriers is that countries from the North and the pharmaceutical companies domiciled there are reluctant to share and transfer knowledge with and to the South because of economic competition and return upon investment reasons. To counter the current static positions, the WHO, in partnership with United Nations Conference on Trade and Development (UNCTAD) and the International Centre for Trade and Sustainable Development, initiated a project on Improving Access to Medicines in Developing Countries through Technology Transfer and Local Production. One of the findings of the project has been that local production has the potential to enhance access to affordable medicines if supported by appropriate and accessible technology (WHO 2011).

Furthermore, the benefits of scientific progress, as opposed to the protection of scientific discoveries and production have not been well explored. The latter has been a topic of thorough discussion and debate under intellectual property rights and law (Besson 2015a, b). This has arguably led to a situation where attention (and legal protection) is typically given to creation of scientific knowledge most likely to benefit the innovators (Yamey and Torreele 2002), while knowledge to address key public health problems of significant magnitude, particularly for poor populations unable to purchase the applications of scientific progress, have been neglected. Or if pursued, the scientific discoveries have been too costly to benefit the majority in need (Yamey 2008). International treaties and agreements on the development, sharing and use of science need to account for the fact that, although the capacity of most LMICs to contribute to scientific progress is more limited than for high income countries, their need to benefit is far greater than that of high-income countries. This imbalance in need versus capacity should be factored into the discourse on intellectual property rights, particularly in terms of impact on access to essential life-saving medicines.

10.6   Intellectual Property Rights and Access to Essential Medicines

The dominant paradigm in scientific development favours the strengthening and protection of Intellectual Property Rights (IPRs) in efforts to encourage and reward innovation. This paradigm was institutionalised after the introduction of patents, which in turn made scientific research more lucrative (Timmermann 2014). A patent is a "government-granted limited property right to exclude others from making, using or selling the patented invention" (Clark et al. 2000).

Patents can have a dramatic impact on access to medicines when they are used to prevent competition. A drug company that holds patents on a medicine has the right to prevent others from manufacturing it and therefore can charge an artificially high price. When a company is selling commodities such as computer components, for example, this might be of no great significance. But when life-saving treatments for diseases such as HIV or cancer become unaffordable to those that need them, the consequences can be – and are – devastating. In many LMICs, where people pay for drugs out of their own pockets and very seldom have health insurance, the high price of medicines becomes a question of life and death.

To advance the protection of intellectual property rights, the Member States of the World Trade Organization (WTO) agreed on the Trade-Related Aspects of Intellectual Property Rights (TRIPS) legal regime that progressively became effective from 1994 onwards (Correa 2007). TRIPS elicited a challenge from developing countries and from Civil Society across the globe, who saw them as an impediment to access to essential medicines. Prior to TRIPS, patent protection on pharmaceuticals was almost non-existent in developing countries and the absence of patents led to the flourishing of generic medicine production in these countries, which significantly lowered the cost of essential medicines. While rewarding innovation is important, it should not occur at the expense of access to the benefits. What is needed is a system for balancing promotion of innovation and access to benefits, which is what the REBSP potentially provides.

For example, in 2001, developing countries initiated negotiations on the interpretations of TRIPS Agreement because they restricted access to drugs for patients with HIV infection, the majority of whom live in developing countries (Correa and WHO 2002). Although the TRIPS agreement itself did not change, a compromise was reached in Doha[3] in the form of a Declaration that clarified that TRIPS should not prevent developing states from dealing with public health crises, that they should not restrict universal access to essential medicines and provided for mechanisms to bypass potential IP obstacles when public health was at stake. This is a case where the right to health was used as a basis to facilitate international trade agreements that would favour access to life-saving medicines, thereby also opening opportunities for realising the benefits of scientific progress.

10.7   Creating an Enabling Environment

In order for scientific progress in TB to thrive, states need to create an enabling environment for research. State duties to meet this right might include those proposed by London, Cox and Coomans (2016): – (i) measures to ensure that researchers have access to infrastructure and equipment to conduct research such as drug development; (ii) adoption of research policies and strategies that foreground research to develop applications for neglected diseases of the poor; (iii) shaping of research funding opportunities to make more attractive research that has lower commercial opportunity; (iv) capacity building of researchers; but also, (v) Public-Private Partnerships to encourage the public sector to contribute to access to treatment; (vi) strengthening collaboration with other countries, especially those contributing to R&D; (vii) reprioritising resources from other sectors such as military to health; and (vi) putting in place more efficient regulatory laws and policies, for providing approval for both research and new drugs.

---

[3] Declaration on the TRIPS Agreement and Public Health, adopted 14 November 2001 by the Ministerial Conference of the World Trade Organisation.

However, for the realisation of the REBSP, there has to be a vehicle or pathways for benefits to be derived from this progress and its applications. Put simply, such a system needs to ensure access to effective treatments for people needing treatment. Where appropriate, the State should make use of the TRIPS flexibilities to develop domestic policies that foster scientific progress in TB as a neglected disease. The case of South Africa and Kenya, which made use of TRIPS flexibilities to enforce compulsory licencing of patents to local producers (Musungu et al. 2006) is an example of the State ensuring access to ARVs. Others efforts can take the form of subsidies for researchers, and tax benefits to encourage research in DR-TB.

10.8   International Cooperation to Improve Access to DR-TB Drugs

International cooperation is integral to the international human rights framework, requiring states to recognise the role of international cooperation in realising human rights globally.. Article 2 of the ICESCR sets out that governments are obligated to "[…] take steps, individually and through international assistance and co-operation, […] to the maximum of its available resources, with a view to achieving progressively the full realization of the rights recognized in the present Covenant […]. The REBSP is not exempted from the international cooperation. Science is too broad to be pursued within the confines of one country, and its benefits, particularly in TB, are far reaching. The new drugs bedaquiline and delamanid have a larger market in LMICs than in high-income countries in which they were developed. Similarly, with ARVs, the market is larger in less-developed countries hit by the HIV epidemic, than it is in the high-income countries, where most R&D takes place. The REBSP, like other social and economic rights, has collective dimensions, in that its realization requires functioning social systems involving population-wide application rather than being a right exercised for any particular individual. This is further underscored by the international dimensions of cooperation across communities and territories. Furthermore, while the REBSP can benefit from international cooperation, it can also be a vehicle to promote such cooperation through the sharing of knowledge and its application.

Some initiatives have been established, proposing how this global ethical responsibility might actually be operationalised in the DR-TB response. One of such initiatives is the 3P project,[4] which seeks to encourage the development of affordable, effective new drugs to treat TB. It makes use of an open collaborative approach to conduct drug research and development (R&D), and recommends some new ways of funding and coordinating the drug research and development process. The

---

[4] https://www.msfaccess.org/sites/default/files/MSF_assets/TB/Docs/TB_briefing_3P-2016_EN.pdf.

process has identified significant weaknesses in the process that hamper development of new regimens and proposes three interventions to address the weakness, these are

- Push funding to finance R&D activities upfront (i.e. through grants);
- Pull funding to incentivise R&D activities through the promise of financial rewards on the achievement of certain R&D objectives (i.e. through milestone prizes)
- Pooling of data and intellectual property to ensure open collaborative research and to ensure fair licensing for competitive production of the final products.

10.9   Conclusion

In conclusion, the REBSP offers great potential for deepening the rights-based approach to health, as well as for enhancing access to effective diagnosis and treatment of DR-TB, and by extension other neglected diseases of the poor, by promoting the development of new drugs, and research in the use of existing regimens in high burden settings. However, the right needs clarity and a universal understanding of the entitlements it confers on right-holders and corresponding obligations of duty bearers. There is also need for clarity on the application of the right beyond national borders, to account for the broad scope of science or global trade-related policies. A thorough conceptualisation of the REBSP will require more than the attention of human rights experts, but also trade experts as well as research and development practitioners.

Despite progress made to advance the right to health, health inequality continues to undermine human rights. For DR-TB, health inequality is apparent at multiple levels; the burden of disease is higher in poorer countries, while access to effective treatment is lower. The moral belief that every human being has the right to health entails a collective global responsibility to those people or countries with poor health outcomes. Therefore, "Everyone with TB should have access to the innovative tools and services they need for rapid diagnosis, treatment and care" (WHO 2015) particularly with DR-TB, ensuring access to products and services will not only benefit patients of TB, but will also benefit global health security. This requires a closer look at how other human rights can, together with the right to health, advance health for all. The REBSP is such a right.

The responsibility to ensure scientific progress in DR-TB and access to such progress cannot lie on most affected countries alone. DR-TB will require a global response and a collective responsibility in an effort to advance the right to health. States' duties to respect, protect and fulfill the REBSP within and beyond their borders is an important vehicle to redress the lack of research into new drug development and repurposing existing drugs for DR-TB and their use in high burden settings. REBSP provides some practical ways in which countries can strengthen their rights based approaches to DR-TB. This includes creating an enabling environment for R&D, promoting public-private partnerships, strengthening international cooperation, prioritizing resources to DR-TB, and implementing more efficient regulatory process for new drugs.

References

Besson, S. 2015a. Mapping the issues. *European Journal of Human Rights* 2015: 403–411.
———. 2015b. Science without borders and the boundaries of human rights: Who owes the human right to science? *European Journal of Human Rights* 2015: 462–486.
CESCR. 2006. *General comment 17: The right of everyone to benefit from the protection of the moral and material interests resulting from any scientific, literary or artistic production of which he or she is the author*. New York: United Nations Economic and Social Council.
Chapman, A.R. 2009. Towards an understanding of the right to enjoy the benefits of scientific progress and its applications. *Journal of Human Rights* 8: 1–36.
Clark, J., et al. 2000. *Patent pools: A solution to the problem of access in biotechnology patents?* Washington, DC: United States Patent and Trademark Office.
Coomans, F., and M.T. Kamminga. 2004. *Extraterritorial application of human rights treaties*. Antwerp: Intersentia.
Correa, C. 2007. *Trade related aspects of intellectual property rights: A commentary on the TRIPS Agreement*. Oxford: Oxford University Press.
Correa, C.M., and WHO. 2002. *Implications of the Doha Declaration on the TRIPS Agreement and public health*. Geneva: World Health Organisation.
Cox, H.S., et al. 2015. The need to accelerate access to new drugs for multidrug-resistant tuberculosis. *Bulletin of the World Health Organization* 93: 491–497.
Daniel, T.M. 2006. The history of tuberculosis. *Respiratory Medicine* 100 (11): 1862–1870.
Dommen, C. 2002. Raising human rights concerns in the World Trade Organization actors, processes and possible strategies. *Human Rights Quarterly* 24 (1): 1–50.
Donders, Y. 2015. Balancing interests: Limitations to the right to enjoy the benefits of scientific progress and its applications. *European Journal of Human Rights* 2015: 486–504.
Evans, C.A. 2011. GeneXpert – A game-changer for tuberculosis control? *PLoS Medicine* 8 (7): e1001064.
Figueroa-Munoz, J.I., and P. Ramon-Pardo. 2008. Tuberculosis control in vulnerable groups. *Bulletin of the World Health Organization* 86 (9): 733–735.
FM't Hoen, E. 2002. TRIPS, pharmaceutical patents and access to essential medicines: Seattle, Doha and beyond. *Journal of International Law* 3: 39–68.
Frick, M. 2016. *2016 report on tuberculosis research funding trends, 2005–2015: No time to lose*. New York: Treatment Action Group.
Kernodle, D.S. 2010. Decrease in the effectiveness of Bacille Calmette-Guérin vaccine against pulmonary tuberculosis: A consequence of increased immune suppression by microbial antioxidants, not over attenuation. *Clinical Infectious Diseases* 51 (2): 177–184.
London, L. 2008. What is a human-rights based approach to health and does it matter? *Health and Human Rights* 10: 65–80.
London, L., et al. 2016. Multidrug-resistant TB: Implementing the right to health through the right to enjoy the benefits of scientific progress. *Health & Human Rights: An International Journal* 18 (1).
Mafukidze, A., et al. 2016. An update on repurposed medications for the treatment of drug-resistant tuberculosis. *Expert Review of Clinical Pharmacology* 9 (10): 1331–1340.
Mann, J., L. Gostin, S. Gruskin, T. Brennan, Z. Lazzarini, and H.V. Fineberg. 1994. Health and human rights. *Health and Human Rights* 1 (1): 6–24.
Mase, S., et al. 2013. Provisional CDC guidelines for the use and safety monitoring of bedaquiline fumarate (Sirturo) for the treatment of multidrug-resistant tuberculosis. *Morbidity and Mortality Weekly Report: Recommendations and Reports* 62 (9): 1–12.

Maskus, K.E. 2002. Regulatory standards in the WTO: Comparing intellectual property rights with competition policy, environmental protection, and core labor standards. *World Trade Review* 1 (2): 135–152.

McFarland, S. 2015. Culture, individual differences, and support for human rights: A general review. *Peace and Conflict: Journal of Peace Psychology* 21 (1): 10.

Musungu, S.F., et al. 2006. *The use of flexibilities in TRIPS by developing countries: Can they promote access to medicines?* Geneva: World Health Organisation.

Ndjeka, N., et al. 2015. Treatment of drug-resistant tuberculosis with bedaquiline in a high HIV prevalence setting: An interim cohort analysis. *The International Journal of Tuberculosis and Lung Disease* 19 (8): 979–985.

Seddon, J.A., et al. 2012. Hearing loss in patients on treatment for drug-resistant tuberculosis. *European Respiratory Journal* 40 (5): 1277.

Sen, A. 2005. Human rights and capabilities. *Journal of Human Development* 6 (2): 151–166.

Shaver, L. 2015. The right to science: Ensuring that everyone benefits from scientific and technological progress. *European Journal of Human Rights* 2015: 411–431.

Skrahina, A., et al. 2016. Bedaquiline in the multidrug-resistant tuberculosis treatment: Belarus experience. *International Journal of Mycobacteriology* 5: S62–S63.

Steingart, K.R., et al. 2006. Sputum processing methods to improve the sensitivity of smear microscopy for tuberculosis: A systematic review. *The Lancet Infectious Diseases* 6 (10): 664–674.

Timmermann, C. 2014. Sharing in or benefiting from scientific advancement? *Science and Engineering Ethics* 20: 111–133.

UNDP. 2012. *MDG report 2012: Assessing progress in Africa toward the Millennium Development Goals*. Addis Ababa: Economic Commission for Africa.

United Nations. 2005. *Economic, social and cultural rights: Handbook for national human rights institutions*. New York: United Nations.

WHO. 2011. *Local production for access to medical products: Developing a framework to improve public health*, 32–36. Geneva: World Health Organisation.

———. 2015. *The end TB strategy*. Geneva: World Health Organisation.

———. 2016. *Global tuberculosis report*. Geneva: World Health Organisation.

———. 2017. *Multidrug- resistant tuberculosis (MDR) TB 2017 update*. Geneva: World Health Organisation.

———. 2018. *Global tuberculosis report*. Geneva: World Health Organisation.

Yamey, G. 2008. Excluding the poor from accessing biomedical literature: A rights violation that impedes global health. *Health and Human Rights* 10: 21–42.

Yamey, G., and E. Torreele. 2002. The world's most neglected diseases. *The British Medical Journal* 325: 176–177.

Zumla, A., et al. 2013. Advances in the development of new tuberculosis drugs and treatment regimens. *Nature Reviews Drug Discovery* 12 (5): 388–404.

Chapter 11
The Right to Refuse Treatment for Infectious Disease

Carl H. Coleman

Abstract One of the central tenets of contemporary bioethics is that mentally competent persons have a right to refuse medical treatment, even if the refusal might lead to the individual's death. Despite this principle, laws in some jurisdictions authorize the nonconsensual treatment of persons with tuberculosis (TB) or other serious infectious diseases, on the grounds that doing so is necessary to protect the safety of others. This chapter argues that, in the vast majority of situations, overriding a refusal of treatment for infectious disease is not justifiable, as the risk to third parties can be avoided by the less restrictive alternative of isolating the patient. At the same time, it rejects the extreme position that the nonconsensual treatment of infectious disease is never appropriate. Instead, it concludes that compelling an individual to undergo treatment for infectious diseases may be ethically justifiable in exceptional situations if a refusal of treatment poses a grave risk to third parties, the treatment is not overly burdensome and has been established to be safe and effective, and less restrictive alternatives, including humanely isolating the patient, are not feasible under the circumstances. The burden should be on those seeking to compel unwanted treatment to demonstrate that these requirements have been met.

Keywords Bioethics · Public Health · Infectious Diseases · Medical Law · Human Rights

11.1 The Right to Refuse Medical Treatment

If there is one unifying concept that runs through the field of bioethics, it is the doctrine of informed consent — i.e., the principle that individuals have the right to make their own decisions about medical treatment after having been informed of the risks, potential benefits, and reasonably available alternatives (Beauchamp and Childress 2013). The principle of informed consent is grounded in two interrelated

C. H. Coleman (✉)
School of Law, Seton Hall University, Newark, NJ, USA
e-mail: carl.coleman@shu.edu

© The Author(s) 2020
E. Jamrozik, M. Selgelid (eds.), *Ethics and Drug Resistance: Collective Responsibility for Global Public Health*, Public Health Ethics Analysis 5,
https://doi.org/10.1007/978-3-030-27874-8_11

justifications. First, by conditioning the imposition of medical treatment on the individual's authorization, informed consent protects persons' right to bodily integrity. Second, by requiring the provision of information needed to evaluate the desirability of proposed treatments, informed consent shows respect for individuals as autonomous decision-makers. The right to refuse medical treatment is "the logical corollary" of the informed consent doctrine: if there were no right to refuse proposed treatments, the process of soliciting consent would be a hollow charade (Cruzan v. Director 1990).

In addition to its ethical foundations, the right to refuse medical treatment is supported by internationally-recognized human rights principles. These principles include the right to security of the person (International Covenant on Civil and Political Rights 1976), the right not to be subjected to torture or to inhuman or degrading treatment (International Covenant on Civil and Political Rights 1976; European Convention on Human Rights 2010), and the right to health (International Covenant on Economic, Social and Cultural Rights 1976). The United Nations Committee on Economic, Social, and Cultural Rights has specifically recognized that the right to health includes the "right to be free from … non-consensual treatment" (United Nations Committee on Economic, Social and Cultural Rights 2000).

Courts have upheld mentally competent individuals' right to refuse medical treatment in a variety of situations, including those in which the refusal of treatment might strike some observers as irrational or unwise. For example, in the case of Fosmire v. Nicoleau, the New York Court of Appeals found that an adult Jehovah's Witness had the right to refuse blood transfusions prior to and immediately after the delivery of her baby, despite her physicians' belief that without the transfusions she was likely to die (Matter of Fosmire v. Nicoleau 1990). While some older judicial decisions in the United States suggested that the right to refuse treatment must be balanced against the state's interests in the preservation of life, the prevention of suicide, and "the maintenance of the ethical integrity of the medical profession" (Annas 1992), there are no recent cases in which any of these state interests has been deemed sufficient to override the refusal of treatment by a mentally competent adult.

However, like most individual rights, the right to refuse treatment is not absolute. First, refusals of treatment may be overridden when the patient lacks the mental capacity to provide informed consent and the imposition of treatment would be in his or her medical best interests (Steele v. Hamilton Cty. Community Mental Health Bd. 2000). This justification for compelled treatment typically arises in the psychiatric context, where a refusal of treatment may be the result of an underlying mental illness rather than the manifestation of a genuinely voluntary choice. The main issues in these situations involve the standards and procedures for determining the patient's mental incapacity (Klein 2012), as well as the appropriate role of family members and other surrogates in making decisions on the patient's behalf (Vars 2008).

Second, the nonconsensual treatment of mentally competent persons may be authorized when doing so is necessary to protect the safety of others. For example, courts have upheld the forcible medication of violent patients in emergency situations "when there is an imminent danger to a patient or others in the immediate vicinity" (Rivers v. Katz 1986). Courts have interpreted the danger-to-others rationale particularly broadly in the context of prisoners. For example, some courts have authorized the forcible feeding of prison inmates on hunger strikes on the theory that a hunger strike can threaten "institutional order and security" (Matter of Bezio v. Dorsey 2013).

Finally, competent individuals may sometimes be compelled to submit to medical interventions as part of the process of investigating or adjudicating criminal responsibility. For example, criminal suspects may be required to undergo physical examinations, which may include the collection of bodily fluids such as saliva or blood (Maryland v. King 2013; Missouri v. McNeely 2013). In addition, courts in the United States have authorized the government to administer medications to render mentally ill criminal defendants competent to stand trial, citing the governmental interest "in bringing to trial an individual accused of a serious crime." However, the forcible medication of criminal defendants is permissible only "if the treatment is medically appropriate, is substantially unlikely to have side effects that may undermine the trial's fairness, and, taking account of less intrusive alternatives, is necessary significantly to further important governmental trial-related interests" (Sell v. United States 2003).

11.2   Existing Approaches to Compelled Treatment for Infectious Diseases

Several jurisdictions have laws that authorize the nonconsensual treatment of persons with serious communicable diseases. For example, in the early 1990s, in response to a dramatic increase in TB cases, New York City amended its health code to authorize the Commissioner of Health to compel TB patients to complete treatment and, if necessary, to detain them during the process (Gasner et al. 1999). Under the law, detention can be authorized for persons with active TB "where there is a substantial likelihood, based on such person's past or present behavior, that he or she can not be relied upon to participate in and/or to complete an appropriate prescribed course of medication for tuberculosis" (N.Y.C. Health Code §11.21(d)(5)). In most cases, treatment under detention is sought only after other less restrictive alternatives, including directly-observed therapy, have already been attempted (Gasner et al. 1999). However, the Department of Health has the discretion to order treatment under detention as a first resort if it concludes attempting outpatient treatment would be futile in light of the patient's history of nonadherence.

Similar laws exist in other jurisdictions. For example, South Africa's National Health Act provides for an exception to the requirement to obtain informed consent to medical treatment in cases where "failure to treat the user ... will result in a serious risk to the public health" (Republic of South Africa National Health Act § 7(1) (d)). Relying in part on this provision, the High Court of South Africa has upheld the involuntary isolation and treatment of patients with multi-drug resistant (MDR) TB (Minister of Health v. Goliath 2008). Similarly, Canadian courts have upheld orders requiring patients with TB to submit to mandatory detention and treatment (Ries 2007), relying on provincial legislation authorizing the board of health to issue orders requiring the "care and treatment" of persons with "virulent" communicable diseases (Ontario Health Protection and Promotion Act § 22(4)(g); Silva 2011). The state of Alabama also has a law authorizing the "compulsory treatment and quarantine" of TB patients, but with an unusual exception: Confined patients are permitted to refuse treatment if they "desire treatment by prayer or spiritual means" (Ala. Code § 22-11A-10). The Alabama law does not permit patients to refuse treatment for reasons unrelated to religion.

In contrast, laws in some jurisdictions provide that patients with infectious diseases can be subject to mandatory detention but not forcible treatment. For example, the state of Minnesota authorizes the involuntary isolation of persons with diseases "that can be transmitted person to person and for which isolation or quarantine is an effective control strategy," but it specifically provides that isolated patients have "a fundamental right to refuse medical treatment" (Minn. Statutes 144.419). Similarly, Israeli law permits the involuntary isolation of persons with TB or other serious infectious diseases, but isolated patients may not be forced to undergo unwanted medical treatment (Weiler-Ravell et al. 2004). The law in Iceland appears to follow a similar approach (Eggertsson 2004).

The Model State Emergency Health Powers Act, a proposed law developed by the Center for Law and the Public's Health, initially included provisions authorizing forcible treatment of patients during an infectious disease outbreak, but after extensive criticism, those provisions were eliminated from the proposal (Annas et al. 2008). The proposed law now provides that "persons who are unable or unwilling for reasons of health, religion, or conscience to undergo treatment" may be subject to isolation (Center for Law and the Public's Health 2001). The implication is that, once isolated, infectious persons may not be subjected to treatment involuntarily.

It is unclear whether compelled treatment for infectious disease would be permitted under international human rights law. On the one hand, some human rights documents cite "the prevention and control of communicable diseases" as a situation in which "coercive medical treatments" could be justified on an "exceptional basis" (United Nations Committee on Economic, Social and Cultural Rights 2000). In addition, the European Court of Human Rights (ECHR) has found that requiring individuals to undergo TB screening by means of a tuberculin skin-reaction test or chest X-ray "can be considered necessary in a democratic society for the protection of health" (Acmanne and others v. Belgium 1984). On the other hand, no human rights tribunal has directly addressed the permissibility of compelling individuals to undergo treatment (as opposed to testing) for infectious diseases. Moreover, citing ECHR decisions condemning the force-feeding of prisoners, a World Health Organization (WHO) mission to the Ukraine concluded that "administering a TB treatment without the consent of the patient is an intrusive major intervention that constitutes a prohibited interference with a person's rights under the ECHR" (Dagron 2016).

Finally, it is worth noting that, although "compulsory vaccination" laws are sometimes cited as support for overriding the right to refuse treatment for patients with infectious diseases (Valenti 2012), those laws do not actually support the forcible imposition of medical interventions over a patient's objection. For example, in the United States, unvaccinated children who have not received an exemption from a state's vaccination requirements may be denied enrollment in the public schools (Barraza et al. 2017), but they will not be forcibly given vaccines to which their parents have objected. In a few other countries, parents can be fined for refusing to vaccinate their children (Reuters 2017), but, again, forcible vaccination does not appear to be authorized. Similarly, in the frequently-cited U.S. Supreme Court case of *Jacobson v. Massachusetts*, which upheld a Massachusetts law requiring individuals to be vaccinated for smallpox, the plaintiff was never actually required to undergo vaccination. Instead, the consequence of his refusal of the vaccine was a fine of five dollars (Jacobson v. Massachusetts 1905).

11.3 Ethical Analysis

As discussed in the preceding section, there are two dominant legal approaches to protecting third parties from the risk of serious infectious diseases: compelling such persons to undergo medical treatment, or isolating them without forcing them to accept treatments to which they object. While the latter approach does not technically override the individual's right to refuse treatment, an offer of treatment that can be refused only by submitting to isolation is inherently coercive. Because "[v]oluntary consent is usually thought incompatible with coercion" (Eyal 2012), both approaches can be seen as exceptions to the general principle that medical treatment requires the patient's voluntary informed consent.

Recognizing an exception to informed consent for patients with infectious diseases can be compared to cases authorizing nonconsensual medical treatment when necessary to protect third parties from violence. As an ethical matter, both situations implicate John Stuart Mill's "harm principle," which provides that individual liberty does not include the right to cause harm to third parties (Brink 2014). As Marcel Verweij explains, "[i]f patients with infectious diseases neglect the treatment they need, this could have harmful implications for others, and that may be reason to overrule the requirements of informed consent" (Verweij 2011).

However, even if the harm principle can justify an exception to informed consent in some situations, the mere possibility of harm, standing alone, is an insufficient basis to justify compulsion. Instead, a full analysis must consider factors such as the likelihood and magnitude of the potential harm, the burdens of compulsion on the individuals affected, the likely effectiveness of the proposed interventions, and the alternatives that are realistically available under the circumstances. This kind of fact-sensitive inquiry is an implicit requirement of consequentialist ethical theories, which require assessing the aggregate balance of potential benefits and harms of any proposed course of action (Sinnott-Armstrong 2015). It is also supported by the human rights principle of proportionality, which requires a "fair balance between the demands of the general interest of the community and the requirements of the protection of the individual's fundamental rights" (Soering v. UK 1989).

Coercive action is most justifiable in cases of highly contagious, life-threatening illnesses, such as MDR-TB. As the likelihood of transmission and/or the consequences of infection lessen, the justification for coercion diminishes as well. The precise point at which a refusal of treatment becomes sufficiently dangerous to justify coercion can be difficult to determine. For example, as Richard Coker asks, "given that the risk of relapse is higher if compliance with [TB] treatment ceases after two months of treatment compared with, say, five months (where, in most cases it is probably very small indeed), should those who fail to comply after two months face the prospect of detention if they fail to comply, whereas those who do so only after five months remain at liberty?" In answering these questions, "one is forced to question issues relating to utility, about how one measures the burden of risk, and utility gains and losses. Objective evidence to support decisions is largely lacking" (Coker 2000).

For many diseases, both the likelihood and magnitude of potential harm exist on a lengthy continuum. With respect to the likelihood of transmission, consider the case of sexually-transmitted infections (STIs). Many persons with STIs pose little or no risk to others because they are either sexually inactive or consistently use barrier protection. However, the level of risk increases if they engage in unprotected sexual encounters with multiple partners. The likelihood of transmission is particularly high for persons whose sexual partners do not recognize the need to take precautions against transmission, such as spouses who may mistakenly assume that their relationships are exclusive. In light of these uncertainties, is the likelihood of harm to unknowing third parties sufficient to justify the use of compulsion? While public health authorities generally favor a voluntary approach to STI testing and treatment (WHO Statement 2012), efforts to use compulsion sometimes occur. For example, under 2017 guidelines from the British Columbia Centre for Disease Control, public health authorities may issue orders compelling HIV-positive individuals to initiate and continue HIV treatment if they engage in high-risk sexual behavior or share needles and/or other drug paraphernalia with other persons and do not disclose their HIV status (British Columbia Centre for Disease Control 2017). Violation of such orders can result in court-ordered detention under the British Columbia Public Health Act, which allows medical health officers "to do anything that the … officer reasonably believes is necessary … to prevent the transmission of an infectious agent" (British Columbia Public Health Act). In 2018, a man from Vancouver was charged under these provisions for allegedly refusing to comply with a medical officer's order to submit to HIV treatment (Proctor). Similarly, a man in Fayetteville, Arkansas was charged in 2016 for a misdemeanor public health violation for refusing to undergo treatment for syphilis (Vendituoli 2016).

Similar uncertainties relate to the magnitude of harm resulting from an individual's refusal of treatment. For example, seasonal influenza typically resolves itself without any long-term consequences, but it can be life threatening for elderly patients or those with compromised immune systems. If a treatment were developed that quickly made patients with seasonal influenza incapable of transmitting the illness, would the magnitude of avoidable harm justify imposing this treatment over a patient's objection? Even seemingly innocuous infections diseases, such as athlete's foot, can cause severe consequences in some subgroups of patients (e.g., persons with diabetes). So far, no one has seriously proposed compelling individuals with athlete's foot to use anti-fungal medications, or to threaten infected persons with detention if they visit a public swimming pool with bare feet during an outbreak. Yet, the idea of invoking the harm principle in this situation might seem perfectly reasonable to a diabetic patient who is forced to undergo a foot amputation after being exposed to an untreated fungal infection.

In determining whether the likelihood and magnitude of harm are sufficient to justify compulsion, some commentators have suggested that the risks posed by an individual's refusal of treatment should be compared to the other risks that individuals typically confront on a day-to-day basis. For example, Mark Cherry argues that compelled treatment for infectious disease can be justified when necessary to prevent "significant and unusual risk" to "non-consenting others." He gives the example of legally-mandated treatment for patients with extensively drug-resistant TB, explaining that "[t]he risks associated with such diseases are significantly greater than the background risks that one generally assumes in daily life" (Cherry 2010).

Yet, while a comparison to the risks of daily life provides a useful framework for analysis, it does not follow that, as long as the risks of refusal are greater than the risks of daily life, the use of compulsion is necessarily warranted. Instead, the existence of a "significant and unusual risk" should be considered a necessary but insufficient criterion for justifying compulsion. In other words, if the risks of treatment refusal are less than the risks of daily life, the use of compulsion should not even be considered; if the risk are greater, compulsion *might* be justifiable, but additional considerations must also be factored into the analysis.

One of these additional considerations is the burden of compulsion on the individuals affected. For example, in upholding Belgium's policy of mandatory skin-tests and chest X-rays for TB, the ECHR pointed to the absence of evidence suggesting that those interventions involved "disadvantages comparable to the former ravages of tuberculosis" (Acmanne and others v. Belgium 1984). The implication was that more burdensome interventions would require a more compelling justification.

Some interventions might be too burdensome to be justifiable under any circumstances. By way of analogy, in refusing to authorize a Caesarean section on a dying woman who had refused the procedure, the District of Columbia Court of Appeals noted that, to perform the procedure, the patient "would have to be fastened with restraints to the operating table, or perhaps involuntarily rendered unconscious by forcibly injecting her with an anesthetic, and then subjected to unwanted major surgery." It concluded that "[s]uch actions would surely give one pause in a civilized society, especially when [the patient] had done no wrong" (In re A.C. 1990). For similar reasons, it is difficult to imagine a scenario in which it would be acceptable to force a patient with an infectious disease to submit to a major surgical procedure, even if doing so were the only way to render the patient non-infectious.

Another consideration is the likelihood that the proposed intervention will be effective in reducing the risk that the disease will be transmitted to third parties. At a minimum, as the World Health Organization emphasizes, overriding an individual's refusal of a medical intervention for the treatment of an infectious disease should not be considered unless the proposed intervention "has proven to be safe and effective and is part of the accepted medical standard of care" (World Health Organization 2016). Likewise, requiring individuals who refuse treatment to undergo isolation would not be justifiable if the disease is already so prevalent in a population that isolating untreated patients is unlikely to make a significant difference in reducing transmission.

Finally, even if an individual's refusal of treatment creates a sufficient risk of harm to others to justify the use of compulsion, the type of compulsion used should be the least restrictive option reasonably available under the circumstances. The "least restrictive alternative" requirement is a well-established principle of human rights jurisprudence. For example, the Siracusa Principles on the Limitation and Derogation Provisions in the International Covenant on Civil and Political Rights provides that any restrictions on human rights must be "strictly necessary" and that "there must be no other, less intrusive means available to reach the same objective" (American Association for the International Commission of Jurists 1985). The United States Supreme Court applies a similar standard in analyzing restrictions on fundamental constitutional rights (Chemerinsky 2015), as does the European Court of Human Rights (Brems and Lavrysen 2015).

In light of this principle, if it is possible to protect the public from harm by isolating the patient, there would be no justification for insisting that the patient, once isolated, submit to medical treatments to which he continues to object. The WHO Ethics Guidance for the Implementation of the End TB Strategy makes this point forcefully:

> While contagious persons with TB who do not adhere to treatment or who are unable or unwilling to comply with infection prevention and control measures pose significant risks to the public, those risks can be addressed by isolating the patient. Patients who are isolated should still be offered the opportunity to receive treatment, but if they do not accept it, their informed refusal should be respected. Forcing these patients to undergo treatment over their objection would require an unacceptable invasion of bodily integrity, and also could put health care providers at risk. Moreover, as a practical matter, it would likely be impossible to provide effective treatment without the patient's cooperation. Nevertheless efforts to convince the patient and re-examine his or her refusal should not be abandoned (World Health Organization 2017).

The Belgian Advisory Committee on Bioethics takes a similar position, finding that "if an MDR patient who does not have any psychiatric problems rendering him incapable of giving his consent refuses a treatment, the health authorities can hold him in isolation to avoid the spread of the disease but cannot force him to receive treatment" (Belgian Advisory Committee on Bioethics 2013).

This position is consistent with the logic of the danger-to-others exception to the informed consent requirement. In particular, cases involving the forcible medication of violent patients emphasize that the use of force is justified only as a temporary measure, "continuing only as long as the emergency persists" (Rivers v. Katz 1986). Once a violent patient has been medicated to a point that he can safely be isolated, there would be no justification for further medication unless another basis for overriding consent, such as mental incapacity, also exists.

Taken together, the considerations discussed above suggest that nonconsensual treatment for infectious disease will rarely be justifiable. Even if the magnitude and severity of harm resulting from a treatment refusal are significant, and even if the proposed treatment is not especially burdensome and is likely to work, the less restrictive alternative of isolating the patient will usually be sufficient to protect the public from harm. While involuntary isolation still involves an element of coercion, in most cases it should be up to the patient to decide whether this option is preferable to submitting to an unwanted medical intervention.

Yet, it would be a mistake to think that the option of isolation will always eliminate the need to consider the appropriateness of nonconsensual treatment. One situation in which isolation may be impractical is when individuals are already living in confined settings, such as in prisons or the military. While isolating a small number of individuals in these settings may be possible, the available space for isolation is likely to be limited. In addition, there may be other persons with a stronger case for access to these limited facilities — for example, prisoners who need to be separated from the general population because they are at heightened risk of violent attack.

Another situation in which isolation may not be a practical alternative is during an epidemic outbreak involving substantial numbers of persons with contagious infections, particularly in low-resource settings with a limited capacity to provide humane isolation facilities. In some cases, voluntary isolation at home may be a reasonable alternative, but this will work only if the patient is willing and able to follow infection control precautions (World Health Organization 2017). Otherwise, if isolation facilities are limited, protecting the public may sometimes require consideration of nonconsensual treatment. Yet, even in these scenarios, compelled treatment would be justified only if a safe and effective treatment exists that would not be unduly burdensome to the patient. If there is such a treatment, it seems unlikely that the number of people refusing it would be high enough to overwhelm available isolation facilities, particularly if public health authorities have engaged in an adequate process of community engagement (World Health Organization 2016). Thus, as a practical matter, the need to invoke this exception may be more theoretical than real.

In all cases, the burden should be on those seeking to compel unwanted treatment to demonstrate that no less restrictive alternatives, including isolation, are realistically available under the circumstances. As the World Health Organization recommends, "objections to diagnostic, therapeutic, or preventive measures should not be overridden without giving the individual notice and an opportunity to raise his or her objections before an impartial decision-maker, such as a court, interdisciplinary review panel, or other entity not involved in the initial decision" (World Health Organization 2016). Such a process provides an important check against abuse and can avoid undermining public trust in the integrity of the public health system.

References

Acmanne and others v. Belgium, No. 10435/83 (European Court of Human Rights 1984).

Ala. Code § 22-11A-10.

American Association for the International Commission of Jurists. 1985. *Siracusa principles on the limitation and derogation provisions in the International Covenant on Civil and Political Rights*. http://icj.wpengine.netdna-cdn.com/wp-content/uploads/1984/07/Siracusa-principles-ICCPR-legal-submission-1985-eng.pdf.

Annas, George J. 1992. *The rights of patients: The basic ACLU guide to patient rights*. Springer.

Annas, George J., Wendy K. Mariner, and Wendy E. Parmet. 2008. *Pandemic preparedness: The need for a public health--not a law enforcement/National Security--Approach*. ACLU. https://www.aclu.org/files/images/asset_upload_file399_33642.pdf.

Barraza, Leila, Cason Schmit, and Aila Hoss. 2017. The latest in vaccine policies: Selected issues in school vaccinations, healthcare worker vaccinations, and pharmacist vaccination authority Laws. *Journal of Law, Medicine and Ethics* 45 (S1): 16–19.

Beauchamp, Tom L., and James F. Childress. 2013. *Principles of biomedical ethics*. 7th ed. New York: Oxford University Press.

Belgian Advisory Committee on Bioethics. 2013. *Opinion No. 55 of 13 May 2013 on the Treatment of Patients with Multidrug- Resistant Tuberculosis from a Public Health Perspective*.

Brems, Eva, and Laurens Lavrysen. 2015. 'Don't use a sledgehammer to crack a nut': Less restrictive means in the case law of the European Court of Human Rights. *Human Rights Law Review* 15 (1): 139–168. https://doi.org/10.1093/hrlr/ngu040.

Brink, David. 2014. Mill's moral and political philosophy. In *The Stanford Encyclopedia of philosophy*, ed. Edward N. Zalta. Metaphysics Research Lab, Stanford University. https://plato.stanford.edu/archives/win2016/entries/mill-moral-political/.

British Columbia Centre for Disease Control. 2017. *Guidelines for medical health officers: Approach to people with HIV/AIDS who may pose a risk of harm to others*. http://www.bccdc.ca/resource-gallery/Documents/Communicable-Disease-Manual/Chapter%205%20-%20STI/MHO-guidelines-PLWH-risk-of-transmission.pdf

British Columbia Public Health Act § 27.

Center for Law and the Public's Health. 2001. *Model State Emergency Health Powers Act*. http://www.publichealthlaw.net/MSEHPA/MSEHPA.pdf.

Chemerinsky, Erwin. 2015. *Constitutional law: Principles and policies*. 5th ed. http://www.wklegaledu.com/aspen-student-treatise-series/id-9781454849476/Constitutional_Law_Principles_and_Policies_Fifth_Edition.

Cherry, Mark J. 2010. *Non-consensual treatment is (nearly always) morally impermissible*. Los Angeles, CA: SAGE Publications Sage CA.

Coker, Richard. 2000. Tuberculosis, non-compliance and detention for the public health. *Journal of Medical Ethics* 26 (3): 157–159.

Cruzan v. Director, Mo. Dept. of Health, 497 US 261 (U.S. Supreme Court 1990).

Dagron, Stéphanie. 2016. *Tuberculosis control and human rights in the National Legislation of Ukraine, report of a Mission 20–24 April 2015*.

Eggertsson, Dagur B. 2004. *Tuberculosis and human rights in Iceland*. http://lup.lub.lu.se/student-papers/record/1554865/file/1563469.pdf.

Eyal, Nir. 2012. Informed consent. In *The Stanford encyclopedia of philosophy*, ed. Edward N. Zalta. Metaphysics Research Lab, Stanford University. https://plato.stanford.edu/archives/fall2012/entries/informed-consent/.

Gasner, M. Rose, Khin Lay Maw, Gabriel E. Feldman, Paula I. Fujiwara, and Thomas R. Frieden. 1999. The Use of Legal Action in New York City to Ensure Treatment of Tuberculosis. *New England Journal of Medicine* 340 (5): 359–366.

"International Covenant on Civil and Political Rights," 1976. http://www.ohchr.org/Documents/ProfessionalInterest/ccpr.pdf.

"International Covenant on EconomicEconomics, Social and Cultural Rights," 1976. http://www.ohchr.org/EN/ProfessionalInterest/Pages/CESCR.aspx.

In re AC, 573 A. 2d 1235 (D.C. Court of Appeals 1990).

Jacobson v. Massachusetts, 197 US 11 (U.S. Supreme Court 1905).

Klein, Dora W. 2012. When coercion lacks care: Competency to make medical treatment decisions and parens patriae civil commitments. *University of Michigan Journal of Law Reform* 45 (3): 561–593.

Maryland v. King, 133 S. Ct. 1958 (U.S. Supreme Court 2013).

Matter of Bezio v. Dorsey, 21 NY.3d 93 (N.Y. Court of Appeals 2013).

Matter of Fosmire v. Nicoleau, 75 NY 2d 218 (N.Y. Court of Appeals 1990).

Minister of Health v. Goliath, No. 13741/07 (High Court of South Africa 2008).

Minn. Statutes § 144.419.

Missouri v. McNeely, 133 S. Ct. 1552 (U.S. Supreme Court 2013).

N.Y.C. Health Code § 11.21(d)(5).

Ontario Health Protection and Promotion Act § 22(4)(g).

Proctor J. Vancouver man charged with ignoring medical health officer's orders. *CBC News*, August 24, 2018.

Reuters. "Italy Passes Law Obliging Parents to Vaccinate Children." May 19, 2017. http://www.reuters.com/article/us-italy-politics-vaccines-idUSKCN18F1J7.

Republic of South Africa National Health Act § 7(1)(d).

Ries, Nola M. 2007. Legal issues in disease outbreaks: Judicial review of public health powers. *Health Law Review* 16 (1): 11–16.

Rivers v. Katz, 67 NY 2d 485 (N.Y. Court of Appeals 1986).

Sell v. United States, 539 US 166 (U.S. Supreme Court 2003).

Silva, Diego S. 2011. *Tuberculosis and persons with severe and persistent mental illnesses: When (treatment) worlds collide?* Munk School Briefings, 2010. Munk School of Global Affairs, University of Toronto.

Sinnott-Armstrong, Walter. 2015. Consequentialism. In *The Stanford encyclopedia of philosophy*, ed. Edward N. Zalta. Metaphysics Research Lab, Stanford University. https://plato.stanford.edu/archives/win2015/entries/consequentialism/.

Soering v. UK, 11 EHHR 439 (European Court of Human Rights 1989).

Steele v. Hamilton Cty. Community Mental Health Bd., 90 Ohio St. 3d 176 (Supreme Court of Ohio 2000).

United Nations Committee on EconomicEconomics, Social & Cultural Rights 2000. *General comment No. 14: The right to the highest attainable standard of health (Article 12 of the International Covenant on Economic, Social and Cultural Rights)*. http://www.nesri.org/sites/default/files/Right_to_health_Comment_14.pdf.

Valenti, Joseph Anthony. 2012. Circumstances when medical treatment may be forcibly imposed despite a patient's explicit refusal: A comprehensive analysis of Pennsylvania law. *Widener Law Review* 18: 27.

Vars, Frederick E. 2008. Illusory consent: When an incapacitated patient agrees to treatment. *Oregon Law Review* 87 (2): 353–398.

Vendituoli, Monica. 2016. Fayetteville man accused of refusing treatment for syphilis. *Fayeteville Observer* 6.

Verweij, Marcel. 2011. Infectious disease control. In *Public health ethics: Key concepts and issues in policy and practice*, ed. Angus Dawson, 100–142. Cambridge/New York: Cambridge University Press.

Weiler-Ravell, D., A. Leventhal, R.J. Coker, and D. Chemtob. 2004. Compulsory detention of recalcitrant tuberculosis patients in the context of a new tuberculosis control programme in Israel. *Public Health* 118 (5): 323–328. https://doi.org/10.1016/j.puhe.2003.10.005.

World Health Organization. 2012. *Statement on HIV testing and Counseling: WHO, UNAIDS re-affirm opposition to mandatory HIV testing.* http://www.who.int/hiv/events/2012/world_aids_day/hiv_testing_counselling/en/.

———. 2017. *Ethics guidance for the implementation of the end TB strategy.* http://apps.who.int/iris/bitstream/10665/254820/1/9789241512114-eng.pdf?ua=1.

———. 2016. *Guidance for managing ethical issues in infectious dsease outbreaks.* http://apps.who.int/iris/bitstream/10665/250580/1/9789241549837-eng.pdf.

Chapter 12
Surveillance and Control of Asymptomatic Carriers of Drug-Resistant Bacteria

Euzebiusz Jamrozik and Michael J. Selgelid

Abstract Drug-resistant bacterial infections constitute a major threat to global public health. Several key bacteria that are becoming increasingly resistant are among those that are ubiquitously carried by human beings and usually cause no symptoms (i.e. individuals are asymptomatic carriers) until a precipitating event leads to symptomatic infection (and thus disease). Carriers of drug-resistant bacteria can also transmit resistant pathogens to others, thus putting the latter at risk of infections that may be difficult or impossible to treat with currently available antibiotics. Accumulating evidence suggests that such transmission occurs not only in hospital settings but also in the general community, although much more data are needed to assess the extent of this problem. Asymptomatic carriage of drug-resistant bacteria raises important ethical questions regarding the appropriate public health response, including the degree to which it would be justified to impose burdens and costs on asymptomatic carriers (and others) in order to prevent transmission. In this paper, we (i) summarize current evidence regarding the carriage of key drug-resistant bacteria, noting important knowledge gaps and (ii) explore the implications of existing public health ethics frameworks for decision- and policy-making regarding asymptomatic carriers. *Inter alia*, we argue that the relative burdens imposed by public health measures on healthy carriers (as opposed to sick individuals) warrant careful consideration and should be proportionate to the expected public health benefits in terms of risks averted. We conclude that more surveillance and research regarding community transmission (and the effectiveness of available interventions) will be needed in order to clarify relevant risks and design proportionate policies, although extensive community surveillance itself would also require careful ethical consideration.

E. Jamrozik (✉)
Monash Bioethics Centre, Monash University, Clayton, VIC, Australia

Royal Melbourne Hospital Department of Medicine, University of Melbourne, Parkville, VIC, Australia

Wellcome Centre for Ethics and the Humanities and The Ethox Centre, Nuffield Department of Population Health, University of Oxford, Oxford, UK
e-mail: zeb.jamrozik@gmail.com

M. J. Selgelid
Monash Bioethics Centre, Monash University, Melbourne, VIC, Australia

© The Author(s) 2020   183
E. Jamrozik, M. Selgelid (eds.), *Ethics and Drug Resistance: Collective Responsibility for Global Public Health*, Public Health Ethics Analysis 5,
https://doi.org/10.1007/978-3-030-27874-8_12

Keywords Antimicrobial resistance · Drug resistance · Asymptomatic infection · Drug-resistant infection · Ethics · Public health ethics

12.1 Introduction

The human body contains more bacterial cells than human cells. We all, thus, "carry" bacteria--especially in our digestive, respiratory, and urogenital tracts – and on our skin (Sender et al. 2016). Bacteria travel with and between human beings, following the daily flux of commuters in cities and crossing the world with international travellers. These 'fellow travellers' are often unsuspected since most of these bacteria cause no overt symptoms (i.e. the human carriers are mostly 'asymptomatic'), and some even have symbiotic benefits for humans.[1] Yet, under certain circumstances (e.g., skin breakdown, bowel surgery, or the use of immunosuppression), some ubiquitous species of bacteria cause 'invasive' disease that may require antibiotic treatment and be life threatening if such treatments are ineffective or unavailable. With increasing use (and overuse) of antibiotics in recent decades, such ubiquitously carried bacteria have in many cases become increasingly resistant to first line (and, in some cases, second line and/or 'reserve') treatments (See Table 12.1). Some have become effectively untreatable with standard antibiotics, and those who develop invasive disease are at high risk of death and/or permanent morbidity (Klein et al. 2007; Tischendorf et al. 2016).

Asymptomatic carriers of resistant ubiquitous bacteria vastly outnumber symptomatic cases (Tischendorf et al. 2016; Smith et al. 2004; Safdar and Bradley 2008) and may carry them unknowingly for months or years (Smith et al. 2004; Carlet 2012; Zimmerman et al. 2013; Marchaim et al. 2007; Kennedy and Collignon 2010), creating risks of disease for carriers as well as risks of transmission to others

Table 12.1 Pathogens frequently carried by healthy individuals

Pathogen	Examples of resistant form(s) of public health importance[b]	Site of carriage
Enterobacteriaceae[a]	CRE (carbapenem resistant enterobacteriaciae)	Digestive tract
Enterococcus faecium	VRE (vancomycin-resistant enterococci	Digestive tract
Clostridium difficile	Vancomycin resistant strains	Digestive tract
Staphylococcus aureus	MRSA (methicillin-resistant S. aureus)	Skin, nose
Streptococcus pneumonia	Penicillin resistant strains	Respiratory tract
Haemophilus influenzae	Penicillin / macrolide resistant strains	Respiratory tract

[a]Enterobacteriaciae include: *Klebsiella pneumoniae, Escherichia coli, Enterobacter spp., Serratia spp., Proteus spp.,* and *Providencia spp, Morganella spp.* [b]Adapted from WHO List of Priority Resistant Pathogens

---

[1] Others can lead to indirect benefits because their presence in the ecological niche of asymptomatic carriage excludes and/or inhibits other more harmful organisms.

(Klein et al. 2007; Tischendorf et al. 2016; Safdar and Bradley 2008; Carlet 2012; Giske et al. 2008). In light of the potential transmission of resistant bacteria from asymptomatic carriers to others, asymptomatic carriage (otherwise referred to as asymptomatic infection, colonization, commensalism, persistence, the carrier state, etc. (Casadevall 2000)) has underexplored ethical implications for public health programs aimed at controlling the spread of drug resistant bacteria.

Though infection control policies often focus on symptomatic cases, in some cases they also apply to apparently healthy individuals (for example, quarantine involves those suspected, but not known, to be infected (Morgan et al. 2017; Kass 2001; Selgelid 2009; Millar 2009)). As more knowledge is gained regarding *community* (as opposed to in-hospital) transmission of drug-resistant bacteria, policy-makers will need to determine appropriate responses to this relatively new set of public health problems in the general community. Policy options could include screening and other kinds of interference with the lives of apparently healthy individuals. When potentially transmissible (resistant) asymptomatic carriage is diagnosed, there may sometimes be an ethical rationale for public health interventions to prevent transmission, and these measures (as well as screening itself) may entail significant burdens[2] for carriers (and others) as well as public health benefits. Beyond screening, interventions could include reporting the diagnosis of asymptomatic carriage to authorities, notification of third parties, monitoring of carriers, restrictions on freedom of movement (e.g. quarantine, isolation, travel bans), exclusion of carriers from working in certain occupations, and/or possibly even requirements for treatment of carriers in certain circumstances (See Table 12.1). Where the rate of carriage of highly resistant (i.e. effectively untreatable) bacteria is increasing and/or where community transmission poses significant risks, such public health interventions could have wide-ranging effects on social norms and the everyday lives of healthy individuals (as was the case for tuberculosis and HIV prior to the availability of effective treatments (Fitzgerald 2007)) resulting in stigma and/or social exclusion, in addition to the direct burdens of complying with public health interventions.

A key ethical question concerns whether or not, or the extent to which, public health decision makers should be especially reluctant to impose public health measures that infringe upon the lives of those who are healthy (as opposed to those who are sick). In this chapter, we summarize current knowledge regarding asymptomatic carriage and community transmission, and argue that (i) beliefs that only those with symptoms pose risks to others (and related views, such as 'microbial determinism' – i.e. the idea that all those who acquire a pathogen will develop symptoms) should be discarded, (ii) policymakers should consider the risks posed by asymptomatic carriers of resistant organisms, and (iii) policy formation should be guided public health ethics frameworks such that the burdens imposed on carriers (and others) are minimized (and/or offset) and proportionate to the public health benefits in terms of risks averted; and (iv) designing proportionate interventions will require, *inter alia*, careful assessments of the risks related to asymptomatic carriage, including through expanded, ethically designed, public health surveillance programs.

---

[2] In this chapter we use the term 'burden' to refer to compromises of people's (especially carriers') liberties and/or wellbeing.

12.1.1   History

In the late nineteenth and early twentieth centuries, pioneering microbiologists such as Robert Koch significantly improved scientific understanding of the microbial agents of infectious disease. In 1890, Koch laid out criteria for inferring causal links between pathogens and disease states, including a requirement that every person with the microbe must show signs of the relevant disease. Only 3 years later Koch realized that this was an error since many people carry pathogenic microbes and can transmit them to others, without themselves showing signs of disease. This insight went against common wisdom at the time and was illustrated in famous cases such as that of 'Typhoid Mary' in New York in 1907: despite being asymptomatic, Mary Mallon transmitted typhoid to many other people through her work as a cook, resulting in several deaths (Soper 1939).

12.1.2   Against Microbial Determinism

Despite this, people might be tempted to think that such cases (of asymptomatic carriage) are exceptions and that the acquisition of potentially pathogenic microbes by a human being (or other animal) will always (or almost always) lead to symptomatic infection (i.e. disease). In some ways, the false view that one's infectious disease status is determined by the pathogens in one's body (which we call 'microbial determinism') is akin to an erroneous view in genetics ('genetic determinism'), according to which phenotype is determined by genotype (de Melo-Martín 2005). Just as particular genetic polymorphisms do not always give rise to particular phenotypes (because many environmental factors as well as cellular and other causal processes are required), acquiring particular pathogens does not always lead to the relevant infectious diseases. Upon acquisition of a pathogen, a complex set of host-pathogen interactions (involving immunological and other causal processes) can lead to a variety of outcomes – e.g. the microbe being eliminated without symptoms, short- or long-term asymptomatic carriage, or symptomatic infection/disease (with rapid or delayed onset)(Casadevall 2000).

12.1.3   Key Drug-Resistant Pathogens

This chapter focuses on the ethical implications associated with a subset of WHO Priority Resistant Pathogens (World Health Organisation 2017a): among those with resistance profiles of public health concern, we concentrate on pathogens that are ubiquitous organisms in the bacterial flora of the human body in healthy individuals (Table 12.1) where – in the majority of carriers, whether the bacteria are resistant or not – they usually cause no symptoms[3] until a precipitating event leads to invasive disease (Tischendorf et al. 2016; Safdar and Bradley 2008; Carlet 2012). Such invasive disease can occur in otherwise completely healthy carriers, although it is more common in those with comorbidities, especially those associated with reduced immune function (e.g. diabetes, HIV etc.). We are also particularly concerned with community as opposed to in-hospital transmission, as the latter has received more analysis elsewhere (Millar 2012), including in this volume.

Many of our arguments may be relevant to other increasingly resistant bacteria, including (i) gastro-intestinal and sexually transmitted pathogens that are associated with a high rate of symptomatic infection when a person is first exposed (e.g. *Campylobacter, Salmonella* (including typhoid*)*, *Shigella,* and *Gonorrhea*) following which only some people will become chronic asymptomatic carriers, and (ii) *Helicobacter pylori*, a less ubiquitous pathogen for which carrier status is associated with a range of clinical severity from no symptoms to mild indigestion to overt peptic ulcer disease and/or stomach cancer.

12.2   The Public Health Problem

With widespread use (and overuse) of antibiotics, the number of asymptomatic carriers of resistant bacteria (henceforth 'asymptomatic carriers') is increasing. It is difficult to characterize the overall carriage rates of the many different types of clinically significant resistant organisms, since the rate of carriage (and the rate of invasive infection) of each varies considerably between populations, and quoted rates will depend on the quality and extent of public health surveillance in different settings (Laxminarayan et al. 2016; Bryce et al. 2016; Bernabé et al. 2017; Nordmann et al. 2011; Schwaber and Carmeli 2013). In any case, *symptomatic* resistant bacterial infections already cause hundreds of thousands of deaths globally per year (although few estimates of the total burden are available) (Laxminarayan et al. 2016; O'Neill 2015) and increase healthcare costs by billions of dollars (Klein et al. 2007; Giske et al. 2008; O'Neill 2015). The true population prevalence of *asymptomatic* carriage of resistant bacteria is often unknown, because not enough community surveillance data are available. Most data come from hospital settings and are biased by the inclusion of a disproportionate number of symptomatic cases and a focus on patients who have contact with healthcare institution(s) as opposed to the wider community (World Health Organization 2014).

Globally, the rate of carriage and/or disease from resistant organisms tracks disadvantage, with higher rates in many low- and middle-income countries (LMICs) (Bernabé et al. 2017). For example, a recent study in a Malaysian hospital found that around 50% of patients screened on arrival to hospital were carrying

---

[3] Emerging microbiome research programs are, however, seeking links between strains of colonization and a wide range of 'subclinical' physical and mental health outcomes.

carbapenem resistant Enterobacteriaciae (CRE)[4] in their digestive tract (Zaidah et al. 2017) – of whom perhaps up to 1 in 6 can be expected to develop invasive disease, with overall mortality among carriers reportedly around 10% (Tischendorf et al. 2016). In general, there is a much higher rate of mortality from resistant infections in poor communities, including among infants – some of whom acquire resistant infections from mothers and/or family members (and/or, for infants admitted to hospital, from staff or other patients) who are asymptomatic carriers (Chan et al. 2013). One 2016 estimate suggested that of the 680,000 annual neonatal deaths due to bacterial infection, the vast majority of which occur in LMICs, around 31% (214,500) were due to resistant infections (Laxminarayan et al. 2016).

Yet the problem is by no means confined to developing regions. A 2016 systematic review of *E. coli*[5] urinary infections among children (with the usual source of such infections being asymptomatic carriage of *E. coli* in the child's digestive tract) found that the highest rates of resistance to first-line penicillin antibiotics occurred in (lower income) non-OECD countries (79.8%), but rates in (higher income) OECD countries were still relatively high (53.4%) (Bryce et al. 2016).

More epidemiological research is urgently needed to quantify the rates of carriage of key pathogens in different populations, as well as the rates of disease among carriers and the rates of transmission from (symptomatic and asymptomatic) carriers to others in different contexts. Novel, less expensive, genomic screening techniques are expected to facilitate such investigations (Kwong et al. 2017).

12.2.1 Antibiotic Use and Drug Resistance

The use of antibiotics that kill sensitive bacteria in the human microbiome inevitably leads to the evolutionary selection of drug resistant bacteria. Despite concerns regarding underuse, the vast majority of drug resistant bacteria in humans arise due to antibiotic overuse and 'appropriate' use (Llewelyn et al. 2017).[6] Importantly, when antibiotics are prescribed/taken to treat one type of (suspected or confirmed) infection, many other bacteria carried in the body are exposed to the same antibiotics, which select for resistant strains by killing sensitive ones. It is mainly these 'off-target' effects that lead to asymptomatic carriage of resistant forms of ubiquitous bacteria (Llewelyn et al. 2017). This means that whether a given prescription for antibiotics is 'appropriate' (e.g. because the patient actually has a symptomatic infection with the pathogen for which she is being treated) or not, each additional dose of antibiotics potentially adds to the burden of resistant bacteria carried in the

---

[4] i.e. highly resistant bacteria predominantly found in the digestive tract, a group considered "critical" (i.e. of highest importance) on the WHO Priority list.

[5] *E. coli* is an important species in the family of *Enterobacteriaciae*, carried primarily in the digestive tract (see Table 12.1 and footnote above).

[6] In this paper we focus on human use, although animal and agricultural use is an important contributor to drug-resistant human infection.

body (Carlet 2012). Furthermore, at the population level, it is otherwise relatively healthy occasional users of antibiotics (taken together) who contribute the most to the prevalence of (asymptomatic carriage of) drug-resistant bacteria, rather than the relatively few, relatively sick individuals whose antibiotic use is more frequent and/ or intensive (Olesen et al. 2018). In any case, although reducing the use of antibiotics is one important policy to reduce carriage of drug-resistant bacteria, this chapter focuses on other potential community interventions that have received less ethical analysis in this context (Bryce et al. 2016; Barbosa and Levy 2000; Bronzwaer et al. 2002).

12.2.2 Transmission

Even those who never use antibiotics can acquire resistant pathogens through (direct or indirect) contact with carriers (Zimmerman et al. 2013; Schwaber and Carmeli 2013; Waters et al. 2004; Paterson 2006). Living in close contact with carriers (Eveillard et al. 2004; Granoff and Daum 1980), hospitalization(Eveillard et al. 2004; Cronin et al. 2017), working in healthcare (Eveillard et al. 2004; Albrich and Harbarth 2008), and travel to countries with high rates of resistant organisms (Kennedy and Collignon 2010) are all risk factors for the acquisition of resistant pathogens via transmission. Increasingly, outbreak investigations have demonstrated transmission networks that connect symptomatic and asymptomatic individuals both within hospitals and in the wider community (Smith et al. 2004; Kwong et al. 2017). Community transmission of some pathogens is relatively well understood; for example, colonization with methicillin resistant *S. aureus* (MRSA) is associated with transmission within families (Eveillard et al. 2004; Fritz et al. 2014; Manian 2003), while the transmission of many other resistant pathogens has been primarily studied in the hospital setting (if at all).

12.2.3 Duration of Carriage

Once an individual becomes a carrier of resistant bacteria, the duration of carriage depends on complex local factors at the site of carriage including competition from other strains or other species of bacteria, as well as the carrier's immune response, and whether (more) antibiotics are used (Andersson and Hughes 2010). Few studies have estimated the average duration of carriage of key resistant bacteria in the general population; studies in returned travellers have suggested a decrease in carriage over months, with most individuals carrying detectable levels of resistant strains for a few months and a minority (around 10%) developing long-term carriage of 6 months or more (Kennedy and Collignon 2010).

12.3 Potential Public Health Responses

Strategies for preventing the transmission of drug-resistant bacteria from asymptomatic carriers include a wide range of potential interventions (Table 12.2). Each of these interventions could involve burdens for carriers (and others, for example family members) – to a greater or lesser degree, depending on the chosen policy and its implementation. Such burdens may include reduced well-being, infringements on privacy,[7] restrictions of other freedoms (including freedom of movement and freedom to decide on the medical interventions to which one will be subjected (Table 12.2)[8]) (Inness 1996), and significant financial costs.[9] Interventions can be more or less coercive ranging from being offered, recommended, or self-enforced to strictly coerced and/or backed by legal sanctions including fines and/or prison terms.

Thus, the design and implementation of infection control policy will inevitably involve ethical tradeoffs (Millar 2009; Millar 2012). The focus of most drug-resistant infection policy regarding asymptomatic carriers has been on healthcare contexts (primarily hospitals), yet – with few exceptions (Millar 2009; Millar 2012) – there has been little explicitly ethical analysis of such policies. There has

Table 12.2 Infection control interventions and potential burdens

Public health interventions that could lead to reduced well-being, privacy infringements, and reductions in other freedoms (e.g. freedom of movement, freedom to decide on medical interventions)	
Intervention Types	Description/examples
Screening	Testing apparently healthy individuals for carriage
Informing carriers of their status	Communicating and explaining the diagnosis of a carrier state
Notifying public health authorities	Requirements for health workers to report the diagnosis of carrier status
Monitoring	Serial (re)testing to assess carrier status
Treatment or decolonization	Skin/nasal decolonization of MRSA; faecal transplant for CRE
Limitations on freedom of movement	Isolation, quarantine, travel bans
Limitations on social practices	Change of occupation, altered norms of social interaction

[7] We consider privacy to be a freedom; see, for example, Inness (1996).

[8] This paper focuses on burdensome interventions for carriers; there are of course a range of other interventions including nudging or minimally-burdensome behavioral interventions (we thank an anonymous reviewer for pointing this out) and interventions to change healthcare worker behaviors such as antibiotic prescribing (which have received some discussion elsewhere).

[9] Costs arise both for the individuals directly affected and, where public health measures are financed through taxation, for the whole community.

been even less ethical analysis of public health infection control interventions to prevent the transmission of drug resistant bacteria in the general community (i.e. outside healthcare settings).

12.3.1 Surveillance, Notification, and Monitoring

Screening is routinely used to detect asymptomatic carriage of resistant bacteria among hospital patients (Morgan et al. 2017; Perencevich et al. 2004; Cooper et al. 2004). In many jurisdictions, hospitals are required to notify or report the diagnosis of asymptomatic carriage to local infectious diseases departments and/or central agencies, resulting in exceptions to the usual right to privacy over one's own health information (Morgan et al. 2017; Inness 1996; World Health Organisation 2017b). In order to form a more accurate estimate of the reservoir of asymptomatic infection in the community, and accurately assess the transmissibility and invasiveness (i.e. propensity to cause symptomatic disease) of a given pathogen, surveillance would ideally go beyond (hospital) patients and include members of the general community such as (but not necessarily limited to) the close contacts of those known to be carriers. Yet this raises questions regarding how policy should address potential scenarios in which large numbers of asymptomatic carriers of highly resistant pathogens are identified. Furthermore, the accuracy of tests used in surveillance has ethical implications especially where, for example, a false positive test result leads to significant burdens for someone who is not actually a carrier of resistant microbes (or where a false negative result provides false reassurance).

12.3.2 Restrictions of Freedom of Movement (Isolation, Quarantine, Travel Bans)

Although isolation and quarantine of asymptomatic carriers and/or their contacts have sometimes been successful in hospitals, whether such measures would be feasible and/or successful in the general community (e.g. where community members are identified as carriers by public health surveillance) remains uncertain. Restricting the freedom of movement of healthy carriers (or those suspected to be carriers) in the general community would plausibly involve significant infringements on individual liberty, and such restrictions might often lead to greater burdens for healthy carriers than for those suffering with the symptoms of the relevant disease (or, for example, confined to hospital for other reasons), as discussed below (See Sect. 12.4.2).

12.3.3 Treatment and Decolonization

Many asymptomatic bacterial infections in certain populations – for example, asymptomatic bacteriuria in most females – do not require treatment because they lead to a low rate of disease in the carrier (Nicolle et al. 2005). In other cases, either because of risks to the carrier or to others, treatment may lead to a net benefit (although not necessarily to the carrier herself). Treatment of asymptomatic *resistant* bacteria is often more difficult and sometimes referred to as decolonization. Decolonization strategies have been used for MRSA, primarily carried on the skin and in the nose – which allows for topical (i.e. non-invasive) bactericidal treatment[10] (Coates et al. 2009). Although decolonization has primarily been used in carriers with recurrent symptomatic infection in order to benefit the carrier herself, they have also been used in family members (and pets) of at-risk patients and healthcare workers in order to prevent harm to others (Albrich and Harbarth 2008; Guardabassi et al. 2004).

Decolonization for organisms carried in the digestive tract is sometimes more invasive than decolonization of the skin. For example, an effective, last-line treatment for resistant *C difficile*[11] is faecal transplantation (i.e. decolonization (Morgan et al. 2015)), whereby the bowel microbiome of the patient is replaced by feces from a healthy donor (Van Nood et al. 2013). There have been recent reports of the successful use of fecal transplantation to clear (symptomatic or asymptomatic) carriage of ubiquitous bowel organisms that have become highly resistant (e.g. CRE), Freedman and Eppes 2014; Manges et al. 2016; Crum-Cianflone et al. 2015). At present, the use of such procedures has been largely confined to unwell patients, including in intensive care units (Carlet 2012). Yet if such strategies prove to be safe, effective and reliable, they could be more widely implemented to address the carriage and transmission of resistant bowel organisms.

12.4 Ethical Issues

Asymptomatic infection raises a number of important ethical issues. In this chapter we focus on ethical considerations related to policy responses to the problem of asymptomatic carriage of resistant strains of ubiquitous bacteria because, as described earlier (i) asymptomatic carriers of such pathogens vastly outnumber symptomatic cases, (ii) carriers are at risk of severe and difficult to treat invasive disease, (iii) carriage of resistant pathogens places others in the community at risk,

---

[10] One limitation of *S. aureus* decolonization is that it uses bactericidal products (albeit with different formulations to other relevant antibiotics) to which the organism can also become resistant.

[11] *C. difficile*: a species of bacteria that is commonly (but not ubiquitously) carried asymptomatically in the bowel and most frequently causes disease in individuals already treated with antibiotics for other conditions.

and (iv) public health interventions to prevent the transmission of resistant bacteria are potentially associated with significant burdens for carriers (and others). Determining the appropriate responses will thus require scientific data regarding the risks involved (e.g. for a given pathogen or resistance mechanism in a given population) as well as moral judgments about the degree to which burdensome interventions to prevent these risks would be justified. An initial question relates to the extent to which the ethical permissibility of a burdensome infection control measure depends on whether contagious carriers are symptomatic as opposed to asymptomatic (i.e. 'healthy'). We begin by arguing for a useful way of applying existing public health ethics frameworks to policy questions, and then give an account of how the principles in such frameworks should be used to inform policy related to asymptomatic carriers.

12.4.1   Applying Public Health Ethics Frameworks

Existing public health ethics frameworks are applicable to the problem of the carriage and community transmission of resistant bacteria by asymptomatic carriers. The principles in different frameworks overlap considerably; Table 12.3 provides a list of relevant principles and examples of how they might be interpreted (based on previous work (Selgelid 2009)). These principles are usually framed as necessary conditions for determining when a given intervention would be justifiable. However, were this to be so, it might seem that certain principles would be difficult to satisfy

Table 12.3 Principles from public health ethics Frameworks

Public Health Ethics Principle	Interpretation/Example
Need for evidence	Evidence of efficacy is needed to justify imposition of potentially burdensome public health interventions.
Least restrictive alternative	Where two interventions are expected to be equally effective, the intervention that involves the least restrictions of liberty should be selected.
Proportionality	The burdens involved in an intervention should be outweighed by public health benefits achieved.
Equity	The intervention should be implemented (and burdens imposed) in an equitable, non-discrimintory manner.
Least harmful alternative	Where two interventions are expected to be equally effective, the intervention that involves the least harms should be selected.
Reciprocity	Those who benefit from public health policies/interventions have a reciprocal duty to assist and/or compensate those on whom burdens are imposed.
Due legal process	Appropriate legal procedures should be followed and individuals should have the right of appeal.
Transparency	Policymaking should be transparent and democratic.

(as necessary conditions) in the context of asymptomatic carriage of drug-resistant bacteria (and/or in many other public health contexts). If the existence of evidence of an intervention's effectiveness is conceived as a necessary condition, for example, then some might think that the current lack of evidence (e.g. of the risks of community transmission of resistant bacteria) precludes the implementation of relevant public health interventions. Such an approach would likely reduce the scope of legitimate public health interventions to a very narrow set (for example, those for which we have very substantial evidence regarding the effectiveness of relevant interventions, and where the attendant burdens are well characterized, etc.). For this and other reasons it is arguably more useful to conceive of such principles as pointing to ethically relevant desiderata the are achieved to a greater or lesser degree (i.e. on a scale) rather than as necessary conditions that are either satisfied or unsatisfied depending on whether some threshold has been crossed (Selgelid 2016).[12] For example, there can be more or less (reliable) evidence regarding the expected public health benefits associated with a given intervention—and a relative lack of evidence (such as the current gaps regarding our knowledge of community transmission of drug-resistant pathogens) might suggest that an intervention should first be instituted as (public health) research (rather than suggesting that it would be ethically unacceptable to implement it at all), and re-evaluated as more evidence comes to light. Furthermore, since there is not likely to be widespread agreement among public health practitioners on any threshold that could be used to characterize a sufficient amount of evidence to justify the implementation of a given intervention (in public health practice, rather than in research), policymakers should consider both evidence and ethical acceptability as matters of degree existing on a scale—the idea being that the more (reliable) evidence one has that an intervention is likely to create a net public health benefit, the more ethically acceptable it would be to implement it in policy, other things being equal.

Likewise the harms of potential interventions should be considered to exist on a spectrum from least to most harmful, liberty infringements on a spectrum from least (e.g., minor and/or short-term) to most restrictive (e.g. major and/or longer duration),[13] and transparency of policy making on a spectrum from least to most

---

[12] For illustration of how such an approach might work in the context of public health policy regarding gain-of-function research see Selgelid (2016).

[13] Although the Least Restrictive Alternative and Least Harmful Alternative principles are usually framed as necessary conditions, there are often cases where policymakers will be uncertain whether one intervention is more effective and/or less restrictive or harmful than another. In such cases of uncertainty, a scalar framing of the principles might run as follows: "The more confidence policymakers have that two interventions are associated with similar expected public health benefits and significantly different restrictions, the more moral considerations would support selecting the less restrictive alternative, other things being equal. The more confidence policymakers have that two interventions are associated with similar expected public health benefits and significantly different burdens, the more moral considerations would support selecting the less harmful/burdensome alternative, other things being equal." However, in a restricted set of cases where policymakers are very certain that one intervention is equally (or more) effective and less restrictive or harmful, interpreting the principles as necessary conditions will yield similar guidance, other

transparent, and so on—the idea being that the ethical acceptability of interventions will be a matter of degree, and a function of the extent to which they are harmful, restrictive, transparent, etc., other things being equal (Allen and Selgelid 2017). There might often be reasonable disagreement about exactly how the estimated degrees of evidence, harmfulness, restrictiveness, transparency, etc. can/should determine what ultimately ought to be done; but, as a starting point, policymakers should consider how well, or poorly, each available intervention fares with respect to the values/concerns highlighted by each principle before making *judgements* regarding what policies/interventions should be implemented. We consider this to be a practical approach to ethically sensitive public health policymaking; in the next section we illustrate such an approach in the context of asymptomatic carriage of drug resistance.

12.4.2 Public Health Intervention for Healthy Carriers

One might think that public health agencies should be more reluctant to interfere with the lives of healthy asymptomatic carriers as opposed to sick individuals who carry, and/or are suffering from clinical infection due to, the same (resistant) pathogens. The proportionality principle may help to explain such intuitions. Firstly, one way in which it may make a difference, ethically speaking, whether a carrier is symptomatic or not when imposing potentially burdensome public health interventions is that asymptomatic carriers are more likely to be living active lives in general society; whereas the more symptomatic one is, the more likely it is that they will be bedbound and/or admitted to a healthcare facility. In the latter kind of case, one's liberty and well-being are in a sense already impaired by illness, and so some restrictions (to prevent transmission of drug-resistant bacteria) may impose few, if any, additional burdens.[14] For example, public health measures limiting the freedom of movement of carriers (See Table 12.2) would be a more significant burden for healthy carriers whereas the isolation of sick individuals is a relatively minor additional burden, since such individuals are more likely to be restricted in their movement (e.g. confined to a healthcare facility) due to illness.[15] Thus, if the risk of transmission from a healthy individual is similar to, or less than, the risk from a sick

---

things being equal. We thank an anonymous reviewer for pressing us on this point. For more on this kind of approach to public health ethics frameworks, see Allen and Selgelid (2017).

[14] Those who are ill may also be more likely to understand and comply with burdensome restrictions because they can more easily perceive that they are infected and a risk to others; in some such cases restrictive measures may even be unnecessary since voluntary measures suffice. We thank an anonymous reviewer for suggesting this point.

[15] Some might worry that the systematic imposition of additional burdens on those who are already (in one sense) badly off due to illness would be inequitable; arguably, this would be particularly worrisome if these burdens were disproportionate (see following discussion of proportionality) to the risks averted which is less likely in cases where (i) the additional burden is small and/or (b) the risk of transmission from sick individual carriers is at least as high as, or higher than, the risk from healthy carriers (i.e. most cases).

individual, the burdens would be higher when imposing a given intervention on healthy carriers. Secondly, there may be cases in which healthy carriers and sick carriers impose different risks of transmission to others. Sick carriers may impose higher risks of transmission because (i) in at least some cases, the degree of symptoms (e.g. related to the resistant bacterial infection in question) is correlated with the risk of transmission[16] (Lerner et al. 2015) or because (ii) hospitalisation for illness places one in (direct or indirect) contact with other patients who are at particularly high risk of acquiring and suffering from (resistant) infection due to medical comorbidities and/or contamination of the shared hospital environment. In cases where such conditions hold, not only would the burdens of certain interventions be lower among sick carriers (as discussed above), but the risks (to others) averted would be greater, meaning that imposing similarly restrictive interventions on healthy carriers would be comparatively less proportionate (because more significant burdens would be imposed to avert lower risks). On the other hand, there might be cases in which healthy carriers impose higher risks on others. For example, healthcare workers or food handlers who carry resistant pathogens might impose increased risks on others because of the nature of their work. In such cases, it may be more justifiable to impose burdensome restrictions on such individuals, because these would be more proportionate to the risks involved (c.f. vaccination of healthcare workers (van Delden et al. 2008)).

12.4.3 Burdens of Interventions and Support for Carriers

As noted previously, public health interventions to control the spread of resistant bacteria in the general community could potentially burden carriers to a greater or lesser degree. They may also involve burdens for family members and contacts of known carriers, as well as direct financial costs for those involved and funding costs of relevant public health policy options (which are borne by the wider community and/or lead to forgoing other opportunities to improve public health)[17]. While the threat of drug-resistant infections as an urgent public health problem might prompt some to consider or propose particularly far-reaching and/or coercive interventions, several other public health ethics principles (in addition to the proportionality principle) can provide useful guidance regarding the ethical reasons to ensure that the burdens (and costs) of an intervention are not only proportionate to the public health benefits but also as low as possible (i.e. without unduly compromising the goals of a policy to reduce risks and thereby lead to public health benefits).

---

[16] for example, urinary frequency and incontinence related to urinary tract infection (e.g. with *E. coli*), diarrhea (e.g. from resistant *Clostridium difficile*), etc. potentially increase the risk of transmission to others.

[17] We thank anonymous reviewers for emphasizing the implications for family members and for pointing out the opportunity cost of public health policies.

First, the least harmful alternative principle holds that interventions must be no more harmful to carriers than necessary (i.e. minimally harmful) – because compromising individuals' well-being requires justification. If two alternative policies/ interventions are expected to be equally beneficial interms of public health protection, then the alternative that is less harmful to carriers should be preferred, other things being equal – See Table 12.3. Second, the reciprocity principle entails that policymakers should consider obligations to assist and support carriers as part of policy implementation so as to reduce or offset the 'net burdens' for those affected. This could include preferential/non-discriminatory access to healthcare, psychological support, assistance with finances or finding other employment, etc.

More broadly, public health agencies should aim to reduce burdens related to stigma through public education campaigns and through informing carriers about important aspects of carriage (for example, that carriage of resistant organisms is frequently not permanent) at the time of diagnosis. Overly burdensome interventions may also lead to perverse incentives for (potential) carriers to avoid diagnosis and/or contact with health authorities, which could lead to greater risks for the carrier and others, undermining the purported public health benefits of a given policy (meaning that the need for evidence includes evidence regarding how well a policy actually works in practice).[18]

12.5   The Need for More Surveillance and Research

Practical ethical deliberations guided by the above principles should always be informed by the best available empirical data on the relevant risks related to carriage of a given pathogen and the expected benefits of an intervention. An initial challenge is that, for many infections transmitted by asymptomatic carriers, the risks are not yet well understood because few data are available. There is thus an ethical imperative for increased public health surveillance and research on such infections, including programs aimed at collecting and analyzing long-term data (i.e. involving monitoring) related to asymptomatic carriage in the general community. This again requires careful ethical consideration, since ethically appropriate surveillance also requires striking a balance between public health goals and individual interests; here, too, public health ethics principles and analysis should help to guide policy formation (World Health Organisation 2017b). Likewise, there is an urgent need for more research regarding the relative effectiveness of different potential public health interventions, as well as qualitative research aimed at better characterizing the burdens experienced by carriers. Being able to draw on such empirical data will only serve to improve public health ethics analysis and, ultimately, public health policy.

---

[18] We thank the participants at the 2018 Brocher workshop "Invisible epidemics: ethics and interventions for asymptomatic carriers of infectious diseases.", in particular Niels Nijsingh and Christian Munthe for helpful discussions of this point.

12.6   Conclusions

All human beings are asymptomatic carriers of bacteria, meaning that 'microbial determinism' is false. The increasing prevalence of carriage of *resistant* strains of ubiquitous bacteria is an urgent public health issue, and many apparently healthy individuals are at risk of resistant infections and risk transmitting such pathogens to others. Deliberations regarding the design and implementation of public health policy should be guided not only by empirical data regarding the health risks of a given resistant strain and the public health benefits of a given intervention (and much more data are needed to clarify these risks and benefits), but also by ethical analysis regarding the justification of burdens imposed on carriers. Principles of existing public health ethics frameworks should help policymakers identify important considerations that have particular implications for the design and implementation of infection control policies regarding asymptomatic carriers. The proportionality principle in particular provides reasons for being wary about imposing potentially burdensome interventions on otherwise healthy carriers of drug resistant pathogens—which is not to say that such interventions would never be ethically appropriate.

Acknowledgement   Reprinted with permission of John Wiley and Sons (Licence 4890151079850); original citation: Jamrozik, E., & Selgelid, M. J. (2019). Surveillance and control of asymptomatic carriers of drug-resistant bacteria. *Bioethics*, 33(7), 766–775.

References

Albrich, W.C., and S. Harbarth. 2008. Health-care workers: Source, vector, or victim of MRSA? *The Lancet Infectious Diseases* 8 (5): 289–301.

Allen, T., and M.J. Selgelid. 2017. Necessity and least infringement conditions in public health ethics. *Medicine, Health Care and Philosophy* 20 (4): 525–535.

Andersson, D.I., and D. Hughes. 2010. Antibiotic resistance and its cost: Is it possible to reverse resistance? *Nature Reviews Microbiology* 8 (4): 260–271.

Barbosa, T.M., and S.B. Levy. 2000. The impact of antibiotic use on resistance development and persistence. *Drug Resistance Updates* 3 (5): 303–311.

Bernabé, K.J., C. Langendorf, N. Ford, J.-B. Ronat, and R.A. Murphy. 2017. Antimicrobial resistance in West Africa: A systematic review and meta-analysis. *International Journal of Antimicrobial Agents*.

Bronzwaer, S.L.A.M., O. Cars, U. Buchholz, et al. 2002. The relationship between antimicrobial use and antimicrobial resistance in Europe. *Emerging Infectious Diseases* 8 (3): 278.

Bryce, A., A.D. Hay, I.F. Lane, H.V. Thornton, M. Wootton, and C. Costelloe. 2016. Global prevalence of antibiotic resistance in paediatric urinary tract infections caused by Escherichia coli and association with routine use of antibiotics in primary care: systematic review and meta-analysis. *BMJ* 352: i939.

Carlet, J. 2012. The gut is the epicentre of antibiotic resistance. *Antimicrobial Resistance and Infection Control* 1 (1): 39.

Casadevall, A. 2000. Pirofski L-a. Host-pathogen interactions: Basic concepts of microbial commensalism, colonization, infection, and disease. *Infection and Immunity* 68 (12): 6511–6518.

Chan, G.J., A.C.C. Lee, A.H. Baqui, J. Tan, and R.E. Black. 2013. Risk of early-onset neonatal infection with maternal infection or colonization: A global systematic review and meta-analysis. *PLoS Medicine* 10 (8): e1001502.

Coates, T., R. Bax, and A. Coates. 2009. Nasal decolonization of Staphylococcus aureus with mupirocin: Strengths, weaknesses and future prospects. *Journal of Antimicrobial Chemotherapy* 64 (1): 9–15.

Cooper, B.S., S.P. Stone, C.C. Kibbler, et al. 2004. Isolation measures in the hospital management of methicillin resistant Staphylococcus aureus (MRSA): Systematic review of the literature. *BMJ* 329 (7465): 533.

Cronin, K.M., Y.S.P. Lorenzo, M.E. Olenski, et al. 2017. Risk factors for KPC-producing Enterobacteriaceae acquisition and infection in a healthcare setting with possible local transmission: A case–control study. *Journal of Hospital Infection* 96 (2): 111–115.

Crum-Cianflone, N.F., E. Sullivan, and G. Ballon-Landa. 2015. Fecal microbiota transplantation and successful resolution of multidrug-resistant-organism colonization. *Journal of Clinical Microbiology* 53 (6): 1986–1989.

de Melo-Martín, I. 2005. Firing up the nature/nurture controversy: Bioethics and genetic determinism. *Journal of Medical Ethics* 31 (9): 526–530.

Eveillard, M., Y. Martin, N. Hidri, Y. Boussougant, and M.-L. Joly-Guillou. 2004. Carriage of methicillin-resistant Staphylococcus aureus among hospital employees: Prevalence, duration, and transmission to households. *Infection Control & Hospital Epidemiology* 25 (2): 114–120.

Fitzgerald, C. 2007. Kissing can be dangerous. The public health campaigns to prevent and control tuberculosis in Western Australia, 1900–1960. *Australian Journal of Social Issues Subscriptions* 42 (2): 273.

Freedman, A., and S. Eppes. 2014. Use of stool transplant to clear Fecal Colonization with Carbapenem-Resistant Enterobacteraciae (CRE): Proof of concept. *Open Forum Infectious Diseases* (1):S65. Oxford University Press.

Fritz, S.A., P.G. Hogan, L.N. Singh, et al. 2014. Contamination of environmental surfaces with Staphylococcus aureus in households with children infected with methicillin-resistant S aureus. *JAMA Pediatrics* 168 (11): 1030–1038.

Giske, C.G., D.L. Monnet, O. Cars, and Y. Carmeli. 2008. Clinical and economic impact of common multidrug-resistant gram-negative bacilli. *Antimicrobial Agents and Chemotherapy* 52 (3): 813–821.

Granoff, D.M., and R.S. Daum. 1980. Spread of Haemophilus influenzae type b: Recent epidemiologic and therapeutic considerations. *The Journal of pediatrics* 97 (5): 854–860.

Guardabassi, L., S. Schwarz, and D.H. Lloyd. 2004. Pet animals as reservoirs of antimicrobial-resistant bacteria Review. *Journal of Antimicrobial Chemotherapy* 54 (2): 321–332.

Inness J.C. 1996. *Privacy, intimacy, and isolation*: Oxford University Press on Demand.

Kass, N.E. 2001. An ethics framework for public health. *American Journal of Public Health* 91 (11): 1776–1782.

Kennedy, K., and P. Collignon. 2010. Colonisation with Escherichia coli resistant to "critically important" antibiotics: A high risk for international travellers. *European Journal of Clinical Microbiology & Infectious Diseases* 29 (12): 1501–1506.

Klein, E., D.L. Smith, and R. Laxminarayan. 2007. Hospitalizations and deaths caused by methicillin-resistant Staphylococcus aureus, United States, 1999–2005. *Emerging Infectious Diseases* 13 (12): 1840.

Kwong, J.C., C. Lane, F. Romanes, et al. 2017. *bioRxiv* 175950.

Laxminarayan, R., P. Matsoso, S. Pant, et al. 2016. Access to effective antimicrobials: A worldwide challenge. *The Lancet* 387 (10014): 168–175.

Lerner, A., A. Adler, J. Abu-Hanna, S.C. Percia, M.K. Matalon, and Y. Carmeli. 2015, 21. Spread of KPC-producing carbapenem-resistant Enterobacteriaceae: the importance of super-spreaders and rectal KPC concentration. *Clinical Microbiology and Infection* (5): 470. e1–70. e7.

Llewelyn, M.J., J.M. Fitzpatrick, E. Darwin, et al. 2017. The antibiotic course has had its day. *BMJ* 358: j3418.

Manges, A.R., T.S. Steiner, and A.J. Wright. 2016. Fecal microbiota transplantation for the intestinal decolonization of extensively antimicrobial-resistant opportunistic pathogens: A review. *Infectious Diseases* 48 (8): 587–592.

Manian, F.A. 2003. Asymptomatic nasal carriage of mupirocin-resistant, methicillin-resistant Staphylococcus aureus (MRSA) in a pet dog associated with MRSA infection in household contacts. *Clinical Infectious Diseases* 36 (2): e26–e28.

Marchaim, D., S. Navon-Venezia, D. Schwartz, et al. 2007. Surveillance cultures and duration of carriage of multidrug-resistant Acinetobacter baumannii. *Journal of Clinical Microbiology* 45 (5): 1551–1555.

Millar, M. 2009. Do we need an ethical framework for hospital infection control? *Journal of Hospital Infection* 73 (3): 232–238.

———. 2012. 'Zero tolerance'of avoidable infection in the English National Health Service: Avoiding the redistribution of burdens. *Public Health Ethics* 6 (1): 50–59.

Morgan, D.J., S. Leekha, L. Croft, et al. 2015. The importance of colonization with Clostridium difficile on infection and transmission. *Current Infectious Disease Reports* 17 (9): 43.

Morgan, D.J., R.P. Wenzel, and G. Bearman. 2017. Contact precautions for endemic MRSA and VRE: Time to retire legal mandates. *JAMA* 318 (4): 329–330.

Nicolle, L.E., S. Bradley, R. Colgan, J.C. Rice, A. Schaeffer, and T.M. Hooton. 2005. Infectious Diseases Society of America guidelines for the diagnosis and treatment of asymptomatic bacteriuria in adults. *Clinical Infectious Diseases*: 643–654.

Nordmann, P., T. Naas, and L. Poirel. 2011. Global spread of carbapenemase-producing Enterobacteriaceae. *Emerging Infectious Diseases* 17 (10): 1791.

Olesen, S.W., M.L. Barnett, D.R. MacFadden, et al. 2018. The distribution of antibiotic use and its association with antibiotic resistance. *bioRxiv* 473769.

O'Neill, J. 2015. Antimicrobial resistance: Tackling a crisis for the health and wealth of nations. The Review on Antimicrobial Resistance. Chaired by Jim O'Neill. December 2014.

Paterson, D.L. 2006. The epidemiological profile of infections with multidrug-resistant Pseudomonas aeruginosa and Acinetobacter species. *Clinical Infectious Diseases* 43 (Supplement_2): S43–S48.

Perencevich, E.N., D.N. Fisman, M. Lipsitch, A.D. Harris, J.G. Morris Jr., and D.L. Smith. 2004. Projected benefits of active surveillance for vancomycin-resistant enterococci in intensive care units. *Clinical Infectious Diseases* 38 (8): 1108–1115.

Safdar, N., and E.A. Bradley. 2008. The risk of infection after nasal colonization with Staphylococcus aureus. *The American Journal of Medicine* 121 (4): 310–315.

Schwaber, M.J., and Y. Carmeli. 2013. An ongoing national intervention to contain the spread of carbapenem-resistant Enterobacteriaceae. *Clinical Infectious Diseases* 58 (5): 697–703.

Selgelid, M.J. 2009. A moderate pluralist approach to public health policy and ethics. *Public Health Ethics* 2 (2): 195–205.

———. 2016. Gain-of-function research: Ethical analysis. *Science and Engineering Ethics* 22 (4): 923–964.

Sender, R., S. Fuchs, and R. Milo. 2016. Revised estimates for the number of human and bacteria cells in the body. *PLoS Biology* 14 (8): e1002533.

Smith, D.L., J. Dushoff, E.N. Perencevich, A.D. Harris, and S.A. Levin. 2004. Persistent colonization and the spread of antibiotic resistance in nosocomial pathogens: resistance is a regional problem. *Proceedings of the National Academy of Sciences of the United States of America* 101 (10): 3709–3714.

Soper, G.A. 1939. The curious career of Typhoid Mary. *Bulletin of the New York Academy of Medicine* 15 (10): 698.

Tischendorf, J., R.A. de Avila, and N. Safdar. 2016. Risk of infection following colonization with carbapenem-resistant Enterobactericeae: A systematic review. *American Journal of Infection Control* 44 (5): 539–543.

van Delden, J.J.M., R. Ashcroft, A. Dawson, G. Marckmann, R. Upshur, and M.F. Verweij. 2008. The ethics of mandatory vaccination against influenza for health care workers. *Vaccine* 26 (44): 5562–5566.

Van Nood, E., A. Vrieze, M. Nieuwdorp, et al. 2013. Duodenal infusion of donor feces for recurrent Clostridium difficile. *New England Journal of Medicine* 368 (5): 407–415.

Waters, V., E. Larson, F. Wu, et al. 2004. Molecular epidemiology of gram-negative bacilli from infected neonates and health care workers' hands in neonatal intensive care units. *Clinical Infectious Diseases* 38 (12): 1682–1687.

World Health Organisation. 2017a. *WHO guidelines on use of medically important antimicrobials in food-producing animals*. Geneva: WHO.

———. 2017b. *WHO guidelines on ethical issues in public health surveillance*. Geneva.

World Health Organization. 2014. *Antimicrobial resistance: global report on surveillance*: World Health Organization.

Zaidah, A.R., N.I. Mohammad, S. Suraiya, and A. Harun. 2017. High burden of Carbapenem-resistant Enterobacteriaceae (CRE) fecal carriage at a teaching hospital: Cost-effectiveness of screening in low-resource setting. *Antimicrobial Resistance & Infection Control* 6 (1): 42.

Zimmerman, F.S., M.V. Assous, T. Bdolah-Abram, T. Lachish, A.M. Yinnon, and Y. Wiener-Well. 2013. Duration of carriage of carbapenem-resistant Enterobacteriaceae following hospital discharge. *American Journal of Infection Control* 41 (3): 190–194.

Chapter 13
Conceptualizing the Impact of MDRO Control Measures Directed at Carriers: A Capability Approach

Morten Fibieger Byskov, Babette Olga Rump, and Marcel Verweij

Abstract Many countries have implemented specific control measures directed at carriers of multidrug-resistant organisms (MDRO) in order to prevent further introduction and transmission of resistant organisms into hospitals and other healthcare related settings. These control measures may in many ways affect the lives and well-being of carriers of MDRO, resulting in complex ethical dilemmas that often remain largely implicit in practice. In this chapter, we propose to conceptualize the impact of MDRO control measures on the well-being of individual carriers in terms of capabilities and functionings. A capabilitarian framework for the ethical treatment of MDRO carriers commits us to conceptualize the harm done to carriers in terms of the impact that MDRO control measures have on what they are able to do or be. Adopting and adapting Nussbaum's list of ten central human capabilities, we present a taxonomy of capabilities and functionings that are normatively relevant for the design and evaluation of MDRO control measures.

Keywords Bioethics · Moral philosophy · Public health · Drug resistance

13.1 Introduction

Antimicrobial resistance (AMR) has been described as one of the major threats to individual and public health (WHO 2014). This threat has justified extensive restrictions on the freedom of individuals (Krom 2011; Littmann 2014; Littmann and

M. F. Byskov (✉) · M. Verweij
Communication, Philosophy, and Technology, Wageningen University & Research, Wageningen, Netherlands
e-mail: morten.byskov@wur.nl; Marcel.verweij@wur.nl

B. O. Rump
The Netherlands National Institute for Public Health and the Environment (RIVM), Bilthoven, Netherlands

© The Author(s) 2020
E. Jamrozik, M. Selgelid (eds.), *Ethics and Drug Resistance: Collective Responsibility for Global Public Health*, Public Health Ethics Analysis 5, https://doi.org/10.1007/978-3-030-27874-8_13

Viens 2015). Many countries have implemented control measures in order to prevent further introduction and spread of MDRO. Some more general, as addressed in Chap. 6 by Gilbert et al., and some more specific, like those targeting the individual who is found to carry an MDRO. Measures directed at MDRO carriers aim to limit the introduction and further transmission of multidrug-resistant organism (MDRO) in hospitals and other healthcare-related settings. The measures vary per micro-organism and include for instance isolation and quarantine; contact precaution; eradication treatment; restrictions in the workplace; refusal of access to important activities; or contact restrictions at the one's family farm. They may in many ways affect the lives and well-being of carriers, resulting in complex ethical dilemmas that often remain largely implicit in practice.

Within the literature, little attention has been paid to how we treat carriers of MDRO, however, and Littmann et al. (2015) includes it as one of four ethical issues that needs further examination when addressing MDRO. In this chapter we aim to start filling this lacuna by proposing to conceptualize the impact of MDRO control measures on the well-being of individual carriers in terms of capabilities and functionings. A capabilitarian framework for the ethical treatment of MDRO carriers commits us to conceptualize the harm done to carriers in terms of the impact that MDRO control measures have on what they are able to do or be. Adopting and adapting Nussbaum's list of ten central human capabilities, we present a taxonomy of capabilities and functionings that are normatively relevant for the design and evaluation of MDRO control measures. Chapter 16 addresses the implications of AMR for child development and adult capabilities.

The chapter is structured as follows: In Sect. 13.2, we shortly present the issue of treating MDRO carriers as an ethical problem before we turn, in Sect. 13.3, to propose a capabilitarian framework for the conceptualization of the impact that MDRO control measures have on the well-being of carriers. In Sects. 13.4 and 13.5, we adapt Nussbaum's list of ten central human capabilities in order to develop a taxonomy of normatively relevant capabilities and functionings in the context of MDRO. In Sect. 13.4, we first present Nussbaum's list of capabilities before we argue that this list needs further specification when applied to the case of MDRO. In Sect. 13.5, we proceed to propose a taxonomy of ethical domains and normatively relevant capabilities and functionings in the context of responsible care for MDRO carriers. In Sect. 13.6, we finally argue that and show how this capabilitarian taxonomy can provide a crucial input to procedures for ethical decision-making on appropriate MDRO control measures.

13.2   The Ethical Treatment of MDRO Carriers: A Neglected Issue

Treating MDRO as an ethical issue is a double-sided coin. On the one side, it involves a concern for public health and how we can ensure that everyone, now and in the future, have access to antimicrobial treatment while minimizing the risk of further spread of MDRO. From this side of the coin, addressing MDRO is primarily an issue of global distributive justice (Littmann 2014; Littmann et al. 2015, 360): how can we distribute antimicrobials in a way that, on the one hand, adequately protects public health by ensuring that everyone has access to antibiotics while, on the other hand, ensuring that antibiotics do not become useless? In the following, however, we shall not primarily be concerned with this distributive question.[1]

The distributive focus has often been accompanied by a discussion of what kinds of control measures we can take to prevent the further spread of MDRO (Selgelid et al. 2009; Coleman et al. 2010): how can we treat carriers in a way that minimizes the risk that they contaminate other individuals? An important element in the fight against MDRO is to adequately treat infections with multi-resistant microbes in patients and to prevent that these persons are re-infected or will infect others with a resistant organism. Due to the threat that MDRO poses to individual and public health (WHO 2014), many countries have implemented specific MDRO control measures in order to prevent further introduction and spread of MDRO. Measures to prevent and control the spread of MDRO may include isolation and quarantine; eradication therapy; restrictions in the workplace; refusal of access to important activities; or contact restrictions with one's family (Verweij and Dawson 2010).

Many of these control measures threaten to seriously affect the lives of individual carriers, however, and as important as such prevention and control is, it may have burdensome implications for infected patients and healthy persons in whom a resistant organism has been colonized: they may feel stigmatized, face restrictions in their work or private life, or might be refused access to certain institutions. For example, in a healthcare context, control measures may mean that surgeons should refrain from operating due to carriership, that infected nurses should not perform patient-related activities, or that we ask infected residents of a nursing home to keep away from social activities.

In some extreme cases it is almost impossible to eradicate the resistant organism and then it may be impossible for the person to return to what used to be his/her normal life. Consider, for example, the case of a medical student who was repeatedly diagnosed as carrier of Methicillin-Resistant Staphylococcus Aureus (MRSA) (Rump 2011; Rump et al. 2016). In line with the MDRO control guidelines, the student was not allowed to be involved in patient-care, which is an implicit part of completing the internships necessary to graduate. Because of this, the student had to

---

[1] For a discussion of different approaches to the distribution of antimicrobials, see especially Anomaly (2010, 2013), Daulaire et al. (2015), Littmann (2014), and Selgelid (2007). For an introduction to and overview of distributive justice in general, see Lamont and Favor (2016).

eventually discontinue his studies. Whether this outcome was indeed necessary or not remains unclear, though, since the risk of further contamination could have been minimized through proper hygiene and guidance.

The consequences of MDRO carriership certainly have the potential to affect the lives and well-being of not only carriers themselves, but also their social connections, such as family members, friends, and colleagues. Yet, it is unclear in what ways MDRO and MDRO carriership affect these individuals. In the remainder of this chapter, we will offer a novel conceptualization of how MDRO control measures can harm carriers and other affected individuals and further reflect on how this conceptualization, and the normatively relevant issues that are thereby revealed, influences the design and evaluation of MDRO control measures. We argue that adopting a capabilitarian framework for the conceptualization of 'harm' done to (potential) carriers can help us make better and more informed decision about what control measures to implement. According to a capabilitarian framework, MDRO control measures may harm individual carriers by negatively affecting their capabilities and functionings.

13.3 A Capabilitarian Framework for Conceptualizing the Impact of MDRO Control Measures

What is the capability approach and how can it be used to conceptualize the (negative) impact that MDRO control measures have on the lives and well-being of individual carriers?[2] Originally conceived by the Indian philosopher-economist Amartya Sen (1979) and further developed by a number of theorists, such as Martha Nussbaum, David A. Crocker, and Ingrid Robeyns, the capability approach is a normative framework for the conceptualization of human well-being (Robeyns 2016a). According to this framework, human well-being should be conceptualized in terms of *capabilities* and *functionings*. Capabilities are the real freedom that people have to do or be certain things, such as falling in love, getting an education, being politically active, riding a bike, reading a book, and so on. Functionings are

---

[2] It is possible that the capability approach can also be used to conceptualize 'harm' within three other related domains of application, which we will not discuss in this chapter. First, we can conceptualize the risk that the spread MDRO poses to the well-being of members of the public in terms of their capabilities and functionings. For a discussion of how to conceptualize public health in terms of the capability approach, see Prah Ruger (2010), Venkatapuram (2011), and Nielsen (2014). Secondly, the capability approach can be used to conceptualize the 'harm' of control measures within infectious disease control in general. While the taxonomy that we provide in Sect. 13.5 may also apply within infectious disease control in general, more research needs to be done in this regard as different capabilities may be relevant in relation to different diseases. Thirdly, the capability approach has been used to conceptualize the idea of 'person-centered healthcare,' what it means to treat patients as persons (Entwistle and Watt 2013).

capabilities that have been realized either by choice or by chance. A person's *capability-set* refers to all the capabilities and functionings that that individual has.[3]

Real freedom in this sense means that there are no restrictions on achieving a particular functioning. Whether or not one has such real freedom crucially depends on certain *conversion factors*. Conversion factors are personal, social, and environmental circumstances that affect the extent to which one can achieve certain doings and beings. For example, whether or not one has the real freedom to be healthy – that is, whether or not one has the capability of achieving the functioning of being healthy – depends on one's physical health, for example the strength of one's immune defense system (personal conversion factors), the extent to which one can rely on family and social relations for care (social conversion factors), and where one lives and whether there are adequate infrastructures, such as accessible health care facilities (environmental conversion factors).[4]

Through the notion of conversion factors, the capability approach captures the fact that human beings are diverse: different people living in different societies would have different needs and capabilities. As we shall see in Sect. 13.5, the different conversion factors are relevant when we consider how MDRO control measures affect the lives and well-being of individual carriers.

The capability approach moves the focus from the means that people have to their ends – what they are able to do or be with these means, such as goods, resources, and formal freedoms. As Sen (1979) argues, this shift in focus is justified because resources and goods alone do not ensure that people are equally able to convert them into doings and beings. Consider, for example, two persons – one disabled, the other able-bodied – with the same amount of resources. According to Sen, the disabled person is disadvantaged relative to the able-bodied person in two regards. First, she is disadvantaged in terms of what she can do or be with her means and resources. She may, for example, be less able to move around because she is confined to a wheelchair. Secondly, she may even be doubly worse off because she only receive the same amount of resources as the able-bodied person, even though she has more expenses in order to correct for her disability, whereas the able-bodied person, *ex hypothesi*, can spend all of her resources to pursue her valued ends. Hence, when evaluating the well-being of individuals, we cannot merely compare the amount of resources that they have without also looking at what they are able to do or be with these resources.

---

[3] Capabilities and functionings can be both positive and negative, as well as neutral (Robeyns 2016b). Positive capabilities are what we consider valuable for someone to do or be. Examples of positive capabilities are good health, adequate nutrition, falling in love, and getting an education. While most applications of the capability approach are primarily concerned with positive capabilities, there are also cases where we want to consider their negative capabilities. When evaluating a person's well-being, for example, it is relevant whether her capability-set include the capabilities to be murdered or raped. Insofar as we, usually, do not consider these capabilities to be valuable, a capability-set that allows for the risks of being murdered or raped would be less valuable than a capability-set that protects the individual from these risks.

[4] This example is adapted from Crocker and Robeyns (2010).

The concepts of capabilities and functionings can help us to better understand how MDRO control measures can 'harm' (potential) carriers of MDRO in terms of how MDRO control measures influence the real freedom that MDRO carriers have to do or be certain normatively relevant things. MDRO control measures can affect the capability-sets of (potential) carriers in at least two ways. First, they may impose certain requirements on (potential) carriers. This is, for example, the case when we subject nurses to strict hygiene regimes or demand that carriers undergo mandatory screenings and eradication therapies. In terms of the capability approach, MDRO control measures thus impose certain doings and beings – that is, functionings – on carriers. Secondly, MDRO control measures can reduce the choices that carriers have to choose from (i.e., the capabilities that they can choose to turn into realized functionings). This is, for example, the case when we place carriers in isolation or ban them from social activities.[5]

Rather than merely focusing on whether or not the autonomy of carriers is being respected (Beauchamp and Childress 2001), by conceptualizing the potential impact of MDRO control measures in terms of capabilities and functionings, we get a broader picture of the many ways in which carriers are affected. In other words, it allows us to move from a singular basis for evaluation, namely in terms of their autonomy, to a multi-dimensional one. The same carrier may be impacted in many different ways by a particular control measure. For example, restricting a resident of a nursing home from participating in the weekly bingo nights not only restricts her capability for participating in social activities, but may also take away an important source of pleasure and happiness or may even lead to stigmatization. Likewise, a particular control measure may impact different carriers in different ways. For example, a child who is at a crucial stage in her social and cognitive development would arguably be negatively affected to a greater extent from being taken out of daycare (even for a short period of time) than a child who is not in this crucial stage of development (Piaget 1971).

Moreover, the capability perspective gives substance to carriers' autonomy: it allows us to identify in which ways MDRO control measures have the potential to (negatively) impact the capabilities of (potential) carriers. We are not merely concerned with the limitation of options that carriers can choose from. Rather, the capability perspective tells us that carriers are concerned with *particular* opportunities for choice (Sen 1991), such as access to day care centers, nursing homes, and physiotherapy; participation in social and leisure activities; opportunities for education and employment; freedom from stigmatization and discrimination; and possibilities

---

[5] That is, there might still be good normative reasons to override this concern for carriers' capabilities, for example out of a concern for the public health. While we do not engage with the discussion on how to weigh the violation of carriers' capabilities against concerns for public health in this paper, do see Sect. 13.6 for an example of how such weighing can take place within an open-ended decision-making framework. The point here is, rather, that there are certain capabilities that are so normatively relevant that we should take them into consideration when deciding on appropriate MDRO control measures – again, even if we do not consider them to have overriding normative status.

for forming and sustaining relationships to friends, family, and pets. Indeed, by employing the capability approach to conceptualize the 'harm' done by MDRO control measures to individual carriers we gain a greater, more in-depth, and more specific understanding of this impact.

As noted, this focus on particular opportunities for choice – rather than freedom or autonomy, in general – moves the discussion away from the singular dichotomy between public health versus the freedom of the individual carrier. The restriction of freedom is not necessarily a bad thing on the capabilitarian view.[6] The restriction of an individual carrier's freedom out of concern for public health is perfectly compatible with the protection of her valued capabilities. What the capability perspective does highlight, though, is that the restriction of some freedoms and opportunities, however *prima facie* insignificant, may affect capabilities that we do find normatively valuable. For example, restricting an MRSA positive child from attending kindergarten for just a few months may not seem like a big deal. However, that restriction may negatively affect a normatively crucial aspect of a person's life, namely the opportunity for a normal social, cognitive, and physical development if the MRSA positive child were, at the moment of isolation, at a crucial stage of her development.

To see how MDRO control measures can (negatively) affect the lives of MDRO carriers, it is crucial to identify what capabilities and functionings that are normatively relevant for carriers in the context of MDRO. In the following two sections, we present a taxonomy of capabilities and functionings that may be normatively relevant when deciding on appropriate control measures. This taxonomy builds on one prominent instantiation of the capability approach, namely Nussbaum's list of ten central capabilities. We first discuss Nussbaum's list in Sect. 13.4 and argue that it needs further adaptation and specification when applied to the context of deciding on appropriate MDRO control measures before we explain the taxonomy in greater detail in Sect. 13.5. In Sect. 13.6, we finally show how this capabilitarian taxonomy can help us make better and more informed decisions when deciding on appropriate MDRO control measures.

13.4 Nussbaum's Ten Central Capabilities: A Starting Point

What capabilities should we be concerned about protecting when implementing certain measures to prevent the spread of MDRO?[7] A good starting point is Nussbaum's influential list of ten central capabilities that, she argues, every

---

[6] See, though, Carter (2014) for a dissenting view.

[7] We have employed what Byskov (forthcoming) refers to as a *synthesizing method* to identify the relevant capabilities. Synthesizing methods compare and reconcile two or more lists of capabilities derived from different theoretical and empirical sources. We have here reconciled Nussbaum's (2000) list of central human capabilities with (i) other lists of relevant normatively domains in healthcare literature, such as Entwistle and Watt (2013) and Huber et al. (2016), (ii) empirical

government should provide for their citizens. While Nussbaum's list is thus derived from a discussion on global justice, it can nevertheless be useful for conceptualizing what kinds of capabilities that are important because it helps us to identify how well-off individuals truly are.[8] The most influential version of Nussbaum's list of capabilities can be found in her book *Women and Human Development* (Nussbaum 2000)[9]:

1. *Life:* Ability to live to the end of a normal length human life, and not to have one's life reduced to not worth living.
2. *Bodily health:* Ability to have a good life, which includes – but is not limited to – reproductive health, nourishment, and shelter.
3. *Bodily integrity:* Ability to change locations freely, in addition to having sovereignty over one's body, which includes being secure against assault (e.g., sexual assault, child abuse, and domestic violence) and the opportunity for sexual satisfaction.
4. *Senses, imagination, and thought:* Ability to use one's senses to imagine, think, and reason in a 'truly human way' informed by an adequate education. The ability to produce self-expressive works and engage in religious rituals without fear of political ramifications. The ability to have pleasurable experiences and avoid unnecessary pain. Finally, the ability to seek the meaning of life.
5. *Emotions:* Ability to have attachments of things outside of ourselves, including being able to love others, grieve at the loss of loved ones, and be angry when it is justified.
6. *Practical reason:* Ability to form a conception of the good and critically reflect on it.
7. *Affiliation:*

    (a) Ability to live with and show concern for others, empathize with and show compassion for others, and the capability of justice and friendship. Institutions help develop and protect forms of affiliation.
    (b) Ability to have self-respect and not be humiliated by others (i.e., being treated with dignity and equal worth). This entails at least protections from being discriminated on the basis of race, sex, sexuality, religion, caste, ethnicity, and nationality. In work, this means entering relationships of mutual recognition.

---

analysis of a database of ethical and practical questions concerning MDRO raised within the Dutch healthcare system as well as (iii) participatory case discussions with practitioners working with infectious disease control. For overviews of the various methods for the selection of capabilities, see Ballon (2013) and Byskov (forthcoming).

[8] Several scholars have taken build on Nussbaum's list and made changes to it, as necessary, when applied in practice (e.g., Alkire 2002). Thus, to be clear, we do not take Nussbaum's list at face value but rather hold that we can compare and specify this list to the particular case of MDRO.

[9] See also Nussbaum (1992, 2011) for similar iterations of her list, albeit based on different normative justifications.

8. *Other species:* Ability to have concern for and live with other animals, plants, and the environment at large.
9. *Play:* Ability to laugh, play, and enjoy recreational activities.
10. *Control over one's environment:*

    (a) Political: Ability to effectively participate in the political life, including having the right to free speech and association.
    (b) Material: Ability to own property, not just formally but materially. Furthermore, having the ability to seek employment on an equal basis and the freedom from unwarranted search and seizure.

Nussbaum's list of central human capabilities provides a good starting point for our attempt to identify what capabilities and functionings that are relevant for evaluating the extent to which MDRO control measures excessively interfere with the lives of MDRO carriers. However, when Nussbaum specifies a list of capabilities she is not concerned with the case of MDRO control measures and the well-being of individual carriers of MDRO but rather with setting out a partial theory of justice. For this reason, when adapting Nussbaum's list to the context of MDRO carriership, we still need to ask (a) whether all items on her list are relevant and (b) to what extent they need to be further specified and/or supplemented by additional capabilities.

First of all, while some of the items on Nussbaum's list may also be relevant for the evaluation MDRO control measures other capabilities are clearly not applicable. For example, while the capability for bodily integrity, seems to be of utmost importance for this discussion, the capability for senses, imagination and thought do not seem to be at stake here. The reason for this is not that being able to use one's senses, imagination, or thoughts are not important human characteristics. Rather, the reason that these capabilities are of little importance in the context of MDRO is that it can be argued that there are no control measures that have the potential to restrict one's use of the senses, imagination, and thoughts. Likewise, it is questionable whether the capability for practical reason – one's ability to form a conception of the good and critically reflect on it – would be thwarted or under threat by any conceivable measure we can take to control MDRO. (However, do note that we suggest to subsume (and expand) the capability for education, which is part of the capability for practical reason, under the capability for life as the capability for proper social, physical, and cognitive development.)

This leaves us the following list of capabilities that we can tentatively assume are relevant for the context of MDRO: Life, bodily health, bodily integrity, affiliation (in both senses), other species, play, and control over one's environment (in both senses). Now, we still need to ask whether these seven items are sufficient for our present purpose. This is so in two ways. We need to ask, first, whether these seven capabilities are comprehensive in the sense that we do not need to add additional capabilities and, second, whether they are sufficiently specified to capture what is at stake in the context of MDRO.

In the first case, are the seven capabilities that we retain from Nussbaum's list sufficient to capture all relevant ethical aspects of the context of MDRO? Do we need to add any further capabilities? In order to answer this, let us first distinguish between Nussbaum's *categories* of capabilities and the more *specific* capabilities that are included within the categories. Thus, for example, the category of 'bodily health' includes the more specific capabilities of adequate health, nourishment, shelter, and reproductive health. Though all of these specific capabilities support the more general categorical capability of bodily health, they can neither be reduced to each other nor to the general category. In other words, the more specific capabilities are distinct capabilities in themselves.

Are the seven categories of capabilities sufficient to capture all ethical aspects of MDRO? In general, the categories on Nussbaum's list seem comprehensive. However, it may be helpful to distinguish carriers' mental well-being from Nussbaum's category of bodily health. Many of the MDRO control measures have little impact on one's physical or bodily health. Even decreases in bodily health – for example, the displeasure caused by eradication therapies – are only temporary. The mental impact, however, may be just as profound and long lasting. Being subject to isolation measures, for instance, is known to increase the levels of perceived stress and anxiety and the stigma of having been a carrier can continue long after carriage has ceased. Thus, the mental impact of MDRO control measures can and should be seen independently from their physical impact. Let us therefore add an additional category, namely *mental health*.

How about the more specific capabilities on Nussbaum's list? Does Nussbaum identify all relevant capabilities to adequately capture what is at stake within the seven general categories in relation to the context of MDRO? Given the particular focus of her own list, Nussbaum naturally leaves off many capabilities that are relevant in the context of MDRO. For example, when deciding on how to treat children and adolescents, a major concern is how the control measure affects their physical and mental development. Prolonged isolation of children in certain age groups may cause setbacks in speech or reading that will disadvantage them later in life. Moreover, Nussbaum does not explicitly address concerns related to healthcare, such as access to timely and effective treatment and protection against intrusive and excessive examinations and therapy.

Nor are Nussbaum's capabilities sufficiently specified to the context of MDRO carriership and MDRO control measures. For example, the way Nussbaum defines sovereignty over one's body (a part of bodily integrity) seems overly abstract. In the context of MDRO, what we mean by bodily integrity and sovereignty concerns not being subjected to unnecessary, intrusive, or excessive examinations and eradication therapies. Likewise, since a large issue in relation to MDRO is how it might contribute to the stigmatization of carriers (Rump et al. 2015), we need to include protection from stigmatization along with the protection from discrimination (a part of the capability for affiliation).

Thus, we can also answer the second question that we asked above, namely whether Nussbaum's capabilities are sufficiently and adequately specified to capture the context of MDRO. There are good reasons to argue that Nussbaum's list needs to be further specified and supplemented with additional capabilities when setting out a taxonomy of normatively relevant capabilities and functionings in the context of how MDRO control measures and carriership may affect the lives and well-being of carriers.

In sum, although Nussbaum's list of ten central capabilities provides a useful starting point for identifying the normatively relevant aspects of how MDRO control measures have the potential to impact the lives of individuals, it still needs to be adapted and specified to this particular context. This is so in several ways: first, some capabilities on Nussbaum's list are irrelevant for the case of MDRO; secondly, Nussbaum's list does not distinguish all relevant capabilities, such as mental health; and, thirdly, Nussbaum's capabilities must be specified to the context of MDRO. In the following section, we proceed to present a taxonomy of normatively relevant capabilities and functionings that we need to take into consideration when deciding on and evaluating MDRO control measures.

13.5   A Taxonomy of Normatively Relevant Capabilities in the Context of Addressing MDRO Carriership

What does a taxonomy of normatively relevant capabilities and functionings look like in the context of MDRO? How can it help us understand what is at stake when deciding on measures to contain the spread of MDRO? Building on Nussbaum's list of central human capabilities, in this section we present a taxonomy that adapts and specifies Nussbaum's list to the particular context of assessing and evaluating MDRO control measures. The taxonomy supplements Nussbaum's list through an analysis of empirical literature and studies on what practitioners and MDRO carriers express as normatively relevant and divides the relevant capabilities into four ethical domains.

Table 13.1 presents a systematic overview of how MDRO control measures can potentially affect the lives and opportunities of individual carriers. The table is divided into three columns, which, from left to right, moves from four general domains of human life (the personal, the social, the institutional, and the environmental) to the more specific capabilities and functionings that are relevant in the context of MDRO.

On the right-hand side of the table, we find a list of the various capabilities and functionings that are (a) normatively relevant for living a decent or flourishing human life, as revealed by Nussbaum's list of capabilities, and (b) specifically relevant within the context of MDRO, as revealed by our empirical analyses.

In the first case, capabilities such as nourishment, shelter, the right to association, and being treated with dignity and equal worth are relevant for human life regardless of whether it involves MDRO or not. In the second case, there are capabilities that only or primarily come become relevant when combined with MDRO, such as protection against stigmatization and pathologization, protection against unnecessary or intrusive examinations and therapy, and the ability to engage in recreational activities.

Table 13.1 Ethical domains and normatively relevant capabilities and functionings for the evaluation of the impact of MDRO control measures on (potential) carriers, partly adapted from Nussbaum (2000). Additional items and specifications in **bold**

Domain	Category	Specific capability
Personal	Life	Not having one's life reduced to not worth living **(especially for elderly)**
		Proper social, physical, and cognitive development (especially for children and adolescents)
	Bodily health	Adequate health **(e.g., to fight off infections)**
		Nourishment
		Shelter
		Reproductive health
	Mental health	**Happiness and peace of mind**
		Self-respect **and self-esteem (e.g., being able not to see oneself as sick or as merely a patient)**
		Protection against internalized pathologization
		Future prospects (e.g., of a speedy recovery)
	Bodily integrity	Sovereignty over one's body **(e.g., not being subjected to unnecessary, intrusive, or excessively costly examinations and intensive eradication therapy)**
		Appearance (i.e., being able to appear in public without shame)
		Freedom of choice and opportunity, both in life and in relation to one's body
		Choice in matters of reproduction
		Protection against internalized pathologization
	Play	Ability to enjoy social and recreational activities

(continued)

Table 13.1 (continued)

Domain		Category		Specific capability
Social and community		Bodily integrity		Protection against assault **(e.g., not being seen as merely a threat), also in the case of relatives of carriers**
		Affiliation	A	Ability to live with others
				Friendship
				Family (incl. reproductive rights)
			B	Being treated with equal dignity and respect
				Social status and prestige
		Play		Ability to **engage and participate in** social and recreational activities
		Control over one's environment A		Right to association
				Ability to form and engage in social relations
Institutional	**(Health) care related**	Bodily health		**Access to adequate (i.e., timely and effective) health care**
		Bodily integrity		Security against assault **in the form of intrusive and excessive examinations and eradication therapy**
		Control over one's environment	A	**Access to just and fair healthcare treatment**
			B	Freedom from unwarranted search and seizure **(e.g., having to pay oneself for excessively expensive examinations)**
		Affiliation	A	Institutions help develop and protect forms of affiliation, **self-respect, and dignity**
			B	Being treated with dignity and equal worth
				Protection from discrimination **and stigmatization**
	Public life	Control over one's environment	A	Ability to effectively participate in political **and public** life (incl. free speech and association)
				Access to just and fair institutions
			B	Ability to seek employment on an equal basis
				Decent working environment (incl. protection against discrimination and abuse)
				Freedom from unwarranted search and seizure **(e.g., having to pay oneself for excessively expensive examinations)**
				Economic security

(continued)

Table 13.1 (continued)

Domain	Category	Specific capability
Environmental	Bodily integrity	Ability to change locations freely/**freedom of movement (incl. the ability to live where one chooses to)**
	Other species	Ability to live with other animals (**i.e., pets and livestock**), plants (**incl. crops**), and the environment at large
	Control over one's environment B	Ability to own property (**e.g., livestock**)

The eight categories of capabilities can, in turn, be relevant within one or more of four domains of human life, identified on the left-hand side of the table: the personal, the social, the institutional, and the environmental. Within the personal domain, MDRO control measures influence the relationship that a carrier has to herself, her own body, and her mental satisfaction. In particular, this includes her bodily health, in the sense of being healthy, well-nourished, and having access to adequate accommodation, her mental health, including being happy, feeling dignified, and being free from stigmatization, and her bodily integrity, most importantly not being subject to excessive and intrusive examinations and eradication therapies.

The social domain concerns individual carriers' relationships to friends and family and the ability to participate in social activities. Human well-being to a large degree depends on well-functioning social relationships, both instrumentally and intrinsically. Not only do we count on friends and family to help us realize certain ends and goals in life; we also attribute intrinsic value to social relationship: we engage in and enjoy social relationships for their own sake and not because they help us fulfill personal goals.

Hence, whenever our social relationships break down it is likely to harm our well-being. Social relationships are especially vulnerable to MDRO and MDRO control measures. Stigmatization and pathologization are social mechanisms by which we respond to perceived threats. In this way, the case of MDRO has a lot in common with the plight of AIDS carriers in the 1980s. However, stigmatization and pathologization are only two ways in which MDRO measures can harm our well-being in a social context. More generally, since we derive pleasure from engaging in social relationships, MDRO measures that restrict the extent to which we can engage in social relations have the potential to lead to a decrease in our well-being.

On the institutional level, we are interested in the carriers' relationship to and standing within institutions, primarily (but not limited to) health care facilities. MDRO is primarily an issue when it comes into contact with a healthcare setting. That is, MDRO is primarily a risk when it comes into contact with already vulnerable individuals who depend on effective antimicrobial treatments for their health and survival. Such individuals are more often found within care facilities, such as hospitals, nursing homes, and rehabilitation centers. Moreover, healthcare settings also provide more fertile breeding grounds for the emergence of multidrug-resistant organism because of the increased exposure to antimicrobials and, hence, the risk that organisms will evolve resistance to these antimicrobials.

Within the environmental domain, we are primarily concerned with carriers' relations with their environments. To what extent, we ask, are carriers able to connect with their environment? Are they able to exercise any control over their environment? We can talk about a person's relationship to their environment in both literal and figurative terms. Literally, we talk about the environment as something that is *there*: a physical presence that we can interact with and influence. In this sense, our relationship with the environment concerns our ability to interact with physical entities such as plants, including flowers, trees, fungi, and so on, as well as animals, including both pets and livestock. In a figurative sense, the environment is a more abstract and indefinite entity. This is so in two ways. First, we can talk about the environment at large, including in the senses of nature and the climate without referring to specific plants or animals. This way of understanding the environment is of little relevance to the context of MDRO. However, secondly, the environment can also be understood as the indefinite but physical space that surrounds us and which we can move around within. In other words, in this sense we understand one's environment as something within which she (can) has control over herself and her choices. Given that two of the primary MDRO control measures – quarantine and isolation – aim to restrict (potential) carriers' ability to move around, this second figurative understanding of the environment is highly relevant to the context of MDRO.

Crucially, a category of capabilities can be specified differently within different domains. For example, the capability for control over one's environment in the context of the social domain concerns one's right and freedom to form social relationships, while in the institutional domain it rather concerns one's institutional status, such as the freedom to participate in political and public life and access to just and fair institutions. Thus, although the different categories of capabilities can be relevant within different domains, the more specific capabilities that they contain depend on the domain.

It is important to stress that the taxonomy here does not make any claims about which capabilities and functionings that cannot be violated by MDRO control measures. Rather, it provides a structural overview of how MDRO control measures may affect the lives and opportunities of individual carriers. We still need to engage in a weighing of the relevant capabilities and functionings in individual cases in order to determine whether they provide overriding normative reasons not to implement a particular control measure. Such weighing would take place on a case-by-case basis because each case includes contextual circumstances that influence what the best course of action would be. Hence, it is not possible to *a priori* determine what control measure (if any) to implement.

However, by offering a taxonomy of relevant domains and capabilities we do make a claim about what is normatively important and relevant when addressing MDRO. First, as argued in Sect. 13.3, MDRO control measures affect carriers in terms of their capability-sets – what they have the real freedom to do or be. Hence, it is claimed that we ought to conceptualize and describe the impact that MDRO control measures have on individual carriers in terms of capabilities and functionings.

Secondly, however incomplete and underspecified, we make a claim about the kinds of capabilities and functionings that are normatively important for (potential) carriers of MDRO and which should be taken into account when deciding on the best course of action. That is, there are good reasons to claim that *these* particular capabilities have the potential to be normatively relevant when dealing with cases of MDRO. There are both normative and empirical reasons for this claim. Normatively speaking, Nussbaum's list of capabilities provides a normative philosophical grounding of the capabilities: these are capabilities that can be subject to an overlapping consensus. Moreover, there is empirical evidence that (some or most of) these capabilities are of relevance to practitioners and carriers when dealing with cases of MDRO in a healthcare setting. The comparison with real-life queries about how to ethically address MDRO – as represented by our database and deliberations with carriers and practitioners[10] – provide empirical basis for the claim that these are the kinds of capabilities that are of concern when deciding on control measures.

How can this taxonomy be implemented in practice to analyze particular cases of MDRO and decide on appropriate control measures? In the final section, we show how our taxonomy can provide an input into ethical decision-making procedures on the appropriate measure to address MDRO carriership.

13.6   Applying the Capabilitarian Taxonomy in Practice

We have in the previous section repeatedly argued that the more general categories of capabilities that Nussbaum identifies – life, bodily health, mental health, bodily integrity, play, affiliation, control over one's environment, and other species – can and should be specified to the particular context of how MDRO control measures impact the lives and freedoms of individual carriers. We further argued that we can and should specify these categories of capabilities differently according to whether they relate to either of four domains of human life, namely the personal, the social, the institutional, and the environmental. While the taxonomy that we have presented in Table 13.1 provides an overview of how the different categories of capabilities can be specified in relation to the different domains, how it contributes to the practice of implementing appropriate MDRO control measures is still unclear.

The above taxonomy can provide a useful input to ethical decision-making procedures on the implementation of MDRO control measures, such as the frameworks developed by Verweij et al. (2012; Krom 2014) or Grill and Dawson (2015). How does the capabilitarian framework help us make decisions about how to address MDRO? How can our taxonomy help us make better and more informed decisions about what kinds of MDRO control measures that are preferable, acceptable, or justifiable? In this section, we briefly consider how the capabilitarian taxonomy can be applied in practice to ethical deliberations on MDRO and what issues that are left unaddressed.

---

[10] See footnote 7.

The capabilitarian taxonomy presented in this chapter is especially useful in two regards. First, it can help professionals better describe cases of MDRO by making explicit what is at stake for the individual carriers and relevant stakeholders. Secondly, it can help us to identify and evaluate possible courses of action by showing how various MDRO control measures may impact the capabilities and functionings of affected carriers and stakeholders. Let us, by way of a case study, briefly show how the taxonomy of normatively relevant capabilities can be put into practice in these two ways.

To illustrate how the capabilitarian taxonomy can be applied in practice, consider, for example, the case of a young, 19-year-old mother with no income or higher education who shares a household with her own mother. The father of her newborn child is unknown or absent and the woman therefore relies on her own mother for economic assistance and care help. However, the grandmother of the child turns out to be MDRO positive and there are concerns that she is a threat to the health of the newborn child. If there is close contact, it is very likely that the grandmother would transmit the resistant organism to the newborn. To make matters worse, the child in case has a heart valve condition and needs to go to the hospital for regular check-ups. Because of the likelihood that the child will become an MDRO carrier if the grandmother is involved in the post-partum care of the child, the hospital insists that the grandmother cannot provide this care or that she should take far-reaching protection measures, such as wearing gowns and masks, that would interfere with the bond between child and grandmother.

How can our taxonomy contribute to the understanding and resolution of this case? What capabilities are at stake in this case? While this case involves a lot of different capabilities within several domains, the primary concern here is the ability to live with others, including family (part 'affiliation A' within the social domain). This capability is restricted not only for the young mother but also for the newborn child as well as the grandmother. However, although we can assume that they value this relationship intrinsically – and hence contributes to the capability of happiness and peace of mind (a part of the capability for 'mental health' within the personal domain) – in this case there are at least two instrumental reasons why this capability is important.

First of all, bonding with relatives may be considered an important part of a child's development (a part of the capability for 'life') and restricting the newborn child's relationship to the grandmother risks harming this development. Secondly, in this case, the social relationship between the young mother and her own mother can also be seen as a proxy for more formal institutional care-relationships. That is, not allowing the grandmother to care for the newborn child is mainly problematic insofar as the mother does not have alternative opportunities for care assistance. Within the taxonomy this is represented by the institutional capabilities for 'bodily health' (access to timely and effective healthcare) and 'affiliation A' (institutions help develop and protect forms of affiliation).

In the described case we have multiple courses of action, which can be employed either independently or in conjunction. How can the capabilitarian taxonomy help us identify and evaluate possible measures? Some of the measures would be directed solely or primarily at the grandmother. First of all, in order to minimize contamination, we could demand that the grandmother undergoes eradication therapy and subsequently attends regular screenings. Secondly, we could demand that the grandmother adhere to a strict hygiene regiment, including the donning of a gown, mask, and gloves when tending to the child. Other measures would be directed at the other stakeholders, in particular the newborn child. For example, thirdly, we could subject the newborn to regular screenings to test for MDRO and, when positive, to eradication therapy.

However, as the case describes, these measures have potentially negative consequences for not only the mother's abilities to engage in social relations and to care for her child, but also the child's well-being, especially in relation to her early childhood development as well as her capability to form an affiliation with her grandmother. From a capability perspective, then, we would do well to look for alternative courses of action that provide better protection of these normatively valuable capabilities.

The analysis of the case from the capability perspective shows that a major issue is that the mother is reliant on *informal care* for her child. Informal care is – usually unpaid – care that is provided by family members or social relations. In contrast, formal care is institutionalized and usually performed by trained professionals. By applying the taxonomy, our analysis shows that the case extends beyond the personal and social domains to reveal a lack of normatively relevant capabilities and functionings at within institutional domain. In the present case, then, a possible solution to the issue could be to increase the access to *formal institutional* healthcare for the young mother and her infant, so that she does not have to rely so much on informal care, thereby avoiding many of the negative consequences that follow from limiting the analysis to focus solely on the informal care-relation between the child, the mother, and the grandmother.[11]

While the proposed course of action in this case might be intuitively clear, in general, a major issue of applying the capabilitarian taxonomy in practice, especially when evaluating the various MDRO control measures, concerns the question of how to weigh different capabilities against each other. That is, we need to ask, when does the reduction of a carrier's capability-set provide an overriding reason to dismiss or provide compensation for a particular control measure? When applying the capabilitarian taxonomy in practice to evaluate different control measures, we should weigh capabilities on at least three levels, namely the intrapersonal, the interpersonal, and the public health level.

The first level at which we need to weight the importance or value of different capabilities against each other when evaluating potential MDRO control measures is at the intrapersonal level. At the intrapersonal level, we ask whether a person is better off within one scenario *as compared to other scenarios*. That is, we can ask, is the person's capability-set more valuable as a result of a particular control

---

[11] A possible objection, also based on the capabilitarian taxonomy, to the proposed solution is that it would negatively affect the grandmother's capability for affiliation with her grandchild.

measure (or combination of control measures) than it would be if we implement another (combination of) control measure(s)?

At the interpersonal level, second, we are concerned with comparing the capability-sets of different stakeholders *within one particular scenario*. Here we should ask: does a particular control measure diminish the value of the capability sets of one or more of the relevant stakeholders to the extent that it outweighs the positive impact on the value of the capability sets of other relevant stakeholders?[12] Finally, thirdly, we should weigh the positive or negative impact to the value of the capability-sets of individual stakeholders against the estimated benefit to public health that the implementation of a particular MDRO control measure has.

It is beyond the scope of this chapter to consider how such a weighing may be done. Since this decision must eventually be made on a case-by-case basis, it must be a subject for further research to set out normative (or pragmatic) principles for the weighing of capabilities. Such principles might include a threshold level of capabilities and functionings: do we really need to compensate someone for a lost job-opportunity if she already has ample opportunity to find alternative employment? Other principles are principles of proportionality and acceptable risk, that can help us determine when a particular MDRO control measure is (dis)proportionate to the harm, conceptualized in terms of capabilities and functionings, that it does to the individual carrier. In this regard, possible connections could, for example, be made between our capabilitarian taxonomy and the approaches of Viens et al. (2009), who set out a principle of reciprocity, Krom (2011), who discusses the shortcomings of the harm principle in infectious disease control, and Grill and Dawson (2015) who propose a value-based approach.

Moreover, the weighing of capabilities and capability-sets should be done in consultation with the relevant stakeholders in order to identify relevant capabilities and their normative weight. This leaves quite a bit of space for professional autonomy in ethical decision-making. It is simply quite impossible *a priori* to determine the normatively relevant capabilities and their relative, normative weight. In this regard, the taxonomy of normatively relevant capabilities and functionings presented in this chapter should be taken as an open-ended and underspecified basis for further deliberation between the various stakeholders (carriers, relatives, professionals, and possibly policy-makers) on a case-by-case basis. Again, how much room to leave for professional autonomy and how exactly to conduct such deliberative exercises must be subject to further research. We have here proposed two promising frameworks for ethical decision-making, namely Verweij et al. (2012; Krom 2014) or Grill and Dawson (2015).

Finally, it might be objected that, while intuitively attractive, the capabilitarian taxonomy presented in this chapter does not add to professional practice on MDRO. That is, it is not clear that the capability perspective would change what professionals already do when addressing cases of MDRO. This objection holds

---

[12] Conversely, this could also be framed as: does a particular control measure increase the value of the capability sets of one or more of the relevant stakeholders to the extent that it outweighs the negative impact on the value of the capability sets of other relevant stakeholders?

that one of the supposed advantages of adopting the taxonomy of capabilities, namely its intuitive appeal, at the same time makes the contribution of this chapter trivial. However, even if the taxonomy largely corresponds to existing practice, there are at least four benefits to making the tacit or implicit assumptions of professional practice explicit through the language of the capability approach and the taxonomy that has been presented in this chapter.

First of all, the taxonomy provides a substantive – yet underspecified and open-ended – view of carriers' well-being. Rather than a person's autonomy, generally speaking, we are, on this view, concerned with protecting carriers' normatively relevant or valuable capabilities and functionings. This allows us, secondly, to provide a structured way of discussing how MDRO control measures impact the lives and well-being of carriers, namely by influencing their normatively relevant or valuable doings and beings. In this sense, the proposed taxonomy could serve as a basis for consultation among professionals and with relevant stakeholders. Third, the capability view presents a multi-dimensional view of the impact that MDRO control measures can have. Different individuals might be impacted in different ways by similar control measures and one individual might be affected in many different ways by a particular control measure. Fourth, by conceptualizing the impact of MDRO control measures in terms of people's capabilities and functionings, it possible to see how affecting one aspect of an individual's life may affect other, less immediately obvious, capabilities and functionings.

In sum, the capability framework to MDRO and the accompanying taxonomy of normatively relevant capabilities does contribute to both the literature on the ethical aspects of MDRO as well as, potentially, to real practice of addressing cases of MDRO.

13.7   Concluding Remarks

In this chapter, we have presented and discussed a capabilitarian conceptualization of how MDRO control measures can (negatively) impact the lives and well-being of individual MDRO carriers. According to the capability approach, we should measure and evaluate this impact in terms of how MDRO control measures (negatively) influence what they are able to do or be. Building on Nussbaum list of central human capabilities, we introduced a taxonomy of normatively relevant capabilities and functionings in the context of MDRO. This taxonomy proposes that measures to contain the spread of MDRO may potentially affect carriers in one or more of four domains human life, namely the personal, social, institutional, and environmental domains. We identified eight categories of capabilities – life, bodily health, mental health, bodily integrity, affiliation, other species, play, and control over one's environment – that can and should be specified differently within the four domains of human life when applied to the context of analyzing how MDRO control measures impact the lives of individual carriers. An overview of this taxonomy can be found in Table 13.1.

The taxonomy, we finally argued, should be used as an ethical input to a decision-making framework when deciding on the best measures to take when dealing with cases of MDRO. As such, the taxonomy is both underspecified and open-ended: it still needs to be expanded and adapted when applied to particular, individual cases of MDRO. It does not, by itself, determine when the infringement of a particular capability or range of capabilities is unjust and should be supplemented with a notion of when the reduction of a carrier's capability-set provides an overriding reason to dismiss a particular control measure.

References

Alkire, S. 2002. *Valuing Freedoms*. Oxford: Oxford University Press.
Anomaly, J. 2010. Combating resistance: The case for a global antibiotics treaty. *Public Health Ethics* 3: 13–22.
———. 2013. Collective action and individual choice: Rethinking how we regulate narcotics and antibiotics. *Journal of Medical Ethics* 39: 752–756.
Ballon, P. 2013. The selection of functionings and capabilities: A survey of empirical studies. *Partnership for Economic Policy (PEP)*.
Beauchamp, T.L., and J.F. Childress. 2001. *Principles of biomedical ethics*. Oxford: Oxford University Press.
Byskov, M.F. forthcoming. Methods for the selection of capabilities and functionings. In *New frontiers of the capability approach*, ed. F. Comim, S. Fennell, and P.B. Anand. Cambridge: Cambridge University Press.
Carter, I. 2014. Is the capability approach paternalist? *Economics and Philosophy* 30: 75–98.
Coleman, C.H., Organization WH, E. Jaramillo, et al. 2010. *Guidance on ethics of tuberculosis prevention, care and control*. Geneva: World Health Organization.
Crocker, D.A., and I. Robeyns. 2010. Capability and agency. In *Amartya Sen*, ed. C. Morris, 60–90. Cambridge: Cambridge University Press.
Daulaire, N., A. Bang, G. Tomson, et al. 2015. Universal access to effective antibiotics is essential for tackling antibiotic resistance. *The Journal of Law, Medicine & Ethics* 43: 17–21.
Entwistle, V.A., and I.S. Watt. 2013. Treating patients as persons: A capabilities approach to support delivery of person-centered care. *The American Journal of Bioethics* 13: 29–39.
Grill, K., and A. Dawson. 2015. Ethical frameworks in public health decision-making: Defending a value-based and pluralist approach. *Health Care Analysis*: 1–17.
Huber, M., M. van Vliet, M. Giezenberg, et al. 2016. Towards a "patient-centred" operationalisation of the new dynamic concept of health: A mixed methods study. *BMJ Open* 6: 1–11.
Krom, A. 2011. The harm principle as a mid-level principle? Three problems from the context of infectious disease control. *Bioethics* 25: 437–444.
———. 2014. *Not to be sneezed at. On the possibility of justifying infectious disease control by appealing to a mid-level harm principle*. Utrecht University.
Lamont, J., and C. Favor. 2016, Winter. Distributive justice. In *The stanford encyclopedia of philosophy*, ed. E.N. Zalta.
Littmann, J. 2014. *Antimicrobial resistance and distributive justice*. Doctoral, UCL (University College London).
Littmann, J., and A.M. Viens. 2015. The ethical significance of antimicrobial resistance. *Public Health Ethics*: phv025.
Littmann, J., A. Buyx, and O. Cars. 2015. Antibiotic resistance: An ethical challenge. *International Journal of Antimicrobial Agents* 46: 359–361.
Nielsen, L. 2014. Why health matters to justice: A capability theory perspective. *Ethical Theory and Moral Practice*: 1–13.

Nussbaum, M. 1992. Human functioning and social justice: In defence of Aristotelian essentialism. *Polit Theory* 20: 202–246.

———. 2000. *Women and human development. The capabilities approach.* Cambridge: Cambridge University Press.

———. 2011. *Creating capabilities. The human development approach.* Cambridge, MA: Belknap Press of Harvard University Press.

Piaget, J. 1971. The theory of stages in cognitive development. In *Measurement and Piaget*, ix, 283. New York: McGraw-Hill.

Prah Ruger, J. 2010. Health capability: Conceptualization and operationalization. *American Journal of Public Health* 100: 41–49.

Robeyns, I. 2016a, Winter. The capability approach. In *The Stanford Encyclopedia of philosophy*, ed. E.N. Zalta. Metaphysics Research Lab, Stanford University.

———. 2016b. Capabilitarianism. *Journal of Human Development and Capabilities* 17: 397–414.

Rump, B. 2011. Stoppen met de studie geneeskunde omwille van MRSA-dragerschap? *Ethiek Infect 19.*

Rump B., M. De Boer, R. Reis, et al. 2015. *Signs of stigma and poor mental health among carriers of MRSA.*

Rump, B., C. Kessler, E. Fanoy, et al. 2016. Case 2: Exceptions to national MRSA prevention policy for a medical resident with untreatable MRSA colonization. In *Public health ethics: Cases spanning the globe*, ed. D.H. Barrett, L.W. Ortmann, A. Dawson, et al., 191–194. Berlin: Springer.

Selgelid, M.J. 2007. Ethics and drug resistance. *Bioethics* 21: 218–229.

Selgelid, M.J., A.R. McLean, N. Arinaminpathy, and J. Savulescu. 2009. Infectious disease ethics: Limiting liberty in contexts of contagion. *Journal of Bioethical Inquiry* 6: 149–152.

Sen, A. 1979. Equality of what? *Tann Lect Hum Values*: 197–220.

———. 1991. Welfare, preference and freedom. *Journal of Econometrics* 50: 15–29.

Venkatapuram, S. 2011. *Health justice. An argument from the capability approach.* Cambridge: Polity Press.

Verweij, M., and A. Dawson. 2010. Shutting up infected houses: Infectious disease control, past and present. *Public Health Ethics*: 1–3.

Verweij, M., A. Krom, and J. Van Steenbergen. 2012. Ethische kwesties in de infectieziektebestrijding. Rijksinstituut voor Volksgezondheid and Milieu.

Viens, A.M., C.M. Bensimon, and R.E.G. Upshur. 2009. Your liberty or your life: Reciprocity in the use of restrictive measures in contexts of contagion. *Journal of Bioethical Inquiry* 6: 207–217.

WHO. 2014. *Antimicrobial resistance global report on surveillance: 2014 summary.* Geneva: World Health Organization.

Chapter 14
A Capability Perspective on Antibiotic Resistance, Inequality, and Child Development

Michael Millar

Abstract Nussbaum's capability theory by drawing attention to multiple determinants of wellbeing provides a rich and relevant evaluative space for framing antibiotic resistance. I consider the implications of antibiotic resistance for child development and adult capabilities. There are common risk factors for childhood growth stunting and the spread of infectious diseases in both antibiotic sensitive and resistant forms. The interaction between infectious diseases, antibiotic resistance and growth stunting illustrates a clustering of disadvantage. The control of antibiotic resistance requires wide-ranging cooperative action. Cooperation is predicated on an expectation of equitable access to effective antibiotics. This expectation is confounded by inequality both in access to antibiotics, and in the risk that available antibiotics will be ineffective. Securing child development (and adult capabilities) requires that inequalities both in access to antibiotics and in risk factors for the dissemination and transmission of antibiotic resistance are addressed. Inequality undermines the cooperative activity that is control of infectious diseases and compounds the threat to the securing of capabilities that arises from antibiotic resistance.

Keywords Antibiotics · Capabilities and growth stunting · Social justice · Mother and child health · Infectious Disease

14.1 Introduction

Antibiotic resistance has been framed as a problem consequent on the lack of development of new antibiotics and overuse of existing antibiotics. How we frame a problem is important in determining our responses to the problem (Tversky and Kahneman 1981). Unsurprisingly solutions to antibiotic resistance have been

M. Millar (✉)
Barts Health NHS Trust, London, UK
e-mail: M.R.Millar@qmul.ac.uk

© The Author(s) 2020
E. Jamrozik, M. Selgelid (eds.), *Ethics and Drug Resistance: Collective Responsibility for Global Public Health*, Public Health Ethics Analysis 5,
https://doi.org/10.1007/978-3-030-27874-8_14

focused on developing new antibiotics, and constraining the use of existing antibiotics (see for example http://drive-ab.eu, and https://www.bu.edu/law/faculty-scholarship/carb-x). Yet, new antibiotics and constraints on use of existing antibiotics can never be a solution to the potentiation of infectious disease transmission (in antibiotic sensitive or resistant forms) consequent on poverty, over-crowding, malnutrition, limited educational opportunity, environmental degradation, poor water quality, inadequate sanitation or conflict.

14.2   Capability Theory

Capability theory has been influential in defining measures of human development, and quality of life, and in the evaluation of the justice of social arrangements. Sen (1999) defined a capability as a 'substantial freedom he or she enjoys to lead the kind of life he or she has reason to value'. Nussbaum emphasises the importance of capabilities for human dignity, and derives entitlements from reflecting on the requirements for equal dignity and respect. She describes necessary conditions for a decently just society, in the form of a set of ten fundamental entitlements for all citizens (Nussbaum 2006). More recently capability theory has been applied to children (Biggeri et al. 2011) and child development (Peleg 2013). Nussbaum and Dixon (2012) have proposed that capability theory can be used to provide theoretical justification, and to justify a degree of special state priority for children's rights – based on the 'unique vulnerability of children to the decisions of others' (Nussbaum and Dixon 2012, p. 575).

Antibiotic resistance has implications for child development, basic capabilities, and the securing of adult capabilities. There is on-going debate about the elements that should be included in a capabilities list that is appropriate for adults (Wolff and De-Shalit 2007) or children (Biggeri and Mehrotra 2011). Children develop in to adults, moving through different developmental stages and capabilities at different ages. Child development both depends on the capabilities of adult carers, and determines the potential for adult capabilities. For the purposes of this chapter I have accepted the list that Nussbaum proposes as appropriate for adults, and that achievement of thresholds of adult capabilities is substantially dependent on child development. The capabilities listed by Nussbaum (2006, p. 154) encompass life expectancy, bodily health & integrity, sense, imagination and thought, emotions, practical reason, affiliation, relations with other species, play and control over one's environment.

14.3   Infectious Disease and Capabilities

The use of antibiotics can be conceptualised as an attempt to try to prevent damage caused by infection to established capabilities (adults) or the potential for capabilities (children). One consequence of the use of antibiotics is antibiotic resistance.

The rate at which antibiotic resistance develops is a function of usage, time, control measures, and the context of use. The availability of effective treatments for infectious disease is a substantial determinant of health. The interactions between health and the determinants of health are complex and not unidirectional, so for example health both determines and is determined by nutrition, education, and social status. The extent of capability fulfilment can be used to define health, while health is required for the fulfilment of capabilities (Venkatapuran 2011, 2013). Uncontrolled infectious disease subverts the achievement of adult capabilities through multiple pathways including through damage to child development. Damage to the capabilities of adult carers also has consequences for child development.

The control of infectious diseases (antibiotic sensitive and resistant) requires fulfilment of multiple entitlements. An entitlement to bodily health requires that an individual is able to have good health, including reproductive health; to be adequately nourished; and to have adequate shelter. The adequacy of shelter is important as a risk factor for the spread of infectious disease, and for damage to child development (see for example Shelter 2006). Nutrition influences both child development and infectious disease susceptibility (Gough et al. 2014). Maternal reproductive health is an important determinant of child development and vulnerability to infectious disease. A capability for senses, imagination, and thought requires education. To secure child development we must secure the capabilities of those who care for them. Nussbaum and Dixon (2012) emphasise that 'the goal remains the full empowerment of all individuals' (p. 578). Maternal education is a key element in assuring healthy child development (see discussion of growth stunting below). Maternal education and parental control of their local environment are necessary elements in protecting children from infectious disease. The entitlements to be able to play and to have relationships with the world of nature can be qualified by adding 'safely'. Children in developing countries may not be able to live and play without exposure to the risk of infectious diseases (antibiotic sensitive and resistant) transmitted as a result of poor sanitation, close proximity to animals with zoonotic infection, and vectors for disease (such as malaria mosquitoes). Infectious disease contributes to impairment of child development through multiple pathways including growth stunting (discussed further below).

The capabilities listed by Nussbaum remind us of important dimensions of the individual experience of infectious disease (such as freedom of movement and engagement in social interactions). The entitlement to affiliation requires that an individual is able to live for and in relation to others, to recognize and show concern for other human beings, and to engage in various forms of social interaction. Bodily integrity requires that an individual can move freely from place to place. These last two entitlements can be breached by restrictions taken to control the spread of infectious disease (antibiotic sensitive or resistant), and by the social consequences of infectious disease, particularly when associated with treatment (antibiotic) resistance (see for example Upshur et al. 2009).

Constraints on freedoms to prescribe, to purchase, to manufacture and formulate, to dispose, to pollute, and to use antibiotics for economic gain can all contribute to the control of antibiotic resistance. Nussbaum's approach does not preclude limitations on freedoms with respect to the use of antibiotics. Nussbaum gives emphasis to the need to limit freedoms when those freedoms adversely impact on the central capabilities. She states that 'no society that pursues equality or even an ample social minimum can avoid curtailing freedom in very many ways, and what it ought to say is those freedoms are not good, they are not part of a core group of entitlements required by the notion of social justice, and in many ways, indeed, they may subvert those core entitlements' (Nussbaum 2011, p. 73). 'In other words, all societies that pursue a reasonably just political conception have to evaluate human freedoms, saying that some are central and some trivial, some good and some actively bad, some deserving of special protection and others not' (Nussbaum 2011, pp. 74–75). Framing the actions, constraints and precautions from a capability perspective also identifies limits to the precautions that we can take and gives priority to actions, which do not undermine capability entitlements. A policy that results in a substantial loss of a capability for some can be challenged from a capability perspective, even if there was an overall benefit. Stigmatisation, isolation, and segregation of individuals to prevent the spread of treatment resistant infection (such as leprosy historically) would not be consistent with a capability perspective while there remain feasible alternative courses of action. The non-availability of effective treatments resulting from antibiotic resistance restricts alternative courses of action. For much of the twentieth century women with leprosy were actively discouraged or prevented from having children. New-born babies of mothers with leprosy were taken from their parents at birth, because otherwise the child would also develop leprosy (see for example International Leprosy Association, History of Leprosy). Capabilities related to childbirth including the opportunities to have and to look after a child were removed. Leprosy has now been controlled to a large extent by the advent and availability of effective antibiotic treatments.

14.4   Human Dignity and Infectious Disease

An emphasis on the importance of human dignity is a substantial element within Nussbaum's capability theory. There is a lack of consensus on how best to define and measure human dignity (Ashcroft 2005). Dignity can be defined positively but also negatively as freedom from sources of humiliation (see Shultziner and Rabinovici 2011). Nussbaum, while acknowledging that dignity is a poorly defined concept, uses human dignity as a touchstone for the selection of capabilities. Nussbaum's list of capabilities is a list of positive entitlements (capabilities) that are necessary for the living of a life with human dignity. Respect for the dignity of others is an important social and political value. Conditions must pertain that reflect that value and these include conditions that foster a sense of personal worth in each person.

There are strong associations between social status, self-esteem and health (Mann 1998; Marmot 2003; Chilton 2006). There are many ways in which dignity interacts with infectious diseases. There are common risk factors for violations of dignity and infectious diseases, for example lack of access to safe toilets contributes to loss of dignity, gender-based violence, and the transmission of infectious disease (see WHO factsheet 392). Inadequate shelter threatens dignity and increases infectious disease risks (Shelter 2006). Some infectious diseases, particularly those that are difficult to treat, such as antibiotic resistant tuberculosis, increase the risk that the dignity of adults will be violated (Upshur et al. 2009). Children with HIV or with a parent with HIV can suffer a substantial loss of self-esteem (Chi and Li 2014). An individual or group with an infectious disease may suffer from stigmatisation, social isolation, and a resultant loss of a sense of self-worth. Social consequences of infection for individuals, groups and institutions also include blame and shame (Sontag 1989). Acquisition and carriage of antibiotic resistant bacteria by individuals while undergoing healthcare can be associated with stigmatisation (Rump et al. 2017). Public health policies designed to support the control of antibiotic resistant bacteria can also lead to stigmatisation of individuals (Ploug et al. 2015). Violations of dignity and damage to self-esteem can increase infectious disease risk through changes in human behaviour. Low self-esteem is associated with sexual risk taking behaviour and sexually transmitted disease (see Byrnes et al. 1999; Ethier et al. 2006).

There are common risk factors for violations of dignity and infectious diseases, violations of dignity increase infectious disease risk, and infectious disease increases the risk that dignity will be violated.

14.5 Clustering of Disadvantage: The Example of Growth Stunting

In discussing actions which lead to the destruction of capabilities Nussbaum states that 'We can certainly agree that capability-destruction in children is a particularly grave matter and as such should be off-limits'. 'Usually situations are not so grave, and thus in many such cases the approach has little to say, allowing matters to be settled through the political process' (Nussbaum 2011, p. 27). Unfortunately capability-destruction is not unusual in many countries. It is estimated that more than 40% of children in lower and middle income countries are at risk of impaired development (Black et al. 2016). A period of particular risk to development is that between conception and 3 years of age. More than 25% of children <5 years of age globally have stunted growth (low height-for-age). Stunting is associated with long-term cognitive and physical impairment (Hair et al. 2015; Noble et al. 2015) and substantial economic consequences for individuals, communities and countries (Horton and Steckel 2013). The WHO Conceptual Framework for stunted growth (Stewart et al. 2013) identifies a range of community and societal contextual factors underlying the causes of stunting of growth of children. These factors include political arrangements, poverty, regulatory frameworks, healthcare systems, beliefs and norms, the status of women, access to safe foods, sanitation, population density, and natural and manmade disasters. Solutions require intervention in multiple sectors with specific emphasis given to the way in which resources are controlled and distributed through the political and economic system, food security, education (particularly of females), water quality, sanitation (and hygiene), ameliorating poverty and vulnerability, and access to healthcare (Casanovas et al. 2013). Nussbaum's capability approach specifies a broad range of entitlements and in so doing is well placed to explicitly accommodate these multiple and complex requirements.

Infections are thought to contribute to the pathogenesis of growth stunting and infections are both more common and more serious in children with stunting. Recently it has been suggested that administration of antibiotics to populations of children could be used to prevent stunted growth (Gough et al. 2014). The proposal to use antibiotics as a population level intervention to mitigate the risk of stunting illustrates the tension between sustaining effective antibiotics while assuring access for those in pressing need. Population level antibiotic interventions have profound implications for present and future generations particularly when the target populations live under conditions of relative deprivation that facilitate the spread of agents of infection in sensitive and resistant forms. Common risk factors for stunting and for the spread of infectious diseases include overcrowding, poor education (particularly maternal), poor nutrition, inadequate sanitation, and poor water quality. Capability insufficiencies (such as poor shelter, threats to bodily health, lack of access to maternal education) contribute both to the transmission of infectious disease and to host susceptibility to disease, and potentially to the burden of antibiotic resistance. There is a clustering of disadvantage in that infectious disease amplifies other disadvantages consequent on capability insufficiencies.

The use of antibiotics in many developed economies has extended beyond the treatment and prevention of life-threatening human infections to include the mitigation of symptoms of self-limiting disease(s), animal husbandry, fish farming, and to allow us to extend the range of medical and surgical interventions including those with limited health benefits such as some forms of cosmetic surgery. In developed countries antibiotics have become a means to a variety of ends, with varying degrees of relationship with mitigation of harm to human capabilities. Despite this there is evidence that the use of antibiotics in developed countries has stabilised or fallen this century (van Boeckel et al. 2014). Most of the recent increase in use of antibiotics has been in rapidly developing countries including Brazil, Russia, India, China and South Africa (BRICS countries). Holland particularly has shown how it is possible for developed countries to have low levels of human antibiotic usage and low levels of antibiotic resistance associated with patients (http://www.ecdc.europa.eu/en/activities/surveillance/EARS-Net/Pages/Index.aspx). By contrast antibiotic resistance is more prevalent now in developing countries with high levels of deprivation than in developed countries (WHO Report 2014). This situation probably reflects the capacity of countries to expend resources on trying to assure socioeconomic conditions (and capability entitlements) such as education, health, nutrition, sanitation, and housing which are significant determinants of the epidemiology of infectious disease. An oft-quoted example of this relationship is that of tuberculosis. The incidence of tuberculosis (TB) and the requirement to treat TB has declined dramatically over recent decades in developed countries and this decline has been attributed to improvements in socio-economic conditions (Comstock 2000). By contrast with developed countries such as Holland the context of use of antibiotics in low to middle income countries frequently involves heightened conditions for the spread of infectious diseases both in antibiotic sensitive and resistant forms.

Antibiotic resistance considered from within a capability perspective draws attention to the importance of the social, political and economic context in determining infectious disease risks. Antibiotic resistance threatens to place an additional burden on communities where growth stunting and infection are already prevalent, because many of the risk factors for growth stunting also determine the risk that antibiotic resistance will spread. Capabilities may be incommensurable but capabilities still interact with each other in contributing to a state of wellbeing. In the real world these interactions may lead to clustering of disadvantage (Wolff and De-Shalit 2007). The interaction of infection with growth stunting in children illustrates the clustering of disadvantage and the importance of addressing the broad range of capability deficiencies. Otherwise preventable childhood stunting has the potential to persist alongside burgeoning levels of antibiotic resistance.

14.6   Capability Thresholds and Inequality?

'The basic claim of my account of social justice is this: respect for human dignity requires that citizens be placed above an ample threshold of capability in all ten of those areas' (Nussbaum 2011, p. 36). In this section I ask if capability thresholds can be achieved while there remains avoidable and substantial inequality in access to effective antibiotics and in the determinants of infectious disease. Inequality can be and often is harmful to human wellbeing (Picket and Wilkinson 2010). Our sense of self-worth, our social status, our wellbeing and inequalities are intertwined (Marmot 2003, 2005). Thresholds of capability may not achievable while substantial inequalities remain. Wolff and De-Shalit (2007, p. 10) define a society of equals as one in which 'disadvantages do not cluster, where there is no clear answer to the question of who is the worst off'. Certainly countries are not equal partners and neither are individuals within countries, when account is taken of the burdens of infectious disease – as exemplified by the contrasting patterns of child development and the clustering of disadvantage.

There is inequality in access to effective antibiotics (Laxminarayan et al. 2016). Pneumonia is still responsible for 1 in 5 deaths of children less than five years old in the world today. Ensuring access of children with pneumonia to antibiotics has been a major objective of the WHO and UNICEF over the last decade (WHO 2013). There is also inequality in the risk of acquisition of antibiotic resistant bacteria. There is increasing evidence of extensive environmental contamination with antibiotics and antibiotic resistant bacteria (Lubbert et al. 2017), particularly from antibiotic manufacturing plants, from use of antibiotics in meat production, from hospitals and urban conurbations (Berendonk et al. 2015). There is increasing use of antibiotics to support animal meat production even in countries with high levels of childhood growth stunting (see Centre for Science and Environment Report 2014). It is estimated that by 2030 the use of antibiotics in livestock production in the US and China will account for 40% of global antibiotic use (Van Boeckel et al. 2015). Antibiotic resistant bacteria, selected in humans and animals, are shared through environmental pollution with both those who can and those who cannot afford to eat meat. Often those with less or no access to antibiotics live in conditions, which promote the spread of antibiotic resistance. These differences are particularly strong in countries where some live in relative affluence in close proximity to slums. Those who are better off have access to the best medical advice, diagnostics for antibiotic resistance, and the latest treatments including antibiotics. When antibiotic resistant bacteria contaminate the environment of people with inadequate sanitation, poor education and with a heightened susceptibility to disease then spread is facilitated. Contamination of the Ganges provides a specific example. While the rich can afford cremation on the Ganges, and pilgrimage to the upper Ganges, the poor cannot. The water of the Ganges has become highly polluted with antibiotic resistant bacteria including a particularly worrying form of antibiotic resistance called New Delhi Metallo-β-lactamase (NDM-1) (Ahammad et al. 2014). Poor people use the water of the Ganges for recreation, washing, and drinking. Poor people with limited access to medical care or antibiotics are exposed to extreme forms of antibiotic resistance through day-to-day activities, with implications for the efficacy of antibiotics, their health and for the health of those around them. There is inequality in the risk of acquisition of antibiotic resistance as well as inequality in access to effective antibiotics (see Note 1).

14.7   International Cooperation, Unequal Partners

Currently the majority of countries collaborate on the control of infectious diseases. One hundred and ninety-six countries have signed up to the International Health Regulations (2005) (developed after the SARS outbreak in 2003), which are designed to control the international spread of infectious disease. The control of epidemic diseases is included in the United Nations Rights Document A/6316 (1996). Over recent years antibiotic resistance has become a focus of international concern. The World Health Organisation (2015) has developed an action plan for antimicrobial resistance, which recognises the need for international collaboration (for example see Section 21 (4), Global Action Plan). There is also broad acceptance in the scientific literature (see for example Institute of Medicine 2010) that control of antibiotic resistance requires international cooperative action.

Unequal relationships between countries can subvert cooperation. For example Indonesia has had outbreaks of avian influenza, which have been associated with high mortality, yet withdrew from cooperation with the WHO Global Influenza Surveillance Network (GISN) in 2007. The position from the Indonesian perspective is described in a journal article as follows – 'Indonesia believes that the world must work in unity against the H5N1 virus infection.... The work must be conducted side by side with mutual trust, transparency and equity as global citizens professionals, taking into consideration the elements of human dignity and solidarity.' 'The avian influenza case in Indonesia has demonstrated once again the unresolved imbalance between the affluent 'high-tech' countries and poor agriculture-based countries. Countries that are the hardest hit by a disease must also bear the burden of the cost of the vaccine, therapeutics and other products, while the monetary and non-monetary benefits of these products go to the manufacturers that are mostly in the industrialised countries' (Sedyaningsih et al. 2008, p. 487). Indonesia and other resource poor countries were expected to participate in cooperating in the control of infectious disease, yet there was a belief that the benefits of cooperation were unequally distributed. Subsequently an agreement was reached which stipulated arrangements for more equitable cooperative arrangements (World Health Assembly 2011).

Another example of an unequal international relationship that relates more directly to the control of antibiotic resistance arises from the Kumarasamy et al. report in 2010 that New Delhi Metallo-β-lactamase-1 was widespread in India. The Indian government considered this name to be 'unfair' and stigmatising, and potentially undermined the burgeoning health tourist market in India (see Pandey 2010). The Editor of the Lancet subsequently described the use of this name as an error and apologized stating that the name had 'unnecessarily stigmatised a single country and city' (Sinhal 2011). This interaction has compromised research on the epidemiology of NDM-1 in India according to the lead author of the Lancet report (Tim Walsh), who named this form of resistance NDM-1. He stated that 'We were banned from India and India had a massive clampdown on sending (biological) strains out' (Sugden 2013). Health tourism is developing fast in India and the stigma associated with the potential acquisition of antibiotic resistant bacteria was characterised by Indian politicians as an international plot to undermine that development. Unfortunately NDM1 is now globally distributed (Berrazeg et al. 2014). Conflict between the Lancet and the Indian government over health policy has continued (Sinhal 2015).

This second story illustrates aspects of the epidemiology of antibiotic resistance. When social conditions are poor then antibiotics quickly become ineffective, antibiotic resistant bacteria can be rapidly spread around the world, and there is a relationship between the sustaining of the functions of antibiotics and other socio-economic 'goods' such as adequate shelter, and clean water. This example also illustrates a relationship between antibiotic resistance, and stigma, which can undermine cooperation in the control of antibiotic resistance. There is considerable inequality in the international influence of the biomedical press with a bias towards developed countries with a strong research base. Combinations of economic inequality, antibiotic prescribing practice norms, and publication practice have marked out India in a way that adds additional disadvantage in a competitive global market for medical tourism.

14.8   A Relational Approach to Capability Inequality

Nussbaum argues that a contractualist account (such as that proposed by Scanlon 1998) based on individuals as moral equals 'is a powerful intuitive way of capturing the idea that human beings are moral equals despite their widely differing circumstances in an unequal world' (Nussbaum 2006, p. 272). Nussbaum states that 'I employ the notion of reasonable rejection, or something very close to it, in articulating my account of political justification' (Nussbaum 2011, p. 89). Her approach 'is a partial account of specifically political entitlements' (Nussbaum 2011, p. 96).

   Scanlon provides an account of why and when we can reject both distributive and non-distributive inequalities (see Scanlon 2003, 2013) which can be applied to Nussbaum's evaluative capability framework. Reasonable rejection and justification to others provide the substantial focus of Scanlon's contractualist approach (Scanlon 1998) (for more discussion of Scanlon's approach see Note 2). Interactions between individuals and groups determine the epidemiology of infectious diseases. Human relations directly or indirectly have a substantial role in the spread of infectious diseases. In addition many of the transactions that determine the use of antibiotics and the consequences of use involve individuals, institutions and nations in dialogue. Examples include healthcare workers (such as doctors) agreeing treatment plans with patients, or healthcare authorities and institutions agreeing antibiotic policies within nations, or nations agreeing approaches to international collaboration on the control of antibiotic resistance. Justification is a key element to these relational interactions and it seems intuitively attractive to start from an acknowledgement of the importance of justification in assuring the validity of principles and agreements. Another attractive feature of Scanlon's approach is a concern with assuring the conditions for self-worth. This is significant both in relation to the epidemiology of infectious disease (as previously discussed) and in relation to a concern with assuring the conditions for a life with human dignity (a substantial concern for Nussbaum's capability theory) (Fitzpatrick 2008).

   For Scanlon everyone counts morally, regardless of race, gender, or where they live. The different reasons for rejecting inequalities are dependent 'on the way that an inequality affects or arises from the relations between individuals' (see Scanlon lecture – Why does inequality matter?). Reasons for rejecting inequalities 'presuppose some form of relationship or interaction between unequal parties' and are based on comparing the differences in the situations of individuals (Scanlon (2004) lecture – When does equality matter?). For Scanlon (2006) reasonable rejection does not depend on rejection of the distribution of goods, but rather it depends on assuring equal respect and fairness. Nussbaum's capability theory provides an evaluative framework for the extent to which equal respect and fairness are achieved, without being itself a complete theory of distributive justice.

   Scanlon argues that 'relief of suffering, avoidance of stigmatising differences in status, prevention of domination by others, and the preservation of conditions of procedural fairness are basic and important moral values' (Scanlon 2003, p. 218) that can give reason to reject inequalities. Inequalities can also be rejected when there are institutional obligations to provide certain benefits in an even-handed way, or 'cases in which individuals, as participants in a cooperative endeavour, have at least a *prima facie* claim to an equal share of the goods which that endeavour produces' (Scanlon 2013, p. 463). The control of antibiotic resistance is a cooperative exercise. Cooperation is predicated on an expectation of a share in access to effective antibiotics. Importantly the scope of these reasons for rejection of inequalities extends beyond national boundaries (for a fuller discussion see O'Neill 2013).

14.9   Inequalities Subvert Capabilities

Inequalities can lead to substantial disadvantage, furthering inequalities in a competitive world, and cluster to give multiple disadvantages. Inequalities can result in stigmatising differences in status, both at the level of the individual and the state as shown by the example of NDM-1 antibiotic resistance described above. Inequalities in access to antibiotics coexist with inequalities in risk factors for infection, and risk factors for antibiotic resistance. These inequalities can and do contribute to social, political and economic disadvantage, as exemplified by the interactions between childhood growth stunting, infectious diseases and antibiotic access.

Wealth determines access to healthcare, medicines (including antibiotics) and healthy living conditions. The less wealthy in many countries have less access to high quality medicines, little access to medical advice, or diagnostic facilities, often live under conditions that may facilitate the spread of antibiotic resistant bacteria, may be exposed to resistant bacteria even before exposure to antibiotics, and are more at risk of infection, and of the serious consequences from infection. Securing child development requires control of infectious diseases in treatment sensitive and resistant forms. There is an unequal distribution of childhood burdens and benefits associated with antibiotics and antibiotic resistance both within and between countries. These inequalities contribute to avoidable suffering (preventable infectious disease), differences in status (growth stunting), and potentially domination of vulnerable children by others. Inequalities subvert the achievement of capability thresholds.

14.10   Addressing Inequalities, Achieving Capability Thresholds

Control of antibiotic resistance is a cooperative enterprise with shared objectives and shared responsibilities. Countries cooperate in the control of infectious disease including antibiotic-resistant agents of infection. If we accept that all of the parties engaged in the control of antibiotic resistance should have an equal *prima facie* claim to access to effective antibiotics then when inequalities exist we can ask if those inequalities can be justified – 'basic structures need to be justified to all who are asked to accept them' (Scanlon lecture – Why Does Inequality Matter?) (see Note 3).

Developed countries have used antibiotics for much longer and in much larger quantities and for a wider range of reasons than most developing nations. India is subject to criticism for insufficient regulation of antibiotic prescribing. Yet, even by 2010, India used half the number of antibiotic units per person compared with the USA (Laxminarayan and Chaudhury 2016). Historically while new antibiotics were regularly coming to the market there was no commercial motivation to constrain antibiotic use because antibiotic resistance was a major justification for using the new antibiotic(s). The profits from antibiotic sales were largely accrued by companies based in the developed world. Marketing decisions were dominated by consideration of profit maximisation with a relatively low priority given to public health. The United Nations has accepted that developed nations should take up a greater part of the responsibility for the control of greenhouse gases than developing countries. 'Common but Differentiated Responsibilities' (UNFCCC 2015) are justified because there are differences in capacity to respond, different priorities, and differences in the historical contribution to the problem of greenhouse gas emissions between developed and developing countries. These differences also apply to antibiotic resistance.

In the World Health Assembly Global Action Plan for Antimicrobial Resistance (2015) Section 21. (3) **Access** states that '*The aim to preserve the ability to treat serious infections requires both equitable access to, and appropriate use of, existing and new antimicrobial medicines*'. Currently despite the high level of international concern there is no international agreement as to what constitutes 'appropriate' use of antibiotics. Equitable access is a long way away when account is taken of the lack of access to antibiotics for the treatment of life-threatening infection in many parts of the world while antibiotics are used extensively without human health benefit in many other parts of the world. I have previously argued that appropriate use is that which prevents some substantial risk of irretrievable harm in patients or their contacts, where a substantial risk is a level of risk which exceeds the range of risks of irretrievable harm that we tolerate in our day to day lives (Millar 2012). Use of antibiotics to support (non-human) animal growth promotion is inappropriate use. Use of antibiotics for animal growth promotion has been reported to be increasing in both developed and developing countries (Van Boeckel et al. 2015), including in areas where human growth stunting is particularly prevalent such as Sub-Saharan Africa. In developing countries people are more likely to live in close proximity to animals, so that the risks that antibiotic resistant bacteria will spread from animals, and that people will be exposed to antibiotics present in their environment, is increased compared with developed countries where the close proximity of farm animals with people is less common. It is strikingly inappropriate for many children to have limited access to antibiotics while antibiotics are being used as animal growth promoters. Developed countries should take a lead in limiting use of antibiotics as animal growth promoters.

The UN has given a substantial place to human dignity in the Sustainable Development Goals (UN 2014). These goals include improvements in education, housing, nutrition, water quality, and sanitation. The scope for international action on antibiotic resistance includes the amelioration of the conditions that potentiate the need for antibiotics and the spread of resistant forms of agents of infection. Improving living conditions (including housing), nutrition, education (for example with respect to risk factors for disease), and other determinants of human dignity such as female empowerment have been shown to reduce the risk of stunting, but these are also important factors in the control of infectious disease transmission and potentially antibiotic resistance. Achieving UN Development Goals (assuming that fairly and honestly set – see Hickel 2017) for 2030 would do much to mitigate the risk factors for both stunting and the transmission of agents of infection in both antibiotic sensitive and resistant forms.

14.11 Conclusions

Nussbaum's capability theory (Nussbaum 2006) provides a rich and relevant evaluative space for framing antibiotic resistance. Securing child development and adult capabilities requires that we address inequalities in access, regulate 'appropriate prescribing', and clarify responsibilities for addressing inequalities in risk factors for the dissemination and transmission of antibiotic resistance. Historical and current patterns of antibiotic use impose a burden of responsibility on developed countries to ensure that their own use is appropriate, and raise questions with respect to the responsibilities of developed countries to address risk factors for antibiotic resistant infection in developing countries. Antibiotics and antibiotic resistant bacteria pollute the environment (particularly when sanitation standards are poor) – compounding the inequality and disadvantage of those without equivalent access to effective antibiotics. Inequality undermines the cooperative activity that is control of infectious diseases and compounds the threat to the securing of child development and adult capabilities that comes from antibiotic resistance.

Note 1
There is some empirical evidence that European countries with more income inequality have higher levels of antibiotic resistance than those with less inequality (Kirby and Herbert 2013). Lack of access to high quality data makes this relationship difficult to study more generally.

Note 2
For Scanlon, reasons are facts (Scanlon 2014, p. 30, note 20), natural (e.g., that you will enjoy some activity) or normative (e.g., a law's being unjust or your having reason to go on living; Scanlon 2014, p. 32). These facts are an essential part of a relation to an agent: consideration (or fact) p is a reason for x to do a in circumstances c. 'Is a reason for' is a four-place relation, R(p, x, c, a) (pp. 31, 37). For Scanlon there are cases where the relative strength of reasons is derived from the

relative amounts of something (such as capabilities), even so he considers it to be a mistake to consider that there must inevitably be a quantitative property which determines the relative strength of reasons (p. 110). 'The strength of a reason is an essentially comparative notion, understood only in relation to other particular reasons' (p. 111). When we judge that certain considerations provide conclusive reasons for (or against) certain actions in certain circumstances, justification comes our understanding of a relationship with other rational beings that we have reason to want, specifically, the relationship of seeing them as beings to whom justification is owed (p. 115). Scanlon acknowledges that 'when we are assessing the justifiability of moral principles we must have reason to appeal to things that individuals have reason to want, and that many of these are things that contribute to well-being intuitively understood.' However, 'we cannot delimit the range of considerations that figure in justification by defining the boundaries of well-being' (Scanlon 1998, p. 140; see Putnam 2008).

Note 3

Scanlon emphasises the importance of probabilities in determining the degree of effort that we make to control risks. 'The probability that a form of conduct will cause harm can be relevant not as a factor diminishing the 'complaint' of the affected parties (discounting the harm by the likelihood of their suffering it) but rather as an indicator of the care that the agent has to take to avoid causing harm'. Scanlon states that '..the cost of avoiding all behaviour that involves risk of harm would be unacceptable. Our idea of 'reasonable precautions' defines the level of care that we think can be demanded: a principle that demanded more than this would be too confining, and could reasonably be rejected on that ground' (Scanlon 1998, p. 209 & pp. 235–6; see Kumar 2016). Scanlon's emphasis on reasons allows the inclusion of morally salient considerations such as responsibility and fairness (Scanlon 1998, p. 243). 'Responsibility of an agent for wrongful conduct, responsibility for creating a situation that gives reason to break a promise, responsibility for engaging in risky conduct that leads to harm and responsibility for misfortune that puts one in need of aid' (Scanlon 1998, p. 244) are all morally salient considerations. The question that arises is 'who has responsibility for 'reasonable precautions' when it comes to antibiotic resistance in developing countries', and what is reasonable?

Acknowledgement I'd like to thank Dr. Andrew Prendergast for raising the issue of growth stunting and the interaction with infectious disease.

References

Ahammad, Z.S., T.R. Sreekrishnan, C.L. Hands, C.W. Knapp, and D.W. Graham. 2014. Increased waterborne blaNDM-1 resistance gene abundances associated with seasonal human pilgrimages to the upper Ganges river. *Environmental Science & Technology* 48 (5): 3014–3020. https://doi.org/10.1021/es405348h.

Ashcroft, R.E. 2005. Making sense of dignity. *Journal of Medical Ethics* 31: 679–682.

Berendonk, T.U., C.M. Manaia, C. Merlin, D. Fatta-Kassinos, E. Cytryn, F. Walsh, H. Bürgmann, H. Sørum, M. Norström, M.N. Pons, N. Kreuzinger, P. Huovinen, S. Stefani, T. Schwartz, V. Kisand, F. Baquero, and J.L. Martinez. 2015. Tackling antibiotic resistance: The environmental framework. *Nature Reviews. Microbiology* 13 (5): 310–317. https://doi.org/10.1038/nrmicro3439.

Berrazeg, M., S.M. Diene, L. Medjahed, P. Parola, M. Drissi, D. Raoult, and J.M. Rolain. 2014. New Delhi metallo-beta-lactamase around the world: An eReview using Google Maps. *Eurosurveillance* 19 (20): pii=20809.

Biggeri, M., and S. Mehrotra. 2011. Child poverty as capability deprivation: How to choose domains of child well-being and poverty. In *Children and the capability approach*, ed. M. Biggeri, J. Ballet, and F. Comim. Basingstoke: Palgrave Macmillan.

Biggeri, M., J. Ballet, and F. Comim, eds. 2011. *Children and the capability approach*. Basingstoke: Palgrave Macmillan.

Black, R.E., C. Levin, N. Walker, D. Chou, M. Temmerman, and for the DCP3 RMNCH Authors Group. 2016. Reproductive, maternal, newborn, and child health: Key messages from disease control priorities 3rd edition. *Lancet* 388: 2811–2824.

Byrnes, J.P., D.C. Miller, and W.D. Schafer. 1999. Gender differences in risk taking: A meta-analysis. *Psychological Bulletin* 125: 367–383.

Casanovas, M.C., C. Lutter, N. Mangasaryan, R. Mwadime, N. Hajeebhoy, A.M. Aguilar, et al. 2013. Multisectoral interventions for healthy growth. *Maternal & Child Nutrition* 9 (Suppl.2): 46–57.

Centre for Science and Environment Report. (2014). www.cseindia.org/node/5487

Chi, P., and X. Li. 2014. Impact of parental HIV/AIDS on children's psychological well-being: A systematic review of global literature. *AIDS and Behavior* 17 (7): 2554–2574.

Chilton, M. 2006. Developing a measure of dignity for stress-related health outcomes. *Health and Human Rights* 9 (2): 208–233.

Comstock, G. 2000. Epidemiology of tuberculosis. In *Tuberculosis: A comprehensive international approach*, ed. L.B. Reichman and E.S. Hershfield, 129–156. New York: Marcel Dekker.

Ethier, K.A., T.S. Kershaw, J.B. Lewis, S. Milan, L.M. Niccolai, and J.R. Ickovics. 2006. Self-esteem, emotional distress and sexual behavior among adolescent females: Inter-relationships and temporal effects. *The Journal of Adolescent Health* 38 (3): 268–274.

Fitzpatrick, T. 2008. From contracts to capabilities and back again. *Research Publications* 14: 83–100.

Gough, E.K., E.E.M. Moodie, A.J. Prendergast, S.M.A. Johnson, J.H. Humphrey, R.J. Stoltzfus, A.S. Walker, I. Trehan, D.M. Gibb, R. Goto, S. Tahan, M.B. de Morais, and A.R. Manges. 2014. The impact of antibiotics on growth in children in low and middle income countries: Systematic review and meta-analysis of randomised controlled trials. *BMJ* 348: g2267.

Hair, N.L., J.L. Hanson, B.L. Wolfe, and S.D. Pollak. 2015. Association of child poverty, brain development, and academic achievement. *JAMA Pediatrics* 169 (9): 822–829. https://doi.org/10.1001/jamapediatrics.

Hickel, J. 2017. *The divide: A brief guide to global inequality and its solutions*. London: William Heinemann Ltd.

Horton, S., and R. Steckel. 2013. Global economic losses attributable to malnutrition 1900–2000 and projections to 2050. In *The economics of human challenges*, ed. B. Lomborg. Cambridge: Cambridge University Press.

Institute of Medicine. 2010. *Antibiotic resistance. Implications for global health and novel intervention strategies workshop summary*, Institute of Medicine (US) Forum on Microbial Threats. Washington, DC: National Academies Press (US).

International Leprosy Association – history of leprosy: leprosy and the family. Available at http://leprosyhistory.org/impact/leprosy-and-the-family. Accessed July 2017.

Kirby, A., and A. Herbert. 2013. Correlations between income inequality and antimicrobial resistance. *PLoS One* 8 (8): e73115. https://doi.org/10.1371/journal.pone.0073115.

Kumar, R. 2016. Risking and wronging. *Philosophy and Public Affairs* 43 (1): 27–51.

Kumarasamy, K.K., M.A. Toleman, T.R. Walsh, J. Bagaria, F. Butt, R. Balakrishnana, U. Chaudhary, et al. 2010. Emergence of a new antibiotic resistance mechanism in India, Pakistan, and the UK: A molecular, biological, and epidemiological study. *The Lancet Infectious Diseases* 10 (9): 597–602.

Laxminarayan, R., and R.R. Chaudhury. 2016. Antibiotic resistance in India: Drivers and opportunities for action. *PLoS Medicine* 13 (3): e1001974.

Laxminarayan, R., P. Matsoso, S. Pant, C. Brower, J.-A. Rottingen, K. Klugman, and S. Davies. 2016. Access to effective antimicrobials: A worldwide challenge. *Lancet* 387: 168–175.

Lübbert, C., C. Baars, A. Dayakar, N. Lippmann, A.C. Rodloff, M. Kinzig, and F. Sörgel. 2017. Environmental pollution with antimicrobial agents from bulk drug manufacturing industries in Hyderabad, South India is associated with dissemination of extended-spectrum beta-lactamase and carbapenemase-producing pathogens. *Infection* 45: 479–491. https://doi.org/10.1007/s15010-017-1007-2.

Mann, J. 1998. Dignity and health: The UDHR's revolutionary first article. *Health and Human Rights* 3 (2): 30–38.

Marmot, M. 2003. Self-esteem and health. *British Medical Journal* 327: 574–575.

———. 2005. Social determinants of health inequalities. *Lancet* 365: 1099–1104.

Millar, M. 2012. Constraining the use of antibiotics: Applying Scanlon's contractualism. *Journal of Medical Ethics* 38 (8): 465–469.

Noble, K.G., S.M. Houston, N.H. Brito, H. Bartsch, E. Kan, J.M. Kuperman, N. Akshoomoff, D.G. Amaral, C.S. Bloss, O. Libiger, N.J. Schork, S.S. Murray, B.J. Casey, L. Chang, T.M. Ernst, J.A. Frazier, J.R. Gruen, D.N. Kennedy, P. Van Zijl, S. Mostofsky, W.E. Kaufmann, T. Kenet, A.M. Dale, T.L. Jernigan, and E.R. Sowell. 2015. Family income, parental education and brain structure in children and adolescents. *Nature Neuroscience* 18 (5): 773–778. https://doi.org/10.1038/nn.3983.

Nussbaum, M. 2006. *Frontiers of justice*. Cambridge, MA: Harvard University Press.

———. 2011. *Creating capabilities: The human development approach*. Cambridge, MA: Harvard University Press.

Nussbaum, M., and R. Dixon. 2012. *Children's rights and a capabilities approach: The question of special priority*, University of Chicago Public Law and Legal Theory Working Paper, No.384. Chicago: University of Chicago Law School.

O'Neill, M. 2013. Constructing a contractualist egalitarianism: Equality after Scanlon. *Journal of Moral Philosophy* 10: 429–461.

Pandey G. 2010. India rejects UK scientists' 'superbug' claim. *BBC News*.

Peleg, N. 2013. Reconceptualising the child's right to development: Children and the capability approach. *International Journal of Children's Rights* 21: 523–542.

Picket, K., and R. Wilkinson. 2010. *The spirit level: Why equality is better for everyone*. New York: Bloomsbury Press.

Ploug, T., S. Holm, and M. Gjerriset. 2015. The stigmatization dilemma in public health policy-the case of MRSA in Denmark. *BMC Public Health* 15: 640. https://doi.org/10.1186/s12889-015-2004-y.

Putnam, H. 2008. Capabilities and two ethical theories. *Journal of Human Development* 9: 377–388.

Rump, B., M. De Boer, R. Reis, M. Wassenberg, and J. van Steenbergen. 2017. Signs of stigma and poor mental health among carriers of MRSA. *The Journal of Hospital Infection* 95 (3): 268–274.

Scanlon, T.M. 1998. *What we owe to each other*. Cambridge, MA: Harvard University Press.

———. 2003. The diversity of objections to inequality. In *The difficulty of tolerance*. Cambridge: Cambridge University Press.

———. 2004. Lecture. When does equality matter? Unpublished paper. Available at http://www.politicalscience.stanford.edu/politicaltheoryworkshop/0607papers/scanlonpaper.pdf. Accessed May 2017.

———. 2006. Justice, responsibility, and the demands of equality. In *The egalitarian conscience: Essays in honour of G. A. Cohen*, ed. Christine Sypnowich, 70–87. Oxford: Oxford University Press.

———. 2013. Reply to Martin O'Neill. *Journal of Moral Philosophy* 10: 462–464.

———. 2014. *Being realistic about reasons*. Oxford: Oxford University Press.

———. (not dated). Lecture. Why does inequality matter? Available at https://www.law.berkeley.edu/wp-content/uploads/2016/01/Why-Does-Inequality-Matter-Chapters-8-9.pdf. Accessed May 2017.

Sedyaningsih, E.R., S. Isfandari, T. Soendoro, and S.F. Supari. 2008. Towards mutual trust, transparency and equity in virus sharing mechanism: The avian influenza case of Indonesia. *Annals of the Academy of Medicine, Singapore* 37: 482–488.

Sen, A. 1999. *Development as freedom*. New York: Random House.

Shelter. 2006. Chance of a lifetime: The impact of bad housing on children's lives. https://england.shelter.org.uk/__data/assets/pdf_file/.../Chance_of_a_Lifetime.pdf. Accessed June 2017.

Shultziner, D., and I. Rabinovici. 2011. Human dignity, self-worth, and humiliation: A comparative legal–psychological approach. *Psychology, Public Policy, and Law* 18 (1): 105. Available at: http://works.bepress.com/doron_shultziner/20/. Accessed May 2017.

Sinhal K. 2011. Lancet says sorry for 'Delhi bug'. *The Times of India*.

———. 2015. British medical journal *Lancet* to take Modi to task for ignoring health sector. *The Times of India*.

Sontag, S. 1989. *AIDS and its metaphors*. New York: Farrar, Straus and Giroux.

Stewart, C.P., L. Ianotti, K.G. Dewey, K.F. Michaelsen, and A.W. Onyango. 2013. Contextualising complementary feeding in a broader framework for stunting prevention. *Maternal & Child Nutrition* 9: 27–45.

Sugden J. 2013. India has lost superbug war. *Wall Street Journal*.

Tversky, A., and D. Kahneman. 1981. The framing of decisions and the psychology of choice. *Science* 211 (4481): 453–458.

United Nations. 1996. *International covenant on economic, social and cultural rights*. Geneva: United Nations.

———. 2014. *The road to dignity by 2030*. Geneva: United Nations.

———. 2015. Framework Convention on Climate Change UNFCCC, Decision 1/CP.21, Adoption of the Paris Agreement, UN Doc. FCCC/CP/2015/10/Add.1.

Upshur, R., J. Singh, and N. Ford. 2009. Apocalypse or redemption: Responding to extensively drug-resistant tuberculosis. *Bulletin of the World Health Organization* 87: 481–483. https://doi.org/10.2471/BLT.08.051698.

Van Boeckel, T.P., S. Gandra, A. Ashok, Q. Caudron, B.T. Grenfell, S.A. Levin, and R. Laxminarayan. 2014. Global antibiotic consumption 2000–2010; an analysis of national pharmaceutical sales data. *Lancet Infectious Diseases* 14: 742–750.

Van Boeckel, T.P., C. Brower, M. Gilbert, B.T. Grenfella, S.A. Levina, T.P. Robinsoni, A. Teillanta, and R. Laxminarayan. 2015. Global trends in antimicrobial use in food animals. *PNAS* 112 (18): 5649–5654.

Venkatapuran, S. 2011. *Health justice*. Cambridge: Polity Press.

———. 2013. Health, vital goals, and central human capabilities. *Bioethics* 27 (5): 271–279.

Wolff, J., and A. De-Shalit. 2007. *Disadvantage*. Oxford: Oxford University Press.

World Health Assembly. 2011. *Report of the open-ended working group of member states on pandemic influenza preparedness: Sharing of influenza viruses and access to vaccines and other benefits*, 64th World Health Assembly A64/8. Geneva: World Health Association.

World Health Organisation. 2005. *Strengthening health security by implementing the international health regulations*. Geneva: World Health Organisation.

————. 2013. Pneumonia still responsible for one fifth of child deaths. http://www.who.int/media-centre/news/releases/2013/world-pneumonia-day-20131112/en/. Accessed May 2017.

————. 2014. *Antimicrobial resistance: global report on surveillance*. Geneva: World Health Organisation.

————. 2015. *Global action plan on antimicrobial resistance*. Geneva: World Health Organisation.

World health Organisation. Factsheet 392. Available at http://www.who.int/mediacentre/fact-sheets/fs392/en/. Accessed May 2017.

Chapter 15
Fairness in the Use of Information About Carriers of Resistant Infections

John G. Francis and Leslie P. Francis

Abstract One standard menu of approaches to the prevalence of anti-microbial resistance diseases is to enhance surveillance, fund research to develop new antimicrobials, and educate providers and patients to reduce unnecessary antimicrobial use. The primarily utilitarian reasoning behind this menu is unstable, however, if it fails to take fairness into account. This chapter develops an account of the fair uses of information gained in public health surveillance. We begin by sketching information needs and gaps in surveillance. We then demonstrate how analysis of information uses is incomplete if viewed from the perspectives of likely vectors of disease who may be subjects of fear and stigma and likely victims who may be coerced into isolation or quarantine. Next, we consider aspects of fairness in the use of information in non-ideal circumstances: inclusive participation in decisions about information use, resource plans for those needing services, and assurances of reciprocal support. Fairness in information use recognizes the ineluctable twinning of victims and vectors in the face of serious pandemic disease.

Keywords Fairness · Data use · Privacy · Surveillance

As many chapters in this volume emphasize, the prevalence of anti-microbial resistant diseases and the comparative paucity of available treatments presents a public health crisis. One standard menu of approaches to this crisis is to enhance surveillance to gain information needed to identify potential disease vectors and to ascertain likely modes of transmission; to fund research to develop new treatments and

J. G. Francis
Department of Political Science, University of Utah, Salt Lake City, USA
e-mail: john.francis@utah.edu

L. P. Francis (✉)
College of Law & Department of Philosophy, University of Utah, Salt Lake City, USA
e-mail: francisl@law.utah.edu

© The Author(s) 2020   243
E. Jamrozik, M. Selgelid (eds.), *Ethics and Drug Resistance: Collective Responsibility for Global Public Health*, Public Health Ethics Analysis 5,
https://doi.org/10.1007/978-3-030-27874-8_15

antimicrobials; and to intervene through education, treatment, and careful stewardship of the existing antimicrobials that retain some efficacy. This combination of approaches is founded primarily in utilitarian reasoning, attempting to achieve the best possible mitigation of the current crisis in the hopes that effective new treatment methods may soon become available.

Such utilitarian reasoning is not entirely stable in practice, however. On the one hand, when the prospects of exposure to untreatable and potentially fatal disease appear imminent, fear may become the overriding reaction to those who are identified as ill. The result may be forms of coercion against people suspected of being vectors of disease that appear prudential in the short term but that are insufficiently grounded in science and potentially counter-productive in the longer term. People may hide to avoid disclosure and deleterious consequences of over-regulation may lead to under-regulation. Recent examples include demands to compel isolation of people believed to have been exposed to Ebola or for banning travel from regions where outbreaks of conditions such as Ebola or Zika have been identified. On the other hand, concerns for victims may generate outpourings of resources for treatment, calls for investment in public health resources in underserved areas, and renewed emphasis on privacy protections. These too may be counterproductive if they result in confusion and waste of resources or multiple conflicting strategies. The upshot may be policies that oscillate between treating people as vectors and treating them as victims but without significant or coordinated progress against the problem of resistance.

Each of these perspectives—victim-hood and vector-hood—is morally important. But in our judgment analysis that is limited to these perspectives is incomplete in its failure to take certain considerations of fairness into account. Our specific focus here is the use of information, but similar points could be made about other types of resources as well. Collection, uses, and access to information, we contend in what follows, must be rooted in the effort to make progress against serious public health problems in a manner that is reasonably fair under the circumstances. This requires not only concern for people as victims and vectors but concerns about how the impact of policies are distributed and foster cooperative connections in both the shorter and the longer term.

15.1 The Important Roles of Information

Traditional public health surveillance methods are both individual and population based. Where particular individuals are concerned, the role of information is primarily to enable strategies to interrupt disease transmission. Case identification, case reporting, contact tracing, treatment if possible, and education and intervention if needed to prevent transmission come to the fore. At every stage, information is critical. If individuals with transmissible disease are unknown or cannot be located, efforts to interrupt transmission will fail. Efforts will also fail if information is not transmitted to those who are capable of acting, whether they be authorities designated to enforce quarantine or isolation or health care personnel equipped to offer treatment or prophylaxis. Education requires information, too, about where to direct educational efforts and what these efforts might contain. Importantly, if people who might suffer exposures are insufficiently informed about the likelihood and seriousness of contagion and the need for precautions, they may unwittingly become infected vectors as well as victims themselves. Such was the case for health care workers during the SARS epidemic of 2003 and for many during the Ebola epidemic of 2014.

Information gleaned in population-level surveillance plays many additional important roles in addressing the problem of anti-microbial resistance. A longstanding recommendation of the WHO, codified in the World Health Regulations that entered into force in 2007 in article 44, is international cooperation in the development of surveillance capacities for the identification of potential global health emergencies of international concern (WHO 2005). Surveillance can help to identify rates of incidence and prevalence of resistant disease. Testing samples can yield information about histories and patterns of disease spread. Samples also can be used to identify biological characteristics of resistant infectious agents that may be helpful in developing methods of treatment or identifying new anti-microbial agents.

Population level surveillance can be targeted to identifying the incidence and prevalence of resistant disease in particular geographical areas. Gonorrhea is an example. There were 78.3 million estimated new cases of gonorrhea worldwide in 2012; the highest number occurred in low-income areas of the western Pacific. Resistant disease has become increasingly prevalent, especially in these areas and among groups such as sex workers and truck drivers (Unemo et al. 2017). Extensively drug-resistant (XDR) gonorrhea cases also have appeared in Spain and in France, although these strains do not appear to have spread, possibly because they are less hardy and so less likely to be passed on. However, significant resistance may not be detected because of "suboptimal antimicrobial resistance surveillance in many settings" (Unemo et al. 2017). A recent international panel reviewing resistant gonorrhea recommends strategies of case management, partner notification, screening (especially of sex workers and men having sex with men), and evidence-based treatment (Unemo et al. 2017); these recommendations are based on surveillance data.

Population-level surveillance information may also be useful in identifying risks associated with providing humanitarian treatment. Over 30,000 young people wounded in the Libyan civil war that began in 2011 were evacuated elsewhere for treatment. Concerns arose that many of these patients were recognized to carry with them resistant organisms—thus bringing along with their needs for treatment risks to other patients being treated in the host facilities (Zorgani and Ziglam 2013). Institutions accepting these patients were informed of this risk so that they could take appropriate precautions. Libya itself was identified as a region with high prevalence of resistant organisms, despite the limited surveillance capacities in that conflict-torn nation. Recommendations included improving surveillance in Libya—which lacks a national surveillance system—and implementation of infection prevention measures in Libyan hospitals.

Surveillance is also used to identify practices that might contribute to the development of resistance. Use of antimicrobials in agriculture is one area of inquiry, although its precise contribution to the problem is not easy to quantify (e.g. Hoelzer et al. 2017). There have been many studies of problematic prescribing practices among physicians in the US (Wigton et al. 2008), Europe (e.g. Jørgensen et al. 2013), Asia (Lam and Lam 2003), and elsewhere (Trap and Hansen 2002), along with efforts to educate physicians about appropriate antimicrobial use.

Ever since the recognition grew that crowds celebrating the return of soldiers from World War I had created a ready opportunity for transmission of the Spanish influenza, epidemiologists have observed the potential health risks of large gatherings that concentrate people together, even for brief periods of time. Examples include music festivals, major sporting competitions, other large festivals, and religious gatherings such as the Hajj or other pilgrimages. The largest estimated gathering is the periodic Kumbh Mela pilgrimage in which Hindus come together to bathe in a sacred river such as the Ganges; over 40 million people, drawn largely from the Indian subcontinent but increasingly international, attend the event (Gautret and Steffen 2016). The largest annual gathering of pilgrims is the Hajj at Mecca which draws over two million people; the Fifth Pillar of Islam is the obligation to undertake the once in a lifetime journey for those who can physically or financially afford to do so. With such great numbers of people together for sustained periods of time, there is a risk of disease outbreaks and the spread of resistant infections. Such events may strain existing sanitation systems or health care facilities if people become ill. Crowding and inadequate facilities contribute to the potential for disease outbreaks (Gautret and Steffen 2016). These events draw people from around the globe and thus may result in the international spread of disease (Gautret and Steffen 2016).

At the same time, many of these events are of great cultural importance and suppression of them is neither a realistic nor a desirable option. There have been extensive discussions of how to address the public health needs of the great numbers of people who undertake pilgrimages or who attend other events that draw great numbers of people together. Vaccination may create herd immunities that reduce risks of disease transmission; for example, for this year's Hajj the Saudi Arabian government is requiring proof of a quadrivalent Meningococcal vaccination in order to receive a visa (Ministry of Hajj 2017). Nonetheless, risks may remain significant for conditions that cannot currently be addressed by vaccination or that are difficult to treat, such as Middle East respiratory syndrome coronavirus (MERS-CoV) or resistant infections. Information too is critical: such well-attended events require imaginative and thoughtful surveillance that informs short-term medical care. Because Saudi Arabia has had the largest number of human cases of MERS-CoV—an estimated 80% (WHO 2017b)—travelers for this year's Hajj are being warned to take extra precautions with respect to sanitation and personal hygiene measures such as handwashing or avoiding direct contact with non-human animals (New Zealand Ministry of Foreign Affairs 2017).

Still other social factors may contribute to the development of resistant disease that can be identified through surveillance. Given the difficulties for women in Saudi Arabia to see physicians without being escorted, it is understandable that in

Saudi Arabia many community pharmacies will dispense antibiotics without a prescription. Zowasi (2016) recommends addressing these issues by increased education especially through social media as to the best approach to respond to the risk of anti-microbial resistant organism.

Still other recommendations about information use involve research on the development of new forms of antimicrobials. According to the most recent review article (Butler et al. 2017), antibiotics "are dramatically undervalued by society, receiving a fraction of the yearly revenue per patient generated by next-generation anticancer drugs." They are in the judgment of these authors an "endangered species,"—but there is some faint encouraging news. WHO and a number of national governments have recently begun to direct attention to the potential threat of resistance and lack of new drugs. Since 2000, five new-in-class antibiotics have been marketed, but these unfortunately only target gram-positive organisms not the gram-negative organisms that are likely to be resistant. Other compounds are also in various stages of the process of clinical trials, but these too are more likely to be active against gram-positive bacteria. In the judgment of the authors of this review article, "the acute positive trend of new approvals masks a chronic underlying malaise in antibiotic discovery and development." Interest in antibiotic development is more likely to be present in smaller biotech companies and in biotech companies located in Europe. The authors conclude: "The only light on the horizon is the continued increase in public and political awareness of the issue." They also observe that with the retrenchment in investment, "we potentially face a generational knowledge gap" and drug development "is now more important than ever."

To address this perilous juncture in antimicrobial research, the Pew Charitable Trust convened a scientific expert group in 2016. The premise of the group was that regulatory challenges, scientific barriers, and diminishing economic returns have led drug companies largely to abandon antibiotic research—yet antimicrobial resistance is accelerating. No entirely new classes of antibiotics useful against resistant organisms have been brought to market that are not derivatives of classes developed before 1984—over 30 years ago. The Pew report advances many explanations for this dismal situation, including importantly the lack of coordinated investment in the relevant basic and translational research. One aspect of the report detailed the major role played by information gaps. Published research is out of date and out of print. Moreover, in today's world of investment in drug discovery, "creating an environment in which data exchange and knowledge sharing are the status quo will be difficult given proprietary concerns and the variety of information types and formats, which may range from historical data to new findings produced as part of this research effort." The Pew consensus is that the following forms of information sharing are needed: a review of what is known about compounds that effectively penetrate gram-negative bacteria, a searchable catalogue of chemical matter including an ongoing list of promising antibacterial compounds, information on screening assays and conditions tested, and an informational database of available biological and physicochemical data. Mechanisms must also be developed for sharing drug discovery knowledge in the area (Pew, pp. 19–20).

In line with Pew, a European antimicrobial resistance project suggests that research is seriously underfunded (Kelly et al. 2015). This group argues that the bulk of the publicly funded research is in therapeutics (63%); among the remainder, 14% of the research was on transmission and only 3% specifically on surveillance. This group also concluded that research is not coordinated and there is little attention to data sharing or sharing of research results. Funding is fragmented, too, with many smaller grants addressing smaller projects independently rather than in a way that builds. This group summarizes: "to conclude, investment at present might not correspond with the burden of antibacterial resistance and the looming health, social, and economic threat it poses on the treatment of infections and on medicine in general. Antibacterial resistance clearly warrants increased and new investment from a range of sources, but improved coordination and collaboration with more informed resource allocation are needed to make a true impact. Hopefully, this analysis will prompt nations to pay due consideration to the existing research landscape when considering future investments."

Additional recommendations from other groups include novel methods for management of resistant disease, such as addressing the intestinal microbiome (e.g. Bassetti et al. 2017); these methods, too, may be furthered by surveillance information as well as information about individual patients.

Analysis of these uses of information from the perspective of vector or victim are, we now argue, incomplete.

15.2 The Vector Perspective

When contagious diseases are serious or highly likely to be fatal and treatments for them are limited at best, fear is understandable. Fear may be magnified if the disease is poorly understood, especially until modes of transmission have been identified. Fear may also be magnified if there are no known effective treatments for the disease, as may be the case for extremely drug resistant infections. It is therefore understandable that proposals may come to the fore that emphasize isolation of those who are known to be infected, quarantine of those who have been exposed, or travel bans from areas of known disease outbreaks. Proposals may even include criminalization of those who knowingly or even negligently take risks of infecting others.

All of these possibilities and more were features of the HIV epidemic. Even as understanding of the disease grew and effective treatment became increasingly available, some of these remain. Criminalization of HIV transmission has not waned, despite the many objections raised to it (e.g. Francis and Francis 2013a, b). Although the US ended its immigration ban on HIV+ individuals in 2010, concerns remain about the risks of undiagnosed infections among immigrant populations in the U.S. (Winston and Beckwith 2011) and some countries (for example, Singapore) continue to ban entry for HIV+ travelers planning stays over thirty days (The Global Database 2017).

As epidemic fears have waxed and waned over recent decades, so have impera-tives for identifying vectors and constraining their activities. These patterns have been apparent for avian influenza, SARS, Ebola, and Zika, among others. The US still bars entry by non-citizens with a list of conditions including active TB, infec-tious syphilis, gonorrhea, infectious leprosy, and other conditions designated by Presidential Executive order such as plague or hemorrhagic fevers (CDC 2017).

Indeed, resistant TB has been a frequent illustration of the vector perspective in operation. Multi-drug resistant tuberculosis is transmissible, difficult to treat, and poses a significant public health problem. Its presence can be identified by methods such as testing of sputum samples. When patients are identified with resistant dis-ease, public health authorities may seek to compel treatment or isolation, especially for patients judged unreliable about compliance with treatment. To avoid transmis-sion, public health authorities have proposed isolating patients who have been iden-tified as infected. Because a course of treatment for TB may take many months—and failure to complete the full course may increase the likelihood of resistant disease—isolation may continue for long periods of time. Controversially, during the early 1990s public health officials in New York isolated over 200 patients identified with MDR TB on Roosevelt Island for treatment out of concern that they would be non-compliant with treatment even when they were unlikely to infect others (Coker 2001).

Perhaps one of the most highly publicized events involving a single patient was the odyssey of Andrew Speaker, a lawyer believed to have extremely resistant TB who eluded authorities as he took airplane flights around the globe in the effort to return home. Speaker's journey created an international scare and calls for travel restrictions. Speaker's lawsuit against the Centers for Disease Control and Prevention alleging violations of the federal Privacy Act, he claimed by revealing more infor-mation than was necessary for public health purposes, was ultimately resolved on summary judgment for the government, largely because the challenged disclosures had been made by Speaker himself. (*Speaker v. U.S. Department of Health and Human Services Centers for Disease Control and Prevention*, 489 Fed. Appx. 425 (2012); Associated Press 2010) But the saga reveals how individuals perceived as threats may be vilified for what were understandable, if unwise, efforts at self-protection.

WHO travel guidelines provide that individuals known to be infected with resis-tant TB should not travel until sputum analysis confirms that they are not at risk of disease transmission (WHO 2017). Evidence is limited, however, about the need for this policy. The most recent literature review suggests that risks of transmission dur-ing air travel are very low and that there is need for ongoing international collabora-tion in contact tracing and risk assessment (Kotila et al. 2016). Blanket travel bans encouraging actions that elude detection may reduce, rather than enhance, this needed collaboration. More subtle policies tailored to need would be preferable, but the fears generated by a focus on fear of vectors may make them unlikely to be developed or implemented.

At best, therefore, the vector perspective is incomplete. Focus on it may be counter-productive, if people hide or try to avoid education. It may encourage expenditures on efforts to identify suspected vectors rather than on evidence based efforts to identify risks of transmission and effective modes of prevention. And, of course, it ignores the plight of victims, to which we now turn.

15.3 The Victim Perspective

People with resistant infections are not only vectors, they are also victims of disease and have ethical claims to be treated as such (Battin et al. 2007). Indeed, it is likely that vectors will themselves be victims, unless they are carriers of the disease in a manner that does not affect them symptomatically.

Concern for victims may take the form of seeking to ease the burdens of constraints such as isolation. A good illustration of the victim perspective in operation is the WHO publication of a pamphlet on "psychological first aid" to those affected by Ebola. The pamphlet is designed to provide comfort to and meet the basic needs of people infected by Ebola and those who are close to them, while maintained the safety of aids workers (WHO 2014). The recommendations rest on the importance of respect for the dignity of those who are suffering amidst disease outbreaks. It also emphasizes the importance of respect for rights such as confidentiality and non-discrimination. The pamphlet is provisional and designed to be updated as knowledge of safety measures improves; this provisional nature is a recognition of the importance of ongoing development of information about how victims' needs can be safely met.

Despite the concern for victims, foremost in the pamphlet's recommendations is safety, both of aid workers and of disease victims, so that no one is further harmed including victims themselves and others close to them. Overall, the pamphlet attempts to counter impulses to come to the aid of victims that may increase transmission risks, such as unprotected contact with those who are ill. But unexplored tensions remain in the document's recommendations. For example: "Respect privacy and keep personal details of the person's story confidential, if this is appropriate" (p. 22). Nowhere does the document discuss when confidentiality is appropriate or what personal details may be revealed and in what ways. Its manifest and important concern for victims is countered by safety but without discussion of how these goals might be implemented together or reasonably reconciled in practice.

The WHO's most recently-adopted strategy for dealing with health emergencies, the Health Emergencies Programme, provides another illustration of concern for victims that may lie in unexplored tension with other values. The Programme urges cooperative methods to meet the immediate health needs of threatened populations through humanitarian assistance while also addressing causes of vulnerability and recovery (WHO 2016). It is a coordinated strategy for emergency response that will move far beyond merely technical help; WHO describes it as a "profound change for WHO, adding operational capabilities to our traditional technical and normative roles" (WHO 2016). It is aimed to provide crisis help, such as to Hurricane Matthew in Haiti or to areas affected by the Zika virus. It requires a major increase in funding devoted to core emergency efforts. Core funding will come from assessed contributions, flexible contributions that the Director-General has discretion to allocate, and earmarked voluntary contributions. But it is clearly under-funded; WHO reported a 44% funding gap as of October 2016, just to meet the program's core capabilities. Moreover, WHO also reported that it has raised less than a third of the funding needed for the WHO Contingency Fund for Emergencies, a fund deployed for the initial 3 months of an emergency before donor funding becomes available (WHO 2016).

The Health Emergencies Programme reflects reactions to the humanitarian disaster of the Ebola epidemic and criticisms of the WHO level of response. The WHO 2016–2017 budget reflects this response as well (WHO 2015). That budget "demonstrates three strategic shifts" (WHO 2015, p. 2). The first is application of the lessons from Ebola especially the need to strengthen core capacities in preparedness, surveillance and response. The second strategic shift is a focus on universal health coverage, which includes enhancing contributions to maternal and child health, speeding progress towards elimination of malaria, and enhancing work on noncommunicable diseases, among other worthy goals. The final strategic shift is towards "emerging threats and priorities"; illustrations of these are "antimicrobial resistance, hepatitis, ageing, and dementia." These are not an obvious group to characterize as "emerging," to the extent that this suggests a developing threat that has not yet become urgent but that may be expected to become so in the near future. Nor are they an obvious group to link together in the same category. This mixture of budgetary priorities suggests is responsiveness to issues raised through consultation with WHO member states, rather than proactive planning.

WHO specific efforts directed to resistance can be characterized as primarily coordination. The WHO website devoted to resistance promotes information sharing and lists research questions and potential funding agencies (WHO 2017a). WHO expresses no judgment about either funding agencies or which of the nearly 100 listed research questions—ranging from research on resistance in day care centers to the biological price that microorganisms pay for resistance—might be fruitfully addressed first or how they might be interconnected.

Concern for victims is surely part of a response to a humanitarian emergency. Responsiveness to urgent health needs is an important goal. Including antimicrobial resistance in a list of "emerging" issues is at least recognition of the problem. But the WHO response to Ebola and the WHO budget overall can be characterized as less than fully set into context in a reasoned way.

Thus, we contend, neither vector nor victim perspectives are adequate. One risks falling prey to fear while the other risks responses that are well-intentioned but that may be difficult to meet or compete with other values in ways that remain underexplored. These perspectives are inevitable and important, but they are each incomplete.

15.4 Fairness in Information Use

In our judgment, a primary difficulty with both vector and victim perspectives is that neither are set into context or seen as interconnected. This section suggests how fairness considerations may help in focusing attention to the most pressing questions to ask about antimicrobial resistance and the directions for surveillance and information use to take.

Fairness entered the philosophical lexicon in discussions about justice as procedural, most famously in John Rawls's "Justice as Fairness" (Rawls 1958). As Rawls initially conceptualized his view, it involved a decision procedure for selecting basic principles of justice in which people were unable to gain unfair advantage. As the debates about Rawlsian justice unfolded, a fundamental issue was whether people with radically different capacities and views of the good life could be expected to accept the results of the decision procedure as formulated. Thus critics raised the concern that people with disabilities might be left out of the decision procedure as "non-contributors" to the practice of justice (Nussbaum 2007; Stark 2007). Critics also pressed the argument that people with radically illiberal conceptions of the good would ultimately destabilize the practice of justice in a Rawlsian ideal society (e.g. Williams 2007a, b). Rawls ultimately accepted the point that proceduralism could not yield a universal theory of justice, pulling back his view to the claim that it only represented a vision of justice for a certain kind of liberal society (Rawls 1993).

But fairness also entered the debates about justice in a more substantive way, especially in bioethics. Norman Daniels (1985), for example, expanded a Rawlsian approach to consider justice in health care. The British idea of a "fair innings," in which the opportunities of each to reasonable health over a normal life span are prioritized, was raised particularly with respect to the distribution of health care resources to the elderly (Bognar 2015; Farrant 2009; Harris 1985; Williams 2007b). Like the metaphor of a level playing field, the fair innings argument comes from sports (Francis 2017). It reflects the idea of everyone having a chance to participate in a game that at least gives them a reasonable opportunity for success. There are four aspects of such opportunity: who plays and whether the rules are constructed to give each an opportunity to win that is reasonable are two. Also important is the balance among opportunities to succeed, so that there aren't consistent tilts in one direction or another, as might be characterized by the further metaphor of leveling the playing field. Finally, attention to the interaction between advantages and disadvantages matters, so that participants are encouraged to continue playing the game rather than dropping out.

Our invocation of fairness as a concept is rooted in the judgment that antimicrobial resistance—or other pressing global public health problems, for that matter—exemplify multiple aspects of non-ideal and partial compliance circumstances. Natural circumstances are less than forgiving; new health threats emerge on a regular basis. Antimicrobial resistance is an ongoing natural challenge to effective therapy for deadly diseases. Social circumstances are imperfect, too: overcrowding, poor sanitation, straitened resources for public health and health care, and cultural practices that increase potential for disease transmission all play roles in the development of resistance. Alexander Fleming, the discoverer of penicillin, warned that the development of resistance was likely, but his warning appears not to have been well heard. Finally, efforts to address antimicrobial resistance are riven with non-compliance: over-prescribing by physicians, over-use of antimicrobials in agriculture, individual failures to take medications as prescribed, and concealment of disease out of fear of discovery and persecution. Because the conditions that give rise to these problems of non-compliance may seem urgent—people seeking anti-microbials are in pain or ill, perhaps gravely; people in hiding from health authori-ties may fear stigmatization or death—they raise in particularly poignant form questions of the extent of obligations under circumstances in which others are not doing what arguably is their fair share (e.g. Stemplowska 2016; Murphy 2000).

Fairness as an ethical concept is especially suited to such imperfect circum-stances. It directs attention to how improvements are distributed. Distributions can be more or less fair, if they distribute benefits and burdens in an increasingly inclu-sive manner (e.g. Francis and Francis 2013a, b). Fairness thus construed is at the heart of perhaps the most influential set of recommendations for ethical pandemic planning, the Canadian *Stand on Guard for Thee* (Toronto Joint Centre 2005). Although much of the discussion of fairness in this document emphasizes inclusive procedures, so that engagement may lead to acceptance of choices as fairly made (e.g., p. 1), the recommendations also contain substantive dimensions. These include fair resource plans for those who fall ill providing necessary services during a pan-demic (p. 11) and assurance that people who are affected by choices are reciprocally supported in a way that they do not suffer "unfair economic penalties" (p. 13). Here, the links between fairness and reciprocity are explicit.

These four aspects of fairness—who is included in the play, what opportunities they have, how these opportunities are balanced, and whether there are elements of reciprocity—can be used to set vector and victim perspectives into context in addressing the gathering and use of information about antimicrobial resistance. Over-emphasizing vectors threatens their opportunities and even possible participa-tion. Overemphasizing victims tilts the field unidirectionally, understandably direct-ing resources to immediate need but without consideration of longer-term consequences. Reciprocity may be the most important of all, creating commitment to workable strategies for addressing resistance when there are difficult choices to be made.

Fear, understood as a threat personal health, is often an ally in persuading people to seek preventive care and to change life styles, or to persuade policy makers to create incentives or penalties for decisions that contribute to poor health. But great fear can also lead to immobility. The real threat posed by the rise of antimicrobial resistance does not seem to be easily addressed by a successful alternative in the view of victims or policy makers. Medical personnel are fearful of not responding to the demands of patients for immediate reductions in pain or suffering at relatively low costs. The scale of the threat posed by rapid rise of antimicrobial resistance may be daunting to policy makers especially as funders of research. The cost of develop-ing ever-new generations of antibiotics seems to suggest a great series of short-term solutions especially as pharmaceutical companies respond to incentives to generate near-term profits. In this context, it is worth recalling how the development of the first antimicrobials contributed to more generally shared benefits: when penicillin became known to people as a wonderful drug it actually helped to speed the adop-tion of the National Health Service in Britain. The popular expectation was health care for all facilitated with the rise of a new generation of low cost wonder drugs and reinforced by low cost vaccinations (Webster 2002). But some of the advan-tages were short-lived, as the costs of pharmaceuticals grew exponentially and inadequate attention was paid to the risks of overprescribing—once again a cautionary reminder of the importance of emphasizing balance rather than one particular perspective such as victimhood. If a promise of sustaining production at lower costs of ever-new generations of antimicrobials from how information is used can offer benefits more widely, then it becomes easier to impose tougher regulations on antimicrobial use that may to some extent stave off the development of resistance.

This approach in terms of fairness directs attention not only to vectors and to victims seen as separate entities. It also directs attention to how they are often, and unpredictably, twinned—given the epidemiology of resistance spread, it is likely to begin within interlaced communities where vectors are also victims. But it also directs our attention to these issues set in distributive context, raising questions such as these: Who is most likely to be affected by resistance? Who will suffer the most severe consequences from resistance? Who is most likely to be disadvantaged by information gained to counter resistance? Who will suffer the most severe disadvantage? Who will benefit from efforts to counter resistance? How can these benefits be spread more inclusively? And, how are the benefits and burdens of addressing resistance intertwined? Are some primarily beneficiaries, while others are primarily burdened? Are there ways to increase reciprocal linkages in these benefits and burdens, so that efforts to counter resistance are accepted and supported more widely? These are the kinds of questions that need to guide how surveillance is deployed in the effort to counter resistance, not vague generalities about the importance of addressing health infrastructure or bromides about the need to increase resources.

References

Associated Press. 2010. Georgia: Suit against disease centers is revived. *The New York Times* (Oct. 22) [online]. http://www.nytimes.com/2010/10/23/us/23brfs-SUITAGAINSTD_BRF.html.

Bassetti, Matteo, Garyphallia Poulakou, Etienne Ruppe, Emilio Bouza, Sebastian J. Van Hal, and Adrian Brink. 2017. Antimicrobial resistance in the next 30 years, humankind, bugs and drugs: A visionary approach. *Intensive Care Medicine*, epub https://link.springer.com/article/10.100 7%2Fs00134-017-4878-x.

Battin, Margaret P., Leslie P. Francis, Jay A. Jacobson, and Charles B. Smith. 2007. *The patient as victim and vector: Ethics and infectious disease*. New York: Oxford University Press.

Bognar, Greg. 2015. Fair innings. *Bioethics* 29 (4): 251–261.

Butler, Mark S., Mark A.T. Blaskovich, and Matthew A. Cooper. 2017. Antibiotics in the clinical pipeline at the end of 2015. *The Journal of Antibiotics* 70: 3024.

Centers for Disease Control and Prevention. 2017. Immigrant and refugee health. https://www.cdc.gov/immigrantrefugeehealth/laws-regs/hiv-ban-removal/final-rule-general-qa.html.

Coker, Richard. 2001. Just coercion? Detention of nonadherent tuberculosis patients. *Annals of the New York Academy of Sciences* 953b: 216–223.

Daniels, Norman. 1985. *Just health care*. 1st ed. New York: Cambridge University Press.

Farrant, Anthony. 2009. The fair innings argument and increasing life spans. *British Medical Journal* 35: 53–56.

Francis, Leslie P. 2017. Promoting equality in and through the paralympics. In *Philosophy: Sport. Macmillan interdisciplinary handbooks*, ed. R. Scott Kretchmar, 245–262. Farmington Hills: Macmillan Reference USA.

Francis, Leslie P., and John G. Francis. 2013a. HIV treatment as prevention: Not an argument for continuing criminalization of HIV transmission. *International Journal of Law in Context* 9 (4): 520–534.

———. 2013b. Informatics and public health surveillance. In *Bioinformatics law: Legal issues for computational biology in the post-genome era*, ed. Jorge Contreras and James Cuticcia. Cleveland: American Bar Association.

Gautret, Philippe, and Robert Steffen. 2016. Communicable diseases as health risks at mass gatherings other than Hajj: What is the evidence? *International Journal of Infectious Diseases* 47: 46–52.

Harris, John. 1985. *The value of life*. London: Routledge & Kegal Paul.

Hoelzer, Karin, Nora Wong, Joe Thomas, Kathy Talkington, Elizabeth Jungman, and Allan Coukell. 2017. Antimicrobial drug use in food-producing animals and associated human health risks: What, and how strong, is the evidence? BMC Veterinary Research 13: 211. https://doi.org/10.1186/s12917-017-1131-3.

Jørgensen, Lars Christian, Sarah Friis Christensen, Gloria Cordoba Currea, Carl Llor, and Lars Bjerrum. 2013. Antibiotic prescribing in patients with acute rhinosinusitis is not in agreement with European recommendations. *Scandinavian Journal of Primary Health Care* 31 (2): 101–105.

Kelly, Ruth, Ghada Zoubiane, Desmond Walsh, Rebecca Ward, and Herman Goossens. 2015. Public funding for research on antibacterial resistance in the JPIAMR countries, the European Commission, and related European Union agencies: A systematic observational analysis. *The Lancet Infectious Diseases* 16 (4): 431–440.

Kotila, S.M., L. Payne Hallström, N. Jansen, P. Helbling, and I. Abubakar. 2016. Review: Systematic review on tuberculosis transmission on aircraft and update of the European Centre for Disease Prevention and Control risk assessment guidelines for tuberculosis transmitted on aircraft (RAGIDA-TB). *Eurosurveillance* 21 (4): 30114. http://www.eurosurveillance.org/ViewArticle.aspx?ArticleId=21357#aff1.

Lam, T.P., and K.F. Lam. 2003. What are the non-biomedical reasons which make family doctors over-prescribe antibiotics for upper respiratory tract infection in a mixed private/public Asian setting? *Journal of Clinical Pharmacy and Therapeutics* 28 (3): 197–201.

Ministry of Hajj, Kingdom of Saudi Arabia. 2017. Saudi Ministry of Health Requirements and Health Matters. http://www.hajinformation.com/main/t20.htm.

Murphy, Liam B. 2000. *Moral demands in nonideal theory*. Oxford: Oxford University Press.

New Zealand Ministry of Foreign Affairs. 2017. Hajj pilgrimage. *Live News* (Aug. 14) [online]. https://livenews.co.nz/2017/08/14/hajj-pilgrimage/.

Nussbaum, Martha C. 2007. *Frontiers of justice: Disability, nationality, species membership*. Cambridge, MA: Harvard University Press.

Rawls, John. 1958. Justice as fairness. *The Philosophical Review* 67 (2): 164–194.

———. 1993. *Political liberalism*. New York: Columbia University Press.

Stark, Cynthia A. 2007. How to include the severely disabled in the contractarian theory of justice. *Journal of Political Philosophy* 15 (2): 127–145.

Stemplowska, Zofia. 2016. Doing more than one's fair share. *Critical Review of International Social and Political Philosophy* 19 (5): 591–608.

The Global Database on HIV-Specific Travel & Residence Restrictions. 2017. Regulations on entry, stay and residence for PLHIV. http://www.hivtravel.org/Default.aspx?PageId=143&Mode=list&StateId=2.

The PEW Charitable Trusts (Pew). 2016. A scientific roadmap for antibiotic discovery. (May 11). http://www.pewtrusts.org/en/research-and-analysis/reports/2016/05/a-scientific-roadmap-for-antibiotic-discovery.

Trap, Birna, and E.H. Hansen. 2002. Treatment of upper respiratory tract infections—A comparative study of dispensing and non-dispensing doctors. *Journal of Clinical Pharmacy and Therapeutics* 27 (4): 289–298.

Unemo, Magnus, Catriona S. Bradshaw, Jane S. Hocking, Henry J. C. de Vries, Suzanna C. Francis, David Mabey, Jeanne M. Marrazzo, Gerard J.B. Sonder, Jane R. Schwebke, Elske Hoornenborg, Rosanna W. Peeling, Susan S. Philip, Nicola Low, & Christopher K. Fairley. 2017. Sexually transmitted infections: challenges ahead. *Lancet Infectious Diseases.* 17(8):e235-e279. 10.1016/S1473-3099(17)30310-9. Epub 9 July 2017.

University of Toronto Joint Centre for Bioethics Pandemic Influenza Working Group (Toronto Joint Centre). 2005. Stand on Guard for Thee: Ethical considerations in preparedness planning for pandemic influenza. http://www.jcb.utoronto.ca/people/documents/upshur_stand_guard.pdf.

Webster, Charles. 2002. *The National Health Service: A political history.* Oxford: Oxford University Press.

WHO. 2005. International Health Regulations, 3rd ed. http://apps.who.int/iris/bitstream/10665 /246107/1/9789241580496-eng.pdf?ua=1.

———. 2014. Psychological first aid during Ebola virus disease outbreaks: Provisional version. http://apps.who.int/iris/bitstream/10665/131682/1/9789241548847_eng.pdf?ua=1.

———. 2015. Programme Budget 2016–2017. http://who.int/about/finances-accountability/ budget/PB201617_en.pdf.

———. 2016. WHO's new Health Emergencies Programme. http://www.who.int/features/qa/ health-emergencies-programme/en/.

———. 2017a. Drug resistance. http://www.who.int/drugresistance/infosharing/en/.

———. 2017b. Middle East respiratory syndrome coronavirus (MERS-CoV) (May). http://www. who.int/mediacentre/factsheets/mers-cov/en/.

Wigton, Robert S., Carol A. Darr, Kitty K. Corbett, Devin R. Nickol, and Ralph Gonzales. 2008. How do community practitioners decide whether to prescribe antibiotics for acute respiratory infections? *Journal of General Internal Medicine* 23: 1615–1620.

Williams, Alan. 2007a. Intergenerational equity: An exploration of the 'fair innings' argument. *Health Economics* 6 (2): 117–132.

Williams, Bernard. 2007b. In *In the beginning was the deed: Realism and moralism in political argument*, ed. Geoffrey Hawthorn. Princeton: Princeton University Press.

Winston, Susanna E., and Curt G. Beckwith. 2011. The impact of removing the immigration ban on HIV-infected persons. *AIDS Patient Care and STDs* 25 (12): 709–711.

World Health Organization (WHO). 2017. International travel and health: Tuberculosis (TB). http://www.who.int/ith/diseases/tb/en/.

Zorgani, Abdulaziz, and Hisham Ziglam. 2013. Letter: Injured Libyan combatant patients: Both vectors and victims of multiresistance bacteria? *Libyan Journal of Medicine* 58: 20325. https://doi.org/10.3402/ljm.v8i0.20325.

Zowasi, Hosam M. 2016. Antimicrobial resistance in Saudi Arabia: An urgent call for an immediate action. *Saudi Medical Journal* 37 (9): 935–940.

Chapter 16
Antimicrobial Resistance and Social Inequalities in Health: Considerations of Justice

Lynette Reid

Abstract Within-country social inequalities in health have widened while global health inequalities have (with some exceptions) narrowed since the Second World War. On commonly accepted prioritarian and sufficientist views of justice and health, these two trends together would be acceptable: the wealthiest of the wealthy are pulling ahead, but the worst off are catching up and more are achieving sufficiency. Such commitments to priority or sufficiency are compatible with a common "development" narrative about economic and social changes that accompany changes ("transitions") in population health. I set out a very simple version of health egalitarianism (without commitment to any particular current theory of justice) and focus on two common objections to egalitarianism. Priority and sufficiency both address the levelling down and formalism objections, but these objections are distinct: giving content to equality (I argue here) places in question the claimed normative superiority of priority and sufficiency. Using examples of the role of antimicrobials in both these trends – and the future role of AMR – I clarify (first) the multiple forms and dimensions of justice at play in health, and (second) the different mechanisms at work in generating the two current patterns (seen in life course narratives and narratives of political economy). The "accelerated transition" that narrowed global health inequalities is fed by anti-microbials (among other technology transfers). It did not accelerate but replaced the causal processes by which current HICs achieved the transition (growing and shared economic prosperity and widening political franchise). The impact of AMR on widening social inequalities in health in HICs will be complex: inequality has been fed in part by tertiary care enabled by antimicrobials; AMR might erode the solidarity underlying universal health systems as the well-off seek to maintain current expectations of curative and rehabilitative surgery and chemotherapy while AMR mounts. In light of both speculations about the impact of AMR on social and global health inequalities, I close with practical and with theoretical reflection. I briefly indicate the practical impor-

L. Reid (✉)
Department of Bioethics, Dalhousie University, Halifax, Canada
e-mail: Lynette.Reid@dal.ca

© The Author(s) 2020
E. Jamrozik, M. Selgelid (eds.), *Ethics and Drug Resistance: Collective Responsibility for Global Public Health*, Public Health Ethics Analysis 5,
https://doi.org/10.1007/978-3-030-27874-8_16

tance of understanding AMR from the perspective of health justice for policy response. Then, from a broader perspective, I argue that the content by which I meet the formalism objection demonstrates that the two trends (broadening within-country inequality and narrowing global inequality) are selective and biased samples of a centuries-long pattern of widening social inequalities in health. We are not in the midst of a process of "catching up". In light of the long-term pattern described here, is the pursuit of sufficiency or priority morally superior to the pursuit of equality as a response to concrete suffering – or do they rationalize a process more objectively described as the best-off continuing to take the largest share of one of the most important benefits of economic development?

Keywords Bioethics · Public health ethics · Antimicrobial resistance · Health inequality · Health justice

16.1   Introduction

What is the significance of antimicrobial resistance (AMR) for health inequalities? This question refers not just to the extent to which AMR might exacerbate or mitigate health inequalities, but to the ways that we should think about health inequalities in order to be adequately sensitive to the justice dimensions of AMR and to the lessons that AMR might offer about the nature of health inequalities as such and their normative status (or the normative plausibility of responses to health inequalities).

The development of AMR in bacteria, viruses, fungi, and parasites detrimental to human health is already a substantial public health concern; it is estimated to cause 700,000 deaths a year globally. The UK government's O'Neill report paints what it calls a worst-case scenario—a post-antimicrobial world in which antimicrobial-resistant infections are collectively the second leading cause of death by 2050, causing 10 million deaths a year and lowering world population by 700 million (O'Neill 2014, 2016). This "worst case scenario" is based on projecting the consequences of AMR for a small selection of conditions and as such may underestimate the potential consequences of AMR (Jamrozik and Selgelid, Chap. 1, this volume).

It is widely acknowledged that the development of AMR will have different effects on high-income, middle-income, and low-income countries. Other chapters in this volume propose or evaluate policies to address and mitigate AMR as these matter for health equity and for global health.

In this chapter, I situate the use of antimicrobials and the growth of AMR in a narrative of population health (Valles 2018), specifically the story of narrowing global (between country) and widening local (within country) health inequalities, and argue that the development of AMR puts pressure on normative commitments to sufficientism and prioritarianism as purportedly more feasible and more normatively satisfying goals for health justice.

Faced with data showing that people who live in one neighbourhood may have a life expectancy 10–20 years lower than people in a nearby neighbourhood, or that the difference in life expectancy between countries is in some cases as high as 35 years, many express a normative worry that social inequality in health and global health inequalities are unjust (Marmot 2015). However, among academics who discuss health justice, few are egalitarians about health. Sufficiency – the attainment of a decent minimum – is often argued to be an appropriate goal; priority for the worst-off is argued to be adequately action-guiding. The question of sufficiency, priority, or equality can be deferred to ideal theory; we need not agree on an ultimate goal in order to act. For example, if we can agree that AMR is likely to hit the worst-off the hardest, as the O'Neill report projects, then sufficientists, prioritarians, and egalitarians alike can focus their attention on alleviating its effects for the worst-off while still remaining faithful to their diverging commitments if they wish.

16.2 Health Inequalities and Health Egalitarianism: Definitions

In this chapter, I use the term "social inequality in health" to refer to inequalities in some health indicator across socioeconomic status (SES) or other category of interest to justice within countries. I will refer to life expectancy or mortality, but metrics of inequality include other key indicators, for example, infant mortality or workplace mortality. I will use "global health inequality" to refer to inequality in such indicators across low-, middle-, and high-income countries (typically referring to LMICs on the one hand and HICs on the other).[1] Granted that there is normatively significant variation in what counts as socioeconomic status and how it is measured, I speak broadly of socioeconomic inequalities in health, while not intending any specific version of the concept as fundamental. (I motivate/explain this below.)

The core commitment of a health egalitarian is that, all things being equal, a society in which persons of different socioeconomic status can lead full lives with minimization and mitigation of health-related disruptions to those lives is more just than a society in which the possibility of a healthy life is patterned by

---

[1] The less cumbersome phrase "health equality" refers simply to the distribution of life years across a population, not correlated with any other variable of interest to justice. For example, Strømme and Norheim recently claim that both global and national health inequalities are shrinking (Strømme and Norheim 2017). They base this claim on a univariate measure of the distribution of a health indicator alone (age at death) within a population: their empirical claim is that the distribution of age at death is becoming more closely clustered around a value rather than spread widely or polarized while their implicit normative claim is that form of health inequality is what matters to justice. Such univariate measures of health inequality, however, capture nothing about who is dying early or late within that narrower distribution – for example, the rich or the poor, the racialized or the non-racialized, those of higher or lower educational attainment, or all social classes equally (Regidor 2004). Life expectancy could cluster around a given value without any change in the gap between the rich and the poor.

socioeconomic status. That is, a society in which service workers, manual labourers, or those excluded from labour market participation live lives as long and healthy as executives or professors is more just than a world in which this is not the case. On a global scale, a world in which such possibilities of health are not patterned by accident of global birth is more just than a world in which that is not the case.

Following Temkin (2000, 2003), I explore the health egalitarian commitment without assuming a specific view on whether this commitment should dominate feasibility considerations or other values that may conflict with the achievement of social equality in health or global health equality. For example, a commitment to cultural self-determination may outweigh a commitment to health egalitarianism in specific cases, granting the moral and political right of a community to pursue economic activity and social organization that exposes it to different supports for and risks to health, such that life expectancy differs from that obtained in urban settings in HICs.

The leveling down objection seems particularly acute for health egalitarians: while it may be acceptable to take financial resources from people for redistribution, can it ever be acceptable to worsen the health of some people in order to achieve some increment of improvement in the health of the less well off? Temkin's approach avoids the common use of this "leveling down" objection (Parfit 1997, 2012) to take egalitarianism off the table. We can explore the normative importance of social equality in health without *ipso facto* committing ourselves to the view that it has such normative weight that it trumps other considerations and values.

16.2.1 Toward a Multi-dimensional Account of Justice, Health, and Equality as a Normative Goal

A health egalitarian faces (at least) three substantial challenges: the leveling down objection (discussed in the previous section), the challenge that health inequalities *qua* natural can only be matters of justice insofar as they cause or are caused by unjust social inequalities, and the challenge of motivating normatively a concern with the abstract and interpersonal measure of inequality. These three challenges – the leveling down objection, the natural difference objection, and the formalism objection – are closely related. For example, one might focus on concrete normative concerns – suffering (and its alleviation) or sufficient health (and its achievement) – and, by adopting prioritarianism or sufficientism, avoid both the formalism and the leveling down objections (as in Anderson (1999, 2010) and other proponents of capabilities (Nussbaum 2006; Venkatapuram and Marmot 2013)). However, these objections are conceptually distinct. Filling out the concrete realities that are reflected in formal measures of inequalities might bolster rather than undermine the case for equality over sufficiency or priority as a normative goal.

In this paper I use examples drawn from antimicrobials and antimicrobial resistance to give content to concerns about social equality in health, showing that the formalist challenge can be met without adopting a sufficientist or prioritarian alternative to egalitarianism. I do not offer or adopt a specific comprehensive theory of justice or even of justice and health; rather, using examples from antimicrobials and antimicrobial resistance, I explore different forms of justice and different relationships between justice and health. These are phenomena that an adequate account of health justice should capture, but what account of health justice in what form will do that adequately is not a question I attempt to resolve here.[2]

Certainly, injustices in domains other than health affect health outcomes; health states lead to injustices in other domains (e.g. fair opportunity, strict equality, reward for merit, reciprocity, respect for privacy or property, etc.). For these reasons, judgements of justice and injustice in other domains inform the normative evaluation of the justice or injustice of social inequalities in health and global health inequalities. But health needs also constitute specific claims for recognition, in response to which we should, as a matter of justice, express respect for equal human worth. That is, health inequalities, in addition to causing and being caused by other unjust inequalities, also constitute and express specific relations of equality and inequality that are of concern to justice.[3]

Furthermore, unequal responsiveness to health needs creates social inequalities that are *sui generis* and that are politically significant in *sui generis* ways. Consider that the inability to access medical treatment that could address a serious health condition creates desperation which may motivate people to enter into social relations they would not otherwise contemplate. For example, in moral reasoning tests, psychological researchers ask participants to evaluate a scenario in which a person needs a medication they cannot afford for their fatally ill spouse. This is the paradigmatic case in which otherwise socially unacceptable actions become reasonable and moral responses to inequality.

My pluralism about forms of justice is not in the first instance motivated by the idea that liberal public policy must avoid commitment where individual moral views diverge. Any claim to neutrality is dubious. Rather, it is an acknowledgement that conceptions of justice and fairness can take different contextually-relevant forms (e.g. fair reward for effort; fair opportunity; recognition of achievement; acknowledgement of and response to need), as can oppression (e.g. economic exploitation; exclusion from wealth or resource base; denial of voice). Inequality or oppression in any of these forms may be produced by other inequalities that themselves raise question of justice, and may produce further such inequalities. As Wolff and

---

[2] This complexity may not be captured by a "spheres of justice" view, where a given domain of human endeavour is governed by a single form of justice (Walzer 2008). Consider that educational policy must be responsive both to the importance of reward for merit and to the importance of equal opportunity for achievement of a reasonable standard of literacy for citizenship and fulfillment. No education system only rewards achievement or only brings the population to whatever minimal or reasonable standard can be achieved by all.

[3] Sen (2002) makes something like this point about the concept of health *equity* as a multi-dimensional concept: there are several forms that equity takes in relation to health and in relation to matters that have bearing on health and all of these must be taken into account in a treatment of health equity.

da-Shalit have argued, injustices that compound in these ways may be of particular concern (Wolff and da Shalit 2007).

Answering the formalist challenge involves contextualization and narrative exploration to evaluate the significance of what is portrayed in health statistics. The relevant narratives include individual lives (life-course narratives) within the context of social structures and broad changes in those structures – i.e. narratives of political economy. Both population level narratives of political economy and individual narratives of life courses contribute to understanding the human and normative significance of broad epidemiological changes in longevity and the social and global equality and inequality in these measures.

16.3 Examples

In this section, I show these multiple relations of health and justice by taking two examples from microbial disease, antimicrobials, and AMR.

16.3.1 Example: AMR, Sex, and Gender

Women's health as measured at the population level is strongly influenced by sexual and reproductive health, and these are in turn conditioned by reproductive autonomy or its absence, as a matter of technology and crucially as a matter of gender politics. (The influence of perinatal experiences on population health measures is substantial; to be discussed below.) At the level of individual life narratives, childbearing is a normative, but of course not a universal, experience for women.

Antibiotics are used prophylactically in childbirth to address the risk of maternal and infant morbidity and mortality from infections. I have been unable to locate in the medical literature discussion of the implications of AMR for obstetric practice, such as discussion of whether prophylactic use constitutes a sustainable form of anti-microbial stewardship, under what conditions it might be abandoned by choice or triaged by need, or what the implications of its waning efficacy would be for women's health. Neither have I been able to locate discussions of the implications of AMR for handling the infections that will be more common if the efficacy of prophylactic antibiotics is lost. Presumably this would have a significant impact on women's reproductive health.

Childbirth is normative for women but not universal – many women give birth and there are cultural expectations whose negotiation characterizes women's lives whether they themselves give birth or not. There are different pathways by which women come through sexual intercourse to childbirth, with greater or lesser degrees of autonomy and under different conditions of gender-related and economic justice and injustice. Sexual assault is both a concern of justice as a criminal matter and a concern of political justice, insofar as it is an expression and a tool of gender-based domination. The sequelae of sexual assault sometimes include health and social effects (e.g. sexual transmitted infections, pregnancy, trauma, altered social standing, and deep effects on current or future intimate relations, exposure to a criminal justice system that may be partly or entirely inadequate to victims' needs, etc.); these have further health and social effects (infection in childbirth, to return to the original example).

Some of the health-related harms of sexual intercourse are harms that we have been able to mitigate with antimicrobials (e.g. gonorrhoea, HIV/AIDS, infections arising from termination, infections arising in childbirth) for the last decades. These will be affected by AMR. Noting substantial disagreement about the moral and political status of sex work, sex workers and/or persons subject to sexual exploitation will be particularly affected by AMR. Some will take this as compounding the injustice of sexual exploitation and others will take it as compounding the injustice of the stigmatization of sex work. As such, AMR will alter the experience of sexual activity and affect concerns of justice in relation to sexual activity.

16.3.2 Example: AMR and Parasitic Infection

Hookworm is a parasitic infection caused by inadequate public health infrastructure (Pilkington 2017). It is virtually unknown in HICs; the United States is an exception to this, which may not be surprising given that it boasts the highest GINI co-efficient (greatest economic inequality) of HICs and captures a low proportion of its national income (30%) for state expenditure (Piketty 2014, pp. 475–6). Anthelmintic resistance (resistance to the anti-parasitic agents used in treatment) is a growing concern for hookworm (Harhay et al. 2010).

The conditions for hookworm arise directly from the legacy of indentured labour on large estates where land was held by a small elite, and so it is strongly conditioned by historical injustices in colonial and slave-holding societies. As a hygienic disease it is tied closely to place, and place is tied closely to identity for many groups subject to colonialism – whether they were brought as slaves to work in a place that is now home or they experienced colonial occupation and seizure of their lands, transforming their status as labourers on those lands.

In turn, hookworm has health sequelae detrimental to participation in the labour market and fulfillment of care responsibilities in the home (McKenna et al. 2017) and so it also has forward-looking intergenerational effects that are of concern for justice and the perpetuation of relations of economic inequality in post-colonial societies. This compounds the historical colonial relations that gave rise to the risk exposure.

These injustices are compounded by the fact that there is little or no investment in research to tackle this problem. Instead, anthelmintic drug development serves the agricultural industry, with the goal of increasing efficiencies in food production for the better-off, rather than serving the medical needs of the worse-off – human beings who live in conditions that make them vulnerable to the parasite (Hu et al. 2013).

16.3.3   Summary of Examples

In these examples, justice concerns and health inequalities are intimately and complexly linked. A given health advantage or detriment can reflect justice or injustice in gender relations, in criminal justice matters, in labour relations, in global economic and colonial relations involving claims to land, resources, and sovereignty – in addition to the common justice-related concerns about the fair distribution of scarce resources within health care.

The forms of stigmatization to which infectious and hygienic disease are subject are paradigmatic and concrete forms of stigmatization. Shame involves the source of the infection (in sexual relations, or in place for historically constituted communities consigned to exposure to hygienic disease) and this extends to shame about the resulting health state, in ways closely tied to disease-related identities, to symbolic and quasi-symbolic stigma (denigration, shunning, isolation, social distancing) and to concrete stigmatization (violation of civil liberties, detention; situating waste disposal on "unproductive" land where this is land occupied by indigenous and poor communities).

AMR has its effects in these fields. As it renders infectious and hygienic diseases less susceptible to treatment, it will exacerbate the ongoing legacies of these historical harms and their forward-reaching effects, in part by compounding the relevant health effects, and in part by the effects of policy responses on marginalized populations. For example, persons exposed to STIs through sex work or sexual exploitation and persons exposed to hookworm through colonialism and its legacy of land marginalization may in turn be subject to increased state surveillance and control, even risk of criminalization, insofar as their health state is seen to pose a threat to others, or insofar as their health state gives them a claim resources that increasingly come to be seen and managed as scarce public or common goods (Smith and Coast, Chap. 17, this volume; Giubilini and Savulescu, Chap. 9, this volume).

In describing the examples, I draw on (not very detailed) meso-level narratives in which justice considerations abound – narratives that involve local manifestations of broader structural relations for specific communities. In these narratives, it is relatively clear how justice considerations relate to health and how health differentials can intrinsically constitute the justice or injustice of relevant relationships. But what connection can we draw from here to the broadest population metrics of health inequalities? What roles have antimicrobials played in these population narratives and what role might AMR play?

16.4   Health Inequalities: The Development Narrative

Recall the 10–20 year within-country differences in life expectancy between high- and low-income neighbourhoods and the 30–40 year differences in life expectancy between HICs and LMICs. Do the differences among groups captured in the population health metrics matter normatively, and are they a concern of justice? What role did antimicrobials play in generating these differences and how might AMR affect these differences?

Now, these formal measures might matter because those at the low end of the distribution are experiencing health deficits which we want to address as a matter of beneficence, or as a reciprocal obligation to supply a decent minimum arising from our economic inter-dependence – or they might matter because the *inequalities* of which they form one extreme matter. It is easy to elicit some kind of normative concern about these inequalities but much more difficult to specify exactly what matters about them as inequalities. Two persons might have different life expectancies because of accident or genetic endowment; it's not obvious why group differences matter. One possible starting place is the thought that if all else were equal the poor would be as healthy as the rich. That is, the poor do not constitute a separate natural kind with different potential to be healthy. If they are differentially unhealthy, it is because we fail to ensure that everyone has access to their potential to be healthy, because incomes are inadequate, housing poor, food of low quality, work hazardous, childbirth ill-supported, and access to care and social support for recovery from ill health limited. This line of thought is similar to the view of that health inequalities matter to justice when they are caused by or cause other injustices, but it maintains at its core the idea that health inequalities matter as such and can constitute injustice.

By analyzing the role of antimicrobials in broad trends for social inequality in health and global health inequality, I argue that understanding the causes and the narrative constitution of these trends renders sufficientism and prioritarianism less attractive as normative stands.

A common account of the broad trends in both global health and economic development in recent decades is that within-country inequalities are widening, but between-country inequalities are narrowing – a narrowing that is particularly dramatic since the 1950s for health and 1980s for economic status. The normative claim attached to this account (implicitly or explicitly) is a prioritarian or sufficientist claim that while things may look grim from the perspective of the middle classes of HICs, from the perspective of those who are truly the worst off, things have never looked better: more and more people in the world are relieved of the worst form of poverty ("absolute poverty") and are achieving a sufficiency of health. In health, the worst off are doing even better: life expectancy is advancing faster than it did when the current HICs built their prosperous economies over the course of the nineteenth and twentieth centuries.

In the examples of the previous section (16.3), I outlined different ways that justice questions might reflect, be reflected by, or otherwise enter into health inequalities, by tying life course narratives to narratives that matter to justice – political and cultural economy, including gender relations. In the same manner, I will approach the significance of population health metrics in relation to justice by exploring the meaning of these metrics in terms of the typical and divergent life courses of the members of a population and in terms of the political and cultural economy of the communities in question (within-country communities and global communities).

The common narrative reflecting the sufficientist or prioritarian reading of narrowing health inequalities is the development narrative (political economy) linked to life course narratives via "transition" theory (demography and epidemiology).

Demographers long ago observed the tendency of populations to pass successively through certain stages, tied to their economic development: populations move from having a high birth rate along with a high death rate (in foraging or pastoral economies), to a more stable and dropping death rate, while the birth rate remains high, resulting in a population explosion (from agricultural to early industrial economies), to a low birth rate-low death rate stage, which we see in the stable or shrinking populations of modern HICs (which depend at this stage on immigration for economic growth). Epidemiologists in the mid-twentieth century linked these demographic changes to patterns of health and disease: the first stage of "pestilence and famine" with its characteristic population swings is followed by a second stage of greater stability but low life expectancy conditioned by infectious and hygienic diseases, initially from the close co-habitation of animals and humans and then the increasingly crowded, eventually urban, living conditions of humans with one another. With improvements in living standards and public health infrastructure, we enter a third stage where chronic ("man-made and degenerative") diseases of later middle age emerge as common causes of death (Omran 1971). A combination of so-called lifestyle changes (e.g. smoking cessation, moderation of red meat consumption) and advanced medical technologies (e.g. cancer treatment) have pushed back these diseases of midlife and created the ongoing extension in lifespan that HICs are now experiencing, which some describe as a "fourth stage" of epidemiological transition (Olshansky and Ault 1986).

These are the economic and epidemiological narratives behind how a nontransparent metric like life expectancy is read in health policy and understood to be a normative concern or a concern of justice. A life expectancy in the 40s reflects substantial maternal, infant, and child mortality caused by infectious and hygienic disease and the lack of empowerment of women to control their sexual and reproductive lives. A life expectancy in the 60s reflects a society in which heart disease and cancer are leading causes of death in later middle life, a phenomenon that emerged over the course of the twentieth century in HICs. A life expectancy in the 80s reflects improved prevention and treatment of these major killers and the emergence of new common forms of dying in advanced old age, e.g. frailty and dementia.

The dramatic narrowing of global health inequalities since the 1950s is called, within this narrative, the "accelerated transition," reflecting the success of the international development agenda in achieving improvements in health status that outstrip the economic and political development of the countries in question – that is, that outstrip both growth in GDP and changes in the franchise and effective political organization of workers to demand the health, education, and safety benefits of modern welfare states. By these means, industrialization and urbanization raised GDP while political change led to a larger proportion of the GDP being invested in the well-being of the population, in a process that involved the empowerment of workers and women, improving population health – although (given the dynamics of democratic pressure) achieving less success in addressing the needs of so-called minorities.[4]

On this narrative, it seems natural to see the changes in global economic and health equalities and inequalities in terms of the idea that some countries got a head start and some lag behind. A common economic belief is that open markets and the transfer of skills and technology will ensure that LMICs continue to advance towards (eventually) "catching up." The common narrative in health is that some (both low and high tech) public health technologies could be transferred in advance of economic development, offering LMICs a leg up in the development process.

16.4.1   Is the Development Narrative True?

Is this development narrative true? Does it do the normative work it claims to do?

One important critique of formal measures of equality and inequality is that they are not transparently related to the underlying realities they measure (King et al. 2012; Harper et al. 2010; Mackenbach 2015, Mackenbach et al. 2016). King et al. point out that the same underlying reality can be represented as a narrowing of absolute inequalities or a widening of relative inequalities, for example. Similarly, the same numeric change may represent different causal pathways or encapsulate different social relations with different significance in terms of justice and injustice and different distributions of well being and suffering. This has implications for the normative concern that these measurements might inspire (which inequalities should we tackle?) and may even raise questions about the reality of abstract interpersonal measures (are inequalities simply imposed on the individual phenomena that, sufficientists and prioritarians argue, should be the object of our moral concern?).

16.4.2   Underlying Realities

The narrative that LMICs are "catching up" suggests that they are achieving what HICs have achieved, but doing so later. However, this suggestion is false. The accelerated narrowing of global health inequalities represents fundamentally different epidemiological patterns, causal pathways, and relations of concern to justice.

---

[4]In the current HICs this process led to the state capturing 40–55% of national income for its spending (Piketty 2014, pp. 475–6).

I described above how a population's life expectancy in a given decade is "read" epidemiologically, and I noted that many different underlying patterns of health and disease can generate the same measure. Substantial infant and early childhood mortality in a subpopulation might depress life expectancy as much as widespread exposure of young adults to interpersonal violence, workplace hazards, tuberculosis, or HIV. Different narratives of political economy can in turn generate these patterns – the sub-population with elevated early childhood mortality may be indigenous or migrant; interpersonal violence may be a matter of warfare or the combination of lack of opportunity, crime, availability of guns, policing, and racialization.

The picture offered by transition theory is simplified (Frenk et al. 1991; Defo 2014). Researchers now emphasize that the same life expectancy in different populations may reflect different realities. They describe the transitions of LMICs as "incomplete" transitions, highlighting counter-transitions within LMICs and even within HICs. For example, LMICs did not leave infectious and hygienic disease behind. Rather, they are taking on the so-called diseases of affluence – diseases that arise from changes in work, nutrition, energy, and transportation – in addition to carrying an on-going burden of infectious and hygienic disease (Santosa et al. 2014; Defo 2014). Specific population groups within HICs experience the re-emergence of infectious and hygienic disease, while the emergence of so-called "diseases of despair" (suicide and substance abuse) may reverse health gains in HICs for some groups (Case and Deaton 2015) or for entire countries (some of the former eastern bloc). Global migration resulting from instability and lack of opportunity in LMICs and the need for low-wage workers particularly in the agricultural sector in HICs brings together the AMR diseases of LMIC and the population health profile of HICs (Suk et al. 2009).

Furthermore, insofar as LMICs have substantially reduced childhood and young adult mortality from infectious and hygienic diseases, these reductions did not come from the developments in political economy that led to such improvements in HICs. In HICs, rising GDP went along with political changes (universal suffrage extending beyond male property owners to labourers and women, and various political movements organized around these identities) and these brought about improvements in determinants of infectious and hygienic disease. In HICs, antibiotics joined and accelerated an existing process of decline in infectious and hygienic disease after the Second World War (Mackenbach 1996). In LMICs, on the other hand, growth in GDP resulting from urbanization has not been distributed or re-distributed and invested to improve living standards and public health infrastructure to the same extent. Political empowerment is limited, in part by the actions of the very same global corporations that bring (some) growth in income to LMICs. In its place, a global network of health philanthropy has delivered effective prevention with vaccines and mosquito nets and treatment with antimicrobials for infectious and hygienic diseases that remain endemic. Antimicrobials played an important role in this so-called "accelerated" transition. The infectious and hygienic diseases that result from crowding and exposure to waste are managed medically e.g. by vaccination for prevention or by antimicrobials for treatment, and not by improved housing for primordial prevention. This is not a lag in economic development or in the uptake of health-related technologies; it is a different path of economic and social change. Even to describe it as "incomplete" is misleading, insofar as the term "incomplete" suggests that LMICs have reached a different point on the same pathway instead of achieving the partial benefits they have achieved from a different trajectory.

This is evident at the level of life course narratives. What a life expectancy of 60 looks like in a current MIC is not what a life expectancy of 60 looked like when HICs reached that stage. Parents in HICs may feel that childhood and adolescence are times of risk and danger, but very few children in HICs experience life-threatening diarrhea or pneumonia and virtually no adolescents and young adults experience TB. (The exceptions to this general picture are in communities within HICs that bear a heavy burden of colonialism such as indigenous communities (Orr 2013 and Møller 2010) and in the globally mobile working class (MacPherson et al. 2009), where migration is also conditioned by colonial histories in addition to current global supply chains.) Parents in LICs with superficially similar mortality figures continue to experience episodes in which their children's lives threatened with diarrhea in infancy – but they now know how to treat it, as few people would have known in the HICs' pre-transition period, and they have antibiotics available for managing severe cases where this is appropriate. Children in MICs do not enjoy a trouble-free childhood, but go through the distress of under-5 pneumonia or adolescent TB, while their parents experience the anxiety of trying to secure the antimicrobials needed to treat these conditions: for 5 million children with pneumonia in LMICs annually, their parents are unable to do this (Laxminarayan et al. 2016). The use of antimicrobials along with other readily transferable technologies sustain different underlying life course narratives in LMICs compared to those that typify HICs.

16.4.3   Underlying Causes

Without a basic understanding of the causal processes that characterize changes in population health and health inequalities, it is not possible to evaluate normative claims about the co-existence of narrowing global health inequalities and the within country widening health inequalities. Sampling changes in a given time period and presenting them as trends (or as trends that are causally linked – with the suggestion that health benefits are, as it were, transferred from the middle classes of HICs to the workers of LMICs) can misrepresent the broad causal picture as it unfolds.

The last 100 years, both globally and within country, constitute one period in a long-term process of *widening* social inequality in health. Broadly speaking, social inequality in health has risen ever since the Middle Ages (perhaps surprisingly), when nobles and peasants seem to have had similar life expectancies (Antonovsky 1967; Bengtsson and van Poppel 2011). The late 19th and early 20th century period – in which social inequalities in health narrowed within those countries that emerged as HICs – was an anomaly.

To relate the trends to transition theory, roughly speaking, we can say that pestilence and famine affect all social classes; as societies transition to agriculture, the wealthy still have surprisingly little protection from the infectious and hygienic diseases that characterize this stage of close co-habitation with animals and increased human crowding. This "agricultural transition" lowers life expectancy for individuals while increasing population size – a qualified form of "improvement" in population health. Urbanization and the initially slow but eventually rapid growth in economic productivity associated with it, by contrast, both raise the life expectancy of the population and increase its size. However, this improvement both for individuals and for populations benefits those of higher SES more than those of lower SES, opening up differences in life expectancy that persist and for the most part continue to grow.

The only period in which social inequalities in health narrowed for a time was the classic period of hygienic and sanitary reform at the end of the 19th and beginning of the 20th centuries – when initial steps to establish safety of the food supply, clean water, and sewage were taken in HICs (Soares 2007). At this point, the happy confluence of political change (increasingly wide suffrage), scientific development (the germ theory of disease), and growth in GDP contributed strongly to this narrowing.

The post-war growth of the welfare state, including systems of universal health coverage, by contrast, has at best slowed the growth of social inequality in health: it has not moved us in the direction of social equality in health (Sreenivasan 2007; Mackenbach 2012; Reid 2016). The better off get more out of the "fourth stage" of epidemiological transition by the differential benefit they derive from programs like universal health coverage. This is not to say that such programs do not promote social equality in health: it is plausible that the poor would have been left yet further behind without them (Reid 2016).

This sampling question is significant for the normative work that the development narrative is supposed to do. The narrative of local divergence (the almost-best-off falling a bit behind) and global convergence (the worst off catching up) relies on sampling HICs just after the one period in which inequalities narrowed (thereby excluding that narrowing) and sampling LMICs at a moment in history that includes the LMIC version of that era in which (in HICs) inequalities narrowed. The underlying dynamic is that as health improves, it improves more for the better off than the worse off, absent a period where we acted on the gains that could be made by providing the basic infrastructure of public health. The broad trend is the same globally and within country: it does not reflect a tradeoff in which we forgo goods for the already pretty-well-off and give them to the worse-off. On the contrary, as we saw in the previous section, we have taken a path to improving life expectancy in LMICs that seems likely not to lead to the same gains. The inference from a given period of narrowing to a broader pattern of a gap being closed is illicit.

16.4.4   Implications for Global AMR Policy

The narrative of catching up ignores longer-term trends and obscures differences in typical and various life courses that cause the life expectancy gains in LMICs. Waning antimicrobial effectiveness will reveal the different underlying causes of these gains.

Current approaches to addressing the role of social determinants of health (SDOH) in the impact of AMR, consistent with the model of global health that gave us the accelerated transition, is medicalized: medical technology transfer again takes the place of political development; inappropriate standards of evidence (the movement towards implementing evidence-based standards in development) direct efforts towards primary prevention and treatment and away from inter-sectoral cooperation on health-related primordial prevention. (See King, Chap. 19, this volume; Silva et al. 2020.) The WHO *Global Action Plan* (2015a) speaks, for example, of "effective prevention of infections transmitted through sex or drug injection as well as better sanitation, hand washing, and food and water safety" as "core components of infectious disease prevention," (§36) and "more widespread recognition of antimicrobial medicines as a public good ... [being] needed in order to strengthen regulation of their distribution, quality, and use" (§41) to address inappropriate antimicrobial use, and the development of awareness and improved veterinary education to address the overuse of antimicrobials in agriculture.

Awareness and education are inadequate to address the struggles of LMICs to promote appropriate antimicrobial use. These countries often make do with half the tax revenues proportionate to GDP or national income that HICs expect: enforcement to tame diversion and over-the-counter sales of antibiotics or the for-profit healthcare sector (a substantial source of poor prescribing – Kuo et al. 2017; Haire, Chap. 3, this volume; Liverani et al., Chap. 5, this volume; Ho and Lee, Chap. 25, this volume) are not free. Neither can LMICs confront the globalized agricultural sector that moves agricultural practices that involve the overuse of antibiotics from the increasingly intolerant regulatory environments of HICs to their permissive regulatory environments. To confront this would in turn require an end to pressure from global supply chains serving HIC-consumers against environmental and labour regulation in LMICs (the justice dimensions of such dependencies are explored in the social connectionist model of Young 2006). Focusing innovation in anthelmintics, for example, on the needs of human beings vulnerable to parasites would change treatment possibilities (in accordance with the call of Shawa et al., Chap. 10, this volume), while reducing or eliminating agricultural use and improving health-related infrastructure – both of which involve political and economic change – would change the distribution of health benefits.

In addition, there is another global structural relationship in inequalities related to antimicrobial use, insofar as antimicrobials are a common or public good, as discussed elsewhere in this volume (Smith and Coast, Chap. 17, this volume; Giubilini and Savulescu, Chap. 9, this volume). LMICs cannot "catch up" in antimicrobial use because HICs are on track to use them up, for their own needs or in service of HIC-led policies that encourage antimicrobial use in LMICs instead of balanced development, high taxation and regulation, and democratic empowerment, the context in which conservative antimicrobial use can be implemented as policy.

The problem of effective regulation and orientation of the pharmaceutical industry towards population needs is not isolated to LMICs. Generic drug shortages in HICs contribute to AMR by driving inappropriate prescribing (Shoham et al. 2016), and every action plan on AMR highlights the failure of industry to invest in new antimicrobials, and not the problem that new antimicrobials, when found, continue to be marketed and deployed in ways that encourage non-beneficial and marginally beneficial use.

16.4.5   AMR and Widening Within-Country Social Inequality in Health

Antimicrobials played a role in the so-called "fourth stage" of epidemiological transition, in which life expectancy has been extended by effective prevention and treatment for cancer and heart disease, among other medical conditions. Insofar as progress in addressing these large sources of disease burden for HICs involves advanced surgical and other tertiary care techniques (in addition to dietary change and tobacco control), this fourth transition relies on antibiotics that make surgical and immune-system suppressing chemotherapeutics possible. Even in countries with universal, comprehensive health care without financial barriers to access (like Canada), innovations in tertiary care provide more benefit to the better-off than the worse-off (Starfield 2011; Asada and Kephart 2007)[5] – contributing to the return to growing social inequality in health despite universal health coverage in contemporary medical care.

At a superficial level, we could speculate that the development of AMR in HICs could undo a line of medical progress that has widened social inequalities in health. But there would be many opportunities for reassertion of the general pattern of the last several centuries, the pattern of the well-off capturing a greater share of benefits and experiencing a lower share of burdens. Moderating the use of antimicrobials to preserve their effectiveness calls on solidarity as a value (Holm and Ploug, Chap. 21, this volume); the implications of waning antimicrobial efficacy will put stress on this same value. For example, the tertiary care interventions of the fourth epidemiological stage are becoming increasingly expensive due to AMR (Smith and Coast 2013; Cosgrove 2006; Teillant et al. 2015). This will continue for some time as those interventions gradually take on an unfavourable intrinsic harm-benefit tradeoff for the individual patient directly involved. It is only at this last stage (where the

---

[5] This is thought to be because the worse-off face barriers to access that are both practical and social (e.g. geographical proximity and social capital that generates attention to the expression of need and referrals); when they do access care, they have fewer resources that would enable them to benefit from it, e.g. the social and income support to enable recovery; furthermore, the care is less suited to their needs, given that complex chronic conditions have a social gradient.

harm-benefit tradeoff shifts) that the loss of this technology would potentially narrow health inequalities by de-implementation of "fourth stage" interventions that have widened social inequalities in health. The effects of specialty and disease interest group advocacy in the intervening period may well result in even greater capture of the resources of the system for the better-off. The increasing cost of supportive care to maintain current surgical interventions while responding to AMR may also feed a political discourse around system sustainability that contributes to eroding the comprehensiveness and depth of universal healthcare coverage. Social inequalities in health could continue to widen as services for those with less voice are eroded to enable the higher cost of coping with AMR in the tertiary care sector that has served higher income persons well.

This is not just a HIC story. The Millennial Development Goals focused on child mortality and mortality of late adolescence/early adulthood (TB; the infectious and hygienic diseases), while the Sustainable Development Goals (SDG) on which the WHO embarked in 2015 focus on developing universal health coverage (UHC) and access to tertiary care – important for delivering the care necessary for chronic diseases and causes of midlife mortality that middle income countries are beginning to experience (WHO 2015b). The expected health transition from a higher burden of infectious and hygienic diseases to a higher burden of chronic, noncommunicable disease has informed this policy move. However, as we saw, MICs in particular are subject to the double burden of emerging chronic noncommunicable diseases alongside persistent infectious and hygienic diseases; they also experience a heavy burden from counter-transitions as new infectious and hygienic burdens arise (Cook and Dummer 2004; Santosa et al. 2014; Defo 2014). There is evidence that the prevalence of hospital-based resistant microbes is inversely associated with national income (Alvarez-Uria et al. 2016), suggesting that LMICs will also bear a heavier burden of hospital-based AMR than HICs, at the same time that they face a heavier burden of community-based resistant diseases. These new health systems will face the same kinds of political struggles about universality, depth, and sustainability that are in play in HICs (Norheim et al. 2014). Given the extent of economic inequalities within these countries, they will face these political struggles without the social solidarity that is somewhat protective of the breadth and depth of UHC in HICs. Indeed, some have speculated that AMR could entirely derail the SDG of UHC (Jasovský et al. 2016).

16.5   Conclusion

The debate between sufficientists and prioritarians on the one hand and egalitarians on the other turns on deep relationships between concerns about the formalism of interpersonal measures and (by contrast) the evident strength of the moral claim of the needs of the worse off. It is true that formal interpersonal measures can mask different underlying realities that would engage different concerns about justice (and other important values). Nonetheless, a fully developed picture of those underlying realities does not necessarily support the case for priority or sufficiency as goals – whether on normative grounds or on grounds of practicality. Perhaps putting the breaks on an underlying reality (the long term trend of growing social inequality in health) is not ultimately feasible: the role that antimicrobials have played in narrowing global health inequalities suggests as much.

The current narrowing of the global gap in life expectancy will not lead to the gap closing if there are fundamental barriers to narrowing the gap built into the path to the current achievement. What looks like a normatively satisfying narrowing of the gap is in reality a period in which a limited version of a familiar (and known to be temporary) reversal of a long term trend is playing out.

In the mid to late 20th century, it was thought that there was a limit to the long-term trend of growing health inequalities – perhaps a natural limit around 75, at which point the better-off would stop gaining life expectancy and then further gains would involve the worse-off (within and between countries) catching up. This has not come about: life expectancy at the top of the scale continues to advance and cautious observers no longer try to predict the upper limit. There is no known end to improvements in the upper limit figure, and so no natural limit to the growth of social inequality in health.

This growing gap is morally complex: Who can object to saving a life of a fellow citizen on the grounds that someone far away does not have access to the same treatment? But our failure to imagine the downside of some having access to lifesaving interventions in their 10th and 11th decades and beyond while others have their prospects for a long and healthy life compromised by social determinants is just that – a failure of imagination.

A century ago, social inequality in health meant that some buried half or more of their children before the age of 5 while others experienced this heartache less frequently. The moral concern we feel when learning of a gap in life expectancy between 40 and 70 is palpable; we are familiar with these narratives and the death of children is a tragedy for which children are not blamed. Social inequality in health now looks quite different. It may mean that some will enjoy long and active retirements approaching the length of their working lives, helping with grandchildren and greeting great-grandchildren, while others manage multiple morbidities related to chronic disease while in their later working lives, juggling this simultaneously with responsibilities to others – perhaps the mental health challenges faced by their children, the substance use entanglements of their partner, and the caregiving needs of their parents who are in turn suffering in their 70s from the cumulative effects of their own chronic conditions as the life narrative legacy of their exposure to the social determinants of ill health. These caregiving responsibilities lead to loss of mobility and skill upgrading, leading to missed work opportunities and, in a technology-driven economy, even loss of work – and so (in turn) to late-life poverty and death shortly after retirement (if retirement is an option at all).

Through the lens of literature, we can look back sympathetically at the concrete social shape that health inequalities took in earlier eras (with Dickens, for example); the form they are taking today are often presented in statistical terms and we have yet to develop a cultural narrative that explores the relations of domination and subordination and shapes the sense of intrinsic moral concern that might be raised by these differences. We are likely to think that individuals bear responsibility for these conditions (utilitarian reformers of the nineteenth century likewise thought those who were worse off responsible for their poor health). Who fundraises by running a race for the cure in their 60s and who is crowd-sourcing to raise money for their own cancer treatment (for travel and loss of work income if that treatment is covered under a universal health system, or for expensive therapeutics not covered in that universal system) – and what micro, meso, and macro relations of power are embodied here?

The *prima facie* moral plausibility of a prioritarian or sufficientist commitment is placed in question by a long-term trend towards a world in which the well-off enjoy a life expectancy – perhaps – double that of the worse-off. Suppose that the situation continued such that the person running for the cure in their 60s would become the centenarian running for the cure, and so on. How long would one remain a sufficientist? At what point would the gap become large enough that an appeal to sufficiency would ring hollow, and closing the gap would become a worthy moral commitment? If AMR makes the tertiary care that has continued the widening of the gap in the fourth stage of epidemiological transition even more expensive, at what cost and opportunity cost will we pursue the continuation of that trend or resist its reversal?

Globally, the response to AMR must move beyond lip service to the most minimal social determinants of health (sanitation and water) and engage with the political and economic process of redistributing resources and power so that governments and professional bodies can effectively regulate pharmaceutical production, distribution, and prescribing across the public and private sectors, and so that the global working class and those living with the legacy of colonization and resource theft (Wenar 2008) – whose housing conditions and social security currently preclude them from benefiting from the primordial prevention of infectious and hygienic disease that most residents of HICs have at this point enjoyed for most of the past century – can survive and even flourish despite the loss of the contribution of antimicrobials to the so-called accelerated epidemiological transition.

References

Alvarez-Uria, G., S. Gandra, and R. Laxminarayan. 2016. Poverty and prevalence of antimicrobial resistance in invasive isolates. *International Journal of Infectious Diseases* 52: 59–61.

Anderson, E. 1999. What is the point of equality? *Ethics* 109 (2): 287–337.

———. 2010. Justifying the capabilities approach to justice. In *Measuring justice: Primary goods and capabilities*, ed. H. Brighouse and I. Robeyns, 81–100. Cambridge: Cambridge University Press.

Antonovsky, A. 1967. Social class, life expectancy and overall mortality. *The Milbank Memorial Fund Quarterly* 45: 31–73.

Asada, Y., and G. Kephart. 2007. Equity in health services use and intensity of use in Canada. *BMC Health Services Research* 7: 41.

Bengtsson, T., and F. van Poppel. 2011. Socioeconomic inequalities in death from past to present: An introduction. *Explorations in Economic History* 48: 343–356.

Case, A., and A. Deaton. 2015. Rising morbidity and mortality in midlife among white non-Hispanic Americans in the 21st century. *Proceedings of the National Academy of Sciences* 112: 15078–15083.

Coast, R.D., and J. Smith. 2020. The economics of resistance through an economic lens. In *Ethics and drug resistance: Collective responsibility for global public health,* ed. E. Jamrozik and M. Selgelid, pp. 281–296. Cham: Springer.

Cook, I.G., and T.J. Dummer. 2004. Changing health in China: Re-evaluating the epidemiological transition model. *Health Policy* 67: 329–343.

Cosgrove, S.E. 2006. The relationship between antimicrobial resistance and patient outcomes: Mortality, length of hospital stay, and health care costs. *Clinical Infectious Diseases* 42 (Suppl 2): S82–S89.

Defo, B.K. 2014. Beyond the 'transition' frameworks: The cross-continuum of health, disease and mortality framework. *Global Health Action* 7: 24804.

Frenk, J., J.L. Bobadilla, C. Stern, T. Frejka, and R. Lozano. 1991. Elements for a theory of the health transition. *Health Transition Review*: 21–38.

Giubilini, A., and J. Savulescu. 2020. Moral responsibility and the justification of policies to preserve antimicrobial effectiveness. In *Ethics and drug resistance: Collective responsibility for global public health*, ed. E. Jamrozik and M. Selgelid, pp. 143–156. Cham: Springer.

Haire, B. 2020. Providing universal access while avoiding antiretroviral resistance: Ethical tensions in HIV treatment. In *Ethics and drug resistance: Collective responsibility for global public health*, ed. E. Jamrozik and M. Selgelid, pp. 37–54. Cham: Springer.

Harhay, M.O., J. Horton, and P.L. Olliaro. 2010. Epidemiology and control of human gastrointestinal parasites in children. *Expert Review of Anti-Infective Therapy* 8: 219–234.

Harper, S., N.B. King, S.C. Meersman, M.E. Reichman, N. Breen, and J. Lynch. 2010. Implicit value judgments in the measurement of health inequalities. *The Milbank Quarterly* 88: 4–29.

Ho, C.W., and T. Lee. 2020. Global governance of anti-microbial resistance: A legal and regulatory toolkit. In Ethics and drug resistance: Collective responsibility for global public health, ed. E. Jamrozik and M. Selgelid, pp. 403–422. Cham: Springer.

Holm, S. and T. Ploug. 2020. Solidarity and antimicrobial resistance. In *Ethics and drug resistance: Collective responsibility for global public health*, ed. E. Jamrozik and M. Selgelid, pp. 347–358. Cham: Springer.

Hu, Y., B.L. Ellis, Y.Y. Yiu, M.M. Miller, J.F. Urban, L.Z. Shi, and R.V. Aroian. 2013. An extensive comparison of the effect of anthelmintic classes on diverse nematodes. *PLoS One* 8: e70702.

Jamrozik, E., and M.J. Selgelid. 2020. Drug-resistant infection: Causes, consequences, and responses. In *Ethics and drug resistance: Collective responsibility for global public health*, ed. E. Jamrozik and M. Selgelid, pp. 3–18. Cham: Springer.

Jasovský, D., J. Littmann, A. Zorzet, and O. Cars. 2016. Antimicrobial resistance-a threat to the world's sustainable development. *Upsala Journal of Medical Sciences* 121: 159–164.

King, N.B. 2020. Technological fixes and antimicrobial resistance. In *Ethics and drug resistance: Collective responsibility for Global Public Health*, ed. E. Jamrozik and M. Selgelid, pp. 311–321. Cham: Springer

King, N.B., S. Harper, and M.E. Young. 2012. Use of relative and absolute effect measures in reporting health inequalities: Structured review. *BMJ* 345: e5774.

Kuo, S.C., S.M. Shih, L.Y. Hsieh, T.Y. Lauderdale, Y.C. Chen, C.A. Hsiung, and S.C. Chang. 2017. Antibiotic restriction policy paradoxically increased private drug consumptions outside Taiwan's national health insurance. *The Journal of Antimicrobial Chemotherapy* 72: 1544–1545.

Laxminarayan, R., D. Sridhar, M. Blaser, M. Wang, and M. Woolhouse. 2016. Achieving global targets for antimicrobial resistance. *Science* 353: 874–875.

Liverani, M., L. Hashiguchi, M. Khan, and R. Coker. 2020. Antimicrobial resistance and the private sector in Southeast Asia. In *Ethics and drug resistance: Collective responsibility for global public health*, ed. E. Jamrozik and M. Selgelid, pp. 78–87. Cham: Springer.

Mackenbach, J.P. 1996. The contribution of medical care to mortality decline: McKeown revisited. *Journal of Clinical Epidemiology* 49: 1207–1213.

———. 2012. The persistence of health inequalities in modern welfare states: The explanation of a paradox. *Social Science & Medicine* 75: 761–769.

———. 2015. Should we aim to reduce relative or absolute inequalities in mortality? *European Journal of Public Health* 25: 185.

Mackenbach, J.P., P. Martikainen, G. Menvielle, and R. de Gelder. 2016. The arithmetic of reducing relative and absolute inequalities in health: A theoretical analysis illustrated with European mortality data. *Journal of Epidemiology and Community Health* 70: 730–736.

MacPherson, D.W., B.D. Gushulak, W.B. Baine, S. Bala, P.O. Gubbins, P. Holtom, and M. Segarra-Newnham. 2009. Population mobility, globalization, and antimicrobial drug resistance. *Emerging Infectious Diseases* 15: 1727.

Marmot, M. 2015. *The health gap: The challenge of an unequal world.* London: Bloomsbury Publishing.

McKenna, M.L., S. McAtee, P.E. Bryan, R. Jeun, T. Ward, J. Kraus, M.E. Bottazzi, P.J. Hotez, C.C. Flowers, and R. Mejia. 2017. Human intestinal parasite burden and poor sanitation in rural Alabama. *American Journal of Tropical Medicine and Hygiene* 97 (5): 1623–1628.

Møller, H. 2010. Tuberculosis and colonialism: Current tales about tuberculosis and colonialism in Nunavut. *Journal of Aboriginal Health* 6: 38–48.

Norheim, O.F. et al. 2014. *Making fair choices on the path to universal health coverage.* World Health Organization. http://www.who.int/choice/documents/making_fair_choices/en/. Accessed 15 Sept 2017.

Nussbaum, M.C. 2006. *Frontiers of justice: Disability, nationality, species membership.* Cambridge, MA: Harvard University Press.

O'Neill, J. 2014. *Antimicrobial resistance: Tackling a crisis for the health and wealth of nations.* https://amr-review.org/sites/default/files/AMR%20Review%20Paper%20-%20Tackling%20a%20crisis%20for%20the%20health%20and%20wealth%20of%20nations_1.pdf. Accessed 15 Sept 2017.

Olshansky, S.J., and A.B. Ault. 1986. The fourth stage of the epidemiologic transition: The age of delayed degenerative diseases. *The Milbank Quarterly* 64 (3): 355–391.

Omran, A.R. 1971. The epidemiologic transition: A theory of the epidemiology of population change. *The Milbank Memorial Fund Quarterly* 4 (1): 509–538.

O'Neill, J. 2016. *Review on antimicrobial resistance.* https://amr-review.org/sites/default/files/160525_Final%20paper_with%20cover.pdf. Accessed 15 Sept 2017.

Orr, P. 2013. Tuberculosis in Nunavut: Looking back, moving forward. *CMAJ* 185: 287–288.

Parfit, D. 1997. Equality and priority. *Ratio* 10: 202–221.

———. 2012. Another defence of the priority view. *Utilitas* 24: 399–440.

Piketty, T. 2014. *Capital in the twenty-first century.* Cambridge, MA: Harvard University Press.

Pilkington, E. 2017. Hookworm, a disease of extreme poverty, is thriving in the US south. Why? *The Guardian*, https://www.theguardian.com/us-news/2017/sep/05/hookworm-lowndes-county-alabama-water-waste-treatment-poverty.

Regidor, E. 2004. Measures of health inequalities: Part 2. *Journal of Epidemiology and Community Health* 58: 900–903.

Reid, L. 2016. Answering the empirical challenge to arguments for universal health coverage based in health equity. *Public Health Ethics* 9: 231–243.

Santosa, A., S. Wall, E. Fottrell, U. Högberg, and P. Byass. 2014. The development and experience of epidemiological transition theory over four decades: A systematic review. *Global Health Action* 7: 59–71.

Sen, A. 2002. Why health equity? *Health Economics* 11: 659–666.

Shawa, R., F. Coomans, H. Cox, and L. London. 2020. Access to effective treatment for drug resistant tuberculosis: Deepening the human rights-based approach. In *Ethics and drug resistance: Collective responsibility for global public health*, ed. E. Jamrozik and M. Selgelid, pp. 157–171. Cham: Springer.

Shoham, S., A.A. Antar, P.G. Auwaerter, C.M. Durand, M.S. Sulkowski, and D.J. Cotton. 2016. Antimicrobial access in the 21st century: Delays and critical shortages. *Annals of Internal Medicine* 165: 53–54.

Silva, J., A.M. Viens, and D.S. Littmann. 2020. The super-wicked problem of antimicrobial resistance. In *Ethics and drug resistance: Collective responsibility for global public health*, ed. E. Jamrozik and M. Selgelid, pp. 423–445. Cham: Springer.

Smith, R., and J. Coast. 2013. The true cost of antimicrobial resistance. *BMJ* 346: f1493–f1493.

Soares, R.R. 2007. On the determinants of mortality reductions in the developing world. *Population and Development Review* 33: 247–287.

Sreenivasan, G. 2007. Health care and equality of opportunity. *The Hastings Center Report* 37: 21–31.

Starfield, B. 2011. The hidden inequity in health care. *International Journal for Equity in Health* 10: 15.

Strømme, E.M., and O.F. Norheim. 2017. Global health inequality: Comparing inequality-adjusted life expectancy over time. *Public Health Ethics* 10: 188–211.

Suk, J.E., D. Manissero, G. Büscher, and J.C. Semenza. 2009. Wealth inequality and tuberculosis elimination in Europe. *Emerging Infectious Diseases* 15: 1812–1814.

Teillant, A., S. Gandra, D. Barter, D.J. Morgan, and R. Laxminarayan. 2015. Potential burden of antibiotic resistance on surgery and cancer chemotherapy antibiotic prophylaxis in the USA: A literature review and modelling study. *The Lancet Infectious Diseases* 15: 1429–1437.

Temkin, L.S. 2000. Equality, priority, and the levelling down objection. In *The ideal of equality*, ed. M. Clayton and A. Williams, 126–161. Basingstoke: Macmillan.

———. 2003. Egalitarianism defended. *Ethics* 113: 764–782.

Valles, S.A. 2018. *Philosophy of population health science: Philosophy for a new public health era*. London/New York: Routledge/Taylor & Francis Group.

Venkatapuram, S., and M. Marmot. 2013. *Health justice an argument from the capabilities approach*. New York: Wiley.

Walzer, M. 2008. *Spheres of justice: A defense of pluralism and equality*. New York: Basic Books.

Wenar, L. 2008. Property rights and the resource curse. *Philosophy & Public Affairs* 36: 2–32.

Wolff, J., and A. da Shalit. 2007. *Disadvantage*. Oxford: Oxford University Press.

World Health Organization (WHO). 2015a. *Global action plan on antimicrobial resistance*. http://www.who.int/antimicrobial-resistance/publications/global-action-plan/en/. Accessed 13 Dec 2018.

———. 2015b. *Health in 2015: from Millennium Development Goals (MDG) to Sustainable Development Goals (SDG)*. https://www.who.int/gho/publications/mdgs-sdgs/en/. Accessed 13 Dec 2018.

Young, I.M. 2006. Responsibility and global justice: A social connection model. *Social Philosophy and Policy* 23: 102–130.

Chapter 17
The Economics of Resistance Through an Ethical Lens

Richard D. Smith and Joanna Coast

Abstract Economics is concerned with the analysis of choice and the efficient use of resources. Markets for antibiotics are heavily affected by their 'public good' nature and the externality that results from their consumption in terms of resistance. The non-excludability and non-rivalry associated with knowledge production in antibiotic development also has implications for the supply of antibiotics. On the demand side there are ethical issues associated with free-riding by consumers, free-riding across nations and free-riding across time. On the supply side, the lack of a pipeline for new antibiotics for the future causes both ethical and economic issues – and from both perspectives, efforts should perhaps focus more on alternatives to antibiotics and adjustments to heath care systems to reduce reliance on antibiotics. Indeed, unlike many areas of health care, where economics and ethical perspectives may differ, antimicrobial resistance is a case where the two perspectives align in terms of ensuring efficient and sustainable development and use of this precious resources. All strategies for dealing with resistance should share the same goals of achieving an optimal balance in the use of antimicrobial agents and explicit consideration of the distributional implications.

Keywords Economics · Externalities · Public-goods · Free-riding · Discounting

17.1 Introduction

Antimicrobial resistance has finally come to the fore on national and international agendas, and is now recognised as a critical threat to public health, modern health systems, and economies (Smith 2015) [Ref: Chap. 1]. Although the broad potential

R. D. Smith (✉)
College of Medicine and Health, University of Exeter, Exeter, UK
e-mail: rich.smith@exeter.ac.uk

J. Coast
Population Health Sciences, Bristol Medical School, University of Bristol, Bristol, UK

© The Author(s) 2020   279
E. Jamrozik, M. Selgelid (eds.), *Ethics and Drug Resistance: Collective Responsibility for Global Public Health*, Public Health Ethics Analysis 5,
https://doi.org/10.1007/978-3-030-27874-8_17

impacts of resistance are clear – that less effective treatments means greater morbidity and mortality, more expensive treatments or hospital care, and more time off work – robust evidence on the extent of likely impacts of resistance on population health, health care and economies remains relatively limited and often contentious (Coast et al. 1996, 2002; O'Neill 2014; Smith and Coast 2013; Wilton et al. 2002). Like climate change, this lack of clarity stems from inherent uncertainty; in the case of resistance, relating to uncertainties around the growth path of resistance over time, the functional form of the relationship between resistance and antibiotic use, and estimates of the direct morbidity and mortality impacts attributable to resistance rather than underlying infection or other causes (Cormican and Vellinga 2012; Courvalin 2008).

However, apart from such uncertainties, a significant issue is the nature of the decision- making around the production and consumption of antibiotics. Although an intrinsically biological phenomenon, the conditions promoting, or mitigating against, the development and spread of antimicrobial resistance are shaped by choices that are made by farmers and vets, doctors and patients, industry and governments, amongst others, concerning what antibiotics to produce, purchase and use (Smith 2015). Economics, at its foundation, is concerned with analysing and evaluating such choices: what and why certain choices are made over others; the conditions under which such choices are made, and how different conditions affect those choices; and, crucially, whether these choices are the 'best' ones (that is, whether they are 'efficient' in the sense that no greater benefit could be gained from another choice that was available).

These choices are determined through some sort of 'market', broadly defined as a place (physical or virtual) where goods are traded. In this case, for example, a market is where patients and doctors will determine whether an antibiotic will be prescribed, or where a vet and a farmer agree on a course of action for sick animals. Regardless of whether money changes hands, this is the point where an antibiotic supplier will engage with the consumer. Markets also underpin whether, what and how much a producer may invest in production of antibiotics, and the price that the payer (the consumer or their agent, often governments or insurers in the case of healthcare) will pay.

However, markets for antibiotics are heavily affected by two forces which are important when considering the development of antimicrobial resistance and strategies to reduce resistance. First, there are significant 'public good' attributes to antibiotics that affect use and distribution. This means that markets, left alone, will 'fail'; that is, they will not result in a socially optimal level of either production or consumption of antibiotics (Smith and Coast 2003). Second, there are 'externality' effects – effects not included in the decision to consume – that are important, and again means that, left alone, the market will 'fail' and not result in a social optimum of antibiotic consumption (Coast et al. 1998).

Together, these two forces drive the under- and/or over-use of antibiotics compared to what is optimal from a societal perspective. This creates major issues for economic efficiency, as different choices would improve overall societal welfare, but also for ethics, in the sense that there will be equity impacts that create problems of distributive (in)justice in access to antibiotics and in the impact of antimicrobial resistance (Coast and Smith 2015).

This chapter describes the core *economic* features of antimicrobial resistance – the public good and externality forces – through an ethics 'lens', focussing especially on the resultant economic and ethical problems of free-riding in both production and consumption. The chapter begins with outlining the core economic perspective on antimicrobial resistance – the market, public goods and externalities – in more detail. From this basis, the chapter then turns to consider the specific issues related to the demand side and then the supply side, focussing on the ethics of the economics, and reflecting on possible strategies to address resistance. In particular, these sections consider the development of appropriate incentives by governments to address free-riding within countries and the development of appropriate incentives by international agencies to address free-riding across countries. Such incentive mechanisms are evaluated both in terms of their value in addressing the economic market- failure and their ethical implications. The chapter concludes with a brief reflection on antimicrobial resistance as a specific case where economic and ethical concerns converge.

17.2   Antimicrobial Resistance and 'The Market'

At its most fundamental, the market for antibiotics seeks to equate demand for them by patients (and farmers or those with pets, although in this chapter we will focus on use in humans only) with supply by providers (doctors, pharmacists or over-the-counter stores for example). Patients demand antibiotics because of the impact they (are believed to have) on a health issue, such as a sore throat or urinary-tract infection. There are well-known problems in the market for healthcare goods and services, including the lack of information held by the patient upon which to decide on whether, or how much, healthcare to demand, the resultant need of the patient to rely on an (in economic terminology) 'agent' (usually a medical professional) to help the patient decide on what to consume, and, often, the separation of the patient from payment through the role of a third-party funder – typically government or insurer (Guiness and Wiseman 2011). Markets in healthcare are, therefore, far from perfect, and thus often subject to government intervention in their provision and/or financing. These problems are also, at least partly, responsible for ethical concerns for distributive justice, as they can leave many individuals with no, or inadequate, access to healthcare. Over time, this has led to increased calls for initiatives to ensure 'universal health coverage'; it is now a major thrust of the World Health Organization and World Bank to deal with general market failures in healthcare and to increase access (World Health Organization 2015), whilst the specific market failures around antimicrobial resistance are dealt with through other agendas, such as the Sustainable Development Goals (Hanefeld et al. 2017).

17.2.1   Externalities

Beyond general issues of failure in healthcare markets, there are two specific features of the market for antibiotics that are significant in considering the generation of antimicrobial resistance, and in establishing policies to contain it (Coast et al. 1998). First, there are externality effects from the consumption of antibiotics. An externality is an effect that is outside of the immediate producer or consumer, and thus does not influence decision-making in the choice about whether to produce or consume that antibiotic. So, for the patient, the direct (expected) benefit of the antibiotic in treating an infection informs his or her decision, as do possible side-effects from taking it, along with any monetary cost associated with its purchase. However, a 'positive externality' that is not generally expected to be taken into account by this patient is the benefit to those individuals who would, in the absence of antibiotics, have been infected by them – that is, there is an external benefit associated with reduced transmission of pathogens. The existence of this positive externality means that, the market will under-provide antibiotics; it might be a good reason for a subsidy or other policy to reduce the private cost to individuals. There is also an ethical issue involved here. There is a responsibility for each individual to consider the impact of their choice about receiving treatment on the wider community – that part of their decision to take treatment should, morally, include consideration of the impact of (reduced) infection on the wider population.

More critically for the discussion here is that there is also a major 'negative' externality from consumption of antibiotics – antimicrobial resistance. It is important to note that this externality is not associated with the production of antibiotics but with their consumption. The effects of antibiotic use, in terms of the sustained effectiveness of that antibiotic for others, or for the same individual in the future, do not influence the costs that must be paid by the patient, nor the benefits to them of taking the antibiotic now. They are therefore unlikely to be considered in the decision to purchase and consume that antibiotic (by either the patient or the health professional acting as their agent). From a societal perspective there is thus an over-consumption of antibiotics. Again, there are distributive issues here around the use of a limited resource knowing that it will generate possible greater ill-health in the future and/or for others.

On balance, the optimal consumption of antibiotics from a societal perspective is a balance of the costs and benefits that accrue directly to the individual concerned, plus these external benefits and costs to society (Coast et al. 1998). In recent years, consensus has been that on balance there is over-consumption – that the private costs and benefits and the positive externality are being severely compromised by not accounting for the negative externality of resistance. Focus has therefore been to see how these external costs of resistance can be internalised in to the decision to consume (in addition to how to increase the supply of effective antibiotics). We look at these strategies – affecting the demand and supply sides respectively – later in this chapter.

17.2.2 Public Goods

Second, there are 'public good' aspects associated with antibiotics [Ref: Chap. 8]. Most goods are what we term as 'private' in nature: their consumption can be withheld until a payment is made in exchange, and once consumed they cannot be consumed again (Woodward and Smith 2003). For example, the consumption of a cake can be withheld from the consumer until the consumer pays the baker a price, and once the consumer has eaten that cake it cannot be eaten again. A private good is therefore considered 'excludable' and 'rival in consumption'. At the other end of the spectrum lie public goods, which are defined as having the opposite characteristics. That is, the benefits, once the good is provided, cannot be restricted and are therefore available to all (i.e. non-excludable), and consumption by one individual does not limit consumption of that same good by others (i.e. non-rival in consumption). A classic example is the service provided by a lighthouse: the warning it provides is available to all who would benefit from it, and one ship's use of it does not limit the ability of other ships to use it. Virtually all public goods are such services or other intangibles, with few, if any, 'commodities' (in the narrow sense of physical objects) meeting these criteria (the exception to this being physical infrastructure, such as sewage systems, which once completed are largely non-rival in consumption, and difficult to exclude people from using).

However, both excludability and rivalry are relative, not absolute, concepts, and there is a scale of both rivalry and excludability. For example, access to public goods in particular may be specific to geography (e.g. conventional television broadcasts, while they broadly satisfy the criteria for a public good, reach only an area defined by the location of transmitters, the strength of signals and topographical constraints) or can be artificially made excludable (such as by using encryption services for satellite broadcasts). These create 'club goods', which are non-rival and are non-excludable to those who can access the 'club', but excludable to those outside of the club. Thus, some people could be (and often are) excluded from the benefits of most theoretically defined public goods through geographic, monetary or administrative prohibition. Similarly, rivalry in consumption may be relative to capacity, particularly in the case of physical infrastructure. For example, if a sewage system has spare capacity its use is non-rival, but as the capacity constraint is approached, its use becomes rivalrous; that person whose use of it causes capacity to be reached has effectively prevented the next person wishing to use it from doing so. Perhaps more usual is that the consumption of a particular good may not prevent others from using it, but simply reduces the benefits available. For example, one person's use of a road does not usually prevent use by others, but the use of the road becomes less beneficial as more and more people use it and the road consequently becomes more congested. This is sometimes termed the 'tragedy of the commons' (Hardin 1968), Chap. 8.

Of key importance, is that 'markets' under-supply public goods. First, non-excludability means that a price cannot be enforced, leading to 'free-riding' (one person can benefit from the actions of another person without reciprocation) and thus there is no incentive for anyone to produce or purchase the good. Second, non-rivalry means that the socially optimal level of consumption is far greater than the level that occurs at the 'market price' (Smith and Coast 2003).

To be clear, antibiotics are not themselves public goods – they can be made excludable, and they are rival in the sense that if I use one dose no one else can use that exact same dose.

However, the knowledge they embody in their development, and the resultant property of reduced infection (bearing in mind the externality properties above) means that they have significant public good attributes associated with them. Once developed, the knowledge required to produce the antibiotic can be made (virtually) freely available as dissemination of it is quick and cheap, and any one firm using it to produce the antibiotic does not prevent another from doing so. We will return to this aspect in the section concerning the supply of antibiotics in light of resistance. It also means that the benefits resulting from the consumption of the antibiotic, of reduced infection, cannot be made excludable – we all benefit from reduced trans-mission of pathogens – and is non-rival – my benefiting from this reduced risk of infection does not diminish you equally benefitting from it. And vice-versa – the reduced effectiveness of antibiotics due to resistance is also non-rival and non-excludable. It is this consumption side to which we now turn.

17.3   The Ethics and Economics of Demand

The characteristics of non-excludability and non-rivalry have important implica-tions for the economics and ethics of the demand/consumption side of the market for antibiotics from a number of perspectives.

17.3.1   Free-Riding by Consumers

Consumers have an incentive to engage in free-riding, meaning that they have no incentive to reduce their own use of antibiotics, but rather to wait for others to do so, Chap. 21. This is perhaps also seen as morally acceptable as each individual consumption of each individual course of antibiotics will only add infinitesimally to the problem of resistance; an individual's personal consumption of antibiotics really is a 'drop in the ocean' relative to the total consumption of antimicrobials (Smith and Coast 2013). There are strong parallels here with climate change of course. Overall, the antimicrobial resistance problem is the accumulation of many millions of decisions by different decision makers for different patients in different health systems and facing different personal, financial and organizational incentives. Economic policies in general aim to induce consumers to internalize the costs of any negative externality, but because of the huge diffusion of the problem in relation to resistance, such policies may be very challenging if even possible at all.

For example, system interventions to address free-riding include taxation (akin to Pigovian taxes or pollution charges in the case of climate change), subsidy, permits or regulation (Coast et al. 1998; Laxminarayan et al. 2010; Smith and Coast 1998) although each of these would have their own distributional implications (Coast and Smith 2015), Chap. 8. Examples of clinical interventions that may address this issue may include better diagnostics (Kolmos and Little 1999; Oppong et al. 2013; Rice 2011) or educational campaigns providing better information (Goossens et al. 2006; Huttner et al. 2010), combined with a gatekeeping role from healthcare providers. Other possibilities include focussing more on assessing option or existence value (the idea that individuals have some personal utility from knowing that the option for them to consume antimicrobials in the future will be retained) (Coast et al. 2006). However, it is more likely that benefits may come from encouraging a greater moral responsibility to avoid resistance within society, such as has been undertaken with climate change and pollution, or as a parallel to stronger regulations, such as with banning smoking in public places, compulsory wearing of seatbelts in cars or drink-driving campaigns.

17.3.2 Free-Riding Across Nations

These issues of individual free-riding are also paralleled by free-riding across macro-settings. Actions taken by current patients have the potential to transmit resistance, through the pathogen this resistance is associated with, across international, cultural and ethnic boundaries. Antimicrobial resistance does not respect regional or national boundaries, with resistant organisms being able to travel from one setting to another just as easily as sensitive ones (Smith and Coast 2002). Globalization has increased the rate at which infectious diseases can travel, and resistances identified in one area are rapidly found in other countries and on other continents. Such spread will be dependent on many epidemiological factors, including for example, socio-demographic factors, density of the population, natural disasters, hygiene levels and so on. Areas with greater poverty may be particularly susceptible to the rapid spread of infection and thus the deleterious effects of resistance.

Transmission of resistance may have particularly severe consequences for economies which already suffer greatly from infectious disease – countries where living conditions are poor and where there are high proportions of individuals with immune-compromise – but it will also have significant economic and health consequences at a global level (Smith and Coast 2002).

Indeed, transmission of resistance across national boundaries compromises measures that individual countries may put in place to try to combat resistance; importing resistance from elsewhere may negate national initiatives, rendering them effectively pointless. Conversely, countries may do little to address resistance within their own borders, relying on the activities of other countries to reduce levels of resistance that come in and out of the country; effectively, individual countries may 'free-ride' on the actions of the wider global community and the containment policies operated in other settings. The potential for resistance to spread across boundaries reduces the incentive for any single country to act against resistance – either because it will appear futile in the face of a global rise in resistance, or because the optimal approach is to await every other country to act first – which means that none will.

The total effort devoted to containing antimicrobial resistance is thus likely to be suboptimal because some nations will rely on the efforts of other nations. This dilemma facing the creation of collective action against antimicrobial resistance at the global level has been highlighted elsewhere (Årdal et al. 2016; Smith and Coast 2002), although it has recently been suggested that international law may provide a means of dealing with the problem of collective action in this context (Hoffman et al. 2015a; Walls and Ooms 2017), Chaps. 24 and 25.

Despite the importance of the global issue in this context, economic (and other) assessments tend to be conducted on a national basis as if systems are closed to the outside world. Opening this up would promote awareness that it is in national self-interest to look to address resistance in other countries, and thus contribute to global initiatives, as well as be open to moral arguments, such as those that drive considerable development funds each year from developed to less developed nations (Molzon et al. 2011).

17.3.3   Time Preference and Discounting: Free-Riding Across Time

The externality effect is also inter-generational, generating significant issues of inter- generational as well as regional equity. Actions taken by current patients will impact on the transmission and emergence of resistance for future generations, but with similar lack of incentive for either individual consumers or nation states to fully account for the impact of their decisions on others.

The relevant economic concept here is that of time-preference and discounting – that people place higher weight on benefits that occur now and costs that occur in the future, thus biasing decisions and choices against future generations (even including themselves in that future generation). The discount rate is the amount at which future years are 'discounted', where a rate of 0 would give equal weight to the current and any future year, and 100% would give total weight to the current year and count future years as worthless. Thus, a higher rate places more weight on the present. Empirical literature finds huge variance in expressed discount rates, but this is overwhelmingly based on personal perspectives (Asenso-Boadi et al. 2008). These, typically high, rates are not consistent with the normative arguments that are considered when intergenerational equity is discussed, and less weighting explicitly given to current generations (Olsen and Richardson 2013; Richardson and McKie 2007). The authors have also previously noted that discount rates with respect to antimicrobial resistance "should reflect collective value judgements and moral issues, rather than just the preference that individuals have for their own consumption over time" (Coast et al. 1996). This is important as high discount rates would result in an almost exclusive focus on the current population, and would probably also mean that the focus would be on interventions intended to reduce transmission of antimicrobial resistance, which has relatively short-term gains, rather than interventions intended to reduce emergence of new resistances, which may be further down the line (but where the total benefits in the absence of discounting may be much greater). This issue is examined in greater detail elsewhere (Coast et al. 2002).

17.3.4 The Importance of Context

On the consumption side, therefore, the focus of policy is on incentives and systems to influence demand through education of patients, restriction of availability through regulation, or greater guidance to providers (Dar et al. 2016). There are huge challenges in this respect, not only of construction, enforcement and evaluation of any single policy, but also across contexts. Policy in most developed countries is concerned with reducing the over-use of antibiotics – their use where there is little or no clinical indication, such as for the common cold, their use where there is marginal clinical benefit (e.g. many minor infections in otherwise healthy people, which may hasten their cure by a few days only) and even use where there may be greater morbidity but the disease is not fatal. In contrast in many low- and middle-income countries the challenges are to increase use, where appropriate, to reduce pools of infection, and to reduce practices of dose splitting, use of sub-standard, out-of-date or counterfeit medicines, and cross use of antibiotics for humans and animals (Review on antimicrobial resistance 2015a).

17.4 The Ethics and Economics of Supply

Externality and public good aspects are not only relevant to consumers, but also to the producers of antibiotics. The supply-chain of antibiotics is long – from original discovery, research and development, which may be in universities, through testing and eventual production by a pharmaceutical company, through wholesalers, health systems or private pharmacy and retailers (with prospects for counterfeit or substandard antibiotics being introduced along the way), Chap. 5 However, critical to this discussion is the core discovery, development and production of the antibiotics; especially that undertaken in the private sector.

It has been noted many times that there has been a distinctly dry pipeline for new antibiotics over recent decades, with no new class of antibiotic being discovered for many years and declining investment by the pharmaceutical industry in the area of anti-infectives (Cormican and Vellinga 2012). What underlies this is the current business model for the development of pharmaceuticals, which is based on a market response to 'free-riding' (Smith et al. 2009).

17.4.1 Free-Riding and Intellectual Property

Basic research and development of new drugs is hugely expensive, but the information that is produced from this research and development activity – such as the chemical formulation of a new antibiotic – is almost costless to disseminate, is very hard (especially in the internet age) to exclude other companies having access to, and is non-rival in the sense that me using this information to produce the drug does not prevent you also using the information to produce the drug. Thus, given the non-excludability of this information, it is socially optimal to encourage wide dissemination and use of it, which can be achieved at almost zero cost once produced. But, the inability to exclude means that a high-price cannot be charged for it to recoup the original investment in its production, and so other producers can 'free-ride' on the back of the one who first produced the information. Since the original producers cannot recoup their costs, let alone gain a profit, then there is no incentive for that information to be produced. Hence, left to this pure market system, there would be no medicines developed (Smith and Coast 2003).

The 'solution' to this situation has been to develop an artificial exclusion system to enable the original producer who discovers and develops the information to charge a higher price to compensate for the research and development costs incurred. This is the 'Intellectual Property Right' system, which in the case of pharmaceuticals is operationalised through a patent (Smith et al. 2009). This system confers a legal exclusion to be placed on others using material that is patented for a specified number of years – usually 20–30. During this time whoever filed the patent, such as the pharmaceutical company, has exclusive rights to use it and can therefore charge a sufficiently high price for the medicine it develops from it to allow it to recoup the costs invested, and to make a profit (Yamabhai and Smith 2015).

This 'solution', however, creates a critical problem for both efficiency and ethics. Due to the non-rival nature of the information – and the non-rival nature of the resultant reduction in infectious disease risk from taking the antibiotic that is based on the information – it is most socially efficient if it is widely disseminated and used. The patent system creates a restriction on that use, since only those who can afford to pay the price now charged can access it, which will be at a level lower than is socially optimal. Ethically, it also means that the information – and resultant antibiotics – are only available at a much higher price than the production cost, with the known consequence that there will be people who can benefit who are being denied that treatment (Mendelson et al. 2016). The 'patent wedge' between cost price and the price charged by the patent holder can typically by seen when the patent period expires and 'generic' versions of branded pharmaceuticals are made available, which are often hundreds of times cheaper (Smith et al. 2009). During the patent period there are clearly therefore poorer people, and countries, that are not able to afford to access these drugs and where ill-health is suffered as a result. For example, the most infamous case is that relating to anti-retroviral therapies for HIV/AIDS which were priced at over US$10,000 per person per year by those producing branded products, but could be produced by generic manufacturers for US$365 per patient per year (Keppler 2013). This is both inefficient and unethical (Laxminarayan et al. 2016). Inequality in access due to price – especially for those in low-income settings – has been the cause of considerable debate more widely with respect to medicines (Trouiller et al. 2002).

There is added complexity in the case of antibiotics, compared to other medicines, in part due to the nature of antibiotics, which are relatively inexpensive compared to other drugs that could be developed and are used for short periods of time (thus reducing the potential profit), and in part due to the development of resistance, as polices elsewhere seek to reduce and restrict use. This is especially the case in high-income countries, which is where the major markets for new antibiotics are based and where companies expect to be able to charge the higher-prices required to support research and development. The combination then of the high development cost, with reduced market volume, has discouraged investment in this area.

17.4.2   Alternative Systems

Over the years there have been growing calls for changes to the patent system as underpinning research and development in pharmaceutical in general, largely from those concerned about the bias towards health needs of high-income countries who are the major markets, resulting in concerns around 'neglected diseases' (Trouiller et al. 2002), Chap. 10. More recently, especially following the report from the Independent Review on Antimicrobial Resistance, chaired by Lord Jim O'Neill, there has been a renewed push for the development of new systems for incentivising investment in new antibiotics, where high-income countries too have considerable need that is not being met (Review on antimicrobial resistance 2015a).

One popular alternative is the de-linking of profitability from sales volumes (Brogan and Mossialos 2016; Review on antimicrobial resistance 2015b; Wise 2015). For example, pre- purchase agreements could be made by governments to guarantee the pharmaceutical company a minimum 'market' in terms of sales, but then the government can choose whether and how those drugs would be disseminated. This builds on other recent work, such as by Chatham House, a think-tank on international affairs, which has outlined alternative business models to similarly change the current financial models for encouraging and supporting research and development in new antibiotics (Kesselheim and Outterson 2010; Outterson 2014; Outterson et al. 2015). There is a lot of work and funding also now going in to research and development by governments themselves to directly develop new antibiotics. For example, The U.S. granted GlaxoSmithKline US$200 million for investment in new antibiotic research and development (GlaxoSmithKline 2013). There are also other initiatives. In the United States for instance there has been an extension to marketing exclusivity, accelerated review, and a relaxation of requirements for approval by the Food and Drug Administration (Hatch 2015). Although this leniency may have increased the development and launch of new antibiotics, there is worrying evidence that such 'fast tracking' may generate considerable adverse effects (Doshi 2015).

Critical, of course, is the basic fact that because resistance to an antibiotic begins as soon as it is developed, new agents can never be the sole solution—and are unlikely to be the most cost-effective or sustainable, Chap. 19. Thus, in addition to looking at encouragement of new drug discovery, another focus has been on initiatives to support the sustainable use of current antibiotics through the development of rapid diagnostic tests to support the more 'appropriate' use of antibiotics (Review on antimicrobial resistance 2015a). In the UK there have been specific funds made available to support diagnostics research and development through the UK Research Councils (Medical Research Council 2014), and the Longitude Prize, where £1 m is to be awarded to whoever first develops a specific rapid – and affordable – diagnostic tool (Nesta 2014). The critical issue here is the last one – affordability. Diagnostic tests face a challenge of affordability in relation to the option to simply take the – currently cheaper – antibiotic. The cost of the test relative to the cost of the antibiotic is critical – if the test costs more than, or even approximates, the antibiotic cost then it is likely to be rational for individuals to simply take the antibiotic and risk it not working – if it does work they save the cost of the test, if it does not work then they may need another drug but have still saved the cost of the test, whereas if they have the test and don't need an antibiotic they may have incurred a cost greater than that of simply taking the antibiotic, and if the test suggests they do need that antibiotic all they have done is add a cost of the test on to what they would have paid in any case. There are challenges especially for this in community settings, rather than hospitals, and in low- and middle-income countries.

Given that a critical feature of antimicrobial resistance is that it occurs naturally for all antibiotics, developing new drugs may win successive battles but not the war. It is possible that genomic developments may produce antibiotics where resistance is not seen, or new ways to tackle pathogens. However, in the absence of those, the most serious issue is not, perhaps, systems for encouraging development in new technologies, but rather how to design health systems that are less reliant on antibiotics. Modern healthcare especially, has been built on the basis that infections can be prevented or treated easily and cheaply. Healthcare has become increasingly technological and invasive, and antibiotics have become integrated in many aspects of care, from prevention of iatrogenic infection in surgery, to women delivering by caesarean section to those having cancer treatment (Smith 2015). From both economic and ethical perspectives efforts should perhaps focus more on alternatives to antibiotics and on adjustments to health systems and care pathways to reduce reliance upon antibiotics.

17.5   Discussion: The Convergence of Ethics and Economics

Far from the 'dismal science', or the often perceived amoral (if not immoral) nature of economics being in conflict with ethics, antimicrobial resistance is a case where the two perspectives are frequently aligned. Many of the economic imperatives for ensuring the efficient and sustainable development and use of antibiotics are in-line with major ethical principles to secure benefits that are equitably available and distributed. In this sense, from an economic and ethical perspective, we are seeking to:

- Develop alternative systems to support the discovery, development and production of new antibiotics that are effective for all major infectious diseases globally
- Ensure that these antibiotics are available to those populations that would benefit from them, wherever they are in the world
- Look for alternatives to reduce reliance on antibiotics – alternatives for the same treatment or prevention of infection, and alternatives to avoid infection and risk of infection and hence requirement for antibiotics or their substitutes

It is also important to note that, at least from an economics perspective, the 'eradication' of resistance is not necessarily a desirable goal. To eradicate resistance entirely – or even to maintain resistance at current levels – would require significant, if not total, reduction in the use of antibiotics, as the use of any antibiotic will lead to the development of some resistance.

Such a goal would imply allowing significant mortality and morbidity to be incurred – far in excess of that caused by the resistance itself. The aim must therefore be to use the available strategies to optimise the balance between the current use of effective antibiotics to treat infection, and thus reduce morbidity and mortality today, and minimising the emergence and spread of resistance to these antibiotics and the consequent increased future morbidity and mortality. This balance depends upon the relative costs and benefits of the positive and negative effects involved. It is determining this balance that is critical, and requires consideration from both an economics and an ethical perspective, as it encompasses significant inter-generational and inter-regional distributional aspects.

Given the diverse contexts of the developed and developing nations, it is likely that optimal strategies will vary considerably across these them, but all strategies should share the same goals of achieving an optimal balance in the use of antimicrobial agents (Hoffman et al. 2015b; Woolhouse and Farrar 2014) and explicit consideration of the distributional implications.

References

Årdal, C., K. Outterson, S.J. Hoffman, A. Ghafur, M. Sharland, N. Ranganathan, et al. 2016. International cooperation to improve access to and sustain effectiveness of antimicrobials. *The Lancet* 387 (10015): 296–307.

Asenso-Boadi, F., T.J. Peters, and J. Coast. 2008. Exploring differences in empirical time preference rates for health: An application of meta-regression. *Health Economics* 17: 235–248.

Brogan, D.M., and E. Mossialos. 2016. Systems, not pills: The options market for antibiotics seeks to rejuvenate the antibiotic pipeline. *Social Science & Medicine* 151: 167–172.

Coast, J., and R.D. Smith. 2015. Distributional considerations in economic responses to antimicrobial resistance. *Public Health Ethics* 8 (3): 225–237.

Coast, J., R.D. Smith, and M.R. Millar. 1996. Superbugs: Should antimicrobial resistance be included as a cost in economic evaluation? *Health Economics* 5: 217–226.

———. 1998. An economic perspective on policy to reduce antimicrobial resistance. *Social Science and Medicine* 46 (1): 29–38.

Coast, J., R.D. Smith, P. Wilton, A.M. Karcher, and M.R. Millar. 2002. Superbugs II: How should economic evaluation be conducted for interventions which aim to reduce antimicrobial resistance? *Health Economics* 11 (7).

Coast, J., R.D. Smith, and M.R. Millar. 2006. Disentangling value: Assessing the benefits of containing antimicrobial resistance. In *The economics of infectious disease*, ed. J.A. Roberts, 201–214. Oxford: Oxford University Press. (Reprinted from: In File).

Cormican, M., and A. Vellinga. 2012. Existing classes of antibiotics are probably the best we will ever have. *British Medical Journal* 344: e3369.

Courvalin, P. 2008. Predictable and unpredictable evolution of antibiotic resistance. *Journal of Internal Medicine* 264 (1): 4–16.

Dar, O.A., R. Hasan, J. Schlundt, S. Harbarth, G. Caleo, F.K. Dar, et al. 2016. Exploring the evidence base for national and regional policy interventions to combat resistance. *The Lancet* 387 (10015): 285–295.

Doshi, P. 2015. Speeding new antibiotics to market: A fake fix? *BMJ: British Medical Journal* 350: h1453.

GlaxoSmithKline. 2013. *GlaxoSmithKline awarded up to $200 million by U.S. Government to develop new antibiotics*. Middlesex: GlaxoSmithKline.

Goossens, H., D. Guillemot, M. Ferech, B. Schlemmer, M. Costers, M. Van Breda, L.J. Baker, et al. 2006. National campaigns to improve antibiotic use. *European Journal of Clincal Pharmacology* 62: 373–379.

Guiness, L., and V. Wiseman. 2011. *Introduction to health economics*. 2nd ed. Maidenhead: McGraw-Hill Education.

Hanefeld, J., M. Khan, G. Tomson, and R. Smith. 2017. Trade is central to achieving the sustainable development goals: A case study of antimicrobial resistance. *BMJ* 358: j3505.

Hardin, G. 1968. The tragedy of the commons. *Science* 162 (3859): 1243–1248.

Hatch, O. 2015. *Promise for antibiotics and therapeutics for health act or the PATH act*. In. www.congress.gov/bill/114th-congress/senate-bill/185?q=%7B%22search%22%3A%5B%22s.1 85%22%5D%7D.

Hoffman, S.J., K. Outterson, J.-A. Røttingen, O. Cars, C. Clift, Z. Rizvi, et al. 2015a. An international legal framework to address antimicrobial resistance. *Bulletin of the World Health Organization* 93 (2): 66–66.

Hoffman, S.J., J.-A. Røttingen, and J. Frenk. 2015b. Assessing proposals for new global health treaties: An analytic framework. *American Journal of Public Health* 105 (8): 1523–1530.

Huttner, B., H. Goossens, T. Verheij, and S. Harbarth. 2010. Characteristics and outcomes of public campaigns aimed at improving the use of antibiotics in outpatients in high- income countries. *Lancet Infectious Diseases* 10: 17–31.

Keppler, H. 2013. *The untold AIDS story: How access to antiretroviral drugs was obstructed in Africa*.

Kesselheim, A.S., and K. Outterson. 2010. Fighting antibiotic resistance: Marrying new financial incentives to meeting public health goals. *Health Affairs* 29 (9): 1689–1696.

Kolmos, H.J., and P. Little. 1999. Should general practitioners perform diagnostic tests on patients before prescribing antibiotics? *British Medical Journal* 318: 799–802.

Laxminarayan, R., I.W.H. Parry, D.L. Smith, and E.Y. Klein. 2010. Should new antimalarial drugs be subsidized? *Journal of Health Economics* 29: 445–456.

Laxminarayan, R., P. Matsoso, S. Pant, C. Brower, J.-A. Røttingen, K. Klugman, and S. Davies. 2016. Access to effective antimicrobials: A worldwide challenge. *The Lancet* 387 (10014): 168–175.

Medical Research Council. 2014. *Tackling AMR theme 2: Accelerating therapeutic and diagnostics development EOI*. In. www.mrc.ac.uk/funding/browse/tackling-amr-theme-2-accelerating-therapeutic-and-diagnostics-development-eoi.

Mendelson, M., J.-A. Røttingen, U. Gopinathan, D.H. Hamer, H. Wertheim, B. Basnyat, et al. 2016. Maximising access to achieve appropriate human antimicrobial use in low-income and middle-income countries. *The Lancet* 387 (10014): 188–198.

Molzon, J.A., A. Giaquinto, L. Lindstrom, T. Tominaga, M. Ward, P. Doerr, et al. 2011. The value and benefits of the international conference on harmonisation to drug regulatory authorities: Advancing harmonization for better public health. *Clinical Pharmacology & Therapeutics* 89 (4): 503–512.

Nesta. 2014. *The challenge: Reduce the use of antibiotics*. In. https://longitudeprize.org/challenge.

O'Neill, J. 2014. *Review on antimicrobial resistance. Antimicrobial resistance: Tackling a crisis for the health and wealth of nations*. London: HM Government; Wellcome Trust.

Olsen, J.A., and J. Richardson. 2013. Preferences for the normative basis of health care priority setting: Some evidence from two countries. *Health Economics* 22 (4): 480–485.

Oppong, R., M. Jit, R.D. Smith, C.C. Butler, H. Melbye, S. Molstad, and J. Coast. 2013. Cost-effectiveness of point-of-care C-reactive protein testing to inform antibiotic prescribing decisions. *British Journal of General Practice* 63 (612): e465–e471.

Outterson, K. 2014. *New business models for sustainable antibiotics*. London: Chatham House.

Outterson, K., J.H. Powers, G.W. Daniel, and M.B. McClellan. 2015. Repairing the broken market for antibiotic innovation. *Health Affairs* 34 (2): 277–285.

Review on antimicrobial resistance. 2015a. *Tackling a global health crisis: Initial steps*.

———. 2015b. *Securing new drugs for future generations: The pipeline of antibiotics*.

Rice, L.B. 2011. Rapid diagnostics and appropriate antibiotic use. *Clinical Infectious Diseases* 52 (Suppl 4): S357–S360.

Richardson, J., and J. McKie. 2007. Economics, political philosophy and ethics: The role of public preferences in health care decision–making. In *Principles of health care ethics*, 2nd ed., 569–576.

Smith, R. 2015. Antimicrobial resistance is a social problem requiring a social solution. *BMJ: British Medical Journal* 350: h1453.

Smith, R.D., and J. Coast. 1998. Controlling antimicrobial resistance: A proposed transferable permit market. *Health Policy* 43: 219–232.

———. 2002. Antimicrobial resistance: a global response. *Bulletin of the World Health Organization* 80: 126–133.

———. 2003. Antimicrobial drug resistance. In *Global public goods for health: A health economic and public health perspective*, ed. R.D. Smith, R. Beaglehole, D. Woodward, and N. Drager. Oxford: Oxford University Press.

Smith, R., and J. Coast. 2013. The true cost of antimicrobial resistance. *BMJ* 346 (f1493): 1–5.

Smith, R.D., C. Carlos, and O. Cecilia. 2009. Trade, TRIPS, and pharmaceuticals. *Lancet* 373 (9664): 684–691.

Trouiller, P., P. Olliaro, E. Torreele, J. Orbinski, R. Laing, and N. Ford. 2002. Drug development for neglected diseases: A deficient market and a public-health policy failure. *The Lancet* 359 (9324): 2188–2194.

Walls, H.L., and G. Ooms. 2017. Innovative use of the law to address complex global health problems: Comment on the legal strength of international health instruments-what it brings to global health governance? *International Journal of Health Policy Management* 6.

Wilton, P., R.D. Smith, J. Coast, and M.R. Millar. 2002. Strategies to contain the emergence of antimicrobial resistance: A systematic review of effectiveness and cost- effectiveness. *Journal of Health Services Research and Policy* 7: 111–117.

Wise, J. 2015. Report calls for $2 bn global fund to kickstart antibiotic development. *BMJ: British Medical Journal* 350: h2635.

Woodward, D., and R.D. Smith. 2003. Global public goods for health: Concepts and issues. In *Global public goods for health: A health economic and public health perspective*, ed. R.D. Smith, R. Beaglehole, D. Woodward, and N. Drager. Oxford: Oxford University Press.

Woolhouse, M., and J. Farrar. 2014. Policy: An intergovernmental panel on antimicrobial resistance. *Nature* 509 (7502): 555–557.

World Health Organization. 2015. *Tracking universal health coverage: First global monitoring report*. Geneva: World Health Organization.

Yamabhai, I., and R.D. Smith. 2015. To what extent are pharmaceutical prices determined by patents? A case study of oncology medicines in Thailand. *Journal of Intellectual Property Rights* 20: 89–95.

Part III
Ethics, Regulation, Governance, and Drug Resistance

Chapter 18
Antibiotics and Animal Agriculture: The Need for Global Collective Action

Jonathan Anomaly

Keywords Drug Resistance · Political philosophy · Moral philosophy · Economics

The use of antibiotics in animal agriculture is steadily increasing, especially in developing countries. The European Union and a handful of developed countries have implemented policies to scale back the use of antibiotics, recognizing its role in the global rise of antibiotic resistance. But many farmers who raise animals live in poor countries without public health regulations, or work for large corporate entities that can move their operations to places with weak regulations. To minimize the careless use of antibiotics around the world, we need multi-lateral coordination between states on some common standards for the use of antibiotics in animals.

18.1 Introduction

Imagine a world in which every time you tied your shoes, you contributed to a process that resulted in the unintended suffering and death of thousands of people you'll never know.[1] In this world, like ours, shoelaces are useful: they save time, are a little cheaper than using Velcro ties, and more convenient than wearing slip-on shoes. But when everyone ties their shoes, lots of people die, and many more suffer.

This is a strange world to imagine, but it is a lot like the world we live in. The culprit isn't tying shoelaces, of course, but consuming factory farmed meat. Factory farms are wicked places – one of the last bastions of legally sanctioned cruelty toward animals. But more than this, they are bad for human health.

---

[1] Parts of the introduction are reprinted from an article that first appeared in *Compass*, the annual magazine of the Kenan Institute for Ethics at Duke University (2017).

J. Anomaly (✉)
University of Pennsylvania, Philadelphia, PA, USA
e-mail: anomaly@upenn.edu

© The Author(s) 2020
E. Jamrozik, M. Selgelid (eds.), *Ethics and Drug Resistance: Collective Responsibility for Global Public Health*, Public Health Ethics Analysis 5,
https://doi.org/10.1007/978-3-030-27874-8_18

Some antibiotics are given to cattle and pigs to marginally speed up their growth. The biological mechanisms through which antibiotics promote growth aren't well understood, but the use of antibiotics to promote growth does seem to work. More importantly, raising animals in densely packed conditions requires a steady dose of antibiotics to prevent infections that would otherwise run rampant.

Like many practices, there are benefits as well as costs: meat from factory farms is cheaper than meat from farms with free-range animals, often about half the price. This is partly because factory farms allow animals to occupy less space, which makes their production cheaper, and this savings is passed on to consumers.

Apart from its obvious benefits, factory farming produces many costs (Anomaly 2015). In this essay, I will focus on the threat that our use of antibiotics in animal agriculture poses for human health. Contrary to popular opinion, the problem is not that antibiotics are passed along from animals to people who eat them, and that this is bad for our health. Instead, the problem is that the more antibiotics we give to livestock, the more we encourage the emergence and spread of antibiotic-resistant bacteria in a microbial environment shared by animals and people (Marshall and Levy 2011; Spellberg et al. 2016).

Like all eukaryotic organisms, people pay a high price for sex: each child only shares half of her genes with each parent. But sexual reproduction seems to confer benefits by increasing variation in the immune system children inherit, thus making it more likely that some of them will survive the onslaught of parasites that continually evolve novel ways of exploiting their hosts (Hamilton et al. 1990). As strange as sex is – each of two independent organisms swapping their genes to create a hybrid – the bacterial equivalent is even kinkier than a San Francisco night club. Bacteria reproduce by cloning themselves, but they evolve throughout their lives by promiscuously swapping genes with other bacteria and by extracting genes from the viruses that parasitize them. This allows them to adapt to new environments quickly: in a lethal environment, a small number of bacteria are likely to have some advantage over the trillions that die. And this advantage comes either from a random genetic mutation, or from the lateral transfer of genes from one bacterium to another.

Some genes allow bacteria to fend off the antibiotics that plants, animals, and other bacteria use to destroy them. Many of these naturally occurring antibiotics have existed for billions of years, as part of an unending evolutionary arms race between host and parasite. Like their naturally occurring cousins, synthetic antibiotics made in a lab usually involve penetrating a bacterial cell wall and disrupting DNA synthesis, or otherwise slowing or stopping bacterial reproduction.

All a bacterium needs to survive an antibiotic is some way to either block the penetration of the chemical with a thick cell wall, degrade it with enzymes, or pump it out if it penetrates its body. Once that happens, it's off to the races. The lucky bacterium multiplies rapidly and spreads its resistance to other bacteria. When new resistant strains of bacteria emerge in animal agriculture, they are passed along to farmers who work with animals, workers who slaughter animals, consumers who eat meat, and people in our more general microbial environment (Laxminarayan et al. 2016).

The average person hosts about 40 trillion bacteria at any given time, and we constantly swap bacteria with each other and with the environment around us (Sender et al. 2016). So even though the overuse of antibiotics tends to affect those closest to the source of resistant bacteria – whether animals or people – over time, strains of bacteria that are resistant to antibiotics can spread through trade and travel among people, and through soil and streams around factory farms. And while reducing the use of antibiotics does tend to reduce resistance, the decline of resistance does not happen immediately, since reservoirs of antibiotic-resistant genes tend to persist in bacterial plasmids for a long time (Andersson and Hughes 2010).

For more than a decade the European Union has banned antibiotics for growth promotion in farm animals, and tried to impose standards that increase animal welfare and reduce the need to use antibiotics. The US has begun to follow suit, driven by consumer demand for antibiotic-free meat, and FDA threats of regulation. But most developing countries are moving in the opposite direction, with explosive growth of antibiotic use in both people and animals in China, India, Pakistan, Egypt, and many sub-Saharan African countries (Van Boeckel et al. 2015).

18.2   Economic Models

The problem of antibiotic resistance is often framed by well-known economic models like the prisoner's dilemma, the tragedy of the commons, or the provision of public goods. All three models are useful in some contexts, but when they are not adequately qualified they can cast shade rather than light on the problem of resistance.

18.2.1   Prisoner's Dilemma

Consider first the prisoner's dilemma (PD). In the original example, we are presented with two prisoners who are suspected of armed robbery, but a District Attorney (DA) who only has enough evidence to prosecute them for the illegal possession of firearms. The prisoners are in separate jail cells, and the DA offers each of them a deal: if you snitch on your accomplice and he stays silent, you'll get off scot free and he'll be executed. If you both stay silent, you'll each get one year in prison. If you both snitch, you'll each get a decade in prison. The payoffs are as follows:

Odin	Loki	
	Silence	Snitch
Silence	1 year/1 year	Death/Freedom
Snitch	Freedom/Death	10 years/10 years

If the accomplices lack friendly feelings for one another, and if neither fears reprisals outside of prison, the rational move for each is to snitch, even if the socially optimal move is for both to stay silent. The PD is interesting because each player acting rationally produces an outcome that is worse for everyone.

The PD is a simple model that is frequently invoked to explain why rational agents act in ways that contribute to air pollution or species extinction even when each person would prefer to breath clean air or preserve biodiversity. Although most of the real-world games the PD is used to illustrate are complicated by the fact that there are more than two players, that players have asymmetric information or poorly formed preferences, and that they face uncertainty about whether (or how many times) the game will be repeated, the simplistic two player model is still of some use in visualizing problems like antibiotic resistance.

Consider the following case. Each carnivore faces the choice to consume meat from factory farmed animals or humanely raised animals free from antibiotics.

Odin	Loki	
	Humanely raised	Factory farmed
Humanely raised	2nd/2nd	4th/1st
Factory farmed	1st/4th	3rd/3rd

The payoff matrix indicates that each person does best by consuming factory farmed meat, that each does worst by consuming humanely raised meat (if the other does not), but that they both do better if they both consume meat from humanely raised rather than factory farmed animals. In the real world, if there were only two consumers and two producers, the effects of Loki's consumption choices would not be big enough to adversely affect Odin's welfare. But when we generalize to hundreds of millions of people, we get a case in which each person marginally increases the probability of antibiotic-resistant bacteria emerging and spreading, but each also saves a bit of money by consuming meat from factory farmed animals. As long as the benefit to each from buying factory farmed meat exceeds the costs associated with the alternative, the model predicts they will continue their socially suboptimal behavior.

There are several limitations of extending a two-person model to a many-person case. First, in the large number case we can treat other people's actions as given, whereas in the small number case we might change their behavior by reasoning with them (Bowles and Gintis 2013). Second, in the large number case we may have to resort to using state power to incentivize socially optimal behavior, whereas in the small number case people are in a better position to create local solutions that exploit social norms and informal punishments to move from the Nash equilibrium to the Pareto optimum (Ostrom 2000).

Similar considerations apply to farmers choosing whether to raise their animals with or without antibiotics, which is a many-person prisoner's dilemma in which most people reason parametrically (taking other's actions as more-or-less given). While there is a growing market for antibiotic-free meat, so that some farmers find it profitable to reject factory farming, most consumers around the world either don't know enough or care enough about the problem to entice farmers to reject antibiotics and raise their animals humanely.

18.2.2   Tragedy of the Commons

Many have argued that our aggregate use of antibiotics – in hospital settings and animal agriculture – is analogous to the misuse of commonly owned resources. In the classic example of a commons tragedy, farmers lack private property rights and are forced to raise animals on a common plot of land. The farmers internalize the benefits from raising animals and selling their meat, but share the costs of grass and soil depletion. Consequently, in the absence of sufficient altruistic restraint, each farmer continues to add animals to the commons up to the point at which the personal benefits equal the personal costs. To the extent that they ignore social costs, farmers add animals even if it makes everyone worse off than they would be if they agreed to a set of enforceable constraints.

Assume, for example, that above some number for each animal added to a common pasture, each farmer will get 10 utility points but the community will lose 20 utility points as the grass becomes overgrazed. If there are 10 farmers, each nets 8 utility points from adding another animal (+10 from selling the meat and $- 2$ from depleting grass and soil), and so they add animals until the commons is ruined. The typical solution to commons tragedies is to privatize plots of land, or (less efficiently) to set up enforceable limits with penalties for exceeding the limits. In small settings, these standards can be enforced by the court of public opinion, assuming farmers care about their reputation in the community. In large settings, standards are usually set by the state, and enforced with penalties for violating laws, or taxes and subsidies that attempt to bring about a socially optimal use of common resources.

Is the use of antibiotics on factory farms a commons tragedy? Some suggest that it is (Hollis and Maybarduk 2015). Others are more cautious, arguing that it depends on assumptions that include how quickly alternative antibiotics and vaccines will be developed, and how accurately we can diagnose infections (McAdams 2017a). Just as there is no such thing as a precise carrying capacity for land (since we can develop chemical fertilizers to increase soil productivity, or genetically engineer animals to more efficiently turn grass into meat), so too there is no specific point at which using more antibiotics necessarily imposes net costs on people.

As with the Prisoner's Dilemma, the commons tragedy model can help us conceptualize the incentives that generate the problem of antibiotic resistance. But it can also be misleading. For example, suppose we develop better diagnostics. Rapid diagnostic tests can make broad-spectrum antibiotics last longer by helping us identify the specific kind of infection plaguing a person or animal so that we can treat it with a narrow-spectrum antibiotic agent (McAdams 2017a). When better diagnostics are available to guide treatment in conjunction with extremely narrow-spectrum agents, David McAdams argues that "greater antibiotic use can in some cases decrease the selective pressure favoring resistant bacteria" (2017a, p. 6). Better diagnostics may also make it more profitable for companies to manufacture and conserve antibiotics if it leads physicians and farmers to more carefully use antibiotics to target specific infections (2017b). Using the wrong antibiotic often fails to treat the relevant infection, and it encourages resistance among all bacteria that the antibiotic affects. Using broad-spectrum antibiotics without a specific diagnosis is like carpet-bombing an entire city in order to kill a few soldiers. To the extent that we can target our enemies with precision strikes, there is less opportunity for collateral damage in the form of resistant strains of bacteria that grow in number as their susceptible compatriots are killed.

In addition to rapid diagnostic tests, the invention of "adjuvants" (supplements that make antibiotics more effective by priming our immune system, or by blocking bacterial resistance) can extend the life of antibiotics (Wright 2016). Rapid diagnostics and effective adjuvants show that the collective consumption of antibiotics does not *automatically* create a commons tragedy. It all depends on how we use antibiotics, and this is in part a function of technology, and the incentives that physicians and farmers face as a result of public policies.

Nevertheless, the careless way in which antibiotics are currently used in animal agriculture outside of Europe probably *is* a commons tragedy. This is because farmers in most countries today simply ignore the social cost of using antibiotics in livestock, and many farmers fail to understand how using antibiotics in agriculture can lead to the rise of bacterial infections in people that are increasingly expensive, difficult, or impossible to treat.

18.2.3   Public Goods

A final model frequently used to describe problems associated with our use of antibiotics requires us to make a distinction. In economics, *private* goods are those that are consumed by individuals in ways that don't involve significant externalities (costs or benefits borne by people external to an economic transaction). For example, when I buy a private good like a cup of coffee or a pair of eyeglasses, the costs or benefits imposed on other people are trivial. *Public* goods, by contrast, are consumed in common, so that we share the benefits of consumption. Public goods can be thought of as non-excludable positive externalities (Cowen 2008), though this is misleading in cases where the public good is experienced as a cost rather than a benefit to those who consume it.

Antibiotics themselves are not public goods, but to some extent the *efficacy* of antibiotics, and efforts made to move us toward the socially optimal use of antibiotics, *are* public goods. Similarly, efforts to eliminate infectious diseases are public goods (Selgelid 2007), since the reduction or eradication of a disease is shared by all people in a region, and potentially all people on the planet. By extension, reducing the reckless use of antibiotics in agriculture is a public good. Although alternative agricultural methods are more expensive, the enormous external costs of drug-resistant diseases that emerge from factory farming almost certainly exceed the benefits of cheaper meat (O'Neill et al. 2015).[2]

Many people, including some economists, equate public goods problems with commons tragedies and prisoner's dilemmas. This is a mistake, although it is understandable since many commons tragedies and public goods problems can be usefully modeled by the prisoner's dilemma. But often public goods are better described as assurance games or coordination games (Hampton 1987), and this is good news for lawmakers and farming associations who are aware of the problem and want to converge on common standards that allow them to make a profit *and* minimize the risk of antibiotic resistance. One problem with preserving global public goods like the efficacy of antibiotics is that most people are unaware of the problem, since each plays a very small role in producing it. In other words, many people who might help preserve or produce public goods are rationally ignorant about the nature of the problem.

18.3 Moral Principles

Ignorance of how the use of antibiotics in agriculture harms human health is *rational* in the economic sense, but it is not necessarily *morally excusable* (Anomaly 2015). Since the problem of AMR is difficult to understand, and since each act of consuming factory-farmed products contributes only imperceptibly to the problem, it makes perfect sense that consumers would ignore the problem and purchase cheap factory-farmed meat, rather than more expensive meat from farms that don't use antibiotics.

But the fact that we can explain consumer ignorance does not absolve consumers of responsibility for contributing to the problem. As information about the private benefits and social costs of using antibiotics in farm animals becomes more widely available, consumers have an increasing responsibility to act on it by changing their purchasing habits and trying to persuade governments to make it harder to purchase meat from animals unnecessarily dosed with antibiotics. Alexander Fleming

---

[2] It may be that *some* use of antibiotics in agriculture is both individually beneficial for animals (who contract infections despite humane and prudent farming practices), and socially beneficial for people (who may be less likely to contract a bacterial infection an animal has). But the growing quantity of antibiotics used in farming today is likely to produce harms that far exceed these benefits.

famously warned that "the thoughtless person playing with penicillin is morally responsible for the death of the man who finally succumbs to infection with the penicillin-resistant organism."[3] One form of "playing with penicillin" is the use of it as a growth promoter on factory farms, or the more common use of it to prevent infections in the cramped and cruel conditions that characterize factory farms.

A more nuanced version of Fleming's admonition requires us to distinguish *actual* harms to discrete people from *probabilistic* harms to actual or potential people. Another way to put the point is to say that the harms of antibiotic resistance are "identity-independent" in the sense that the victims of AMR cannot be known ahead of time and, in some cases, are not yet born. While a single farmer (or consumer) misusing antibiotics can create or encourage a resistant strain that spreads to other people, generally the prevalence of resistant bacteria in the environment depends on how all of us act. By acting in ways that create genetic pollution in our microbial environment, we make it a little more likely that someone will suffer or die of a previously treatable infection.

Many other pollution problems are structurally similar to antibiotic resistance. For example, each of us drives to work and produces the social costs of pollution and traffic congestion as a byproduct, while experiencing the private benefit of an enjoyable ride in our own car. There are also *social* benefits when each person drives, if driving contributes to a more efficient workforce that creates better goods at lower cost. Suppose the social costs of air pollution and traffic congestion exceed the individual benefits of driving. A common response is to impose a price on driving by taxing fuel or charging user fees to encourage the efficient use of roads and the atmosphere. The underlying moral principles are that we should pay in proportion to the amount we contribute to the problem, and that if anyone's liberties to pollute are restricted, then all us should face the same restrictions (Gaus 1999, p. 197).

Similar arguments have been made for taxing antibiotics in medicine and agriculture to discourage low-value use (Kades 2005; Anomaly 2013). But antibiotic resistance is much more complicated than air pollution or traffic congestion: in some cases we may want to *subsidize* rather than *tax* the use of antibiotics when people who can't afford them are likely to spread infectious diseases to others (Selgelid 2007). Apart from taxes and subsidies, there is a vast literature on how to harness intellectual property rights, prescription requirements, basic science research funding, and shared surveillance to control the problem (O'Neill et al. 2016).

What I now want to argue is that without more coordination between states, the problem of antibiotic resistance in agriculture will likely get worse, with dire consequences for human health in the coming century.

---

[3] http://www.fda.gov/NewsEvents/Speeches/ucm427312.htm

18.4   Global Coordination

The provision of global public goods like conserving antibiotics and reducing infectious disease raises two problems: the *free rider* problem occurs when individual consumers, farmers, or states seek the gains of limited antibiotic use without paying the costs; the *assurance* problem occurs when each is willing to pay the cost of reducing unnecessary use, but lacks the assurance that others will abide by policies that constrain our collective use of antibiotics.

The first problem is difficult to overcome to the extent that self-interest dominates the actions of farmers in a market or of politicians in a government. But there is some evidence that most consumers who understand the problem are willing to pay higher prices for meat from animals not given antibiotics (Spellberg et al. 2016). Moreover, if people really understood the problems factory farms create they would likely be willing to pay significantly more for meat, since most people support taxes (or costly regulations) when they are reasonably sure the tax will be used to discourage the problems associated with pollution (Kallbekken et al. 2011).

Agricultural producers are also likely to be willing to comply with standards that limit antibiotic use provided other firms are also forced to internalize the cost of similar regulations or taxes. The fact that the assurance problem is often more serious than the free rider problem in trying to elicit cooperation in many public goods games (Bowles and Gintis 2013) is good news for those who worry about the feasibility of states setting mutually beneficial standards.

Part of the problem with antibiotics in agriculture is that as transportation costs decline, the market for animal meat becomes increasingly global: animal feed is produced in one country, animals are raised in another, and then meat is exported to a third country. Since producers in many countries are now in a position to operate industrial animal farms, unless all states set standards that limit antibiotic use, producers will tend to migrate to countries with the weakest regulations. There is already some evidence that this "race to the bottom" is happening as Chinese farms are producing meat in factory farms that use more confinement and antibiotics than farms in other countries. In fact, just as the US is beginning to move away from factory farming due to consumer demand and threats of regulation by the US Food and Drug Administration, many of the most populous developing countries – including China, India, and Brazil – are embracing factory farming (O'Neill et al. 2015).

A well-designed trade treaty between major exporters and importers of meat should recognize the problem of "leakage," which occurs when one country sets relatively high environmental standards, and allows other countries with weaker standards to increase the production of similar goods in ways that simply changes where the pollution is emitted (Barrett 1999). In other words, any treaty worth implementing cannot reward free-riding countries whose firms are permitted to externalize the costs of their production, while firms in other countries internalize the costs of complying with policies that would make everyone better off if countries complied with them.

A second feature of an effective treaty to limit antibiotics in agriculture is a minimum participation clause to assure prospective signatories that unless a sufficient number of nations sign on, they will not be forced to pay additional production costs (Barrett 1999). This feature solves the assurance problem for firms and nations that are willing to comply with stricter production practices provided enough others do to produce the global benefits associated with restricting antibiotic use.

A third feature of any multi-lateral agreement to restrict antibiotic use is that it would need to be flexible enough to allow countries to achieve collective goals in different ways. For example, some experts advocate setting targets for the *per capita* quantity of antibiotics that can be administered to animals. According to the British Review on Antimicrobial Resistance, pork producers in Denmark (the first country to ban antibiotics as growth promoters) use about 50 mg of antibiotics per kg of livestock in the country (O'Neill et al. 2015, p. 2).

A flexible treaty would take something like this number as a benchmark that all countries must meet, but it would allow countries to achieve the relevant goal in different ways: by taxing antibiotics, placing a cap on total use, restricting antibiotics by requiring veterinary oversight, or some combination of these policies. Antibiotics deemed especially important for human use should probably be banned for use in agriculture by all countries. But what often matters is the *quantity* of antibiotics used, not just the *kind*. This is especially true because plasmids that confer antibiotic resistance can be transferred between bacteria of different species, and can reduce the efficacy of different drugs than those administered by farmers (Marshall and Levy 2011).

One advantage of imposing "pollution taxes" or user fees on antibiotics in agriculture is that, unlike regulations, governments have strong incentives to enforce them. Governments can use the revenue raised from taxes to finance vaccination programs that minimize the need to administer antibiotics. They might also fund basic science research that aims to develop new vaccines and diagnostics for infectious diseases, and to develop entirely new treatments like genetically engineered bacteriophage viruses (Bikard et al. 2014).

Taxing socially costly activities like using antibiotics in agriculture also incentivizes farmers to find alternative ways to produce meat that minimize antibiotic resistance. These alternatives may include increasing the roaming space animals have, and decreasing the stress they face when forced to live in extreme confinement. A more promising alternative is to create "in vitro meat" made in a lab from embryonic stem cells. This would avoid the need to raise animals at all, thus reducing untold amounts of suffering and potential public health problems.

Finally, any agreement to restrict antibiotic use should be attractive enough for each participating country to be willing to *enforce* it. It is likely that offering benefits for compliance will be more effective than simply threats of sanctions for noncompliance. For example, it is in the interest of all nations that each nation monitors the outbreak and spread of infectious diseases, as well as novel patterns of antibiotic resistance. But sometimes only wealthier states have the budgets and technology to accomplish this. By sharing information and technology with developing countries, wealthier countries can both signal goodwill and deliver tangible benefits to other countries they wish to comply with stricter controls on antibiotic use. This may act as a positive incentive for poor countries to do their part, even if the threat of sanctions for non-compliance with collectively beneficial restrictions is also important.

Each nation faces its own challenges, including an electorate that is unlikely to fully understand the social benefits and costs of antibiotics, and factory farmers who are unlikely to welcome regulations that impose new costs on them. Governments can justify spending some money to ease the transition from factory farming techniques to alternatives that produce better consequences for the same reason they can justify compensating taxi cab drivers who were required to buy a costly permit from the state to drive a taxi, but who are now forced to compete with companies like Uber, whose drivers did not have to pay for the right to operate as a taxi service. In fact, if relatively wealthy governments offer temporary assistance to domestic firms to transition away from factory farming, and to relatively poor governments to comply with new restrictions, the move away from the reckless use of antibiotics may be easier to induce, and more fair from the standpoint of global distributive justice.

References

Andersson, Dan, and Diarmaid Hughes. 2010. Antibiotic resistance and its cost: Is it possible to reverse resistance? *Nature* 8: 260–271.

Anomaly, Jonathan. 2013. Collective action and individual choice. *Journal of Medical Ethics* 39: 752–756.

———. 2015. What's wrong with factory farming? *Public Health Ethics* 8 (3): 246–254.

Barrett, Scott. 1999. Montreal vs Kyoto. In *Global public goods: International cooperation for the 21st century*, ed. Inge Kaul et al. Oxford: Oxford University Press.

Bikard, David, et al. 2014. Exploiting CRISPR-Cas nucleases to produce sequence-specific antimicrobials. *Nature Biotechnology* 32: 1146–1150.

Bowles, Samuel, and Herbert Gintis. 2013. *A cooperative species: Human reciprocity and its evolution*. Princeton: Princeton University Press.

Cowen, Tyler. 2008. *Public goods. The concise encyclopedia of economics*. 2nd ed. http://www.econlib.org/library/Enc/PublicGoods.html.

Gaus, Gerald. 1999. *Social philosophy*. London: M.E. Sharpe Publishing.

Hamilton, William, et al. 1990. Sexual reproduction as an adaptation to resist parasites. *Proceedings of the National Academy of Science* 87: 3566–3573.

Hampton, Jean. 1987. Free rider problems in the production of collective goods. *Economics and Philosophy* 3: 245–273.

Hollis, Aidan, and Peter Maybarduk. 2015. Antibiotic resistance is a tragedy of the commons that necessitates global cooperation. *Journal of Law, Medicine, and Ethics* 43: 33–37.

Kades, Eric. 2005. Preserving a precious resource: Rationalizing the use of antibiotics. *Northwest University Law Review* 99: 611–675.

Kallbekken, Steffen, et al. 2011. Do you not like Pigou, or do you not understand him? *Journal of Environmental Economics and Management* 62 (1): 53–64.

Laxminarayan, Ramanan, et al. 2016. Antibiotic use and resistance in food animals. *The Center for Disease Dynamics, Economics, and Policy*. https://cddep.org/sites/default/files/india_abx_report.pdf. Accessed 1 May 2017.

Marshall, Bonnie, and Stuart Levy. 2011. Food animals and antimicrobials: Impacts on human health. *Clinical Microbiology Reviews* 24 (4): 718–733.

McAdams, David. 2017a. Resistance diagnosis and the changing epidemiology of antibiotic resistance. *New York Academy of Sciences*. 1388 (2017): 5–17.

———. 2017b. Resistance diagnosis and the changing economics of antibiotic discovery. *New York Academy of Sciences*. 1388 (2017): 18–25.

O'Neill, Jim, et al. 2015. Antimicrobials in agriculture and the environment: Reducing unnecessary use and waste. *The Review on Antimicrobial Resistance*. http://amr-review.org/.

———. 2016. Tackling drug-resistant infections globally: Final report and recommendations. *The Review on Antimicrobial Resistance*. http://amr-review.org/.

Ostrom, Elinor. 2000. Collective action and the evolution of social norms. *Journal of Economic Perspectives* 14 (3): 137–158.

Selgelid, Michael. 2007. Ethics and drug resistance. *Bioethics* 21 (4): 218–229.

Sender, Ron, et al. 2016, August 19. Revised estimates for the number of human and bacterial cells in the body. *PLOS Biology*. https://doi.org/10.1371/journal.pbio.1002533.

Spellberg, Brad, et al. 2016. *Antibiotic resistance in humans and animals*. Discussion paper for the National Academy of Medicine. Accessed May 2017. http://www.nam.edu/antibiotic-resistance-in-humans-and-animals.

Van Boeckel, T.P., et al. 2015. Global trends in antimicrobial use in food animals. *Proceedings of the National Academy of Sciences* 112 (18): 5649–5654.

Wright, Gerard. 2016. Antibiotic adjuvants: Rescuing antibiotics from resistance. *Trends in Microbiology* 24 (11): 862–871.

Chapter 19
Technological Fixes and Antimicrobial Resistance

Nicholas B. King

Abstract A 'technological fix' reduces the negative impact of a problem without addressing its underlying political, economic, or social causes. This chapter examines antimicrobials' central role in both the modern faith in technological fixes in medicine, and critiques of over-reliance on technological interventions that produce unintended consequences. The enduring appeal of technological fixes is rooted in their promise to provide simple, efficient, measurable, and effective solutions to complex problems; but this practically is purchased at the price of eliding important distributive concerns.

Keywords Ethics · Public health · Infectious diseases · Social policy · History of science

19.1 Introduction

In October, 2003, The Bill and Melinda Gates Foundation (BMGF) announced the first 14 'Grand Challenges in Global Health,' as part of an initiative to stimulate scientific research on diseases affecting less-developed nations. Grand Challenge #10 asked for applications that would help "discover drugs and delivery systems that minimize the likelihood of drug resistant microorganisms." In the intervening years, combatting antimicrobial resistance has continued to be a priority for the BMGF, with at least 50 grants funded through initiatives aimed at 'Creating Drugs and Delivery Systems to Limit Drug Resistance,' and identifying 'Novel Approaches to Characterizing and Tracking the Global Burden of Antimicrobial Resistance.' Funded projects include efforts to crowdsource surveillance of antimicrobial resistance (AMR), investigating the role of metabolic pathways in *M. tuberculosis*

N. B. King (✉)
McGill University, Montreal, QC, Canada

© The Author(s) 2020
E. Jamrozik, M. Selgelid (eds.), *Ethics and Drug Resistance: Collective Responsibility for Global Public Health*, Public Health Ethics Analysis 5,
https://doi.org/10.1007/978-3-030-27874-8_19

bacteria as a target for drug therapy, and identifying chemical entities that would act as 'selection inverters' to limit or reverse the development of AMR.

Given the problem's scale and complexity, the importance of developing techno-logical solutions to improve surveillance and response to AMR might seem unas-sailable. However, soon after the initial announcement of the BMGF Grand Challenges, historian Anne-Emanuelle Birn published a sharp critique in *The Lancet*. Birn argued that, "in calling on the world's researchers to develop innova-tive solutions targeted to 'the most critical scientific challenges in global health', the Gates Foundation has turned to a narrowly conceived understanding of health as the product of technical interventions divorced from economic, social, and political contexts"(Birn 2005, p. 515). In the area of AMR, Birn argued that:

> Three more Grand Challenges addressing drug resistance and the development of cures for latent and chronic infections are also short-sighted. While access to effective immunothera-pies for HIV/AIDS, tuberculosis, and other ailments has been appropriately termed a human right, integration of treatment with the well- established social and economic com-ponents of prevention surely merits at least one Grand Challenge. Here what deserves care-ful consideration are the factors associated with both HIV and multiple drug resistant tuberculosis: typically a combination of deprived social conditions—poor nutrition, over-crowded and unsanitary housing, economic insecurity, and inadequate health-care ser-vices—which lead to disease and can inhibit the taking of a complete course of medication (Birn 2005, p. 516).

Noting that "it is easy to be seduced by technical solutions"(Birn 2005, p. 516), Birn questioned the BMGF's assumption that the most pressing global health prob-lems can be solved through the application of innovative scientific and technologi-cal solutions that target proximate causes of disease, while ignoring more distal political, economic, or social determinants of health. She concluded with a call for integrating both narrow technological and wider social interventions, and cautioned that "the longer we isolate public health's technical aspects from its political and social aspects, the longer technical interventions will squeeze out one side of the mortality balloon only to find it inflated elsewhere"(Birn 2005, p. 518).

Birn's general argument is commonplace in contemporary public health. These arguments frequently employ a critique of 'technological fixes' (or in Birn's lan-guage, 'technical solutions') – attempts to solve problems through technological innovations or interventions, without attending to the social, economic, or political factors that may have contributed to the development of the problem in the first place. Over-reliance on the promise of technological fixes has long been controver-sial in public health, particularly since the widespread recognition of the importance of the social determinants of health. Critics contend that too much emphasis is placed on 'downstream' technical solutions to health problems whose root 'upstream' causes are not amenable to technological intervention.

In this chapter, I will use the idea of a 'technological fix' as a window into the moral economy of responses to AMR, paying special attention to questions of dis-tributive justice. Responding to antimicrobial resistance – which results from both a lack of appropriate medical technology and a complex web of behavioral, social, political, and economic factors – invites but also troubles the easy distinctions at the heart of the technological fix idea, and confronts us with difficult decisions regarding how best to distribute social and economic resources.

19.2 Technological Fixes

The idea of a technological fix arose within a context of postwar optimism that advances in science and technology could provide cheap and effective solutions to seemingly intractable social problems. While there were important precursors, one of the most important early popularizers of this view was Alvin Weinberg, a physicist and longtime director of the Oak Ridge National Laboratory. Weinberg had long advocated philosophical reflection and social awareness among scientists, particularly about the implications of 'Big Science.' During the 1960s, in a series of conversations with colleagues, he developed the idea that at least some pressing problems that were largely seen as intrinsically social might be amenable to simple technological interventions (Johnston 2018), culminating in an influential 1966 speech, "Can Technology Replace Social Engineering?" which was printed in a number of periodicals, including the *Bulletin of the Atomic Scientists.*

Weinberg argued that the causes of many pressing social problems of the day were intractably complex, and interventions targeting those causes are infeasible or impossible. He thus advocated remedies that can reduce or eliminate the harms generated by these problems, without addressing those causes:

> There is a more basic sense in which social problems are much more difficult than are technological problems. A social problem exists because many people behave, individually, in a socially unacceptable way. To solve a social problem one must induce social change – one must persuade many people to behave differently than they behaved in the past. One must persuade many people to have fewer babies, or to drive more carefully, or to refrain from disliking blacks. By contrast, resolution of a technological problem involves many fewer individual decisions (Weinberg 1991, p. 42).

Weinberg identified technologies that had already 'fixed' social problems, including the intra-uterine device (a 'one-shot' form of birth control which requires a minimum of individual motivation), development of safer cars (which reduces traffic deaths without having to improve driver competence), and even the hydrogen bomb (a strong disincentive to war that does not require greater tolerance or understanding). He proposed additional fixes, such as providing air conditioning to low-income households (to improve immediate personal comfort and potentially reduce rioting during hot summer months, without having to address the underlying social or economic causes of riots), and employing nuclear-powered desalination (which would eliminate water shortages without changing water consumer behavior).

While Weinberg was an admitted optimist and a tireless promoter of technological fixes, in his 1966 speech he struck a conciliatory note with potential critics: Noting that "technological solutions to social problems tend to be incomplete and metastable, to replace one social problem with another," he conceded that "we technologists shall not satisfy our social engineers, who tell us that our Technological

Fixes do not get to the heart of the problem; they are at best temporary expedients; they create new problems as they solve old ones" (Weinberg 1991, pp. 47–8). Answering the question posed in his title, Weinberg concluded that "technology will never *replace* social engineering…It is only by cooperation between technologist and social engineer that we can hope to achieve what is the aim of all technologists and social engineers – a better society, and thereby, a better life" (Weinberg 1991, p. 48).

Weinberg's hedging notwithstanding, his ideas about technological fixes met with immediate and sustained criticism; as early as 1970 the Oxford English Dictionary included wholly negative or ironic definitions of the term (Rosner 2004). Critics have mounted both practical and ideological objections. On the practical side, technological fixes provide at best short-term relief from the deleterious symptoms of social problems, but fail to address their root causes; at worst, they oversimplify intrinsically complex problems which they then fail to solve, and may introduce side effects that are worse than the problems they intend to solve. In a larger sense, the search for technological fixes is fundamentally reactionary in nature, encouraging society to seek cheap, quick, and partial solutions to holistic problems, while turning a blind eye to fundamental social and distributive injustices.

The transparent pursuit of technological fixes has had few defenders as vocal as Weinberg, and many detractors. This does not mean that the idea has fallen out of favor. Indeed, continued criticism is a sign of the enduring appeal of the idea of a technological fix, even if the term itself is not commonly used. For example, Evgeny Morozov's 2013 book, *To Save Everything, Click Here: The Folly of Technological Solutionism*, criticizes the 'there's an app for that' mentality of technological 'solutionism' in the internet age. Echoing earlier criticisms, Morozov argues that efforts to use internet-based technology to solve complex social problems – such as combatting obesity through exercise-tracking apps rather than food regulation – encourage us to define problems in terms of the technologies that might solve them, and discourage broader consideration of macro-level interventions (Winograd 2013).

19.3   Technological Fixes and Health

Technological fixes have been the subject of intense and sustained debate in the areas of public health and medicine. Advances in medical technology (e.g. vaccines, therapeutic drugs, surgical techniques, medical devices such as x-rays) play an undeniable role in diagnosing and curing disease, relieving suffering, increasing life expectancy, and improving population health. This fact alone would seem to justify continued investment of the BMGF Grand Challenges variety. Yet, beginning around the same time that Weinberg began evangelizing for technological fixes, critics began to question a similar logic underlying increasing investment in medical technology.

First, critics questioned whether medical advances played as great a role in health improvements as commonly believed. The classic statement of this critique is

Thomas McKeown's 1976 book, *The Role of Medicine: Dream, Mirage, or Nemesis?* McKeown argued that changes in political economy – including rising standards of living and better sanitation and nutrition – rather than specific therapeutic or preventive efforts, were responsible for the massive mortality declines of the ninteenth and early twentieth centuries (McKeown 1979). While the McKweown thesis has been hotly contested (Szreter 1988; Fairchild 1998; Colgrove 2002), it is now widely accepted that the contribution of technological innovation to historical improvements in health and longevity has been greatly overstated.

Critics also pointed to the unintended consequences of medical technology. The classic statement (and the 'Nemesis' of McKeown's title) is Ivan Illich's 1975 book, *Medical Nemesis: The Expropriation of Health*. Illich criticized medicine's role in the production of 'iatrogenic' illness, side effects of medical interventions that might often be worse than the original pathology they were intended to treat. He lamented the progressive "medicalization of life," which left individuals and societies ill-equipped to contend with normal aspects of human life including birth, pain, and death, as anything other than technologically-mediated pathology. And he objected to modern medicine's "radical monopoly" on all aspects of health and disease, diverting economic and intellectual resources from other attempts to improve the human condition.

Finally, critics contended that the pursuit of medical technology promoted a narrow approach to health and illness, rather than the environmentally and socially holistic approach that is actually necessary. In his 1959 book, *Mirage of Health: Utopias, Progress, and Biological Change*, René Dubos – a microbiologist who as early as 1942 had predicted the emergence of bacterial resistance (Moberg 1996) – argued that attempts to create a world free of disease are bound to fail, "because paradise is a static concept while human life is a dynamic process" (Dubos 1996, p. 281). Modern medical technology is ill-equipped to address the evolutionarily dynamic biological and social worlds, and single-minded pursuit of technological innovation is a fool's errand.

In the years since, the arguments mounted by McKeown, Illich, and Dubos have resonated with a wide variety of critics of technological fixes in clinical care, modern medicine, and public health. They have found most purchase among proponents of the social determinants of health, who contend that the majority of the world's inequalities in health result from political, economic, environmental, and social injustices, rather than lack of technological innovation. As the World Health Organization's 2008 report, *Closing the Gap in a Generation*, argues:

> Traditionally, society has looked to the health sector to deal with its concerns about health and disease. Certainly, maldistribution of health care – not delivering care to those who most need it – is one of the social determinants of health. But the high burden of illness responsible for appalling premature loss of life arises in large part because of the conditions in which people are born, grow, live, work, and age. In their turn, poor and unequal living conditions are the consequence of poor social policies and programmes, unfair economic arrangements, and bad politics. Action on the social determinants of health must involve the whole of government, civil society and local communities, business, global fora, and international agencies. Policies and programmes must embrace all the key sectors of society not just the health sector (WHO Commission on Social Determinants of Health 2008, p. 1).

19.4 Technological Fixes in the Context of Antimicrobial Resistance

It is no accident that Dubos developed his theories of microbial ecology, and his broader critique of the 'mirage' of modern medicine, within the context of his early work on bacterial resistance (Moberg 1996). Antimicrobials appear at first glance to be a technological fix par excellence, in the sense that they are a simple and effective technical solution to a complex problem with biological, social, political, and economic determinants. Indeed, a recent article critiquing the pursuit of technological fixes in dementia cited antibiotics as the classic case of a technological fix with unintended consequences (Jongsma 2017).

One could in fact argue that the apparent success of antibiotics is in large part responsible for a broader faith in technological fixes in medicine in the developed world. In 1967, during an oft-cited speech before American public health officials at the White House, U.S. Surgeon General William H. Stewart declared that it was time to close the book on infectious diseases and turn attention towards chronic health problems (Garrett 1994, p. 33). Stewart's optimistic claim reflects a perennial American faith in the power of biomedical science to conquer disease, but it is also evidence of a significant transformation in the burden of disease during the past century. In what is often referred to as the "epidemiologic transition," the proportion of deaths caused by infectious disease in the United States (and other industrialized nations) declined precipitously (Omran 1983). In 1900, infectious disease was responsible for 797 deaths per 100,000 population; by 1980, this figure had dropped to 36 deaths per 100,000 population. As mortality from infectious disease climbed, death rates for heart disease, various cancers, stroke, and accidents held steady or increased. In a dramatic reversal, chronic diseases replaced infectious ones as the leading killers in the United States. In 1900, 40% of deaths in the U.S. were caused by the eleven major infectious diseases (pneumonia, influenza, and tuberculosis alone accounting for more than 25% of all deaths), 16% by the three major chronic diseases (heart disease, cancer, and stroke), and 4% by accidents; by 1973, only 6% of deaths were caused by infectious diseases, 58% by chronic diseases, and 9% by accidents (Armstrong 1999).

The social and institutional ramifications of the epidemiologic transition were profound. Deluged with dramatic stories of patients snatched from the brink of death by antibiotics such as streptomycin and penicillin (dubbed "yellow magic" by Reader's Digest in 1943), Americans increasingly credited biomedical science for reducing the threat of infectious disease. This accelerated the transformation (or "narrowing"), underway since the early part of the century, of the focus of public health, from broad preventive measures towards clinical medicine, screening and the early detection of disease (Rosenkrantz 1974; Tomes 1998). This transformation was accompanied by changes in federal health expenditures as well: between 1950 and 1959, federal grants-in-aid declined from $45 million to $33 million, while funding for clinical and laboratory research jumped from $28 million in 1947 to $186 in 1957 (Fee 1994). The redirection of public and private funding from public health towards biomedical research would continue to accelerate into the 1980s, when it would be exacerbated by the dismantling of public health infrastructures in general under the pressure of Reagan-era budgetary constraints.

Even as much of the institutional apparatus for addressing infectious disease was being dismantled, the public and the medical profession increasingly mirrored Stewart's optimism regarding the threat of infectious disease. As historian Nancy Tomes notes, "With the array of drugs and vaccines available by 1965, the need to guard against contact infection understandably relaxed. Americans quickly came to believe that with a few soon-to-be-cured exceptions, modern medicine and public health had 'conquered' epidemic disease. Young physicians in the 1960s were advised, 'Don't bother going into infectious diseases,' and to concentrate on cancer or heart disease instead" (Tomes 1998, p. 254) Americans increasingly came to see infectious disease as a thing of the past, and they were bolstered in this confidence by the lack of significant epidemics of communicable disease – a confidence only briefly shaken by the HIV/AIDS epidemic, at least until the discovery of antiretro-viral drugs.

Yet one can argue whether antibiotics truly 'fixed' the problem of infectious disease. As already noted, while they are often given full credit for the decline of infectious disease in Europe and North America, antibiotics were introduced after the majority of mortality declines that are more correctly attributed to rising stan-dards of living, better nutrition, and basic public health preventive measures such as sanitation, vaccination, vector control, and provision of clean water (Szreter 1988; Cutler and Miller 2005). Moreover, social and economic factors continue to be major determinants of infectious disease (Semenza 2016). Even in high-income countries with well-functioning health care systems and access to a full range of antibiotics, poverty, social marginalization, and food and water quality continue to play a significant role in the incidence of infectious disease (King 2003; Semenza 2010). It remains to be seen whether, from a population perspective, investment in additional antimicrobials is the most effective or efficient way to reduce the burden of infectious disease.

Whether or not antibiotics 'fixed' infectious disease, their introduction generated the unintended consequence of antimicrobial resistance. While it might seem logi-cal that a 'technological' problem (antimicrobials) demands a 'technological' solu-tion (more antimicrobials), as the other chapters in this volume illustrate, the causes of and solutions to antimicrobial resistance are more complicated. As with infec-tious disease, the proximate cause of antimicrobial resistance is biological – patho-gens develop resistance under the selective pressure exerted by use of antimicrobials – but underlying this proximate cause is a range of more distal deter-minants. These include, among others:

- Physician behavior – e.g., prescribing antibiotics for conditions caused by viruses (Sprenger 2015); engage in suboptimal practices such as use of monotherapy rather than combination therapy; and incorrect drug administration routes (Struelens 1998).
- Patient behavior – e.g., failure to complete full course of treatment; and self-medication, particularly in countries where antimicrobials are available over the counter.
- Poor hygiene, sanitation, and infection control in hospitals and other healthcare facilities, leading to cross-infection with multiple strains of bacteria (Aiello 2006; Struelens 1998).
- Widespread use – including misuse and overuse – of antimicrobials in agriculture (Levy 2014; Ventola 2015).
- Lack of development of new antimicrobials, which are less profitable than drugs for chronic conditions that are generally more expensive and used for much longer periods of time (Brown and Wright 2016).

Novel antibiotics will be introduced into the same social, political, and economic contexts that have contributed to the development of antibiotic resistance in the first place. Addressing these contexts would require, among other interventions: changing the behavior of physicians and patients to encourage appropriate stewardship of antimicrobials; instituting and enforcing new sanitation and infection control protocols at healthcare facilities across the globe; reforming an agricultural system dependent upon cheap antibiotics; and changing the incentive structure of the pharmaceutical industry to ensure that need rather than profit drives drug research.

Faced with a wide and seemingly insurmountable range of determinants, focusing on the development of new lines of antimicrobials through targeted research grants and incentive programs is attractive. Yet such a narrow pursuit would, ultimately, amount to little more than layering technological fixes on prior technological fixes. As Dubos noted a half-century ago, "Granted the obvious usefulness of sanitary practices, immunological procedures, and antimicrobial drugs, it does not necessarily follow that destruction of microbes constitutes the only possible approach to the problem of infectious disease, nor necessarily the best"(Dubos 1996, p. 53).

19.5 Technological Fixes and Distributive Justice

Technological fixes have an enduring appeal. Part of this is practical: they appear to provide simple, efficient, measurable, and effective solutions to complex problems. However, the presumed practicality of these fixes conceals underlying moral and political assumptions, with important ramifications for distributive justice.

One of Weinberg's precursors was Richard L. Meier, a chemist and technological optimist who advocated the use of technological systems as a means of eliminating poverty and other social ills (Johnston 2018). It is worth noting an important distinction between the little-known precursor and the widely-known successor: while Meier saw technology as a means of remedying distributive injustices, Weinberg saw it as a means for minimizing the harms of those injustices without addressing them. Indeed, in his influential 1966 essay, Weinberg's primary example of a successful technological fix is the advances in mass production of goods that "enable[d] our capitalistic society to achieve many of the aims of the Marxist social engineer without going through the social revolution Marx viewed as inevitable" (Weinberg 1991, p. 43).

The distinction between Meier and Weinberg illustrates an often-hidden ethical component of the technological fix approach. Technological fixes generally reward the rich. This is partially by design: fixes are appealing precisely because they avoid social engineering that might upset the status quo, including extant inequalities. They require no hard questions about the justness of current distributional procedures or outcomes, and they often expressly promise to reduce the negative consequences of extant inequalities without addressing the inequalities themselves.

The concentration of benefits among the rich is also a consequence of the global economy of health research. Health-related technological innovations are most likely to occur in wealthy institutions – top research universities, multinational pharmaceutical companies and medical device manufacturers, highly-capitalized biotechnology startup companies – concentrated in high-income countries. Initiatives that target technological innovation, from BMGF research grants to targeted incentives for pharmaceutical development, thus overwhelming benefit wealthy individuals and institutions in wealthy countries. While it is true that these investments may eventually benefit others, in the short-term they simply circulate resources from rich donors to rich institutions in the global north, often in the name of benefitting the least advantaged.

Technological fixes may thus be 'practical' insofar as they provide an easily identifiable, targeted, and efficient strategy for distributing resources; but this practically is purchased at the price of eliding important distributive concerns. In the context of antimicrobial resistance, we would do well to resist the urge to unthinkingly pursue technological fixes, lest they contribute to precisely the inequalities that underlie the spread of infectious disease and the continued generation of antimicrobial resistance.

Acknowledgments The author thanks Joy Tseng for research assistance in the preparation of this chapter.

References

Aiello, Allison E., Nicholas B. King, and Betsy Foxman. 2006. Antimicrobial resistance and the ethics of drug development. *American Journal of Public Health.* 96(11):1910–1914.

Armstrong, Gregory L., Laura A. Conn, and Robert W. Pinner. 1999. Trends in infectious disease mortality in the United States during the 20th century. *Journal of the American Medical Association* 281:61–66.

Anne-Emanuelle, Birn. 2005. Gates's grandest challenge: Transcending technology as public health ideology. *Lancet* 366:514–519.

Brown, E.D., and G.D. Wright. 2016. Antibacterial drug discovery in the resistance era. *Nature* 529(7586):336–343.

Colgrove, James. 2002. The McKeown thesis: A historical controversy and its enduring influence. *American Journal of Public Health.* 92(5):725–729.

Cutler, David. 2005. And miller, Grant. The role of public health improvements in health advances: The twentieth-century United States. *Demography* 42(1):1–22.

Dubos, Rene. 1996. *Mirage of health: Utopias, Progress, and biological change.* New Brunswick: Rutgers University Press.

Fairchild, Amy L., and Gerald M. Oppenheimer. 1998. Public health nihilism vs. pragmatism: History, politics, and the control of tuberculosis. *American Journal of Public Health.* 88:1105–1117.

Fee, Elizabeth. 1994. Public health and the state: The United States. In *The history of public health and the modern state*, ed. Dorothy Porter, 224–275. Amsterdam: Editions Rodopi B.V.

Garrett, Laurie. 1994. *The coming plague: Newly emerging diseases in a world out of balance.* New York: Farrar, Strauss and Giroux.

Johnston, Sean F. 2018. Alvin Weinberg and the promotion of the technological fix. *Technology and culture.* 59(2).

Jongsma, Karin R., and Sand, Martin. 2017. The usual suspects: Why techno-fixing dementia is flawed. *Medicine, health care, and philosophy.* 20(1):119–130.

King, Nicholas B. 2003. The influence of anxiety: September 11th, bioterrorism, and American public health. *Journal of the History of Medicine and Allied Sciences.* 58(4):433–448.

Levy, Stuart B., and Marshall, B. 2014. Antibacterial resistance worldwide: Causes, challenges and responses. *Nature Medicine.* 10(12 Suppl):S122–S129.

McKeown, Thomas. 1979. *The role of medicine : Dream, mirage, or nemesis?* Princeton, N.J.: Princeton University Press.

Moberg, Carol L. 1996. Rene Dubos: A harbinger of microbial resistance to antibiotics. *Microbial Drug Resistance* 2(3):287–297.

Omran, Abdel R. 1983. The epidemiologic transition theory. A preliminary update. *Journal of Tropical Pediatrics.* 29:305–316.

Rosenkrantz, Barbara. 1974. 'Cart before horse': Theory, practice, and professional image in American public health, 1870-1920. *Journal of the History of Medicine.* 55–73.

Rosner, Lisa. 2004. *The technological fix: How people use technology to create and solve problems.* London: Routledge.

Semenza, J.C., J.E. Suk, and S. Tsolova. 2010. Social determinants of infectious diseases: A public health priority. *Euro Surveillance* 15(27):1–3.

Semenza, J.C., E. Lindgren, L. Balkanyi, L. Espinosa, M.S. Almqvist, P. Penttinen, et al. 2016. Determinants and drivers of infectious disease threat events in Europe. *Emerging Infectious Diseases.* 22(4):581–589.

Sprenger, M. 2015. How to stop antibiotic resistance? Here's a WHO prescription.. Retrieved from http://www.who.int/mediacentre/commentaries/stop-antibioticresistance/en/

Struelens, M.J. 1998. The epidemiology of antimicrobial resistance in hospital acquired infections: Problems and possible solutions. *BMJ* 317(7159):652–654.

Szreter, Simon. 1988. The importance of social intervention in Britain's mortality decline, c. 1850-1914: A re-interpretation of the role of public health. *Social History of Medicine.* 1:1–37.

Tomes, Nancy. 1998. *The gospel of germs: Men, women, and the microbe in American life.* Cambridge, MA: Harvard University Press.

Ventola, C.L. 2015. The antibiotic resistance crisis: Part 1: Causes and threats. *Pharmacy & Therapeutics.* 40(4): 277–283.

Weinberg, Alvin M. 1991. Can Technology Replace Social Engineering. In *Controlling Technology: Contemporary Issues,* ed. W.B. Thompson. Buffalo, NY: Prometheus Books.

Winograd, Terry. (July 1, 2013). What's wrong with technological fixes?: Boston review.

WHO Commission on Social Determinants of Health. 2008. *Closing the gap in a generation: health equity through action on the social determinants of health.* Final report of the Commission on the Social Determinants of Health. Geneva, Switzerland: World Health Organization, Commission on Social Determinants of Health.

Chapter 20
Tackling Anti-microbial Resistance: An Ethical Framework for Rational Antibiotic Use

Jasper Littmann, Annette Rid, and Alena Buyx

Abstract To reduce the effect of antimicrobial resistance and preserve antibiotic effectiveness, clinical practice guidelines and health policy documents call for the "rational use" of antibiotics that aims to avoid unnecessary or minimally effective antibiotic prescriptions. In this paper, we show that rational use programmes can lead to ethical conflicts because they place some patients at risk of harm – for example, a delayed switch to second-line antibiotics for community-acquired pneumonia is associated with increased fatality rates. Implementing the rational use of antibiotics can therefore lead to conflicts between promoting patients' clinical interests and preserving antibiotic effectiveness for future use. The resulting ethical dilemma for clinicians, patients and policy makers has so far not been adequately addressed. We argue that existing guidance for acceptable risks in clinical research can help to define risk thresholds for the rational use of antibiotics. We develop an ethical framework that allows clinicians and policy-makers to evaluate policies for rational antibiotic use in six practical steps.

This paper is an extended version of: Littmann J, Rid A, Buyx A: Tackling anti-microbial resistance: ethical framework for rational antibiotic use. European Journal of Public Health 2018; 28(2): 359–363. A fuller defense of our framework is provided in: Rid A, Littmann J, Buyx A: Evaluating the risks of public health programs: rational antibiotic use and antimicrobial resistance. Bioethics 2019; 33(7): 734-748.

J. Littmann
National Institute of Public Health, Centre for Antimicrobial Resistance, Oslo, Norway

Institute for Ethics and History of Medicine, Technical University, Munich, Germany

A. Rid
Department of Bioethics, The Clinical Center, National Institutes of Health, Bethesda, MD, USA

A. Buyx (✉)
Institute for Ethics and History of Medicine, Technical University, Munich, Germany
e-mail: a.buyx@tum.de

© The Author(s) 2020   321
E. Jamrozik, M. Selgelid (eds.), *Ethics and Drug Resistance: Collective Responsibility for Global Public Health*, Public Health Ethics Analysis 5,
https://doi.org/10.1007/978-3-030-27874-8_20

Keywords Antimicrobial resistance · Medical ethics · Rational use of antibiotics · Research ethics · Rationing · Acceptable risk

20.1   Introduction

Antimicrobial resistance (AMR) has been described as one of the major threats to public health in the twenty-first century and sparked concerns about the possibility of a post-antibiotic era, in which many infections are no longer treatable (WHO 2014; General Assembly of the United Nations 2016; Davies et al. 2013). Such a development would not only have detrimental effects on patient care, but is also likely to significantly affect public health, agriculture, as well as economic and national security (PCAST 2014; O'Neill 2014; Adeyi et al. 2017; OECD 2018).

Over the past few years, the challenge of AMR has increasingly been recognised by policy makers. The World Health Organization (WHO) developed a global action plan on AMR (WHO 2011, 2015), the United Nations General Assembly passed a declaration on the issue (General Assembly of the United Nations 2016) and the G20—a forum of the world's major economies—has placed AMR on its agenda (Bundesministerium für Gesundheit 2011). In addition, considerable investments have been made to strengthen and incentivise the development of new antibiotics (Boucher et al. 2013; Drive-AB-12 2018; Rex 2014; Outterson et al. 2016a, b).

Since one of the major causes of AMR is the overprescription and overconsumption of antibiotics, comprehensive response plans to preserve antibiotic effectiveness do not focus solely on the development of new drugs. They also include a wide range of measures, including surveillance, infection control and, crucially, the promotion of the "rational" use of antibiotics (WHO 2014). Rational use programmes primarily focus on avoiding unnecessary prescriptions of antibiotic treatments, for example in the context of viral respiratory infections. However, they also include practices that involve delaying or withholding access to antibiotics that are known to be beneficial. These practices can place some patients at risk of harm and thereby lead to conflicts between two important ethical goals: promoting patients' clinical interests and preserving antibiotic effectiveness for future use.

The ethical conflict between these goals has so far not been adequately addressed. In this paper, we describe programmes for rational antibiotic use and show when they require placing some patients at risk of harm for the benefit of managing the existing pool of effective antibiotics. We argue that ethical guidance on acceptable risks in clinical research can help to develop acceptable programmes for rational antibiotic use. Based on a recognized ethical framework for risk and risk-benefit evaluations in clinical research, we propose a novel framework that allows clinicians and policy-makers to evaluate when rational antibiotic use is (or is not) ethically justified.

20.2   Rational Use of Antibiotics

The promotion of "rational" antibiotic use – sometimes referred to as "prudent" or "appropriate" use – is a key component of many response plans to AMR (Bundesministerium für Gesundheit 2011; UK Department of Health 2013; Public Health Agency of Canada 2014; The White House 2014; Australian Department for Health 2015). The WHO provided the most widely cited definition of the rational use of medicines in general, which includes the rational use of antibiotics. According to WHO, this requires that "patients receive medications appropriate to their clinical need, in doses that meet their own individual requirements, for an adequate period of time, and at the lowest cost to them and their community" (WHO 1985). Elsewhere, the WHO defined appropriate antibiotic use as the "cost-effective use of antimicrobials which maximizes clinical therapeutic effect while minimizing drug-related toxicity and the development of antimicrobial resistance" (WHO 2011). Most concise is the definition provided by the Alliance for the Prudent Use of Antibiotics (APUA), which defines prudent use as "the right drug for the right condition for the right amount of time" (Wilson and Tan 2010).

Notwithstanding some differences in meaning, these definitions all delineate types of antibiotic use that are clearly "irrational"—that is, obvious instances of overuse or misuse where antibiotics are prescribed or taken even though they are not the appropriate treatment. For example, antibiotics are an appropriate treatment for "strep throat", which is caused by a bacterium susceptible to certain antibiotics. However, they are not an appropriate treatment for most sore throats, which are caused by viruses against which antibiotics are not effective.

There is a wide array of interventions designed to reduce antibiotic prescribing, and evidence suggests that restrictive measures, which limit the availability of antibiotics, have a higher chance of success than information campaigns or educational interventions that teach prescribing clinicians about AMR (Charani et al. 2011; Davey et al. 2013). In many instances, such restrictions appear to be a promising and effective way of reducing antibiotic use. For instance, recent research has shown that antibiotic prescribing in primary care can be restricted substantially, without a negative effect on clinical outcomes, through a practice called delayed prescribing (Schuetz et al. 2009; Spurling et al. 2013; Little et al. 2014). Here patients for whom antibiotics are unlikely to be an appropriate treatment—say most patients with sore throats—receive a prescription that they can use at the pharmacy if their symptoms do not improve on their own within a few days. Receiving such a prescription gives patients reassurance, even though only few end up filling it. A Cochrane review concluded that delayed prescribing offers no additional risks to the patient, while offering similar patient satisfaction when compared to immediate prescribing (Spurling et al. 2013).

So understood, rational use of antibiotics is ethically justified because it promotes patients' clinical interests while also preserving antibiotic effectiveness for future use. Not only are present patients spared the side effects of unnecessary treatment, avoiding such treatment also curbs the spread of antibiotic resistance and thereby helps to ensure that existing antibiotics remain effective to treat future patients. Indeed, because rational use so understood promotes the interests of present as well as future patients, clinicians and policy-makers are ethically required to ensure that patients use "the right drug for the right condition for the right amount of time" (Wilson and Tan 2010).

20.3   Ethically Challenging Instances of Rational Use

In addition to avoiding unnecessary antibiotic use, rational use programmes embrace practices that restrict access to antibiotics even when treatment would reduce a given patient's health risk (Millar 2012). One example for this are programmes that ask clinicians to factor resistance thresholds in the community when they suggest antibiotic treatment to their patients. For example, clinical practice guidelines for treating community-acquired pneumonia recommend the use of macrolides—a particular class of antibiotics—unless 25% or more of the relevant pathogens are resistant to macrolides in the given local community (Mandell et al. 2007). Only then should clinicians switch to more effective second-line drugs (Mandell et al. 2007).[1] When considering the entire patient population, this is a sensible recommendation, since it is to be expected that macrolides—the first line of treatment—remains effective in most patients if resistance is below 25%. However, a delayed switch to second-line drugs negatively affects clinical outcomes in some patients, notably those with severe pneumonia. Some authors estimate that a similar guideline has carried a mortality risk of more than 1% for all patients in the past (Daneman et al. 2008). Rational use practices of this type effectively delay or withhold a proven beneficial antibiotic in order to curb the spread of antimicrobial resistance. This places some present patients at risk for the sake of future patients.

In these and comparable cases, the rational use of antibiotics confronts clinicians with a trade-off between promoting their present patients' best clinical interests and helping to maintain effective antibiotic treatments for future patients (Kollef and Micek 2014). This trade-off seems particularly stark for several reasons (Littmann et al. 2015; Littmann and Buyx 2018). First, it is difficult to measure the contribution of any individual course of antibiotics to the emergence of AMR in a population, so that estimates about the benefits of rational use programmes are relatively uncertain. It seems likely, however, that delaying or withholding any given antibiotic treatment makes no more than a very small contribution to curbing AMR and, in some cases, no contribution at all. For example, a patient might develop resistant

---

[1] Since the time of writing, an updated version of the consensus guidelines has superseded the cited work (Metlay et al. 2019).

bacteria but die before transmitting these to other people or the environment. As a result, rational use programmes that involve delaying or withholding antibiotic treatment are neither necessary nor sufficient for addressing AMR. Moreover, the benefits of preserving antibiotics may accrue to unknown people in the potentially distant future. By contrast, the benefits of antibiotic treatment for the present patient are much easier to measure and can therefore be predicted with greater certainty. Antibiotic treatment also has immediate benefits that accrue to known or "identified" persons. These differences readily explain why a typical clinician who is tasked with delaying or withholding antibiotics from her present patients with the goal of tackling AMR would feel an acute ethical conflict (Cohen et al. 2015).

Bioethicists have started to discuss this ethical conflict (Selgelid 2007; Battin et al. 2009; Millar 2012; Oczkowski 2017; Littmann and Buyx 2014; Littmann et al. 2015). However, at both the clinical and policy level, there is no ethical guidance on when rational use programmes that involve delaying or withholding antibiotics and hence compromise the best clinical interests of present patients are acceptable. Moreover, many policy documents lack explicit discussion of any ethical challenges, and those that include ethical considerations are narrowly disease- and pathogen-specific (Bundesministerium für Gesundheit 2011; WHO 2011; Davies et al. 2013; Andreasen 2014). What is therefore needed is a more specific analysis of whether and when it is acceptable to restrict antibiotics use with the goal of curbing AMR, even when doing so poses risks to present patients and thereby compromises their best clinical interests.

20.4 The Analogy to Clinical Research

To address this question, it is helpful to consider more generally whether and when it is justified for clinicians not to promote the best clinical interests of their present patient. Standard professional guidance takes a relatively restrictive stance on this issue. For example, the World Medical Association states that clinicians owe their patients "complete loyalty", and that they may place the interests of others above those of the patient only in "exceptional situations" (Williams 2015). However, closer analysis reveals many clinical practices that place the interests of other patients, individuals or society above the interests of a given present patient (Wendler 2010). Consider, for example, that clinicians routinely give vaccinations not to protect the present patient, but to maintain herd immunity; that clinicians transplant kidneys from healthy individuals to patients with renal failure in order to enable these patients to live a better and longer life; and that clinicians allow younger colleagues to gain experience and perform procedures for the first time, even though they know that this will result in higher complication rates for the present patient. Importantly, these practices are not only frequent, but ethically justified provided that a number of conditions are met (Wendler 2010). This finding raises the possibility of evaluating the ethical acceptability of rational use programmes for antibiotics in comparison to other practices that involve compromising the interests of individual patients for the benefit of others.

For such comparisons to be both sound and useful, it is essential to identify comparator practices that are relevantly similar to rational use programmes for antibiotics and widely considered to be acceptable (Sunstein 1993). Moreover, detailed ethical guidance on how to evaluate the given comparator practice should ideally exist. Clinical research fulfils all three desiderata, as the following paragraphs demonstrate.

Clinical research is a subset of research with human participants that focuses on evaluating methods to prevent, treat or cure illness and disease, or on generating the knowledge necessary to develop such methods. One of the key characteristics of clinical research is that investigators expose participants to some risks for the potential benefits of unknown patients in the potentially distant future. Clinician-investigators routinely perform research procedures that do not promote the best clinical interests of the present patient-participant, but serve solely to address research questions. In fact, the majority of clinical trials involve procedures that have no prospect of clinical benefit for participants but help to answer important research questions. For example, most clinical trials involve additional blood draws, biopsies or imaging procedures that would not be performed as part of routine clinical care, but serve to test the safety and efficacy of an investigational drug. Moreover, phase 1 trials with healthy individuals by definition have no prospect of clinical benefit. It is widely considered acceptable for clinician-investigators to perform these "net risk" procedures and trials—and thereby deviate from the general clinical norm of promoting the present patient's best clinical interests—because clinical research generates knowledge with the potential to improve the health or care of future patients (WMA 2013; CIOMS 2016). That is, the potential benefits of clinical research for future populations are widely considered to justify that investigators expose patient-participants to some level of net research risk.

Importantly, the potential benefits for future patients—or the so-called "social value" of the research—is generally seen as the fundamental justification for exposing participants to net research risks, not the participants' informed consent. Participants' consent is, of course, relevant for determining what levels of net research risk are acceptable in socially valuable research. Only small levels of net risk are acceptable when participants cannot give their own informed consent, for example in research involving children, or when obtaining participants' consent is not feasible, for example in research on large datasets that were originally collected for clinical (and not for research) purposes. By contrast, greater net risks are acceptable when participants give their voluntary and informed consent. However, it is generally accepted—and for good reason in our view—that research without social value is not justified even when the research poses low net risks and participants consent (Wendler and Rid 2017). This underscores that the fundamental justification for exposing participants to net research risks lies in the potential benefits of clinical research for future patients.

Restricting antibiotic use with the goal of curbing AMR is relevantly similar to clinical research in this important respect. Just like in research, clinicians expose the present patient to some risks of harm for the potential benefits of unknown patients in the potentially distant future. Moreover, what level of risk is acceptable arguably depends on whether the present patient consents to delaying or foregoing antibiotic treatment that would be in her best clinical interests. In clinical research, it is not certain that any given clinical trial generates findings that make a significant contribution to improving the health or care of future patients and populations. Moreover, the enrolment of any single patient-participant is uncertain to make a significant difference to the value of any given trial. Yet without sufficient overall participation, trials—and clinical research more generally—would fail. Rational use programmes that involve delaying or withholding antibiotics are similar in this respect. It is not only uncertain that such programmes will make a significant contribution to addressing AMR, but restricting a single patient's access to antibiotics may not lead to a measurable effect on the overall level of AMR. Yet reducing antibiotic prescriptions overall is a key component of strategies for curbing AMR and ensuring that bacterial infections can still be treated effectively in the future (Costelloe et al. 2010; Davies et al. 2013). Given this relevant similarity between clinical research and rational use programmes that delay or withhold antibiotics, and considering that it is widely accepted for clinician-investigators to impose some level of net risk on patient-participants for the potential benefit of future patients (Selgelid 2007; Kollef and Micek 2014), it should also be acceptable to impose comparable levels of risks on patients in order to preserve antibiotic effectiveness for future patients.[2]

Moreover, there is a long-standing and nuanced debate about acceptable risk in clinical research. This does not only mean that our judgments about acceptable risks are relatively considered in the research context, but also that detailed ethical guidance in this area exists (WMA 2013; CIOMS 2016; Emanuel et al. 2008). Clinical research therefore fulfils the three desiderata for sound and helpful comparisons formulated above: it is relevantly similar to rational use programmes that delay or withhold antibiotics for the sake of curbing AMR; it is widely considered an acceptable exception from the general principle that clinicians should act in the best interests of the present patient; and detailed ethical guidance on how to evaluate research risks, including net research risks, exists. This suggests that judgments about acceptable risk in clinical research can inform the ethical evaluation of rational use programmes that impose risks on present patients with the goal of curbing AMR and safeguarding antibiotic effectiveness for future patients (Rid et al. 2019).

20.5 An Ethical Framework for Evaluating Rational Antibiotic Use

For clinician-researchers to justifiably impose risks on patient-participants in clinical research, a number of conditions must be met. Specifically, the research must be scientifically valid and socially valuable; the risks to participants have to be

---

[2] There are, of course, relevant dissimilarities between clinical research and rational use programmes that involve delaying or withholding antibiotics. However, as we argue elsewhere in more detail, the similarities between the two clearly dominate the dissimilarities (Rid et al. 2019).

minimised and reasonable in relation to the potential clinical benefits for them and/ or the social value of the research; and, when participants cannot or do not consent, any net risks to them should be no greater than minimal or, in cases of compelling social value, no greater than a minor increase over minimal risk (WMA 2013; CIOMS 2016).

To operationalise these conditions, one of the authors of this paper (AR) has developed a systematic framework for evaluating research risks and potential benefits based on prominent research ethics guidelines and literature (Rid and Wendler 2011). This framework can be usefully adapted to evaluate the risks of rational use programmes that involve delaying or withholding antibiotics from patients who could clinically benefit (Appendix). The framework is most useful for evaluating rational use programmes that do not require clinicians to obtain the patient's informed consent for restricting antibiotics. In particular, the minimal risk threshold for research without informed consent (Rid 2014) provides much needed orientation regarding acceptable levels of unconsented to net risks in the public health context. However, the framework could also help to evaluate rational use programmes that would require the patient's informed consent. Here, the debate about upper net risk limits in research with competent consenting adults is particularly informative (London 2006; Miller and Joffe 2009).

The framework sets out six steps for evaluating rational use programmes that involve delaying or withholding antibiotics. A hypothetical programme to restrict antibiotic use for lower respiratory tract infections (LRTIs) helps to illustrate how these steps would work in practice (Table 20.1, Ethical framework for rational antibiotic use).

The first step—*ensure and enhance the programme's social value*—requires decision-makers to develop rational use programmes based on sound evidence and policy methods. This is to ensure that all programmes are rigorously developed and judged to be both feasible and effective. For example, the available data should suggest that delaying or withholding the given antibiotics for LRTIs has the potential to help curb AMR. Data should also be sufficient to estimate the risks of restricting the antibiotics for patients; if there is too much uncertainty regarding how likely patients will suffer harm, relevant data should first be gathered in a research study. For programmes that meet these conditions, step one also requires conducting (and/or reviewing) observational research on their implementation. For example, how does restricted antibiotics use for LRTIs affect prescription rates, clinical outcomes, and antimicrobial resistance in the community? Such research serves to confirm that the given programme is indeed effective, while also expanding the existing knowledge of AMR. Evidence from accompanying observational research should inform regular re-evaluations of the programme.

The second step—*identify the programme interventions*—requires considering how the care of patients under the given rational use programme would differ from recommended standard care. Of course, the key difference would be that antibiotics are delayed or withheld under clearly specified circumstances. However, programmes might also include supplementary, non-routine interventions that aim to reduce the risks of restricting antibiotic use. Importantly, these non-routine

Table 20.1 Ethical framework for rational antibiotic use

Step	Elements	Example
1. Ensure and enhance the policy's social value	1. Ensure that restricting antibiotics use for the given condition is based on sound evidence and policy methods	Available data allow estimating the clinical and public health impact of restricted antibiotics use for lower respiratory tract infections
		Policy to restrict antibiotics use for lower respiratory tract infections is rigorously developed (policy analysis, stakeholder consultation, etc.) and judged to be feasible and effective
	2. Ensure that restricting antibiotics use for the given condition passes a minimum threshold of social value	Restricted antibiotics use for lower respiratory tract infections has the potential to address antimicrobial resistance, a major public health problem
	3. Enhance the knowledge to be gained from the policy and use it to refine the policy	Initiate research[a] to evaluate how restricted antibiotics use for lower respiratory tract infections affects prescription rates, clinical outcomes, antimicrobial resistance etc. and update policy as new evidence comes in
		Evaluate how the experience with restricting antibiotics use for lower respiratory tract infections can inform the management of antibiotics use for other conditions (e.g. urinary tract infection)
2. Identify the policy interventions	1. Identify the policy intervention and any supplementary (non-routine) interventions to protect patients	Policy intervention: restricted antibiotics use for lower respiratory tract infections
		Supplementary interventions: provide relevant information to patients in writing (e.g. signs of worsening condition, next steps in this situation), provide 1-time respiratory therapy session to manage symptoms
		Exclude routine clinical interventions from analysis: provide relevant information verbally, provide supportive measures for lower respiratory tract infections (e.g. mucolytics, inhalation, pain medication)
	2. Ensure that each supplementary intervention is essential for protecting patients	Could consider whether written patient information is essential

(continued)

Table 20.1 (continued)

Step	Elements	Example
3. Evaluate and reduce the risks to patients	1. Evaluate the risks of the policy intervention and each supplementary intervention	Restricting antibiotics use for lower respiratory tract infections poses risks (e.g. of increased complications such as pneumonia or empyema), which may require antibiotics and/or hospitalisation)
		Providing relevant information to patients in writing (rather than verbally) can cause mild anxiety
	2. Reasonably reduce the risks	Ensure that routine clinical interventions to treat lower respiratory tract infections are implemented (e.g. supportive measures)
		Ensure that serious harms, should they occur, are adequately managed at no financial cost to the patient (e.g. antibiotics and/or hospitalisation) and patients receive compensation for any lasting serious harms
		Actively monitor or exclude patient groups from policy who are at increased risk of complications or other serious harms (e.g. immunodeficiency)
		Instruct clinicians to exercise judgment in individual cases
		Reassure patients about written information sheet
4. Evaluate and enhance the potential benefits for patients	1. Evaluate the potential clinical benefits of the policy intervention and each supplementary intervention	Restricting antibiotics use for lower respiratory tract infections spares patients the risks of treatment (e.g. diarrhea) 1-time respiratory therapy session helps to manage symptoms
	2. Enhance the potential clinical benefits	Consider complementing 1-time respiratory therapy session with educational material
5. Evaluate whether the interventions pose net risks	1. Determine whether the risks of each individual intervention exceed the intervention's potential clinical benefits (implies net risks): Would an *informed clinician* who is committed solely to promoting the patient's clinical interests recommend the intervention?	*Informed clinician* would not recommend withholding antibiotic treatment for lower respiratory tract infections or measures to manage the resulting risks (e.g. written information, active monitoring): net risks
		Informed clinician would recommend respiratory therapy session: no net risks
	2. Determine whether the unit of policy and supplementary interventions pose net risks: would an *informed clinician* recommend the unit of interventions?	*Informed clinician* would not recommend unit of interventions: potential clinical benefits of respiratory therapy session do not outweigh the risks of withholding antibiotic treatment and measures to manage the resulting risks

(continued)

Table 20.1 (continued)

Step	Elements	Example
6. Evaluate whether the net risks are justified by the policy's social value	1. Determine the level of cumulative net risk posed to patients	Relatively small level of net risk from withholding antibiotic treatment and measures to manage the resulting risks
	2. Determine whether the policy's cumulative net risks fall within the general range of acceptable net risk: could judge in light of net risk limits in biomedical research, notably the "minimal risk"" threshold in research without informed consent for rational use programmes that do not require clinicians to obtain the patient's informed consent and upper net risk limits in research for rational use programmes that require informed consent	Net risks from withholding antibiotic treatment for lower respiratory tract infections arguably falls within the range of acceptable net risk in clinical research (i.e. the net risks arguably fall below the minimal risk threshold)
	3. Evaluate whether the given level of cumulative net risk is proportionate to the social value of implementing the policy: would an *ideal social arbiter* recommend the policy?	*Ideal social arbiter* may recommend the policy, given the importance of curbing antimicrobial resistance and the arguably minimal net risks to patients

The framework is adapted from an existing framework for risk-benefit evaluations in biomedical research (Rid and Wendler 2011; see Appendix for additional details). Example: hypothetical policy to restrict antibiotics use (delayed or no prescription) for lower respiratory tract infections without obtaining the patients' informed consent

[a]Research to evaluate the impact of restricted antibiotics use for a given condition has to be judged based on standard ethical criteria for research (CIOMS 2016; WMA 2013)

interventions can carry their own risks. For example, if antibiotics for LRTIs were delayed or withheld, patients might be informed about signs of pneumonia or empyema (i.e. potential more serious complications of an initial infection) in writing, rather than verbally as in standard practice, so as to ensure that they know when to seek expert advice and thereby reduce risks to them. However, receiving written information can itself cause concern among patients. It is therefore essential to identify all programme interventions in order to comprehensively evaluate the risks that the given rational use programme poses to patients. As part of this step, the need for any supplementary interventions should also be scrutinized with a view to avoiding unnecessary risks. For example, do patients really require written information?

Step three—*evaluate and reduce the risks to patients*—requires evaluating the risks of all programme interventions, including those of any supplementary interventions, based on the available evidence. Moreover, all reasonable measures must be taken to reduce the risks to patients. For example, if written information about pneumonia and/or empyema is considered necessary, how can this information be conveyed in ways that reduce anxiety and stress for patients? Similarly, rational use programmes that involve delaying or withholding antibiotics need to ensure that patients who suffer serious harms are adequately treated at no financial cost. In the case of restricting antibiotics for LRTIs, for instance, it must be guaranteed that any collections of pus in the pleural cavity (i.e. empyemas) are promptly treated for free. Patients with any lasting serious harms should also receive compensation. Furthermore, patient groups at increased risk of experiencing serious harm—for example, patients with immune deficiency—should either be actively monitored or excluded from the programme. Guidelines should also instruct clinicians to exercise their judgment in individual cases.

Step four—*evaluate and enhance the potential benefits for patients*—requires evaluating based on the available evidence to what extent patients could benefit clinically from the given rational use programme. For example, not receiving antibiotics for LRTIs means avoiding their side effects and reducing the risk of infection with *Clostridium difficile*—which is associated with antibiotic use and sometimes leads to a life-threatening infection of the colon—or resistant infection in future. Moreover, non-routine supplementary interventions can have potential clinical benefits for patients. For instance, if a one-time respiratory therapy session is introduced to help manage patients' symptoms, this should be considered when evaluating the risks of the overall programme. Any supplementary interventions with potential clinical benefits for patients should also be targeted at groups who would benefit most, especially when they cannot be provided to all patients. For example, the one-time respiratory therapy session might be focused on patients who suffer from chronic pulmonary problems (provided they are not excluded from the programme for being at increased risk).

Step five—*evaluate whether the interventions pose net risks to patients*—requires judging, again based on the available evidence, whether the risks of each individual intervention exceed the intervention's potential clinical benefits. If the answer to this question is yes, then the given intervention is said to pose "net risks" to the patient. One way of making this determination is to ask whether an "informed clinician" who is committed solely to the promoting the patients' best clinical interests would recommend that they undergo the intervention in question (Rid and Wendler 2011). If the clinician would recommend the intervention, it promotes the patients' clinical interests and thus does not pose net risks. This implies that the intervention's risk-benefit profile is acceptable and—assuming the requirements of the previous steps are satisfied—needs no further evaluation. If the clinician would be indifferent, then undergoing the intervention neither undermines nor promotes the patients' clinical interests. Provided that including the intervention in the rational use programme is necessary to ensure its safety (step 2), this suggests that it has an acceptable risk-benefit profile. If the clinician would advise against the intervention, then it poses net risks that require further evaluation (step 6). For example, an informed clinician who is committed solely to promoting the patients' best clinical interests would not recommend delaying or withholding antibiotics for LRTIs (at least where the expected benefit of antibiotic treatment outweighs the expected harms of adverse effects of antibiotics outlined above). This means that this intervention poses net risks whose acceptability needs to be considered further.

Finally, step six—*evaluate whether the net risks to patients are justified by the programme's social value*—serves precisely this purpose. It first requires determining the cumulative level of net risk of the given rational use programme by adding the net risks of all its interventions. For example, restricting antibiotics for LRTIs likely poses a relatively small level of cumulative net risk from delaying or withholding antibiotic treatment and providing patients with written information in order to manage the resulting risks. Evaluators must then judge whether the programme's cumulative net risks fall within the general range of acceptable net risk to patients and, if the answer is yes, whether the risks are proportionate to the social value of implementing the programme.

With regard to the general range of acceptable net risk, there is a long-standing debate about this issue in clinical research that can inform judgements about the acceptability of rational use programmes. Debate has been most intense in the context of research involving participants who cannot give their own informed consent, such as children or patients with dementia. Many ethical guidelines and regulations, as well as many research ethicists, endorse a "minimal risk" threshold for such research in order to protect participants from excessive net research risks (Emanuel et al. 2008; WMA 2013; CIOMS 2016). The precise interpretation of this threshold remains contested. However, a recent analysis of the existing literature identifies several "lessons learned" that are equally pertinent for the evaluation of rational use programmes (Rid 2014). Specifically, the most convincing definitions of minimal risk refer to the two basic components of research risk, likelihood and magnitude of research harm. Further, these definitions distinguish different magnitudes of harm (e.g. small, moderate, serious) and set approximate likelihood thresholds for each magnitude of harm, where the likelihood thresholds are anchored with numeric information (e.g. 1 per 100,000) derived from risk comparisons. For example, if the risks of participating in a charity soccer game are considered acceptable, and this activity is thought to be relevantly similar to clinical research, data on the risks of playing soccer can be used to anchor likelihood thresholds for different magnitudes of research harm. At present, comparator risks are best specified as the risks that average, healthy, normal individuals in different age groups face in riskier but still acceptable activities that are directed at benefiting others, such as playing soccer for charitable purposes in children. In practice, this implies that the most convincing definitions of the minimal risk threshold equate minimal net research risks with very low likelihoods of serious and moderate harm and modest likelihoods of small harm (Rid 2014). Using this threshold as a preliminary guide for evaluating the acceptability of rational use programmes, net risks such as those reported for delaying antibiotics for LRTIs would seem to qualify as minimal. Specifically, if the two main complications—empyema and pneumonia—are classified as moderate harms, the up to 10% likelihood that they will occur (Schuetz et al. 2009) is broadly consistent with the likelihood of moderate harms observed in appropriate comparator risks (Rid et al. 2010).

Similar arguments have been advanced in the debate about upper risk limits in research with competent consenting adults. Specifically, several authors suggest that these upper risk limits should be delineated in comparison to appropriate comparators risks for competent adults, such as emergency medical assistance (London 2006; Miller and Joffe 2009). In rational use programmes that require clinicians to obtain informed consent for delaying or withholding antibiotics, the range of acceptable net risks could be evaluated based on what levels of risk are considered acceptable in research involving competent consenting adults. However, all current rational use programmes that we are aware of proceed without the patients' informed consent (though individual doctors may discuss prescribing decisions with their patients and seek to understand patients' preferences), so that the minimal net risk threshold is more pertinent. Moreover, although some ethical guidelines allow a "minor increase" above minimal risk in research without consent when it is not possible to gather the necessary data in a less risky manner and the social and scientific value of the research is compelling (CIOMS 2016), the minor increase over minimal risk threshold remains underspecified in the research ethics literature. We therefore would not recommend applying it systematically in the context of rational use programmes.

With regard to judging whether the net risks of a given rational use programme are proportionate to the social value of implementing it, one way of making this judgment is to ask whether an "informed and impartial social arbiter" would recommend the programme in question after carefully considering the risks and potential benefits for all affected parties and giving everyone's claims fair consideration, while treating like cases alike in similar areas of policy (Rid and Wendler 2011). This test provides no more than a heuristic to guide evaluators' judgment; however, the idea of a social arbiter helps to ensure that rational use programmes are evaluated just like other programmes or policies that impose some level of risk on individuals for the sake of realising an important social good—within clinical medicine, but also beyond. In the case of delaying or withholding antibiotics for LRTIs, the (arguably) minimal net risks that this programme involves for patients would seem proportionate to the importance of preserving antibiotic effectiveness and the potential of this programme to help curb AMR. Moreover, given that net risks are minimal, the programme could proceed without obtaining the patients' informed consent.

20.6 Potential Objections

Critics might argue that there are less invasive strategies for curbing the spread of AMR that we should exhaust before delaying or withholding antibiotics in clinical medicine, such as eliminating unnecessary antibiotic treatments, reducing antibiotic use in farming, and investing more, and more effectively, into research on new antibiotics. These other strategies should indeed be pursued with urgency, and there is little reason to believe that adopting rational use programmes would undermine them. However, given that AMR is now recognized as one of the major public health threats of the twenty-first century, it is essential to consider all approaches that could delay or prevent AMR in an ethically justifiable way. As this paper argues, rational use programmes that involve delaying or withholding antibiotics should become part of the conversation.

It might also be objected that few such programmes would satisfy the minimal risk threshold known from the context of clinical research. At this point, it is difficult to respond to this objection because few reliable data exist on patient outcomes when antibiotics are delayed or withdrawn. However, we have identified at least one case—restricting antibiotic use for LTRIs—where reasonable estimates suggest that the minimal risk threshold would be respected. As more data become available, more cases could follow. Moreover, our proposed ethical framework would also be helpful for evaluating rational use programmes that pose more than minimal risks to patients and therefore require obtaining their informed consent to participate. However, as mentioned, all current rational use programmes we are aware of proceed without obtaining the patients' informed consent, and we suspect this will stay the same in future. To be effective, rational use programmes need to be implemented as widely as possible—and in practice this is likely to be feasible only when patients are not asked to give their informed consent to having antibiotic treatment withheld or delayed. Given the considerable uncertainty about the social value of rational use programmes in terms of curbing AMR, and considering that such programmes are generally rolled out without requiring that clinicians obtain patients' informed consent, we would argue that rational use programmes should pose no more than limited net risks to patients until more evidence supports their social value.

Finally, some might argue that the present argument rests on an ethical framework for risk-benefit evaluations in clinical research that is not universally accepted. In particular, prominent research ethicists hold that clinician-investigators should generally not delay or withhold established effective treatments from patient-participants in clinical trials (Weijer and Miller 2004). This argument is based on the idea that clinician-investigators continue to have clinical obligations in the research context, which make it unacceptable for them to act against the best clinical interests of patient-participants by depriving them of proven effective interventions. And, it might be argued, if it is not acceptable to delay or withhold proven interventions in the research context, then it is not acceptable to do the same in rational use programmes. However, it is important to see that this latter position is not universally accepted in research ethics. This is because it cannot explain why it is justified for clinician-investigators to pose any net research risks to patient-participants—for example, by performing research-specific blood draws or biopsies that have no prospect of clinical benefit. After all, these interventions by definition do not promote the best clinical interests of patient-participants (Miller and Brody 2003). The ethical framework for risk-benefit evaluations that forms the basis of the present paper avoids this problem. It justifies why some level of carefully evaluated net risk can be imposed to patient-participants, namely because clinical research has important social value. Moreover, the framework applies the same standards for acceptable net risk to all research interventions, whether or not net risks result from delaying or withholding effective treatments or from performing research procedures without a prospect of clinical benefit (Rid and Wendler 2011). Thus, the ethical framework for risk-benefit evaluations in clinical research used here, while not universally accepted, is arguably more defensible than the existing alternatives.

20.7   Conclusion

For clinicians, rational use programmes can pose ethical conflicts when they involve delaying or withholding antibiotics and thereby compromise the clinical interests of their present patients for the sake of preserving antibiotic effectiveness for future patient populations. Such exceptions from the general norm that clinicians should always act in the best interests of the present patient need to be carefully developed and managed. In this paper, based on a comparison to clinical research, we presented an ethical framework that enables an explicit and transparent evaluation of the net risks that rational use programmes can pose to patients in order to address AMR. Because these evaluations require judgments about complex empirical facts and normative questions, it is essential that rational use programmes be developed transparently and with the involvement of patients, clinicians and other relevant stakeholders. This involvement would not only improve the quality of deliberations, but also help to ensure the fairness and legitimacy of the resulting policies and their successful implementation.

Finally, the rational use of antibiotics can only be one part of a comprehensive strategy to address the danger of AMR; other measures have to be pursued with equal urgency, including better infection control, comprehensive strategies to reduce antibiotics use in animals, and the development of new antibiotics.

Acknowledgments We thank Jonathan Grant, Harald Schmidt, David Wendler, James Wilson and Peter West-Oram, as well as the editors of this volume, Euzebiusz (Zeb) Jamrozik and Michael Selgelid, for comments on earlier versions of this manuscript. AB's and JL's work was supported by the Deutsche Forschungsgemeinschaft (grant BU 2450/2-1).

Disclaimer This work was completed as part of Annette Rid's official duties as an employee of the NIH Clinical Center. However, the opinions expressed are the authors's own. They do not represent any position or policy of the National Institutes of Health, Public Health Service, or Department of Health and Human Services.

Appendix: Ethical Framework for Rational Antibiotic Use

Clinical research			Restricted antibiotic use		
Step	Elements	Example	Step	Elements	Example
1. Ensure and enhance the study's social value	1. Ensure the study methods are sound	Study uses recognized scientific methods	1. Ensure and enhance the policy's social value	1. Ensure that restricting antibiotics use for the given condition is based on sound evidence and policy methods	Available data allow estimating the clinical and public health impact of restricting antibiotics use for lower respiratory tract infections. Policy to restrict antibiotics use for lower respiratory tract infections is rigorously developed (e.g. policy analysis, stakeholder consultation) and judged to be feasible and effective
	2. Ensure the study passes a minimum threshold of social value	Study addresses a valuable question: liver cancer is a serious disease and few treatments exist		2. Ensure that restricting antibiotics use for the given condition passes a minimum threshold of social value	Restricted antibiotics use for lower respiratory tract infections has the potential to address antimicrobial resistance, a major public health problem
	3. Enhance the knowledge to be gained from the study	Could make specimens available for other research		3. Enhance the knowledge to be gained from the policy	Initiate research* to evaluate how restricted antibiotics use for lower respiratory tract infections affects prescription rates, clinical outcomes, antimicrobial resistance etc. and update policy as new evidence comes in. Evaluate how the experience with restricting antibiotics use for lower respiratory tract infections can inform the management of antibiotic use for other conditions (e.g. urinary tract infection)

(continued)

Clinical research			Restricted antibiotic use		
Step	Elements	Example	Step	Elements	Example
2. Identify the research interventions	1. Determine which interventions address the study question(s) or protect subjects1	Investigational drug, blood draws, CT scan, liver biopsy (with ultrasound)	*2. Identify the policy interventions*	1. Identify the policy intervention and any supplementary (non-routine) interventions to protect patients	Policy intervention: restricted antibiotics use for lower respiratory tract infections Supplementary interventions: provide relevant information to patients in writing (e.g. high fever and other signs of a worsening condition, steps to take in this situation), provide 1-time respiratory therapy session to manage symptoms Exclude routine clinical interventions from analysis: provide relevant information verbally, provide supportive measures for lower respiratory tract infections (e.g. mucolytics, inhalation, pain medication)
	2. Ensure research interventions are likely to yield important and non-duplicative information	Could consider whether a liver biopsy is important for answering the study question(s)		2. Ensure that each supplementary intervention is essential for protecting patients1	Could consider whether written patient information is essential

| *3. Evaluate and reduce the risks to participants* | 1. Evaluate the risks of each research intervention | Liver biopsy poses a 8-35 per 100,000 risk of hematothorax | *3. Evaluate and reduce the risks to patients* | 1. Evaluate the risks of the policy intervention and each supplementary intervention | Restricting antibiotics use for lower respiratory tract infections poses risks of i.e. increase of complications like pneumonia or empyema, which may require antibiotics and/or hospitalisation) Providing relevant information to patients in writing (rather than verbally) might cause mild anxiety |
| | 2. Reasonably reduce the risks | Could consider whether liver biopsy is safer with 1-on-1 monitoring | | 2. Reasonably reduce the risks | Ensure that routine lower respiratory tract infections to treat condition are implemented (e.g. supportive measures) Ensure that serious harms, should they occur, are adequately managed at no financial cost to the patient (e.g. antibiotics and/or hospitalisation) and patients receive compensation for any lasting serious harms Actively monitor or exclude patient groups from policy who are at increased risk of complications or other serious harms (e.g. immunodeficieny) Instruct clinicians to exercise judgment in individual cases Reassure patients about written information sheet |

(continued)

Clinical research			Restricted antibiotic use		
Step	Elements	Example	Step	Elements	Example
4. Evaluate and enhance the potential benefits for participants	1. Evaluate the potential clinical benefits of each research intervention	Only investigational drug: may reduce cancer	**4. Evaluate and enhance the potential benefits for patients**	1. Evaluate the potential clinical benefits of the policy intervention and each supplementary intervention	Restricting antibiotics use for lower respiratory tract infections spares patients the risks of treatment (e.g. diarrhea) / 1-time respiratory therapy session helps to manage symptoms
	2. Enhance the potential clinical benefits	Could focus on advanced liver cancer patients		2. Enhance the potential clinical benefits	Could focus on patients with chronic pulmonary problems
5. Evaluate whether the interventions pose net risks	1. Determine whether the risks of each individual intervention exceed the intervention's potential clinical benefits (implies net risks): Would an *informed clinician* who is committed solely to promoting patients' clinical interests recommend the intervention?1)	*Informed clinician* would recommend investigational drug: no net risks / *Informed clinician* would not recommend blood draws, CT scan, liver biopsy: net risks	**5. Evaluate whether the interventions pose net risks**	1. Determine whether the risks of each individual intervention exceed the intervention's potential clinical benefits (implies net risks): Would an *informed clinician* who is committed solely to promoting the patient's clinical interests recommend the intervention?	*Informed clinician* would not recommend withholding antibiotic treatment for lower respiratory tract infections and measures to manage the resulting risks (e.g. written information, active monitoring): net risks / *Informed clinician* would recommend respiratory therapy session: no net risks
				2. Determine whether the unit of policy and supplementary interventions pose net risks: would an *informed clinician* recommend the unit of interventions?	*Informed clinician* would not recommend unit of interventions: potential clinical benefits of respiratory therapy session do not outweigh the risks of withholding antibiotic treatment and measures to manage the resulting risks
6. Evaluate whether the net risks are justified by the potential benefits of other interventions	1. Determine whether any of the interventions stand in a relation of strict scientific necessity or unity (unit of interventions)	Blood draws, CT scan, and investigational drug (presumed true for present purposes)	*N/A*	*Reason: The policy and supplementary interventions always stand in a relation of strict "policy necessity" (unit of interventions) because the supplementary interventions are considered essential for protecting patients' interests (step 2). The first element of this step is therefore unnecessary. The second element has been included in the previous step (step 5).*	
	2. Determine whether any units of interventions pose net risks: would an *informed clinician* recommend the unit(s) of interventions?	*Informed clinician* would recommend the above unit of interventions (stipulates that the potential clinical benefits of the investigational drug offset the drug's own risks and the risks of the blood draws and CT scan)			

7. Evaluate whether the remaining net risks are justified by the study's social value			*6. Evaluate whether the net risks are justified by the policy's social value*		
	1. Determine the level of cumulative net risk in the study: add any absolute, relative, indirect, and excess net risks	Considerable level of cumulative net risks from the liver biopsy		1. Determine the level of cumulative net risk posed to patients	Relatively small level of net risk from withholding antibiotic treatment and measures to manage the resulting risks
	2. Determine whether the study's cumulative net risks fall within the general range of acceptable net risk: could judge in light of net risk limits in biomedical research, notably the "minimal risk"[a] threshold in research without informed consent for rational use programmes that do not require clinicians to obtain the patient's informed consent and upper net risk limits in research for rational use programmes that require informed consent	Cumulative net risks probably fall within the general range of acceptable net risk for research with competent participants		2. Determine whether the policy's cumulative net risks fall within the general range of acceptable net risk: could judge in light of risk limits in biomedical research	Net risks from withholding antibiotic treatment for lower respiratory tract infections arguably falls within the range of acceptable net risk in clinical research and practice (i.e. the net risks arguably fall below the minimal risk threshold)
	3. Evaluate whether the given level of cumulative net risk is proportionate to the social value of conducting the study: would an *ideal social arbiter* recommend the study?	*Ideal social arbiter* would recommend the study (presumed true for present purposes)		3. Evaluate whether the given level of cumulative net risk is proportionate to the social value of implementing the policy: would an *ideal social arbiter* recommend the policy?	*Ideal social arbiter* may recommend the policy, given the importance of curbing antimicrobial resistance and the arguably minimal net risks to patients

The table shows how an existing framework for risk-benefit evaluations in clinical research (Rid and Wendler 2011) is adapted for the purposes of evaluating the risks of restricted antibiotic use. Example for clinical research: hypothetical phase 2 study of an investigational treatment for liver cancer, involving administration of the investigational drug, a series of blood draws, a CT scan, and a liver biopsy (taken without modifications from [Rid and Wendler 2011]). Example for restricted antibiotic use: hypothetical policy to restrict of antibiotic use for lower respiratory tract infections without obtaining the patients' informed consent

[a]Research to evaluate the impact of restricted antibiotics use for a given condition should be judged based on standard ethical criteria for research (CIOMS 2016; WMA 2013)

References

Adeyi, O. et al. 2017. Drug-resistant infections: A threat to our economic future (vol. 2): final report. Washington, DC. World Bank Group. http://documents.worldbank.org/curated/en/323311493396993758/final-report.

Andreasen, M. 2014. *Arbejdspapirer om antibiotikaresistens (Working papers on antibiotic resistance)*. Det Etiske Råd (Danish Ethics Council): Copenhagen.

Australian Department of Health. 2015. Responding to the threat of antimicrobial resistance: Australia's first national antimicrobial resistance strategy 2015–2019. Canberra.

Battin, M.P., L.P. Francis, J.A. Jacobson, et al. 2009. *The patient as victim and vector: Ethics and infectious disease*. New York: Oxford University Press.

Boucher, H.W., G.H. Talbot, D.K. Benjamin, et al. 2013. 10 × '20 Progress—Development of new drugs active against gram-negative bacilli: An update from the Infectious Diseases Society of America. *Clinical Infectious Diseases* 56: 1685–1694.

Bundesministerium für Gesundheit. 2011. Deutsche Antibiotikaresistenz-Strategie [German antibiotic resistance strategy]. Berlin.

Charani, E., R. Edwards, N. Sevdalis, et al. 2011. Behavior change strategies to influence antimicrobial prescribing in acute care: A systematic review. *Clinical Infectious Diseases* 53: 651–662.

Cohen, G., N. Daniels, and N. Eyal. 2015. *Identified versus statistical lives: An interdisciplinary perspective*. Oxford: Oxford University Press.

Costelloe, C., C. Metcalfe, A. Lovering, et al. 2010. Effect of antibiotic prescribing in primary care on antimicrobial resistance in individual patients: Systematic review and meta-analysis. *BMJ* 340: c2096.

Council for International Organizations of Medical Sciences, World Health Organization. 2016. International Ethical for Health-related Research Involving Humans. Geneva. https://cioms.ch/wp-content/uploads/2017/01/WEB-CIOMS-EthicalGuidelines.pdf. Accessed 26 Nov 2018.

Daneman, N., D.E. Low, A. McGeer, et al. 2008. At the threshold: Defining clinically meaningful resistance thresholds for antibiotic choice in community-acquired pneumonia. *Clinical Infectious Diseases* 46: 1131–1138.

Davey, P., E. Brown, E. Charani, et al. 2013. Interventions to improve antibiotic prescribing practices for hospital inpatients. *Cochrane Database of Systematic Reviews* 30 (4): CD003543.

Davies, S.C., J. Grant, and M. Catchpole. 2013. *The drugs don't work: A global threat*. London: Penguin.

Drive-AB. 2018. Driving reinvestment in research and development for antibiotics and responsible antibiotic use. http://drive-ab.eu. Accessed 26 Nov 2018.

Emanuel, E., C. Grady, R.A. Crouch, et al. 2008. *The Oxford textbook of clinical research ethics*. New York: Oxford University Press.

General Assembly of the United Nations. 2016. High-level meeting on antimicrobial-resistance. New York. http://www.un.org/pga/71/event-latest/high-level-meeting-on-antimicrobial-resistance/. Accessed 26 Nov 2018.

Kollef, M.H., and S.T. Micek. 2014. Rational use of antibiotics in the ICU: Balancing stewardship and clinical outcomes. *JAMA* 312: 1403–1404.

Little, P., M. Moore, J. Kelly, et al. 2014. Delayed antibiotic prescribing strategies for respiratory tract infections in primary care: Pragmatic, factorial, randomised controlled trial. *BMJ* 348: g1606.

Littmann, J., and B. Buyx. 2014. Antibiotikaresistenz. Ethische Aspekte einer drängenden Herausforderung. *Ethik in der Medizin* 27 (4): 301–314.

Littmann, J., and A. Buyx. 2018. Rationaler Antibiotikaeinsatz als ethische Herausforderung. *Bundesgesundheitsblatt*. https://doi.org/10.1007/s00103-018-2716-0.

Littmann, J., A. Buyx, and O. Cars. 2015. Antibiotic resistance: An ethical challenge. *International Journal of Antimicrobial Agents* 46 (4): 359–361.

Littmann, J., A. Rid, and A.M. Buyx. 2018. Tackling anti-microbial resistance: Ethical framework for rational antibiotic use. *European Journal of Public Health* 28 (2): 359–363.

London, A.J. 2006. Reasonable risks in clinical research: A critique and a proposal for the integrative approach. *Statistics in Medicine* 25 (17): 2869–2885.

Mandell, L., R. Wunderink, A. Anzueto, et al. 2007. Infectious Diseases Society of America/American Thoracic Society: Consensus guidelines on the Management of Community-Acquired Pneumonia in adults. *Clinical Infectious Diseases* 44: 27–72.

Metlay, J.P., G.W. Waterer, A.C. Long, et al. 2019. Diagnosis and treatment of adults with community-acquired pneumonia. An official clinical practice guideline of the American Thoracic Society and Infectious Diseases Society of America. *American Journal of Respiratory and Critical Care Medicine* 200 (7): e45–e67.

Millar, M. 2012. Constraining the use of antibiotics: Applying Scanlon's contractualism. *Journal of Medical Ethics* 38: 465–469.

Miller, F., and H. Brody. 2003. A critique of clinical equipoise. Therapeutic misconception in the ethics of clinical trials. *The Hastings Center Report* 33 (3): 19–28.

Miller, F.G., and S. Joffe. 2009. Limits to research risks. *Journal of Medical Ethics* 35: 445–449.

Oczkowski, S. 2017. Antimicrobial stewardship programmes: Bedside rationing by another name? *Journal of Medical Ethics* 43 (10): 1–4. https://doi.org/10.1136/medethics-2015-102785.

O'Neill, J. 2014. Review on antimicrobial resistance. Antimicrobial resistance: tackling a crisis for the health and wealth of nations. London.

Organisation for Economic Co-Operation and Development (OECD). 2018. Stemming the superbug tide: Just a few dollars more, OECD Health Policy Studies, OECD Publishing, Paris, https://doi.org/10.1787/9789264307599-en. Accessed 26 Nov 2018.

Outterson, K., J.H. Rex, T. Jinks, et al. 2016a. Accelerating global innovation to address antibacterial resistance: Introducing CARB-X. *Nature Reviews Drug Discovery* 15: 589.

Outterson, K., S.J. Hoffman, A. Ghafur, et al. 2016b. International cooperation to improve access to and sustain effectiveness of antimicrobials. *Lancet* 387: 296.

President's Council of Advisors on Science and Technology (PCAST). 2014. *Report to the President on combating antimicrobial resistance*. Washington: PCAST.

Public Health Agency of Canada. 2014. Antimicrobial resistance and use in Canada: a federal framework for action. Ottawa.

Rex, J.H. 2014. ND4BB: Addressing the antimicrobial resistance crisis. *Nature Reviews Microbiology* 12: 231–232.

Rid, A. 2014. Setting risk thresholds in research: Lessons from the debate about minimal risk. *Monash Bioethics Review* 32 (1–2): 63–85.

Rid, A., and D. Wendler. 2011. A framework for risk-benefit evaluations in biomedical research. *Kennedy Institute of Ethics Journal* 21 (2): 141–179.

Rid, A., E.J. Emanuel, and D. Wendler. 2010. Evaluating the risks of clinical research. *Journal of the American Medical Association* 304 (13): 1472–1479. https://doi.org/10.1001/jama.2010.1414.

Rid, A., J. Littmann, and A. Buyx. 2019. Evaluating the risks of public health programs: Rational antibiotic use and antimicrobial resistance. *Bioethics* 33 (7): 734–748.

Schuetz, P., M. Christ-Crain, R. Thomann, et al. 2009. Effect of procalcitonin-based guidelines vs standard guidelines on antibiotic use in lower respiratory tract infections: The prohosp randomized controlled trial. *JAMA* 302: 1059–1066.

Selgelid, M.J. 2007. Ethics and drug resistance. *Bioethics* 21: 218–229.

Spurling, G.K., C.B. Del Mar, and L. Dooley et al. 2013. Delayed antibiotics for respiratory infections. In: Cochrane database of systematic reviews, ed: Cochrane, 30(4): CD004417.

Sunstein, C. 1993. On analogical reasoning. *Harvard Law Review* 106: 741–791.

The White House. 2014. National strategy for combating antibiotic-resitant bacteria. Washington.

UK Department of Health. 2013. UK five year antimicrobial resistance strategy 2013 to 2018. London.

Weijer, C., and P. Miller. 2004. When are research risks reasonable in relation to anticipated benefits. *Nature Medicine* 10 (6): 570–573.

Wendler, D. 2010. Are physicians always obligated to act in the patient's best interests? *Journal of Medical Ethics* 36 (2): 66–70.

Wendler, D., and A. Rid. 2017. In defense of the social value requirement for clinical research. *Bioethics* 31 (2): 77–86.

Williams, J. 2015. World Medical Association: Medical ethics manual, 3rd Edition 2015. Ferney-Voltaire Cedex: World Medical Association. https://www.wma.net/wp-content/uploads/2016/11/Ethics_manual_3rd_Nov2015_en.pdf. Accessed 26 Nov 2018.

Wilson, M., and M. Tan. 2010. *Raising awareness for prudent use of antibiotics in animals: Position paper of the global alliance for the prudent use of antibiotics (APUA).* Rome: APUA.

World Health Organization (WHO). 1985. *The rational use of drugs, report of the conference of experts.* Geneva: WHO Press.

———. 2011. *Global strategy for containment of antimicrobial resistance.* Geneva: WHO Press.

———. 2014. *Antimicrobial resistance – global report on surveillance.* Geneva: WHO Press.

———. 2015. *Worldwide country situation analysis: response to antimicrobial resistance.* Geneva: WHO Press.

World Medical Association (WMA). 2013. Declaration of Helsinki: Ethical principles for medical research involving human subjects. Helsinki. https://www.wma.net/policies-post/wma-declaration-of-helsinki-ethical-principles-for-medical-research-involving-human-subjects/. Accessed 26 Nov 2018.

Chapter 21
Solidarity and Antimicrobial Resistance

Søren Holm and Thomas Ploug

Abstract The concept of solidarity has received increasing attention in discussions about public health interventions, both as a possible justification for such interventions and as a possible motivating factors for individual action. This chapter provides an analysis of whether thinking through a lens of solidarity is likely to be helpful in devising strategies and policies to combat antimicrobial resistance. It first provides a critical overview of recent accounts of solidarity and argues that solidarity must be understood as a group based concept. It then applies this conception of solidarity to individual use of antibiotics through a case study of the antibiotic treatment of moderate and severe acne where it is argued that solidarity based thinking is valuable within a context of shared decision-making. Issues of policy making are then discussed and it is argued that basing a policy on solidarity on the one hand constrains the methods chosen to pursue public health goals, but that on the other hand solidarity may provide a strong and durable motivation to comply with such a policy. The limits of solidarity are explored in the final section and it is concluded that 1) the concept of solidarity does have an important role to play in thinking about public health, 2) considerations of solidarity can help us shape the goals and methods of public health policies in the area of antibiotics, and 3) that it is likely that solidarity may also be helpful in thinking through other contentious issues in public health.

Keywords Antibiotic resistance · Bioethics · Collective action · Solidarity

S. Holm (✉)
Centre for Social Ethics and Policy, School of Law, University of Manchester, Manchester, UK

Center for Medical Ethics, Faculty of Medicine, University of Oslo, Oslo, Norway
e-mail: soren.holm@manchester.ac.uk

T. Ploug
Centre for Applied Ethics and Philosophy of Science, Department of Communication, Aalborg University, Copenhagen, Denmark
e-mail: ploug@hum.aau.dk

© The Author(s) 2020
E. Jamrozik, M. Selgelid (eds.), *Ethics and Drug Resistance: Collective Responsibility for Global Public Health*, Public Health Ethics Analysis 5,
https://doi.org/10.1007/978-3-030-27874-8_21

21.1   Introduction

Antibiotic resistance in bacteria, viruses and other pathogenic microorganisms is an increasing public health problem. The treatment of infectious diseases depends on the availability of effective antibiotics and modern surgery is only possible because post-operative infections can be effectively treated. It is therefore of the utmost importance to devise policies that can minimise and delay the emergence of resistance, as well as policies that promote the development of new classes of antibiotics.

But developing such policies and ensuring compliance with them is a complicated problem. It is complicated by (1) the fact that the development of resistance is not only caused by misuse or overuse of antibiotics, (2) the possibility of resistant strains and resistance genes to spread rapidly in our globalised and interconnected world, and (3) the close connection between health system deficiencies in resource poor environments and the development of resistance (Daulaire et al. 2015).

Any use of antibiotics can lead to the selection of genetic variants that confer resistance to that antibiotic or class of antibiotics in microorganisms. Although proper use of antibiotics can minimise the emergence of such genetic variants it cannot completely prevent it, and there is some evidence that genes conferring resistance can be present even before the antibiotic is developed and used (Rolo et al. 2017a, b). Patients shed microorganisms during their treatment (e.g. approximately 1×10^{11} bacteria per gram of faeces), and the antibiotics themselves are also excreted in urine and faeces leading to selection for antibiotic resistance in bacteria in the sewer systems of health care institutions (Hansen et al. 2016).

This means that antibiotic resistance cannot be prevented merely by preventing misuse and overuse of antibiotics. In some circumstances we will also need to restrict proper use. We may, for instance have to restrict the use of an effective antibiotic against a known microorganism which is sensitive to that antibiotic in contexts where most of the cases of the illness in question are self-limiting. The cost of this will be that many patients will experience illness for longer than if they had been treated, and that some whose illness turn out not to be self-limiting may suffer more significant effects.

In this paper we will analyse to what extent the concept of 'solidarity' can help in (1) guiding personal and professional decision-making about the use of antibiotics, (2) designing and deciding on proper policies for minimising and delaying the emergence of resistance, and (3) maintaining support for and promoting compliance with anti-resistance policies. Solidarity is an old concept, but has recently experienced a resurgence in public health ethics.

The focus of the paper is on the use of antibiotics in human health care, and primarily on policies that aim to control the use of antibiotics in health care. There are significant issues concerning how to properly incentivise the development of new antibiotics and how to control the marketing of antibiotics, but these issues of 'industrial policy' are better analysed through the concepts of justice and injustice. Antibiotic use in veterinary medicine and in agriculture more generally, e.g. as growth promoters in animal husbandry is also outside the scope of this paper. These practices contribute very significantly to the development and maintenance of antibiotic resistance, but we simply do not have space to analyse them in depth within the solidarity framework.[1]

21.2 Solidarity and Public Health

Solidarity is an old concept with roots in both moral theology and socialist/social democratic political philosophy. Both the authors of this paper are from Denmark and up until the mid-1990s 'solidarity talk' was common in Danish political discourse and it is still common within the organised labour movement.

The concept of solidarity has recently been revived in public health ethics by Prainsack & Buyx and by Jennings & Dawson (Dawson and Jennings 2012, Jennings 2015, Jennings and Dawson 2015, Prainsack and Buyx 2011, 2016; see also Baylis et al. 2008).

Prainsack & Buyx develop their conception of solidarity from the bottom up in their important book-length exploration of the concept, building on the concept of the solidaristic act defined in the following way:

> Solidarity is an enacted commitment to carry 'costs' (financial, social, emotional or otherwise) to assist others with whom a person or persons recognise similarity in a relevant aspect. (Prainsack and Buyx 2016, p. 52)

Based on this definition they then distinguish three 'tiers' of solidarity:

1. Interpersonal solidarity
2. Group solidarity
3. Contractual, Legal or Administrative Norms (or perhaps better 'institutionalised solidarity')

And they define group solidarity as:

> … solidarity comprises manifestations of a *shared* commitment to carry costs to assist others with whom people consider themselves bound together through at least one similarity in a relevant respect (e.g. a shared situation, characteristic, or cause). (Prainsack and Buyx 2016, p. 55, emphasis in original)

Other important features of their account are that they see solidarity as an inherently symmetrical relation, and thus distinct from asymmetrical relations like charity (p. 67); and that they claim that solidarity is a hybrid descriptive-normative concepts and that it is therefore "… suited neither to be framed nor applied in the way clearly deontic concepts such as human rights, or justice, can be." (p. 93)

We agree with their analysis in many respects. It is very useful because it shows that although the concept of solidarity has a history which links it historically to socialist and social democratic political philosophy, it can be developed and used

---

[1] We do not even have space to consider whether or not human networks of solidarity can and should be extended to also encompass some or all animals.

for bioethical purposes without relying on any explicit or implicit socialist premises.[2] However, the commitment to develop the conception from the bottom up, from the individual, isolated solidaristic act leads to some problematic conclusions.

The first problem is that it may not make sense to define a specific class of solidaristic acts without reference to a motive grounded in group solidarity. Acts of beneficence or charity are often based on a recognition of 'similarity in a relevant aspect' between the agent and the beneficiary of the act, e.g. the ability to suffer or the ability to experience a certain kind of welfare benefit or the simple fact that the beneficiary is recognised as being a fellow human being; and beneficial acts based in justice are often built on the recognition that the agent and the recipient are similar in the relevant sense of being moral agents subject to obligations of justice.

The second problem is that it is far from obvious that relations of solidarity are, or have to be inherently symmetrical. Agents may perform solidaristic acts for other persons in the full knowledge that they can perform these acts because they have more power or more resources than those they assist, and that it is unlikely that they will need reciprocation. But, if the acts are motivated by solidarity and not by, for instance pity or a desire to do good works, then they should count as within the scope of solidarity. And even if the acts are performed with the expectation that they will be reciprocated if and when needed, that does not entail that the acts or the relationship is symmetrical at the time when the act is performed. If I, based on a motive of solidarity donate money to the striking workers during a long term strike, I may have an expectation that if ever I was participating in a long term strike someone would, based on solidarity donate money to me and my fellow strikers, but that would not necessarily be an expectation that the money would come from the very same workers who are now on strike, and it could still involve the realisation that I am currently much better off than they are. A strict symmetry requirement would also entail that solidarity with future generations beyond the lifetime of the moral agent would be ruled out, simply because whereas I can do something for future generations, they cannot do anything for me.[3]

It also cannot be definitional for acts of solidarity that they must involve net costs. Walking in the 1st of May parade with the other members of the blacksmiths' labour union was, at least in years when the weather was good, a great source of joy for one of our grandfathers, but it was also an act of solidarity and identification with the union, its members and its causes, as well as with the larger network of labour unions.

Jennings & Dawson has a more traditional understanding of solidarity as a group concept and identify three relational dimensions of solidarity:

> The fundamental gesture or posture of solidarity is *standing up beside*. This posture has three relational dimensions: standing up *for*, standing up *with*, and standing up *as*. (p.32, emphasis in original)

---

[2] Whether that is sufficient to ensure that those socialist historical connotations will not be activated in the minds of readers or listeners when the term 'solidarity' is used is a different question.

[3] Except, perhaps keeping my good reputation alive.

We stand up for when we "assist or advocate for the other" (p 36). We stand up with when our solidarity requires us to enter "into the lifeworld of the other"; and we stand up as when there is "a yet stronger degree of identification between the agents of solidaristic support and the recipients of such support." (p. 37). These three dimensions are not fundamentally different types of solidarity and they all draw their justification from the same conception of solidarity, but they point to different levels of engagement and action. It is, for instance uncontroversial that health care professionals should in many circumstances stand up for the patient groups they serve, but their obligation to stand up with these patient groups only get activated in certain contexts, and it is perhaps even rarer that they are required to stand up as with a patient group. But, the obligation to stand up as may be activated in more circumstances for health care professionals if the issue that requires solidaristic action is an issue affecting other health care professionals. In our analyses below of the implications of solidarity for individual and group actions and societal policies in relation to antimicrobial resistance we use Jennings & Dawson's three dimensions to indicate the level of engagement and commitment necessary for a particular kind of solidarity based action or policy to be likely to be effective.

21.3 Solidarity and Antimicrobial Resistance

Antimicrobial resistance is a problem that may potentially affect each and every one of us. It may currently be a more significant problem in some areas of the world, but it is not geographically containable. In order to reduce the rate at which antimicrobial resistance develops and spreads we also all have to be involved, since there is no use of antibiotics which does not promote resistance to some small degree.

We thus have a coordination problem. Each of us will in each instance where antibiotics can be prescribed personally benefit or potentially benefit by using antibiotics more than is optimal seen from the point of an optimal, overall balance between use and development of resistance. Or to put it the other way around, slowing down the development of resistance will require each of us sometimes to suffer when that suffering could have been potentially reduced by the prescription of antibiotics.

Coordination problems can, as discussed in other chapters of this book be solved in a variety of different ways, e.g. by changing the choice architecture and/or incentive structure. But some of these more structural solutions may either be illiberal in the sense that they reduce the option space available to individuals, even if none of the individual choices have any harmful consequences as individual choices, or they may create perverse incentives to try to circumvent restrictions in the choice architecture. It is therefore worth considering whether there are alternative options.

Can considerations of solidarity help to solve these coordination problems in a more constructive way, and what is required for solidarity to be activated?

If we look at Jennings & Dawson's three relational dimensions of solidarity it seems to be the case that even persons who have not realised that antimicrobial resistance is a problem for themselves can come to see that they ought to stand up for others in relation to antimicrobial resistance. They might, for instance decide to stand up for victims of extensively drug-resistant tuberculosis (XDR-TB) in sub-Saharan Africa, because they recognise some connection between themselves and the victims, even though they do not see themselves in any way threatened by XDR-TB.

However, once persons have realised that they share in the problem of antimicrobial resistance the two more intensely relational dimensions of solidarity can be activated, i.e. they can come to the realisation that they have to stand up with or as those who are threatened by antimicrobial resistance.

This does, potentially have different implications for ordinary persons /patients and for health care professionals. Ordinary persons can stand up with others in a number of ways in this context, they can advocate for policy changes, for research investment and for increased health care; and they can stand up as and take responsibility for their own use of antibiotics and encouraging others to do the same.

Health care professionals have a wider range of actions available to them because of their roles as gatekeepers to antibiotics, experts in public debates, and trusted advisors to individual patients. But this also means that they have to bring two different perspectives to bear, their personal perspective and their professional perspective; and that they can be part of potentially different networks of solidarity (e.g. solidarity with other health care professionals in areas of the world where resistance is very prevalent).

It is important to note that activating solidarity in the context of antimicrobial resistance does not necessarily have to be based solely on the realisation that I and others share this particular similarity or vulnerability. Groups within which we feel and enact solidarity are often sustained by multiple, complexly intertwined perceived similarities (e.g. a national, regional or professional identity that engenders group solidarity is not just about one shared feature), and it may well be that the chance of successfully engendering 'antimicrobial solidarity' is much larger if it takes place within already existing networks of solidarity. Single similarity solidarity may in some instances be strong, but multi-similarity solidarity is often stronger, partly because it receives support from a multiplicity of motivating factors.

It is likely that the most important and effective kind of solidaristic identification and action in relation to antimicrobial resistance is if enough people stand up as and take responsibility for their own use of antibiotics. What are the necessary conditions for this kind of solidarity to be engendered and for it to spread within already existing networks of solidarity? The first necessary precondition is, as mentioned above the realisation that antimicrobial resistance is not only a problem for others, it is also a problem for me and for others like me. However, for this to result in more than a sentiment of solidarity or a solidarity based call or activism for others to do something, I also need to know that there is something I can do personally in relation to the use of antibiotics. Many people in the industrialised world have probably internalised three key messages about antibiotics during their own upbringing, and later as parents:

1. Don't ask for antibiotics unnecessarily, e.g. when there is no bacterial infection
2. Take the antibiotics if they are offered by the doctor
3. Take the full course of antibiotics that is prescribed

1. and 3. were traditionally backed up by the claim that this will help prevent the development of resistance, and 2. by the general idea that if a doctor offers a treatment option it must be good for you. However, 3. which has been a mainstay of antibiotic folk-knowledge for generations has now been repudiated by a group of scientist in a recent high profile paper where they claim that taking the full course actually promotes the development of resistance (Llewelyn et al. 2017).

So, what knowledge do people now need to have in order to be able to act effectively in this context? They should still not ask for antibiotics when they are not necessary, but the concept of 'not necessary' may have to be further explicated so that we all understand that many bacterial infections are self-limiting and therefore not in need of antibiotic treatment, and also that pursuing treatment even when it is technically effective may bring about bad consequences for others to which we stand in a relation of solidarity.

Let us briefly analyse a specific case study of antibiotic use through the lens of solidarity, i.e. the use of antibiotics in the treatment of acne (Acne vulgaris) in teenagers and adolescents. Acne is a common disease and moderate and severe acne can lead to permanent scarring of the skin, as well as social and psychological problems in those affected by the condition (Hassan et al. 2009; Kellett and Gawkrodger 1999; Murray and Rhodes 2005). Oral antibiotics have for decades been used when topical treatments have been tried but turn out not to be effective and before the prescription of systemic isotretinoin or other retinoic acid analogues, since the retinoic acid analogues have many and potentially severe side-effects (Nagler et al. 2016). There has, however been an increasing realisation that whilst this use of oral antibiotics is effective in many cases, it also leads to increasing antimicrobial resistance (Dreno et al. 2014; Sinnott et al. 2016). What implications do solidarity in terms of standing up with and as have for patients and doctors in this scenario? We will first focus on the patient-doctor encounter and then move on to the question of policy.

In the patient-doctor encounter the enactment of solidarity is a matter for personal and potentially shared decision-making. The doctor can and should advise the patient on the benefits and drawbacks of antibiotics and of other available treatments and the patient should make a decision based on an evaluation of how the different treatment options will affect his or her life and life plans. What solidarity brings to the table are two things, it provides the doctor with a license (and potential obligation) to appeal to the patients solidaristic motivations and it creates an obligation for the patient to consider the choice through the lens of solidarity and not merely as an individual decision. This does not mean that the patient has to refuse antibiotics, or that the doctor has not to mention them as one of the treatment options. Acne is not a trivial disease, the side-effects of other available treatments are significant, and it is only the patient who can judge whether the sacrifice of forgoing antibiotics is outweighed by the solidaristic public health benefits in terms of reducing the development of antibiotic resistance. It is also important to note that the doctor's license to appeal to solidaristic motivations is not generated by the doctor's professional role but by the fact that he or she stands with the patient in a mutual relation of solidarity and that they both recognise this.

Things are more complicated at the policy level. The 'spontaneous' enactment of solidarity that we would wish to crystallise in policy arise in a situation where the doctor interpellates the patient as a person within a network of solidarity and encourages the patient to consider and enact solidarity in his or her decision-making, and where the patient enacts solidarity in a nuanced way by giving proper weight to the public health consequences of using antibiotics in his or her specific circumstances. This is tricky to implement as a policy and the main risks are that the doctor will be seen not as someone who stands with the patient and interpellates him or her in a network of solidarity, but as an agent of the state or the public health system, and that the patient will therefore not engage in solidaristic decision-making but perceive the situation as one where he or she is simply being asked or forced to sacrifice personal interests for some abstract conception of the public good. To avoid this problem the justification given for a policy, the way it is communicated to doctors and patients, and the way in which it is being implemented in health care will have to foreground solidarity considerations as the driving force behind the policy. A policy of simply banning the use of oral antibiotics for the treatment of acne is, for instance unlikely to sustain a solidarity based motivation for compliance.

Consideration of the implications of solidarity can also guide the choice of methods used to pursue public health goals in other areas. It has for instance been suggested in the literature that the active stigmatisation of the overweight and obese is an acceptable public health intervention, if there is evidence that it is effective (Callahan 2013); and there have been suggestions that targeted stigmatisation of farmers whose pigs carry multi-resistant bacteria is acceptable or perhaps even appropriate.[4] But, if there are some obese people or some pig farmers within my circles of solidarity a choice of pursuing active stigmatisation to further public health goals becomes potentially problematic seen from the perspective of solidarity. These are people with whom I share important things and with whom I work on common projects and not people who should be harmed or ridiculed. And, given that standing up with or as them requires me to understand their lifeworld I may well come to understand why they act as they do. More generally considerations of solidarity enjoin us to choose supportive methods in our public health policies, and only resort to proscription when absolutely necessary.

---

[4] For a rejection of this view on other grounds than solidarity see Ploug et al. 2015.

21.4 Solidarity as a Motivational Factor

We have argued above that considerations of solidarity can be useful in the design of policies to minimise and delay antibiotic resistance. But, building on the case study we need to consider solidarity as a motivating factor for action in more detail. We do not want to become embroiled in the interminable discussion about whether moral considerations necessarily motivate, or whether they only motivate when combined with a suitable desire. And, working with a concept like solidarity we do not need to become embroiled in that debate because realising that one has an obligation derived from or in solidarity straddles the cognitive/affective divide. Being in a relation of solidarity with a group is not a purely cognitive matter, but also a matter of felt identification. You don't walk in the parade on the eighth of March, or participate in a 'Reclaim the Streets' action purely on the basis of a dispassionate, rational assessment of your ethical obligations. You do it because you identify with the cause and with the group pursuing the cause.

This motivating force of solidarity lead to and support individual solidaristic action, but it can also be an important factor in engendering support for policy change and maintaining support for policies once they are implemented. However, the mere fact that a policy builds on or crystallises an already existing, or a developing solidarity based practice does not automatically mean that the support will transfer from the practice to the new context where what was previously a freely chosen action is now prescribed and must be performed (or proscribed and no longer available).

In the design of policies we therefore need to be careful not to lose the connection to solidarity as a motivating factor. The policy in a sense has to continue to speak to people and call them out in the language of solidarity and interpellate them as active participants in networks of solidarity.

This has implication both for how policies should be designed and how they should be communicated. We need to understand which networks of solidarity the relevant groups of citizens are embedded within, and design policies that appeal to those already existing networks.

21.5 The Extent and Limitations of Solidarity

There are a number of possible criticisms of the line of argument which has been pursued above. Here we want to look at two of them which are especially relevant in the context of antibiotic resistance since it is a global problem that needs global solutions.

The first possible criticism is that because we develop solidarity as a group concept, there will be some individuals and groups that are outside the solidarity group, that are Other than the ones included in our solidarity and who will be disadvantaged by solidarity in the in-group. The second criticism is that it is implausible to think that the solidarity of ordinary people could ever have a global scope (Holm 1993). Saints may aspire to global solidarity, but most of us restrict our solidarity to smaller and more local groups. Taken together these two criticisms imply that solidarity is an unsuitable, and perhaps even pernicious concept on which to base a solution to a global public health problem.

We take the second criticism to be empirically true. Few people manage to engender a state of 'solidarity with the people of the world'[5] in themselves (*pace* West-Oram and Buyx 2017). But perhaps that is not what is needed for solidarity to have global reach and be useful in supporting public health policies and initiatives. A global reach of local solidarity can come about in two ways. The simplest way is if local networks of solidarity are exhaustive in the sense that everyone is a member of at least one local network of solidarity, and if the results of solidarity in all, or perhaps just most of these networks contribute to a similar aim, i.e. in the present context the conservation of effective antibiotics through a reduction in the development of resistance. In this scenario a positive global outcome can be achieved by purely local action. Each local network of solidarity does not have to pursue an aim which is precisely identical to that of other networks. The second way in which local solidarity can have global reach is when we have local, but overlapping circles of solidarity that reach in an unbroken chain from people in the affluent north to people in the impoverished south. This may be sufficient to sustain the public health policies globally. In such a situation, persons may for instance accept policies restricting the use of antibiotics based on their solidarity with their fellow citizens in a particular country or region, but most of the people in that circle of solidarity which is defined by local citizenship will also be in other overlapping circles of solidarity based on other identifications like language, gender, profession etc. etc. Some of these overlapping circles may reach directly from the north to the south, there may for instance be solidarity among nurses or teachers or solidarity based on shared history. Others may be longer and involve more linking circles of solidarity but may never the less still be effective.

The first criticism also contains an element of truth, but it is perhaps not as damaging to the use of solidarity as it initially seems. It is undoubtedly true that there are some circles of solidarity that are partly constituted by identifying a particular group of others as 'the enemy', or as the radical Other. Communist solidarity among workers for instance posits the violent struggle between the working class and the capitalist class as an inevitable historical fact. However, not all forms of solidarity have to work in that way, and most do not. Danish identity and solidarity among Danes, for instance originally relied on distinguishing Danes from Swedes and Germans and seeing those two groups as enemies. But, although Swedes and Germans are still the Other and instantiate all of the traits that are non-Danish and therefore not-ours, they are no longer the enemy. So whereas solidarity almost inevitably expresses a form of partiality, which is partly what distinguishes it from thin

---

[5] 'Solidaritet med verdens folk' is a common slogan in left wing politics in the Nordic countries

conceptualisations of justice or utility maximisation, it may be a perfectly benign type of partiality.[6]

21.6 Conclusion

In this paper we have argued for three main conclusions. First, that the concept of solidarity has an important role to play in analysing how individuals and groups can and should act in relation to the threat of antimicrobial resistances and that it is an important counterweight to pure self-interest. Second, that considerations of solidarity can help us to shape both the goals and the methods of public health policies in order to make them long-term socially sustainable. And, third by implication that given that solidarity is helpful when thinking through one thorny issue in public health ethics and policy, it is likely to be helpful in other areas of public health ethics as well and that it therefore warrants continued attention from the public health (ethics) community.

We have also analysed the problems in moving from a voluntary practice sustained by solidarity to an official policy crystallising that practice as prescriptions and proscriptions and shown that careful design and communication is necessary not to lose the motivating force of solidarity when formulating policy.

References

Baylis, F., N.P. Kenny, and S. Sherwin. 2008. A relational account of public health ethics. *Public Health Ethics* 1 (3): 196–209.

Callahan, D. 2013. Obesity: Chasing an elusive epidemic. *Hastings Center Report* 43 (1): 34–40.

Daulaire, N., A. Bang, G. Tomson, J.N. Kalyango, and O. Cars. 2015. Universal access to effective antibiotics is essential for tackling antibiotic resistance. *The Journal of Law, Medicine & Ethics* 43 (S3): 17–21.

Dawson, A., and B. Jennings. 2012. The place of solidarity in public health ethics. *Public Health Reviews* 34 (1): 65–79.

Dreno, B., D. Thiboutot, H. Gollnick, V. Bettoli, S. Kang, J.J. Leyden, A. Shalita, and V. Torres. 2014. Antibiotic stewardship in dermatology: Limiting antibiotic use in acne. *European Journal of Dermatology* 24 (3): 330–334.

Hansen, T.A., T. Joshi, A.R. Larsen, P.S. Andersen, K. Harms, S. Mollerup, E. Willerslev, K. Fuursted, L.P. Nielsen, and A.J. Hansen. 2016. Vancomycin gene selection in the microbiome of urban Rattus norvegicus from hospital environment. *Evolution, Medicine, and Public Health* 2016 (1): 219–226.

Hassan, J., S. Grogan, D. Clark-Carter, H. Richards, and V.M. Yates. 2009. The individual health burden of acne: Appearance-related distress in male and female adolescents and adults with back, chest and facial acne. *Journal of Health Psychology* 14 (8): 1105–1118.

---

[6] We will here assume without argument that any complete system of ethics for human agents must incorporate various types of partiality.

Holm, S. 1993. Solidarity, justice and health care priorities. In *Solidarity, justice and health care priorities*, Health service studies, ed. Z. Szawarski and D. Evans, vol. 8, 53–64. Linköping: Linköping University Press.

Jennings, B. 2015. Relational liberty revisited: Membership, solidarity and a public health ethics of place. *Public Health Ethics* 8 (1): 7–17.

Jennings, B., and A. Dawson. 2015. Solidarity in the moral imagination of bioethics. *Hastings Center Report* 45 (5): 31–38.

Kellett, S.C., and D.J. Gawkrodger. 1999. The psychological and emotional impact of acne and the effect of treatment with isotretinoin. *British Journal of Dermatology* 140 (2): 273–282.

Llewelyn, M.J., J.M. Fitzpatrick, E. Darwin, C. Gorton, J. Paul, T.E.A. Peto, L. Yardley, S. Hopkins, and A.S. Walker. 2017. The antibiotic course has had its day. *BMJ* 358: j3418.

Murray, C.D., and K. Rhodes. 2005. Nobody likes damaged goods: The experience of adult visible acne. *British Journal of Health Psychology* 10 (2): 183–202.

Nagler, A.R., E.C. Milam, and S.J. Orlow. 2016. The use of oral antibiotics before isotretinoin therapy in patients with acne. *Journal of the American Academy of Dermatology* 74 (2): 273–279.

Ploug, T., S. Holm, and M. Gjerris. 2015. The stigmatization dilemma in public health policy-the case of MRSA in Denmark. *BMC Public Health* 15 (1): 640.

Prainsack, B., and A. Buyx. 2011. *Solidarity: Reflections on an emerging concept in bioethics*. London: Nuffield Council on Bioethics.

———. 2016. *Solidarity in biomedicine and beyond*. Cambridge: Cambridge University Press.

Rolo, J., P. Worning, J.B. Nielsen, R. Bowden, O. Bouchami, P. Damborg, L. Guardabassi, et al. 2017a. Evolutionary origin of the staphylococcal cassette chromosome mec (SCCmec). *Antimicrobial Agents and Chemotherapy* 61 (6): e02302–e02316.

Rolo, J., P. Worning, J.B. Nielsen, R. Sobral, R. Bowden, O. Bouchami, P. Damborg, et al. 2017b. Evidence for the evolutionary steps leading to mecA-mediated β-lactam resistance in staphylococci. *PLoS Genetics* 13 (4): e1006674.

Sinnott, S.-J., K. Bhate, D.J. Margolis, and S.M. Langan. 2016. Antibiotics and acne: An emerging iceberg of antibiotic resistance? *British Journal of Dermatology* 175 (6): 1127–1128.

West-Oram, P.G., and A. Buyx. 2017. Global health solidarity. *Public Health Ethics* 10 (2): 212–224.

Chapter 22
Justifying Antibiotic Resistance Interventions: Uncertainty, Precaution and Ethics

Niels Nijsingh, D. G. Joakim Larsson, Karl de Fine Licht, and Christian Munthe

Abstract This chapter charts and critically analyses the ethical challenge of assessing how much (and what kind of) evidence is required for the justification of interventions in response antibiotic resistance (ABR), as well as other major public health threats. Our ambition here is to identify and briefly discuss main issues, and

N. Nijsingh (✉)
Centre for Antibiotic Resistance Research (CARe), University of Gothenburg, Gothenburg, Sweden

Department of Philosophy, Linguistics and Theory of Science (FLoV), University of Gothenburg, Gothenburg, Sweden

Institute of Ethics, History and Theory of Medicine, Ludwig Maximilians Universität, Munich, Germany
e-mail: niels.nijsingh@med.uni-muenchen.de

D. G. J. Larsson
Centre for Antibiotic Resistance Research (CARe), University of Gothenburg, Gothenburg, Sweden

Department of Infectious Diseases, Institute of Biomedicine, The Sahlgrenska Academy, University of Gothenburg, Gothenburg, Sweden
e-mail: joakim.larsson@fysiologi.gu.se

K. de Fine Licht
Centre for Antibiotic Resistance Research (CARe), University of Gothenburg, Gothenburg, Sweden

Department of Technology Management and Economics, Chalmers University of Technology, Gothenburg, Sweden
e-mail: karl.licht@sp.se

C. Munthe
Centre for Antibiotic Resistance Research (CARe), University of Gothenburg, Gothenburg, Sweden

Department of Philosophy, Linguistics and Theory of Science (FLoV), University of Gothenburg, Gothenburg, Sweden
e-mail: christian.munthe@gu.se

© The Author(s) 2020
E. Jamrozik, M. Selgelid (eds.), *Ethics and Drug Resistance: Collective Responsibility for Global Public Health*, Public Health Ethics Analysis 5,
https://doi.org/10.1007/978-3-030-27874-8_22

point to ways in which these need to be further advanced in future research. This will result in a tentative map of complications, underlying problems and possible challenges. This map illustrates that the ethical challenges in this area are much more complex and profound than is usually acknowledged, leaving no tentatively plausible intervention package free of downsides. This creates potentially overwhelming theoretical conundrums when trying to justify what to do. We therefore end by pointing out two general features of the complexity we find to be of particular importance, and a tentative suggestion for how to create a theoretical basis for further analysis.

Keywords Antibiotic resistance · Public health ethics · Precautionary principle · Complexity

22.1   Antibiotic Resistance

Antibiotic resistance is emerging as one of our largest global challenges: more and more bacterial infections[1] are becoming increasingly impervious to antibiotics, which increases morbidity, mortality and societal costs around the world.

The evolutionary principle that drives ABR is relatively simple: when populations of bacteria are exposed to an antibiotic, strains that have acquired resistance to the drug (through mutations or through uptake of genetic material) are favoured over the sensitive ones. The emergence of ABR on a macro-scale is, however, notoriously complex.[2] One reason is that ABR is a global phenomenon with a variety of causes on different levels and in different contexts, some of which are poorly understood.

The most obvious cause of ABR is the use of antibiotics in humans, especially when antibiotics are used inappropriately (e.g. when overly broad antibiotics are used, or when a patient has no benefit from antibiotic treatment). The use of antibiotics in animals, both for treatment and prevention of disease and for growth promotion, also contributes to the problem.[3] Some bacteria have the ability to colonize both humans and domestic animals, and mobile genetic elements, such as resistance plasmids, often move across bacterial species. Hence, there are no firm barriers that

---

[1] We will limit ourselves here to antibiotic resistance. Antibiotic resistance is a sub-category of antimicrobial resistance, which also includes drug resistance in viruses, fungi and other microorganisms than bacteria.

[2] World Health Organization (2014). *The evolving threat of antimicrobial resistance: options for action*. Geneva, Switzerland: World Health Organization.

[3] Anomaly, J. (2020). Antibiotics and Animal Agriculture; The Need for Global Collective Action. In Ethics and Drug Resistance: Collective Responsibility for Global Public Health. Springer, Cham.

separate the microflora of animals from that of humans. The external environment is another source of resistance, both as a transmission route for certain pathogens, for example through faecal contamination of water, and as a source for resistance genes that over time are recruited from harmless bacteria into pathogens, assisted by a selection pressure from antibiotics.[4] The need to take into account the interconnection between humans, animals and the external environment is often referred to as a "One Health perspective".[5]

Clearly, there is an urgent need to address all of these causes of ABR and implement interventions at different sites and different levels of organization. However, as we will see, securing the evidence required to establish both the effectiveness and the risks of such interventions, comes at a moral price. This raises in a straightforward manner the question of what the criteria of evidence should be for the various interventions that aim to fight ABR. This question links the ethical justification of ABR interventions to debates around the ethics of risk and precaution. In other words, all ABR interventions pose the challenge of what quality of evidence for what balance of risks and possible benefits is required for such an intervention to be justified.

22.2 Precaution

The notion of precaution is central to much public health and environmental thinking. Specifically, when faced with complex and potentially extremely threatening phenomena such as a pandemic, global warming or pollution, it makes sense both to act in response to them even if there is a lack of evidence, but also to proceed with caution when enacting precautionary measures to mitigate or prevent damage.

Scholars of the *Precautionary Principle* (PP) have worked to express this intuition more clearly, resulting in a generic criterion of justified decision-making and policy arrangements that can be expressed in the following way:

> ... in the face of an activity that may produce great harm, we (or society) have reason to ensure that the activity is not undertaken, unless it has been shown not to impose too serious risks.[6]

This criterion expresses three basic things: First, the idea that uncertain major threats may provide reason for action.[7] Second, the contention that whatever such

---

[4] Bengtsson-Palme, Johan, and DG Joakim Larsson (2015). Antibiotic resistance genes in the environment: prioritizing risks. *Nature Reviews Microbiology* 13.6: 396–396.

[5] Boden, L. & D. Mellor. (2020). Epidemiology and ethics of antimicrobial resistance in animals. In Ethics and Drug Resistance: Collective Responsibility for Global Public Health. Springer, Cham.

[6] Munthe, Christian (2016). Precautionary principle. In: Ten Have (ed.) *Encyclopedia of global bioethics*. Dordrecht: Springer International Publishing.

[7] Compare also: "uncertainty should not be a reason for inaction in the face of serious environmental threats". Daniel Steel calls this idea the ´meta-precautionary principle´. Note that the vagueness of this procedural meta-criterion allows PP to be applied in a large number of different contexts

actions are taken must not themselves impose too serious risks or new uncertain major threats, and, third, that we are required to *demonstrate reasons* both why responses to threats are motivated and why apparent threats may be accepted. The criterion expresses a generic formula, within which more specific PP *versions*, or specific precautionary policy suggestions, must fit in order to be justified. There are thus various ways to flesh out the idea that we have reason to take precautions in the face of major, but uncertain threats. As a version of PP is specified, it further delineates what can properly be considered *responsible* decision-making in such contexts, not least regarding what is required more precisely to satisfy the requirement of demonstrating reasons for whatever precautionary action is suggested.[8]

One basic assumption underlying PP is that there is a moral price to exposing people to risk, as well as to proceed with activities in the face of uncertain risks. However, there is also always a price to any precautionary intervention that aims to clarify uncertainties and to prevent or mitigate risks: these will always claim resources that could have been used for other worthwhile purposes, create risks of their own, and delay or stop possibly valuable activities. For that reason, suggestions for precautionary action need to be subjected to precautionary scrutiny too, and to be justified it needs to be demonstrated that they incur an acceptable price and level of precaution. Particularly in systemically complex situations, the emergence of risks and uncertainties on various levels raises complications concerning how to balance the type and severity of the various harms and uncertainties involved.

A version of PP has to set *standards* concerning when precautionary action is required, and what is required of it in order to be responsible. Daniel Steel has recently explained this in terms of a 'tripod', consisting of a knowledge condition, a harm condition, and a suggested precautionary action.[9] Variations of how this tripod is construed will affect the price of precaution, as well as the level of precaution enacted. A PP version thus needs to specify for (1) any suggested precautionary action, (2) what threat is sufficiently serious for such action to be defensible, and (3) what degree of uncertainty is acceptable for it. For example, in order to, say, justify taking expensive precautionary measures to curb ABR (1), there needs to be a scientifically plausible model (3) in which failure to introduce these measures leads to significant economic or health damage (2). Whether or not in a specific case the model leading to harm is ´plausible´ and the damage is ´significant´ of course requires further elaboration. In any justifiable specification of the 'tripod', it is necessary to balance in a responsible way the need for precautionary action against the price of precaution.

Although details vary among authors, critical debate on what it takes to justify a PP version has led to a reasonably broad consensus on some minimal desiderata. These regard that a sound PP must not balance its required level and price of

---

and on different levels of organization. See Steel, Daniel (2014). *Philosophy and the precautionary principle*. Cambridge University Press.

[8] Munthe, Christian (2011). *The price of precaution and the ethics of risk*. Dordrecht: Springer.
[9] Steel (2014).

precaution *arbitrarily* (but according to a general principle that applies equally to all cases), that it needs to avoid so-called *precautionary paradox*, and that principle for responsible balancing of precautionary level and price must be *proportional*.[10]

PP is *arbitrary* when it offers no good, generalizable reasons why a specific course of action is acceptable or not. If an appeal to PP is used to recommend intervention 1, but to prohibit intervention 2, it should be able to meaningfully distinguish between the two measures and show how these are relevantly different. Note that the requirement to avoid arbitrariness also excludes treating the *status quo* with special regard: the fact that things are currently done in certain way is not in itself an argument for doing it that way.[11] It also means that whatever requirements are set by the specification of the tripod in a PP, these apply both to uncertain threats in order to justify precautionary action, *and* to the uncertainties of these actions themselves.

This links to the need to avoid 'precautionary paradox'. PP can lead to paradox in two related ways: Either its requirements are so strong that it tends to ban all options in most situations, thereby undermining any capacity to guide decision-making.[12] Or it issues inconsistent prescriptions by requiring and banning one and the same option.[13] It has been a theme among critics to point out how simplistic versions of PP may easily become paradoxical in any or both of these ways.[14]

The desideratum of proportionality follows from both of these requirements. In order to avoid paradox and arbitrariness, a justified version of PP must present a principle of responsible balancing of what level of precaution is required and what price of precaution is acceptable to pay that applies equally to all situations, as well as to all options in such situations. Any plausible version of PP will thus offer principled grounds for comparing suggested precautionary interventions, or the acceptance of an uncertainty or a threat, to alternative options in a unified manner. Such a version will express an allegedly morally responsible way of balancing the required level and price of precaution. To justify a specific precautionary action in a situation, it is therefore necessary to point how such a PP version supports it. As different situations vary with regard to what options are available, what stakes in terms of threats and prospects these actualise, and what knowledge is available with regard to these factors, one and the same precautionary intervention may therefore be

---

[10] Munthe, 2011, 2016; Munthe, C. (2017). *Precaution and Ethics: Handling risks, uncertainties and knowledge gaps in the regulation of new biotechnologies.* Berne: Swiss Federal Office for Buildings and Publications and Logistics (FOBL); Steel, (2014).

[11] However, there may be good instrumental reasons to be cautious when implementing change in a situation of great uncertainty. We will return to that point later.

[12] What Munthe (2011, ch. 2) has called *decisional paralysis*.

[13] What Steel (2014) terms *inconsistency*.

[14] Holm, Søren, and John Harris (1999). "Precautionary principle stifles discovery." Nature 400.6743: 398–398. Sunstein, Cass R (2005). *Laws of fear: Beyond the precautionary principle.* Cambridge University Press. McKinney, WJ, & Hill, HH (2000). Of sustainability and precaution: The logical, epistemological, and moral problems of the precautionary principle and their implications for sustainable development. *Ethics and the Environment,* 5: 77–87.

justifiable in some situations, but not in others. This regards also what quality and type of information about risks and effectiveness we require, and how much of further investigation to mitigate uncertainties is needed in the light of that. Precautionary requirements will therefore be gradual rather than absolute, and context-dependent rather than rigid. Different situations will justify different levels of precaution, and different prices of precaution to attain such levels.[15]

This regards not least the option of postponing a specific intervention in order to gather more evidence to ensure its effectiveness and responsibility. Possibly, this is the most common type of precautionary measure, familiar from standard regulation of drugs and the introduction of novel biotechnology.[16] It is also easy to see how this type of precautionary action may often be justified on the basis of a defensible version of PP. However, knowledge is never perfect, and the option to further update the basis of information for assessing the effectiveness and riskiness of an intervention is ever present. So, when do we know enough? How much time and resources should we spend on making sure that what we do in order to invoke responsible precautionary response to dangers and uncertainties will not in fact worsen the situation from a precautionary standpoint by invoking an unjustifiable price of precaution? This is a distinct ethical issue that becomes a particular challenge in the face of complex and drastic public health threats, such as ABR, where the price of delaying interventions is obvious, and costs and new risks of conducting research are salient. If we wait, the ABR problem continues to grow and increasingly threatens to overwhelm us, and if we experiment with interventions this will usually create new uncertainties and risks of harm. At the same time, unproven interventions may both escalate the ABR problem, and expand it to include severe policy failures. This takes us to the question of how these stakes, and options of collecting (or not collecting) evidence, should be assessed and evaluated.

22.3   Evidence

Traditionally, guidelines for evidence basing and research in the area of medicine confine themselves to clinical trials of biomedical interventions, focusing mostly on the immediate somatic effects on individual patients.[17] At the same time, as mentioned, ABR (and most other public health) interventions greatly surpass that area, and mostly occur outside of immediate therapeutic action (although sometimes intended to affect it, e.g., those interventions that regard antibiotic prescription practices). However, the recently revised guidelines by the Council for International Organizations of Medical Sciences (CIOMS) allow for a broader conception of

---

[15] Munthe, (2017).

[16] Munthe (2011), p. 97. See also Munthe, (2017).

[17] World Medical Association (2014). "World Medical Association Declaration of Helsinki: ethical principles for medical research involving human subjects." *The Journal of the American College of Dentists* 81.3: 14

'health research', including the study of any intervention that aims to change health-related behaviour on both individual and institutional levels.[18] This more inclusive conception clearly and significantly leaves room for the gauging of the proper amount of evidence for public health interventions.

A starting point for this type of assessment is the recognition of the fact that all health research – not only on biomedical interventions – imposes risks on research subjects, while the projected aim is to gather more knowledge in the interest of science or society.[19] A central tenet of research ethics therefore is that health research either has to plausibly benefit the research subject, or the societal benefit needs to be very large. In the new CIOMS guidelines, the latter is explicitly recognized in terms of the "social value" that may be attained by an intervention.[20] Furthermore, considerations of promoting trust towards health professionals and the complexity of the ethical issues involved provide arguments to treat health research with a certain amount of caution.[21] To this, we may add the precautionary considerations related in the preceding section: while a public health threat may be major and acute, any intervention meant to mitigate or prevent it may instead make it worse, or produce structural side effects that undermine other types of social goods. Therefore, the CIOMS frame is helpful to understand the question of evidence in public health interventions, such as the interventions aimed at fighting ABR.[22] To establish whether the evidence is sufficient, we have to chart the types of harm and uncertainty for various interventions in order to determine whether the expected (social) value of the intervention outweighs the value of postponing the use of a new intervention to collect more solid information about it.

This challenge is well illustrated by debates over suggested interventions in public health emergencies, such as Ebola.[23] When, in 2014, the West African Ebola epidemic was finally recognized as a global threat, it was suggested to prevent

---

[18] Council for International Organizations of Medical Sciences (CIOMS) (2016). International Ethical Guidelines for Health-Related Research Involving Humans. Geneva, Switzerland: Council for International Organizations of Medical Sciences. http://www.cioms.ch (accessed July 28, 2017). Munthe, C., Nijsingh, N., de Fine Licht, K., & Joakim Larsson, D. G. (2019). Health-related Research Ethics and Social Value: Antibiotic Resistance Intervention Research and Pragmatic Risks. Bioethics, 33(3), 335–342.

[19] Wilson, James, and David Hunter (2010). "Research exceptionalism." *American Journal of Bioethics* 10.8: 45–54.

[20] CIOMS (2016).

[21] Wilson and Hunter (2010).

[22] Attena, Francesco (2014). "Complexity and indeterminism of evidence-based public health: an analytical framework." *Medicine, Health Care and Philosophy* 17.3: 459–465.

[23] National Academies of Sciences, Engineering, and Medicine (2017). *Integrating Clinical Research into Epidemic Response: The Ebola Experience*. National Academies Press. An even more recent example is the Zika epidemic. See Edwards, Sarah JL (2016). "The precautionary paradox and Zika." *Research Ethics*: 178–181.
and Glenza, J. "Zika virus: Floridians fear 'Pandora's box' of genetically altered mosquitos." *The Guardian*, August 14, 2016: https://www.theguardian.com/us- news/2016/aug/14/florida-keys-zika-virus-genetically-modified-mosquitoes (accessed July 25, 2017).

further harm by 'fast tracking' new vaccines and experimental drugs, thus relaxing the demands of evidence required to introduce new medication.[24] This suggestion was countered by public health officials, who argued that the epidemic should rather be controlled by means of proven public health policies, such as proper hygiene, surveillance and quarantine.[25] Another issue that was debated was whether randomized clinical trials could be justified in the context of an epidemic and the extent to which genuine equipoise could be presumed. In part, the answers to these questions depend on the relative risk to which the affected communities were exposed, in another part it depends on how we assess the gravity of uncertainties underlying the assessment of these risks and how we value the importance of acting on good evidence in view of those uncertainties. While the question on the evidence of interventions to counter ABR is similar to such debates, the issue of ABR also raises a new set of worries and topics for discussion. Specifically, whereas the Ebola crisis was unexpected and presented an acute emergency, ABR is – for now – slowly emerging, albeit foreseen, but nevertheless posing a major and growing public health threat. Already a substantial amount of morbidity and mortality is attributed to ABR, but this number is likely to keep growing in a way well known to us.

22.4   Justifying Interventions

Since the causes are varied, the fight against ABR takes place in different arenas. In this section, we distinguish between various groups of interventions. The first set concerns the development of new types of (or alternatives for) antibiotics. Second, we consider interventions that target the access to antibiotics by individuals. Third, various interventions aim to establish a larger degree of surveillance. Last, we bring together various institutional measures to attack the environmental health side of the ABR problem, such as the use of antibiotics in animals, as well as emissions of antibiotics. In accordance with the broad notion of 'health research' introduced in the former section, these interventions span a wide array of different actions and policies. As a consequence, we will consider many different levels and types of intervention; both on the scale of an individual patient–doctor interaction, as well as on the level of macro-economic interventions, institutional regulation and global health treaties. Varied though these interventions may be, they all share the characteristic of aiming to help curbing – or otherwise fighting – ABR. To what extent it can be demonstrated that they are effective in that regard, and to what extent they pose risks of their own, determines whether they can be responsibly introduced.

Not all interventions in the fight against ABR are new. In fact, a number of important interventions intended to counter ABR belong to the classic public health

---

[24] Geisbert, Thomas W. (2015). "Emergency treatment for exposure to Ebola virus: the need to fast-track promising vaccines." Jama 313.12: 1221–1222.

[25] Rid, Annette, and Ezekiel J. Emanuel. (2014). "Ethical considerations of experimental interventions in the Ebola outbreak." *The Lancet* 384.9957: 1896–1899.

repertoire: screening, surveillance, quarantine, hygiene, and so on. Although they are not always uncontroversial, these interventions have been thoroughly tested and proven effective. Unfortunately, however, they will not suffice in addressing the problem of ABR.[26] New methods will need to be explored, which raises the question how to determine which intervention is preferable; which offers greater relative benefit, and which poses fewer relative risks? The answer to that question depends on the evidence available to assess the various interventions. We have no ambition here to be complete in listing the possible, but aim to illustrate and map some major complexities that arise when balancing the level of precaution against the price of precaution.

22.4.1 Biomedical Interventions

A fundamental problem in managing and fighting ABR is the lack of appropriate biomedical interventions. One aspect of this is *the lack of truly new antibiotics*. Although there is some progress in the development of novel antibiotics that affect Gram-positive bacteria (bacteria with a single outer cell wall),[27] innovation for Gram-negative bacteria that has reached the market has for decades consisted only in variations of the same.[28] In part, this can be attributed to the fact that developing new antibiotics is relatively unappealing from a business point of view. Therefore there is a widely recognized and urgent need to encourage academia and pharmaceutical companies to develop new antibiotics, and to facilitate their introduction.[29]

So-called *expedited programs* to this effect have been launched, for example by the US Food and Drug Administration (FDA).[30] Interventions included in such programs are priority review, accelerated approval, and fast track (which can be combined).[31] By promising a swifter, simplified and/or more relaxed process for licencing new therapies, such options both offer incentives to industry to invest in

---

[26] O'Neill, Jim (2014). Antimicrobial resistance: tackling a crisis for the health and wealth of nations. Review on antimicrobial resistance.

[27] Wright, Gerard. (2015). "Antibiotics: An irresistible newcomer." *Nature* 517.7535: 442–444.

[28] See, e.g., University of Illinois at Urbana-Champaign. "Antibiotic breakthrough: How to overcome gram-negative bacterial defenses." *ScienceDaily*. www.sciencedaily.com/releases/2017/05/170510132012.htm (accessed July 6, 2017); WHO 2015.

[29] World Health Organization, WHO. *Global priority list of antibiotic-resistant bacteria to guide research, discovery, and development of new antibiotics*. Geneva: World Health Organization, 2017. Online access: http://www.who.int/medicines/publications/global-priority-list-antibiotic-resistant-bacteria/en (accessed July 28, 2017)

[30] U.S. Department of Health and Human Services. *Guidance for Industry. Expedited Programs for Serious Conditions – Drugs and Biologics*. Washington: USDHHS, 2014. Online access: https://www.fda.gov/downloads/drugs/.../ucm358301.pdf *(accessed July 7, 2017)*. See for a more elaborate discussion: Munthe, C., & Nijsingh, N. (2019). Cutting red tape to manage public health threats: An ethical dilemma of expediting antibiotic drug innovation. Bioethics, 33(7), 785–791.

[31] https://www.fda.gov/forpatients/approvals/fast/ucm20041766.htm (accessed July 28, 2017)

the development of new antibiotics, and speed up the introduction of successful fruits of such endeavours. Interventions of this sort appear attractive, considering the potential damage that lack of development and delay could cause, motivating a lower acceptable price of precaution than in the case a "normal" drug development context. At the same time, entirely new classes of antibiotics imply elevated uncertainties regarding effect and side effects, pointing to a need for *more* caution, and motivating a higher price of precaution. In addition, problems with regards to the control of prescription, use and transmission imply further uncertainties regarding the benefits of "expediting" new antibiotics. In particular, it creates a stark tension between the overall aim of ABR research and the needs of patients burdened by resistant infections. If a new compound is introduced in a setting where the mentioned problems have not been mastered, resistance, though inevitable, can be expected to develop faster. As a result, there is a relative public health benefit to delay the discovery and introduction of new antibiotics while addressing the problems of ensuring responsible use, and mitigating transmission of resistance. Still, earlier introduction may save lives and reduce morbidity of individuals. Therefore, it is less clear whether expedited programs for the introduction of new antibiotics should be at the top of the priority list. Unless they are combined with effective measures to control usage and transmission, they introduce graver uncertainties of both negative side effects for patients, and of having the overall aim of managing the ABR problem undermined. Below, we will comment on interventions to manage this complexity of the ABR challenge.

Another aspect of this challenge is that, if resources are concentrated to this effect, it may be possible to develop drugs to take in order to mitigate plasmid-mediated transmission of resistant bacteria from one patient to others. These could be taken by patients with resistant infections, but also patients who take antibiotics where this treatment may otherwise give rise to local resistance. This is a possible intervention that is still in a very early stage of exploration,[32] which means there will be a long and expensive path to any possible actual treatment. At the same time, there is an obvious risk that no such success awaits at the other end – creating a severe uncertainty with regard to the actual worth of incentive schemes aimed at effecting such focused research and development endeavours. In addition, any successful treatment of this sort will create an ethical challenge in terms of exposing patients to the risk of side effects of the treatment without any sort of potential somatic benefit for these same patients. If it is successful, it will have an important general primary preventive effect of great public health value in the face of the ABR problem. If the introduction of such a drug would be "expedited", this will at the same time increase the risk and uncertainty regarding negative side effects concentrated only to those people taking the drug. Weighing these stakes has to be a part of striking the balance between what the acceptable price of precaution is to be when comparing incentive schemes.

---

[32] See, e.g., Buckner, Michelle, Maria Laura Ciusa, and Laura JV Piddock. (2018). "Strategies to combat antimicrobial resistance: anti-plasmid and plasmid curing." *FEMS microbiology reviews.*

Controlling use and transmission is less essential regarding therapeutic interventions where resistance development is not an apparent threat. Phage therapy (the therapeutic use of viral strains to attack bacteria) might fall into that category.[33] The efficacy and safety of phage therapy has not been proven to the stage where it would fulfil current guidelines in Europe and the USA. However, it might be that these guidelines do not quite suffice in assessing the responsible introduction of phage therapy, e.g. since individually designed cocktails may be required for each patient, creating an impediment for designing controlled trials. Thus, phage therapy, or other innovative solutions that do not (as new antibiotic compounds) feed immediately into the ABR development problem, may be a better target for "expedited programs" from an ABR standpoint – at least while we lack effective means to control use and transmission. On the other hand, accepting the higher degree of uncertainty, means lowering the level of required precaution, which may harm patients severely if experimental treatments turn out to be unsuccessful.

A more general challenge posed by all types of expediting program interventions, is that they may inadvertently create incentives that give rise to negative dynamics regarding drug development. The basic problem is that any expedited program creates an incentive for industry to re-direct their research efforts in a way that shapes studies to be less stringent and clinically relevant than what they would otherwise have been. A well-known example of this is the acceptance of surrogate outcome variables (an essential part of accelerated approval interventions), which makes it economically attractive for companies to run studies measuring only these, meaning that there will be a structural dynamic change of clinical research efforts into paths with less potential or without demonstrating actual clinical value. Similarly, so-called compassionate use programs have recently come under fire for creating a structural incentive for industry to move more and more drug development out of the default review process, thus creating a generally decreasing level of safety and elevated uncertainty regarding effect. To be sure, expediting programs partly aim at having industry thus allocate their efforts and resources, however, if there is a structural negative dynamic on the general effectiveness of new drugs, this must be viewed as a relevant downside. For that reason, policy makers may want to consider other solutions to the issue of drug development, such as rewarding pharmaceutical companies for developing new antibiotics, for example with exclusivity extensions, buyouts and entry prizes.[34] Each of these interventions has the potential to offer incentives to the pharmaceutical companies, but also to pose risks to society

---

[33] De Vos, Daniel & Pirnay, Jean-Paul (2015). "Could viruses help resolve the worldwide antibiotic crisis?" *AMR Control*, 110.

[34] Seth Seabury Neeraj Sood (2017, May 18). Toward A New Model For Promoting The Development Of Antimicrobial Drugs. *Health Affairs Blog*: http://healthaffairs.org/blog/2017/05/18/toward-a-new-model-for-promoting-the-development-of-antimicrobial-drugs/(accessed July 28, 2017); Morel, Chantal M., and Elias Mossialos. "Stoking the antibiotic pipeline." *BMJ: British Medical Journal (Online)* 340. Jim O'Neill (2014) has also suggested a ´pay or play´ principle, where pharmaceutical companies are required to either contribute or pay a fine.

and individuals, for instance a risk of social backlash.[35] There are, of course, also large uncertainties regarding whether or not such actions would be money well spent.

Returning to our main question concerning the evidence required for justifying various interventions that aim to offer incentives to develop new medicine, we see that a trade-off has to be made not just between individual and public interest, but also between various levels of uncertainty, and risks of structural negative dynamic effects. If faster development of antibiotics comes at the price of faster emergence of ABR for those same drugs, this raises the question on how to appreciate the urgency of the matter. In particular, it demands that we weigh current ABR against possible future ABR and the likelihood of developing alternatives for which ABR development is not an issue. There is both a danger of being retrospectively overly restrictive in the use of antibiotics when an alternative to the current drugs is found, as well as a danger of complacency based on the false reliance on such an alternative. At the same time, we need to weigh into the balance the apparent but uncertain risk of incentive schemes being structurally counterproductive.

22.4.2 Prescription Practices

Since the individual use of antibiotics is an important driver of ABR, interventions aiming to control the distribution of antibiotic drugs to individual patients are an important part of ABR policy. The proposed interventions include mandating prescription policies (in those countries where this is not already the case), various limitations to the type of antibiotics that are made available and improved access where antibiotics are currently lacking.

It is a received wisdom that the prescription system is an effective way of controlling the use of drugs. At the same time, the effectiveness on the system may vary, depending on numerous factors. For instance, antibiotic prescription practices across European regions vary considerably, linked to varying levels of institutional corruption.[36] Such structural challenges can be assumed to multiply in countries where there is no system or culture of effective prescription for antibiotics. Given the widespread acceptance of the over-the-counter availability of antibiotics in such societies, not only among citizens, but also medical professionals, and sometimes policy makers, there is a recognised uncertainty as to the real impact of trying to create or toughen up such regulation.[37]

---

[35] Munthe et al. (2019).

[36] Rönnerstrand, Björn, and Victor Lapuente. (2017). "Corruption and use of antibiotics in regions of Europe." Health Policy 121.3: 250–256.

[37] Radyowijati, Aryanti, and Hilbrand Haak (2003). "Improving antibiotic use in low-income countries: an overview of evidence on determinants." Social science & medicine 57.4 (2003): 733–744. Dreser, Anahí, et al. (2012). "Regulation of antibiotic sales in Mexico: an analysis of printed media coverage and stakeholder participation." BMC public health 12.1: 1051.

We meet here with a type of uncertainty that is entirely about how societies may react to attempted institutional change.[38] Weighing into the mix economic, cultural and institutional factors of relevance, a more incremental change seems preferable. It provides opportunity to attend to the interests of various stakeholders, as well as taking the time for a society to adjust, in order to ease both the passing of regulation, and its effective implementation. However, that requires quite a bit of knowledge of such mechanics of overarching social change, and also uses time itself as a factor. This raises the question of how long is long enough to attempt establishing social change, and how much effort should be spent on securing the understanding of how to make such attempts work. Facing the ABR challenge, how high should the price of precaution due to delaying prescription regulative action be allowed to rise while attending to such uncertainties?

For countries where a reasonably effective prescription practice is in place, unless a patient is critically ill, the first choice of antibiotics is often not the latest, most potent formula (with still limited resistance problems). Therefore, antibiotics prescribed usually bring a greater risk that the treatment will not cure the infection due to resistance. At the same time, this practice serves to protect the future integrity of "last line antibiotics" by minimising their use and thereby inhibiting the evolutionary drive towards resistance to them. Most commentators describe the payoff of these interventions in terms of *public good*, whereas risks of implementing them are considered to be carried by single *individuals*.[39] However, matters are slightly more complex than that. First, although there is agreement that this intervention does delay resistance for broad-spectrum compounds, the magnitude of the effect is still uncertain. Second, since broad-spectrum antibiotics are more likely to drive resistance in the individual patient's own gut flora,[40] there is also a chance of individual benefit linked to prescription practice.

We thus face a complex trade-off situation, where individual risks of suffering untreated infections must be balanced against the uncertain prospect that patients are protected against being infected by resistant bacteria, at the same time as the question remains whether this mix of risk and uncertain benefit for some individuals can be justified by a social benefit of uncertain magnitude. This also raises the question how much effort should be spent on making sure that the right balance is struck, for example by straightening out some of the important uncertainties.

In any case, agreeing that such a practice is indeed justified does not end the problem. We must also ask what intervention would actually effectively address it.

---

[38] The risk of incentive schemes for drug creation to produce unintended negative dynamics via their effect on industry in the former section also belongs to this type.

[39] Littmann, Jasper, and A. M. Viens. (2015). "The ethical significance of antimicrobial resistance." *Public health ethics* 8.3: 209–224.

[40] This is phenomenon can be observed in urinary tract infections, for example. As the normal non-resistant invading bacteria of this flora are exterminated by the treatment, a very fertile living space is created for bacteria that are resistant against the drug used. Costelloe, Ceire, et al. (2010). "Effect of antibiotic prescribing in primary care on antimicrobial resistance in individual patients: systematic review and meta-analysis." Bmj 340: c2096.

One idea, of course, is to make professional prescription guidelines for doctors to use. However, this introduces the uncertainty that doctors may fail to apply them, e.g., due to patient pressure, economic counter-incentives, or the mere inertness of habit. To address that, there is the option of allowing professionals less choice, for example by requiring application to a higher instance and proof of due cause for having a prescription green lighted. Such an intervention could consist of several levels of requirements, and for some antibiotics regular doctors may be stripped of all prescription rights. At the same time, being able to leave professional discretion to doctors in individual cases also has its value, and the more of rigid restriction is built into an intervention, the bigger the risk that individuals are harmed due to lack of (timely) access to treatment. However, rigid regulatory interventions clearly avoid the uncertainty with regard to the overall aims of delaying resistance development, as well as avoiding harmful individual prescriptions. To make this trade-off, it would be of great value to know more about the social dynamics creating the uncertainty around the effectiveness of prescription interventions, as well as how these might be complemented by additional institutional changes to mitigate the pressure on doctors from patients, and to remove economic counter-incentives.[41] On the other hand, as we delay action, or apply overly cautious interventions with uncertain effectiveness while making sure what more exact variant would be best, the price of precaution is allowed to go up in terms present prescription practices being allowed to continue.

This precautionary challenge is further complicated by the fact that there is an instrumental value to fine-tune prescription interventions so that treatment of infection is optimized also under a restrictive prescription practice. The reason for this is that increased persistent infection can be expected to increase the future demand and consumption of antibiotics, thereby accelerating rather than mitigating resistance development in the long run.[42] Depending on what current prescription practices look like in specific societies, this may mean that an optimal prescription intervention should not only decrease prescription, but in some cases leave it as it is, and in yet other even *improve* the access to antibiotics. Considerations of fairness may add further reasons to a similar effect, and also the need of securing the legitimacy of any policy in this area. After all, of what interest is the issue of ABR to anyone who is barred from accessing appropriate antibiotics in the first place? This further complicates the uncertainty about what exact intervention would be most effective. But it also adds a basic source of uncertainty with regard to how the moral stakes should be balanced in a measure of effectiveness. A sound precautionary solution therefore needs to acknowledge the latter point when striking the balance between ensuring a desired level of precaution at an acceptable price of precaution, and allow both considerations of health promotion and fair distribution of the population health.

---

[41] As such changes may involve drastic reform to entire health care and health insurance systems, the knowledge required is quite advanced and complicated to collect.

[42] Daulaire, N., et al. (2015). "Universal access to effective antimicrobials: an essential feature of global collective action against antimicrobial resistance." *Journal of Law, Medicine & Ethics* 43.2.

22.4.3 Surveillance

The fight against ABR also requires enhanced possibilities of diagnosis and surveillance of resistant bacteria. Better diagnostic methods are in themselves unobjectionable as increased speed, precision and readiness in determining the cause of an infection limits the danger of squandering antibiotics. However, there will be trade-offs between increasing speed, increasing precision and financial costs and possibly the intrusiveness of the sample taking. Because of this, as well as the general uncertainties befalling any new measurement tool, there is once again a challenge to decide how much support for the reliability and validity of a new diagnostic tool has to be secured in order to start implementing it. The balance here, as before, includes assessing the value of more firm knowledge against the price of precaution in terms of delaying tools that may also offer opportunities for better surveillance in the face of the ABR threat. But it also includes the complications as more speedy introduction will tend to increase one or the other of well-known downsides to such interventions.

These complications become especially challenging as resistant infections or even carriership may often actualise restrictive communicable disease management measures, such as compulsory isolation, quarantine or mandatory life-style restrictions. The implied tension between individual interest and collective good is particularly salient when this involves asymptomatic carriers, who have nothing to benefit from being institutionalized.[43] The issue is further complicated when we consider the possibility of false positives, where patients are wrongly identified as carrying resistant bacteria. As in the former section, this also links to a risk of undermining the legitimacy of ABR policies. Thus, while speedy introduction of diagnostic methods certainly has its potential upsides, it will increase uncertainties of a sort that in other areas are often taken to undermine health surveillance programs.

Attempting to strike these several balances, we face the general problem of having the right idea concerning the moral stakes involved and a sound notion of what price of precaution to allow. In addition, increased complexities of how to assess the quality of available and attainable evidence for ambitious and complex public health interventions add another layer of uncertainty.[44]

---

[43] Weinstein, Robert A., Daniel J. Diekema, and Michael B. Edmond. (2007). "Look before you leap: active surveillance for multidrug-resistant organisms." Clinical Infectious Diseases 44.8: 1101–1107. Nijsingh, N., Juth, N., Munthe, C., "The Ethics of Screening", in: Quah, Stella R. International encyclopedia of public health. Academic Press, 2016. Nijsingh, N., Munthe, C., Lindblom, A., & Åhrén, C. (2020). Screening for multi-drug-resistant Gram-negative bacteria: what is effective and justifiable?. Monash bioethics review.

[44] Attena, (2014).

22.4.4 Environment and Animals

A wide variety of ABR interventions relate to attempts to curb the emission of anti-biotics in the environment[45] and their use in animals. We have grouped these together because of the potential risks of the interventions, which seem mostly economical. For example, attempts to enhance transparency of pharmaceutical companies[46]or banning of the use of antibiotics as a growth enhancer,[47] or taxing consumer prod-ucts emanating from ABR driving practices, such as meat production,[48] do not have direct health risks for humans. Compared to the possible economic damage of such interventions, the health risks of ABR may seem to clearly win out. However, there is still much uncertainty concerning the role of non-human use and pollution in the establishing of ABR and economic cost of interventions carry their own set of sec-ondary risks and uncertainties, which might be substantial indeed as the incurred costs become more significant.

One obvious uncertainty regards the effectiveness of systems of surveillance and control of emission rates in production or compound use in farming. These will include uncertainties and imprecisions of technical methods, but even more institu-tional uncertainties of the sort we have already discussed related to prescription and surveillance interventions. As already observed, straightening these uncertainties out includes coming to grips with very complex social circumstances, and may require quite a lot of time and resources.

On top of this, macro-economic ABR interventions targeting environmental emission may have adverse effects in themselves, both socially and economically, for instance, by discouraging pharmaceutical business and thereby restricting access to drugs generally. Consequently, we may legitimately ask which interventions are necessary, or reasonable, and which are disproportional, given the uncertain effects of current practices. Should, for instance, pharmaceutical companies be required to monitor and make sure antibiotics emission from manufacturing are very low?[49] Or should regulation rather target the pricing of products, adding tax or extra cost in the procurement of drugs by public national health services? Or should some other

---

[45] Pruden, Amy, et al. "Management options for reducing the release of antibiotics and antibiotic resistance genes to the environment." *Environmental health perspectives* 121.8 (2013): 878.

[46] Larsson, DG Joakim, and Jerker Fick. "Transparency throughout the production chain—a way to reduce pollution from the manufacturing of pharmaceuticals?." *Regulatory Toxicology and Pharmacology* 53.3 (2009): 161–163. Nijsingh, N., Munthe, C., & Larsson, D. J. (2019). Managing pollution from antibiotics manufacturing: charting actors, incentives and disincentives. Environmental Health, 18(1), 95.

[47] Laxminarayan, Ramanan, Thomas Van Boeckel, and Aude Teillant. (2015). "The economic costs of withdrawing antimicrobial growth promoters from the livestock sector.".

[48] Giubilini, Alberto, et al. (2017). "Taxing Meat: Taking Responsibility for One's Contribution to Antibiotic Resistance." *Journal of Agricultural and Environmental Ethics* 30.2: 179–198.

[49] Bengtsson-Palme, Johan, and DG Joakim Larsson. (2016). "Concentrations of antibiotics pre-dicted to select for resistant bacteria: Proposed limits for environmental regulation." *Environment International* 86: 140–149.

institutional intervention to similar effect be chosen, for example having high emissions in production reduce the calculated health benefit in the context of health technology assessment? In all these cases, should the link between detected emission rates and such incentives be proportionally or more rigidly designed? If the former, according to what formula or proportionality, if the latter, on what grounds should thresholds be set? And what institutional arrangement would be effective to have whatever intervention is chosen to be effectively implemented? Similar questions appear with regard to interventions aimed at creating incentives to reduce the use of antibiotics in farming.[50]

All of this makes for a considerable difficulty in assessing the proper balancing of the level of precaution and its acceptable price. Surely, the urgency of mitigating major environmental practices that fuel antibiotic resistance development is a priority. However, to find the right way of doing this requires quite a bit of very complex knowledge, and behind this need lurks the very real risk that a more speedy introduction of interventions is not only sub-optimal, but actually makes the problem worse. For instance, implementing any of the regulative interventions mentioned may mainly have the effect of having pharmaceutical and food production relocating to areas where the regulative situation is even worse. Or secondary effects, e.g., in the form of drastically increased food prices may both undermine the legitimacy of ABR policies and create a public health threat of its own. On the other hand, it is well known that it takes considerable time to have large-scale operations such as drug production and farming change their longstanding ways, and in the light of that, applying interventions to address the environmental side of the ABR challenge is paramount.

22.5   Discussion

We have assessed interventions with regard to how much and what kind of evidence is needed when evaluating and implementing interventions in response to antibiotic resistance, a public health threat of immense proportions. The notion of responsible precautionary decision-making provides a basic and strong reason to act in response to this threat. However, determining *what* response to go for introduces complex problems of balancing what level of precaution to aim for and what price of precaution to pay, actualising much more difficult ethical challenges than what is often acknowledged.

We end this exploration by briefly addressing two issues that emerge when we consider the evidence for interventions that aim to fight antibiotic resistance, leading into a final broad suggestion for future analyses to build on.

---

[50] Silley, Peter, and Bernd Stephan. (2017). "Prudent use and regulatory guidelines for veterinary antibiotics—politics or science?." *Journal of applied microbiology* 123.6: 1373–1380.

First, there is the sheer size of the possible consequences of increased ABR. For example, there is a real question whether standards of treatment and diagnosis in research ethics and clinical ethics may need revision in light of a public health threat as significant to global wellbeing as ABR. Although one should be wary to discard too easily the frameworks that have proven to be of value throughout the years, the possible disruptive effects of ABR raise the issue to what extent these standards can be maintained, given the range of difficult choices we might face. At the same time, we have seen that many of the uncertainties posed by ABR interventions are not so much about having risks of undesirable side effects as such are typically conceived of when evaluating pharmaceuticals. Rather, the important uncertainties are about risks of outright counterproductivity due to social psychological and institutional dynamics, where apparently promising attempts to counter ABR may instead lure us into political, economic or psychological dead ends from which we are unable to get out. Social processes are typically slow, variably inert and intractable, which means also that they may be very difficult to reverse, and that doing so may require a lot of time. Given that interventions on all of the mentioned levels are probably necessary to reduce the risk of emerging resistance and that they are to a large extent interrelated, the standards of evidence should be set from an integrated, One Health perspective.

This connects to the general observation that methods in response to ABR have to intervene on a variety of different levels, from the everyday practice of physicians to those affecting global structures. Interventions worthy of consideration therefore involve a myriad of different types and degrees of uncertainty and risks, which are also unevenly distributed across people, societies and time. Assessment of the evidence thus needs to consider a multi-layered mosaic of uncertainties and ethical dilemmas regarding the short- and long-term trade-off between individual interests and public health aims. This regards especially the issue of how much and what evidence to collect regarding the effectiveness of interventions, and their potential long-term legitimacy.

These considerations may drive one to despair whether a responsible, measured approach to the issues at play here is at all feasible. One way of moving forward in the light of these considerations is to acknowledge that there are good – moral *and* precautionary – reasons to cut the Gordian knot: just as there is a question of how much to amass and ponder evidence and the proper resolution of ethical dilemmas, we must not get stuck forever in the precautionary conundrum. Moving along such a path, one primary consideration is then to assess the relative importance of avoiding harm and risk of harm that result from otherwise apparently effective intervention packages, while avoiding the pull of the enormity of the ABR challenge to lure us into policy deadlocks. In doing so, the complexity and close connections of the various risks offer strong grounds for putting the reversibility of potential adverse consequences at centre stage. Related to debates on the ethics of precaution, this links to different proposals on how to limit the scope of a precautionary principle, e.g., in terms of *de minimis* risk, and, more specifically, to the importance of avoiding irreversible negative outcomes. This is not to say that this is a generally plausible solution for all precautionary decision-making, but the peculiar complex challenges of assessing evidence for ABR interventions seem to add reasons for the fittingness of an approach that prioritises reversibility.[51]

[51] This research was supported by the UGOT Challenges Initiative at the University of Gothenburg; and by the Swedish Research Coucil, VR, contract no. 2014–40, for the Lund-Gothenburg Responsibility Project.

Chapter 23
Antimicrobial Footprints, Fairness, and Collective Harm

Anne Schwenkenbecher

Abstract This chapter explores the question of whether or not individual agents are under a moral obligation to reduce their 'antimicrobial footprint'. An agent's antimicrobial footprint measures the extent to which her actions are causally linked to the use of antibiotics. As such, it is not necessarily a measure of her contribution to antimicrobial resistance. Talking about people's antimicrobial footprint in a way we talk about our carbon footprint may be helpful for drawing attention to the global effects of individual behaviour and for highlighting that our choices can collectively make a real difference. But can we be morally obligated to make a contribution to resolving a collective action problem when our individual contributions by themselves make no discernible difference? I will focus on two lines of argument in favour of such obligations: whether a failure to reduce one's antimicrobial footprint is *unfair* and whether it constitutes wrongdoing because it is *harmful*. I conclude by suggesting that the argument from collective harm is ultimately more successful.

Keywords Political philosophy · Ethics · Public health · Antimicrobial resistance · Collective action problems

23.1 Introduction

Anti-microbial resistance and a decline in anti-microbial efficacy are urgent collective action problems. Who should act on this problem? According to the World Health Organisation's recommendations, concerted action on this issue requires efforts from a diverse array of actors: patients, drug prescribers and dispensers, hospitals, policy makers, and food producers (WHO 2001: 68–70, see also Littmann and Viens 2015).

A. Schwenkenbecher (✉)
Murdoch University, Perth, Western Australia
e-mail: A.Schwenkenbecher@murdoch.edu.au

© The Author(s) 2020
E. Jamrozik, M. Selgelid (eds.), *Ethics and Drug Resistance: Collective Responsibility for Global Public Health*, Public Health Ethics Analysis 5,
https://doi.org/10.1007/978-3-030-27874-8_23

In this chapter I explore the idea of an 'antimicrobial footprint' and discuss whether or not individual agents are under a moral obligation to reduce theirs. Importantly, I am not suggesting that reducing our antimicrobial footprints by way of individual behavioural change is the best or most efficient way of decelerating antimicrobial resistance, since that is an empirical question. However, given that the WHO identified individual agents such as patients and prescribers as agents of change, it seems that individuals' moral obligations deserve some discussion, which is why I will focus on those in this chapter. But before I do so, let me briefly point to another way in which individual agents are implicated in anti-microbial resistance: as consumers of products from animal industries. Notably, the aforementioned WHO report treats the implications of our aggregate meat consumption as an issue for regulation, but not one for individual behavioural change. In contrast, my argument includes individual consumer choices amongst the options individuals have for addressing antimicrobial resistance.

I will focus on two lines of argument for moral obligations to reduce one's antimicrobial footprint: whether a failure to reduce it is *unfair* and whether it constitutes wrongdoing because it is *harmful*. I conclude by suggesting that the argument from collective harm is ultimately more successful.

23.2 Antimicrobial Resistance as a Collective Moral Action Problem

Antimicrobial resistance is a collective action problem in that it is the result of many different agents' activities, it can only be solved by the concerted efforts of many different agents, and it seems rational for individual actors to free-ride because individual behavioural change (if taken in isolation) is neither responsible for the problem's occurrence nor could it ever remedy the problem.

Crucially, too, antimicrobial resistance is the *inevitable* result of using antimicrobials and thereby selecting microorganisms that are resistant to our drugs. Resistance will eventually emerge to any antimicrobial agent we use. This means that resistance as such is an effect that has to be factored into the 'good' that specific antimicrobials provide. To put it differently, it is only a matter of time for any antimicrobial drug to lose its efficacy. To undermine the public good of antimicrobial efficacy is to reduce overall efficacy and to produce resistance at a faster-than-necessary rate. Some have warned that we might be in danger of losing this public good altogether one day – a worst-case scenario, which we are currently capable of preventing. In order to do that, we need to slow down the process of emerging resistance through a more limited and more considerate use of such drugs.

But who is meant by 'we'? Unsurprisingly, many call for global regulation or even the socialization of the use of antibiotics in order to delay the erosion of this good (Smith and Coast 2002; Anomaly 2010). And no doubt, regulators, policy-makers and industry leaders must be at the forefront of restricting the use of antimicrobials in a way that secures their continued efficacy.

But what about ordinary people – individual agents who consume antibiotics either directly (as patients) or indirectly (as consumers of animal products) or who prescribe them (as medical doctors)? The 'general community' was identified by the WHO as a target of intervention (WHO 2001). The assumption behind that seems to be that individual members of the general community *can* jointly reduce resistance. If that is the case, does it follow that we *ought to* do something about reducing resistance?

One of the starting points for answering this question is to establish what causal relationship obtains between our use of antibiotics and emerging resistance. According to WHO authors, "the relationship between use and resistance is not a simple correlation" when it comes to antimicrobials. "Paradoxically, underuse through lack of access, inadequate dosing, poor adherence and sub-standard antimicrobials may play as important a role as overuse" (WHO 2001: 15).

Further, it is not simply the case that those who are *causally* responsible for antimicrobial resistance are automatically *morally* responsible. That is, knowing how a problem came about, or which agent(s) caused it, does not necessarily tell us which agent(s) can be blamed for its occurrence or even who should fix it. *Retrospective moral responsibility* is often used synonymously with *moral blameworthiness*. The focus in this chapter will be on *prospective – or forward-looking – moral responsibility* in the sense of having a moral obligation to act or to bring about a certain outcome.

Clearly, any answer to the question of prospective moral obligations must be based on empirical data concerning which actions will really make a difference to antibiotic resistance. One of the great difficulties for making the case for moral obligations to change individuals' behaviour lies in the fact that no individual (human) agent's actions will make a measurable or perceptible difference to solving the problem. It is an issue on which only the aggregation and combination of countless individual actions and enduring behavioural change will have a real impact.

Both common-sense morality and traditional moral theory often struggle in dealing with collective moral action problems – cases where what is wrong or right cannot be determined by looking at individuals and their actions in isolation, but where instead these must be considered in conjunction or in aggregation. Increasingly, scholars are making an effort to rethink traditional ethical approaches with a view to better account for collective agents, actions and effects (May and Hoffman 1991; French and Wettstein 2006; French and Wettstein 2014; Hess et al. 2018). One of the early attempts to do so will be discussed further down: Derek Parfit proposed that we re-think our 'moral mathematics'. According to Parfit, we need to revise our notions of wrong and right, harm and benefit regarding aggregate effects, where individual actions only make a significant difference in conjunction with countless actions of others (Parfit 1984). It is easy to see that such collective moral action problems abound: Apart from anti-microbial resistance, climate change, and overfishing are cases that come to mind.

So who should act on these problems? The most obvious response would be to point to states and state agents and the need for new policies and regulation. And there is no doubt that such agents are in principle best suited for dealing with such complex large-scale problems. But there is a role for 'ordinary citizens' where governments fall short of doing what is required. Where climate change mitigation is concerned, for instance, the combined actions of individual agents can make a significant contribution to closing the so-called emissions gap, that is, the gap between the emission reductions countries have currently committed to and the reductions required for limiting global warming to a maximum of 2 °C (Dietz et al. 2009; Ostrom 2010; Wynes and Nicholas 2017).

Whether or not anti-microbial resistance is a problem that can be fixed or improved through the aggregate effect of individual behavioural change by patients and doctors, consumers and producers is ultimately an empirical question. But, in line with the *WHO Global Strategy for Containment of Antimicrobial Resistance,* I will proceed on the assumption that collectively individual actors can make a significant difference. Can this ground an obligation for patients, doctors, consumers and producers to make a joint effort towards reducing the use of antibiotics? In the following, I will re-assess some of the philosophical arguments defending ascriptions of individual obligations in combating collective action problems. I will introduce the idea of an 'antimicrobial footprint' and discuss whether not contributing to the public good of antimicrobial efficacy is unfair and whether or not it constitutes harmful behaviour. I will conclude by suggesting that not reducing your antimicrobial footprint (where it is possible for you to do so at an acceptable cost) is potentially wrong because it is harmful (even if your individual actions as such make no difference to antimicrobial resistance).

23.3 Antimicrobial Footprints

Let me start by introducing a new concept: that of an *antimicrobial footprint.* An individual agent's antimicrobial footprint would result from the extent to which her actions are causally linked to the use of antibiotics. The idea mirrors that of a carbon footprint, a measure which – however imperfect (Wright et al. 2011) – reflects the amount of greenhouse gases released into the atmosphere as a result of individuals' actions. Importantly, it links global effects to individual behaviour and highlights that our choices can collectively make a real difference. It may be a helpful tool, then, to start talking about our antimicrobial footprint in a way we talk about our carbon footprint.

With regard to antibiotics, a person's antimicrobial footprint would not necessarily be a measure of her contribution to resistance, but merely of her overall direct and indirect use. Direct use would involve using such drugs as a patient, prescriber or agricultural producer. Indirect use would involve the consumption of goods from animal industries that were produced by overusing antimicrobials. Our diet, then, plays a major role in accelerating resistance (Giubilini et al. 2017) (see also the chapter by Anomaly "Antibiotics and Animal Agriculture). It is important to note, though, that underuse of antibiotics also causally contributes to resistance, not just overuse.

To reiterate, the anti-microbial footprint is – just like one's carbon footprint – an imperfect measure. As mentioned above, the causal links between our use of antimicrobials and resistance are not always straightforward. But the concept as such draws attention to an important fact – that every single one of us is causally and morally implicated in the problem of antimicrobial resistance.

Note further that – just like with our carbon footprint – our antimicrobial footprint will differ depending on our needs and circumstances. If we live in a climate which forces us to heat or cool our dwellings during major parts of the year in order to be healthy and safe then our carbon footprint will necessarily be greater than that of a person living in a milder climate. Likewise, if we suffer from health conditions that require the use of antibiotics we will necessarily have a greater antimicrobial footprint. Reducing our carbon footprint as well as reducing our antimicrobial footprint must not involve unacceptable cost.

But just like in the case of greenhouse gas emissions, there are many instances where we can reduce our antimicrobial footprint at an acceptable cost. First, research shows that patients often ask for such drugs (and are prescribed such drugs) when it would not have been necessary (WHO 2001, see also chapter by Oakley). If doctors can avoid prescribing such drugs and patients stop insisting on them where they are not needed this can make a significant difference for the better.

Another way to reduce one's antimicrobial footprint at an acceptable cost (and with numerous co-benefits such as improved health) is to become vegetarian (or vegan) or at least to have a meat-reduced diet (or else to resort to game and fish caught in the wild). This is a factor that is missing from many public debates concerning antimicrobial resistance and also missing from the WHO report (2001) mentioned earlier.

Let us assume for the sake of argument that a reasonable way of promoting the idea of antimicrobial footprint reductions can be found – one which does not unduly jeopardize individuals' health and which promotes reductions that are truly effective. Do we have *moral obligations* to reduce our antimicrobial footprint? Why would anyone have such an obligation? The question is a serious one: by themselves, none of our individual antimicrobial footprint reductions would make a difference to local, or regional, let alone global antimicrobial efficacy. I call this the *impotence objection*, or the *no-effect-view*. The issue is a familiar one: can we be morally obligated to make a contribution to resolving a collective action problem when our individual contributions make no discernible difference? The view that we cannot be obligated to perform an action if it makes no discernible positive difference to a morally desirable outcome seems to be entailed by standard individualist act-consequentialism. The discussion of obligations to contribute to collective endeavours even where our individual actions make no perceptible difference is ongoing (Parfit 1984; Cullity 1995; Kagan 2011; Nefsky 2011; Schwenkenbecher 2014; Spiekermann 2014; Pinkert 2015).

I will not rehearse all positions here, nor even the main ones, but instead focus on two solutions that appear particularly interesting and suitable to the kind of problem we are faced with and which move outside the standard act-consequentialist framework: the argument from unfairness and the argument from collective harm (for a different argument based on solidarity, see chapter by Holm and Ploug "Solidarity and Antimicrobial Resistance"). Most importantly, these solutions avoid the problem of impotence or imperceptible effects by locating the wrongness of failing to contribute somewhere other than in the effects of one's individual actions.

23.4 The Argument from Unfairness

The first of these arguments is about fairness: Under certain conditions, it is unfair not to contribute to schemes that we benefit from, regardless of the immediate effect of our free-riding, that is, regardless of whether or not we undermine the scheme or make people worse off by defecting. According to Garrett Cullity's *Principle of Fairness* (Cullity 1995), if a person receives benefits from a scheme that satisfies the following conditions, it is unfair of her not to meet the requirements the scheme makes on those enjoying its benefits:

(i) The practice of participation in the scheme represents a net benefit for her;
(ii) Similarly, this practice does not make most others worse off either;
(iii) She is not raising a legitimate moral objection to the scheme. (p. 18f, paraphrased)

According to Cullity, the free-rider's unfairness lies in giving herself *objectionably preferential treatment* in such cases. The benefits she seeks to gain from free-riding "only exist because others who seek them take it upon themselves to contribute toward their production". In other words, her choice to free-ride is motivated by the benefits that others provide, while she grants herself the privilege of enjoying those benefits without providing them (1995: 22–23).

In a later paper, Cullity specifies that unfair actions are failures of *appropriate impartiality* (Cullity 2008). Judgments about fairness and unfairness concern actions for which one particular way of being impartial is morally required (2008: 3). "Unfairness requires not just that the impartiality you fail to display would have been appropriate, but that it is the appropriate way of doing what ought to be done, as it ought to be done." (2008: 5). Cullity gives the following general description of what is common to unfair actions:

"Not Φ-ing is unfair when:

(i) something ought, all things considered, to be done;
(ii) doing it as it ought to be done requires a form of impartiality;
(iii) Φ-ing is the appropriate form for that impartiality to take; and
(iv) the failure of appropriate impartiality can contribute to a non-instrumental explanation of the failure to do what ought to be done." (ibid.)

According to Cullity, then, what matters for assessing the wrongness of free-riding is not only whether there is an action that ought to be performed (or an outcome to be produced or a scheme to be implemented) but that there is a specific way in which this ought to be done, which requires people to apply some kind of impartial rule, rather than look to their own advantage. Doing "what ought to be done as it ought to be done" (ibid.) requires that individuals do not exempt themselves from contributing. That is, out of the two imperatives that bind agents in such cases – the imperative to produce the collective good and the imperative of distributive (or procedural) justice – the free-rider violates the latter even where she cannot be said to clearly violate the former (because she does not jeopardize the collective outcome with her defection alone).

How does this relate to our specific problem of antimicrobial footprint reductions? Let us assume that Cullity is correct in claiming that the above features characterise unfair actions. Is failing to reduce one's antimicrobial footprint unfair? In order for that to be true, it would have to be the case that reducing or limiting antimicrobial resistance is something that all-things-considered ought to be done. Such a claim implies that it can be done at an acceptable overall cost. I think we can safely assume that both are the case.

But what about doing it *as it ought to be done*? Is reducing our individual antimicrobial footprint the method by which we ought to combat anti-microbial resistance? Cullity rejects the idea that whenever a group ought to collectively act or produce a good, individual group members ought to be doing something to produce that good: "That would have odd implications for collective actions to which no one is contributing" (2008: 11). He thinks that it is not unfair if I do not unilaterally pursue a goal if there is no collectively agreed method for pursuing it (ibid.). Defecting (or exempting yourself from contributing to a collective good) is only unfair if there is such a method.

According to Cullity, a collectively agreed method for addressing a collective action problem is in place where the required course of action was decided in a fair procedure. He makes two qualifications though: first, that sometimes decisions produced by fair procedures can be bad and therefore need not to be respected. Second, that we may sometimes be obligated to respect the outcomes of procedures that though not perfectly fair are good enough. Unfortunately, Cullity does not specify what it means for a procedure to be good enough.

It is not possible here to have a detailed discussion on fair (or good enough) procedures for deciding on the production of collective goods. Regulation and legislation – where they result from legitimate democratic procedures – should arguably count as such. What is crucial for Cullity's procedural condition is the underlying rationale: that in order for a collective scheme to have legitimacy, in the sense that it gives individual agents binding reasons for playing their role therein, such a scheme must have been produced in the right way. If that is the case, then we as individual agents can be bound by rules (including laws) that are not of our own making and that we would in fact not have chosen ourselves. But these clarifications do not help with our current enquiry, since our focus is precisely on actions that are not called for by regulation and legislation, but on voluntary individual behavioural change that might be necessary while regulation and legislation fall short of reining in the problem.

This is the point where the fairness argument in favour of reducing our antimicrobial footprint crumbles, I believe. It is quite unclear what kind of method or procedure would count as fair where aggregate individual behavioural changes to reduce our use of antimicrobials are concerned. Would it be enough for such changes to have been recommended by an authoritative, politically neutral global body such as the WHO or other expert panels? According to WHO, its *Global Strategy for Containment of Antimicrobial Resistance* report is the result of expert consultation, workshops and consensus meetings. It is doubtful that this is the kind of procedure Cullity had in mind. Moreover, even though the panel has made recommendations for individual behaviour change, it has not in fact proposed an outright 'scheme' for individual participation with clearly defined roles and contributory actions. For both of these reasons, it does not constitute the kind of collective agreement that gives potentially binding reasons to individual agents. In sum, the argument put forward by Cullity cannot support the idea that individuals ought to take on a share in reducing antimicrobial resistance as a matter of fairness.

A different and more promising approach might be built on an argument that antibiotic overuse or misuse is a way of wronging others in that it harms those who suffer its consequences. This argument relies on a notion of 'collective' harm – a relatively new concept that is increasingly gaining traction.

23.5 The Collective Harm Argument

According to Elizabeth Cripps (2011), individual agents can be collectively responsible for harm brought about by their aggregate individual actions in some cases:

> a person becomes one of a group collectively responsible for harm once her contribution exceeds the amount such that, were everyone contributing only to that level, there would be no harm (p. 181)

In order for a person to be thus responsible for harm, certain conditions have to be met:

1. "individuals acted in ways which, in aggregate, caused harm, and which they were aware (or could reasonably be expected to have foreseen) would, in aggregate, cause harm (although each only intentionally performed his own act);
2. they were all aware (or could reasonably expected to have foreseen) that there were enough others similarly placed (and so similarly motivated to act) for the combined actions to bring about the harm; and
3. the harm was collectively avoidable: by acting otherwise (which they could reasonably have done), the individuals making up the putative group could between them have avoided the harm." (pp. 174f)

The crucial point to be noted is that in order to be *weakly collective responsible* (as Cripps puts it) for harm, individuals need to *know* (or be in a position to foresee) two things: (i) that if enough other people did what they do it will cause harm, and (ii) that there are enough other people doing what they do.

Whether or not a large enough number of people are in this position vis-à-vis antimicrobial resistance is an empirical question. However, I suspect that these epistemic conditions are not met when it comes to our antimicrobial footprint. The problem of antimicrobial resistance has much less presence in the media and public discourse than the problem of climate change and carbon footprint reductions, for instance.

Cripps' criteria are clearly modelled on Derek Parfit's (1984) conditions for collectively doing wrong or harming others. He, too, relies on an epistemic condition that is – currently – unlikely to be met where antimicrobial resistance is concerned:

> (C12) When (1) the outcome would be worse if people suffered more, and (2) each of the members of some group could act in a certain way, and (3) they would cause other people to suffer if *enough* of them act in this way, and (4) they would cause these people to suffer *most* if they *all* act in this way, and (5) each of them both **knows** these facts and **believes** that enough of them will act in this way, then (6) each of them would be acting wrongly if he acted in this way. (p. 81, my emphasis in bold)

According to both Cripps and Parfit, then, we only act wrongly if we know about the effects of our own antimicrobial overuse or misuse and we are aware that enough others are engaged in this practice. Consequently, public awareness campaigns would make it the case that Cripps' and Parfit's conditions are met. Public knowledge – which obtains where most people know some proposition to be true and most people know that most people know – would turn harmless actions into harm. But still, on their accounts there is – currently – no harm or wrongdoing committed by many if not most of those who contribute to antimicrobial resistance. Also, for Cripps, weakly collective responsibility does not imply that any individual has direct duties to avert the (aggregate) harm. Instead such duties fall to the group, first and foremost. That is, even if we were collectively responsible for antimicrobial resistance we would not be required to individually reduce our antimicrobial footprint on her account.

Let me now turn to Judith Lichtenberg, who combines the unfairness argument and the argument from aggregate harm (2010): If we knowingly contribute to harms that "depend on the joint effects of many people's actions" (p. 568) we accept that if a sufficient number of other persons act in the same way, these harms will occur. She thinks that to do so is wrong because it means to act *unfairly*: "In the case of aggregate harms, doing the right thing involves an appeal to the unfairness of acting inconsistently with how one thinks others ought to act." (2010: 569). As I understand Lichtenberg, contributing to aggregate harms is not *intrinsically* wrong, but is wrong because it cannot be justified in rule-consequentialist terms or by way of universalizing. Similar to Cullity, she argues that the wrongness lies in exempting oneself from a rule that one should accept as morally optimal.

Note that Lichtenberg's account is more demanding than Cripps' and Parfit's because it does not have as strong a knowledge condition. For the wrongness of contributing it does not matter whether or not an individual agent knows that enough others will perform the same action and harm will be thus caused in aggregation. It suffices for the individual to know that collectively we should adopt a rule prohibiting such actions. This is a more demanding account because it seems to require us (*pro tanto*, at least) to individually refrain from doing what is collectively suboptimal. As I understand it, Lichtenberg's rule, if applied to antimicrobial footprint reductions, would imply that avoidable antimicrobial overuse and misuse are instances of harming, from which we (*pro tanto*) ought to abstain.

In response, one might argue that to demand – as Lichtenberg appears to do – that we individually do our part in a pattern that is collectively optimal is too strong a requirement. After all, sometimes it may be right to do what is collectively suboptimal if no one else does what is collectively optimal and our individual 'sacrifice' would be pointless. However, note that if Lichtenberg's proposal is safe from this objection as long as it is understood as generating pro tanto obligations to avoid contributing to collective harm, that is, obligations that can be overridden by other, more important obligations. If the collective defection rate is too high, my *pro tanto* obligations may simply fail to become all-out obligations. That is, if not enough others contribute, I may not have an all-things-considered obligation to avoid collective harm.

23.6   Conclusion

In this chapter I discussed arguments in favour of a moral obligation to reduce one's individual antimicrobial footprint. Despite the intuitive appeal of this idea, there exists no simple, straightforward defence of an obligation to change our individual behaviour. High levels of collective awareness and a genuine collective willingness to address the problem of anti-microbial resistance appear to be important preconditions for motivating (all-out) obligations for individuals to reduce their antimicrobial footprint. It is one of the most frustrating aspects of collective action problems that it is precisely the publicly known lack of commitment to resolving them which seems to sustain and justify a (further) lack of commitment for all those who could potentially resolve it.

Acknowledgements   The author gratefully acknowledges the financial support received from the Oxford Martin School for this project as a Visiting Fellow to the program "Collective Responsibility for Infectious Diseases" (April – July 2017) as well as the extremely valuable feedback from colleagues in the Oxford Martin School and the Uehiro Centre for Practical Ethics.

References

Anomaly, J. 2010. Combating resistance: The case for a global antibiotics treaty. *Public Health Ethics* 3 (1): 13–22.
Cripps, E. 2011. Climate change, collective harm and legitimate coercion. *Critical Review of International Social and Political Philosophy* 14 (2): 171–193.
Cullity, G. 1995. Moral free riding. *Philosophy and Public Affairs* 24 (1): 3–34.
———. 2008. Public goods and fairness. *Australasian Journal of Philosophy* 86 (1): 1–21.
Dietz, T., G.T. Gardner, J. Gilligan, P.C. Stern, and M.P. Vandenbergh. 2009. Household actions can provide a behavioral wedge to rapidly reduce US carbon emissions. *Proceedings of the National Academy of Sciences* 106 (44): 18452–18456.
French, P.A., and H.K. Wettstein, eds. (2006). Special Issue: Shared intentions and collective responsibility, Midwest Studies in Philosophy 30.

———, eds. (2014). Special issue: Forward-looking collective responsibility, Midwest Studies in Philosophy 38.

Giubilini, A., P. Birkl, T. Douglas, J. Savulescu, and H. Maslen. 2017. Taxing meat: Taking responsibility for one's contribution to antibiotic resistance. *Journal of Agricultural and Environmental Ethics* 30 (2): 179–198.

Hess, K.E., V.E. Igneski, and T.L.E. Isaacs. 2018. *Collectivity: Ontology, ethics, and social justice*. London: Rowman & Littlefield International.

Kagan, S. 2011. Do I make a difference? *Philosophy & Public Affairs* 39 (2): 105–141.

Lichtenberg, J. 2010. Oughts and cans. *Philosophical Topics* 38 (1): 123–142.

Littmann, J., and A.M. Viens. 2015. The ethical significance of antimicrobial resistance. *Public Health Ethics* 8 (3): 209–224.

May, L., and S. Hoffman. 1991. *Collective responsibility: Five decades of debate in theoretical and applied ethics*. Rowman & Littlefield: Savage.

Nefsky, J. 2011. Consequentialism and the problem of collective harm: A reply to Kagan. *Philosophy & Public Affairs* 39 (4): 364–395.

Ostrom, E. 2010. Polycentric systems for coping with collective action and global environmental change. *Global Environmental Change* 20 (4): 550–557.

Parfit, D. 1984. *Five mistakes in moral mathematics. Reasons and persons*. Vol. 1, 55–83. Oxford: Clarendon Press.

Pinkert, F. 2015. What if I cannot make a difference (and know it). *Ethics* 125 (4): 971–998.

Schwenkenbecher, A. 2014. Is there an obligation to reduce one's individual carbon footprint? *Critical Review of International Social and Political Philosophy* 17 (2): 168–188.

Smith, R.D., and J. Coast. 2002. Antimicrobial resistance: a global response. *Bulletin of the World Health Organization* 80: 126–133.

Spiekermann, K. 2014. Small impacts and imperceptible effects: Causing harm with others. *Midwest Studies In Philosophy* 38 (1): 75–90.

World Health Organisation. (2001). WHO global strategy for containment of antimicrobial resistance.

Wright, L.A., S. Kemp, and I. Williams. 2011. 'Carbon footprinting': Towards a universally accepted definition. *Carbon Management* 2 (1): 61–72.

Wynes, S., and K.A. Nicholas. 2017. The climate mitigation gap: Education and government recommendations miss the most effective individual actions. *Environmental Research Letters* 12 (7): 074024.

Chapter 24
Global Health Governance and Antimicrobial Resistance

Belinda Bennett and Jon Iredell

Abstract This chapter analyses the challenges and the adequacy of existing frameworks to provide a strong foundation to support global responses to antimicrobial resistance. Calls for global responses are indicative of a growing global commitment to seeking practical means of tackling the growing problem of antimicrobial resistance. While antimicrobial resistance is often conceptualised as an emergency, the application of the International Health Regulations, designed to govern responses to public health emergencies of international concern, remains unclear. Furthermore, there may be challenges for countries in developing and resourcing national approaches to address antimicrobial resistance. Clarity and agreement around definitions of key concepts related to antimicrobial resistance will also be essential to antibiotic stewardship and development of policy in this area. Finally, improvements to health systems as a result of the Sustainable Development Goals may help to support improvements in public health and may play a role in global strategies to address antimicrobial resistance.

Keywords Antimicrobial resistance · International Health Regulations · Sustainable Development Goals

24.1 Introduction

Antimicrobial resistance (AMR) has been described as 'a global health crisis,' (Review on Antimicrobial Resistance 2015) 'a slowly emerging disaster' (Viens and Littman 2015), and 'a complex global public health challenge' (World Health

B. Bennett (✉)
School of Law, Queensland University of Technology, Brisbane, Queensland, Australia
e-mail: belinda.bennett@qut.edu.au

J. Iredell
Department of Infectious Diseases, Westmead Hospital and University of Sydney, Sydney, NSW, Australia

© The Author(s) 2020
E. Jamrozik, M. Selgelid (eds.), *Ethics and Drug Resistance: Collective Responsibility for Global Public Health*, Public Health Ethics Analysis 5, https://doi.org/10.1007/978-3-030-27874-8_24

Organization 2014: xix). With AMR raising the possibility that we could soon be living in a 'post-antibiotic era' (World Health Organization 2014: ix) and with recognition that AMR 'threatens the sustainability of the public health response to many communicable diseases, including tuberculosis, malaria and HIV/AIDS' (World Health Assembly 2014), there is increased international dialogue around the need to find practical global solutions for the looming crisis (Laxminarayan et al. 2013). Others have argued for the need for AMR strategies to address equitable *access* to medicines; *conservation* through stewardship and appropriate use of antimicrobials; and *innovation* to ensure the development of the new antimicrobials, as well as the importance of a binding legal framework addressing these issues (Hoffman, Outterson, Røttingen et al. 2015; Hoffman, Røttingen and Frenk 2015). Proposals have included a new international treaty (Hoffman, Outterson, Røttingen et al. 2015; Anomaly 2010), the establishment of an intergovernmental panel (Woolhouse and Farrar 2014), a range of policy options to support action on AMR (Hoffman, Caleo, Daulaire et al. 2015), and the use of the International Health Regulations as an existing framework for managing global health risks (Wernli et al. 2011). Such is the complexity of the challenge posed by AMR that national and regional approaches alone are unlikely to be effective solutions (Littman et al. 2020). Yet the complex nature of AMR presents considerable challenges for global health governance, and with recent global health crises revealing shortcomings in global health governance, there are clearly significant challenges associated with utilising existing frameworks to address the growing problem of AMR. This chapter analyses these challenges and the adequacy of existing frameworks to provide a strong foundation to support global action on AMR.

24.2   The Rise of Drug Resistance

Severe infection long ago overtook heart attacks in terms of likelihood of hospital admission in developed countries (Seymour et al. 2012). The ability to manage it is essential for intensive care, major surgery and transplantation services that are increasingly routine. Up to 20 million people each year are treated with mechanical ventilation in intensive care (Adhikari et al. 2010). Infection complicates about half of ICU admissions and is an important cause of death in the critically ill (Vincent et al. 2009).

Antibiotics are a cornerstone of treatment of severe infection, with the risk of death increasing in the presence of antibiotic resistance (Review on Antimicrobial Resistance 2016a). In response to the steady rise in antibiotic resistance that began soon after the first introduction of antibiotics and has been quickly accelerating (Davies et al. 2013; Laxminarayan et al. 2013; Review on Antimicrobial Resistance 2016a: 10), experts call for international co-ordination of public policy solutions that include (i) better antibiotic controls in industry, agriculture and medicine (antimicrobial stewardship), and (ii) better surveillance and containment (infection control) (Howard et al. 2013; Laxminarayan et al. 2013; Piddock 2012; Spellberg et al. 2008).

24.3 A Global Approach to AMR?

With the recognition of the problems posed by growing AMR have come calls for a global approach to limiting the spread of AMR and its associated risks. In 2015 the World Health Assembly adopted a global action plan on antimicrobial resistance (World Health Organization 2015a; Ho and Lee 2020) and called on Member States to develop national actions plans on AMR and to 'mobilize human and financial resources through domestic, bilateral and multilateral channels in order to implement plans and strategies in line with the global action plan' (World Health Assembly 2015). The global action plan included five strategic objectives:

(1) to improve awareness and understanding of antimicrobial resistance; (2) to strengthen knowledge through surveillance and research; (3) to reduce the incidence of infection; (4) to optimize the use of antimicrobial agents; and (5) to ensure sustainable investment in countering antimicrobial resistance (World Health Organization 2015a: 1)

The World Health Assembly resolution was followed in 2016 by a declaration by the high-level meeting of the United Nations General Assembly on antimicrobial resistance. The declaration of the high-level meeting included committing to work nationally, regionally and globally to address AMR, and to mobilize funding and resources to support the development and implementation of national plans (United Nations General Assembly 2016).

These calls for global action on antimicrobial resistance indicate a growing global commitment to seeking practical means of tackling the growing problem of antimicrobial resistance. As Wernli et al. have pointed out, this growing focus on AMR 'also indicates that AMR has transformed into a global governance priority, which requires international co-operation' (Wernli et al. 2017: 1). The policy discourses around AMR are also complex however, with Wernli et al. identifying five key policy frames that are evident in the debates around AMR, with each of these frames providing insights into AMR and the interdependencies between these policy frames: (i) 'AMR as a healthcare issue'; (ii) 'AMR as a development issue,' reflecting the burden of infectious diseases in low- and middle-income countries; (iii) the relationship between AMR and innovation, which recognises the importance of research and development for new antibiotics; (iv) AMR as a global health security issue; and (v) AMR and the role of a One-Health approach to addressing AMR (Wernli et al. 2017).

International cooperation will clearly play a key role in responding to AMR (Årdal et al. 2016). However, recent global health crises have revealed difficulties in ensuring effective responses to emerging public health crises, leaving the role of global health governance in leading responses to AMR unclear. The remainder of this chapter identifies four key areas that will present challenges for a global response to AMR: (i) the challenge of fitting responses to AMR within other global health governance frameworks that support urgent responses to health crises; (ii) the challenge of developing national and regional capacities to identify and respond to AMR; (iii) the challenge of developing common understandings and definitions of AMR; and finally (iv) the challenge of fitting AMR within the global priorities identified by the Sustainable Development Goals.

24.4   The International Health Regulations, Global Health Governance and AMR

Adopted to manage the international spread of infectious diseases, the International Health Regulations (2005) (IHR) (World Health Organization 2016a) could possibly provide a framework for improved surveillance and reporting of AMR. However, there has been debate over whether the scope of the IHR could include evolving events such as AMR or whether the IHR are more properly focused on emergency situations (Wernli et al. 2011; Kamradt-Scott 2011). It has been argued that AMR may, at least in some instances, fulfil two of the four IHR criteria for a 'public health emergency of international concern' (PHEIC) i.e. those relating to the seriousness of the public health impact of an event, and the potential for international spread of disease (Wernli et al. 2011: 3), with the remaining two IHR criteria being that the event is 'unusual or unexpected,' and that there is 'a significant risk of international travel or trade restrictions' (World Health Organization 2016a: Annex 2). It is certainly the case that the IHR provide a strong normative scaffold for the shared expectations about how both states and the international community should respond to the international spread of disease (Davies et al. 2015), as well as clear processes for reporting progress.

While patterns of AMR vary between countries (Review on Antimicrobial Resistance 2014), AMR certainly poses major risks to public health globally (World Health Organization 2015a). The focus of the IHR on managing global risks to public health, and the inclusion of consideration of whether there is 'a significant risk of international spread' as one of the criteria for determining whether events that may constitute a PHEIC should be notified to WHO (World Health Organization 2016a: Annex 2), provide a global perspective to the risks posed by spread of AMR. As it is now more than a decade since the revised IHR took effect in 2007, the IHR provide a familiar and well-established framework for the global community and, it has been argued that it may be possible to build on existing frameworks rather than building new ones (Wernli et al. 2011: 5). Yet these factors do not necessarily make the IHR an ideal mechanism for addressing AMR (Kamradt-Scott 2011), even though, with the global initiatives referred to above, global health governance seems set to play a key role in addressing AMR.

Despite the promise that initially heralded the adoption of the revised IHR in 2005 (Fidler and Gostin 2006), recent global health crises have revealed significant shortcomings in contemporary global health governance including, inadequate capacity building of national health systems, a failure by countries to comply with the IHR 2005 through the imposition of trade and travel restrictions contrary to WHO recommendations, and a lack of funding to support an effective global response to global public health emergencies (Gostin et al. 2015, 2016; Moon et al.

2015; Ottersen et al. 2016). Furthermore, the suitability of the IHR as a mechanism for addressing AMR remains unclear, with the primary focus of the IHR being on the global capacity to deal with global public health emergencies (Kamradt-Scott 2011). With the IHR's focus being on acute emergencies, rather than issues requiring sustained response over a prolonged period, their utility as a framework for coordinating global responses to AMR appears less certain (Wernli et al. 2017: 5; Kamradt-Scott 2011).

While a new treaty may provide a clear mechanism for addressing the challenges posed by AMR (Anomaly 2010; Hoffman, Outterson, Røttingen et al. 2015), as has been noted, 'Reaching such an ambitious legal agreement will take leadership, skill, and perseverance from a wide range of actors' (Rochford et al. 2018: 1977). A treaty would also need to balance stewardship of antibiotics and innovation with recognition of lack of access to antibiotics for many people, particularly in low and middle income countries (Padiyara et al. 2018: 3). Furthermore, as discussed below, the issue of capacity building remains an enduring challenge, particularly in low resource countries.

24.5 Building Capacity

A further difficulty associated with using the IHR as a mechanism for addressing AMR is in the challenges faced by many countries to meet their IHR core capacity requirements (Davies et al. 2015: 126-132; World Health Organization 2015b). The IHR require Member States to develop core capacities within their national health systems to 'detect, assess, notify and report events' that are within the scope of the IHR (Article 5). These national capacities required by the IHR will provide an important step in building surveillance capacities for AMR (Wernli et al. 2011). Countries were required to achieve the core capacity requirements of the IHR by 2012 i.e. within 5 years of the IHR coming into force, with some extensions possible, but many countries still had not achieved their core capacity requirements (World Health Organization 2016b). Although global collaboration (Goff et al. 2017) and regional networks may provide an additional mechanism for capacity building, particularly in developing regions (Bennett and Carney 2017), the challenge of building strong public health systems remains a key one for global health (Gostin et al. 2016; Moon et al. 2015).

In addition, although the World Health Assembly has urged its Member States to develop national action plans for antimicrobial resistance (World Health Assembly 2015), the results of 2013–2014 WHO survey showing that few countries reported having comprehensive national AMR plans (World Health Organization 2015c: 1), suggests that there may be considerable work to be done in many countries. More recently, a self-assessment survey of countries and their progress in developing and implementing national AMR plans also shows that considerable work remains, with only 60% of countries having a national multisectoral action plan on AMR, although a further 33% had plans in development (World Health Organization et al. 2018: 7). In addition, although 125 countries reported having awareness raising activities about AMR risks and human health, only 36 had campaigns in animal health (World Health Organization et al. 2018: 12). While 103 countries reported having a national surveillance system in humans, only 41 counties had systematic data collection in animals, and most countries have no surveillance system in place in the plant and environment sectors (World Health Organization et al. 2018: 15).

One key difficulty is that although the IHR require countries to develop their national capacities to deal with potential global public health emergencies that arise within their borders, poorer countries may have difficulty in finding the resources to support such capacity building (Gostin and Katz 2016: 276-277; Davies et al. 2015: 126-132), leaving many countries unable to achieve the capacities required by the IHR. There is renewed interest in developing sustainable mechanisms for global health financing, including financing of global public goods such as addressing antimicrobial resistance (Ottersen et al. 2017; Mendelson et al. 2016; Moon et al. 2017). Clearly there is a need for investment to address shared vulnerabilities in global health (Gostin 2017). Global commitments and the development of effective financing mechanisms will be key to the development of effective responses to AMR. Without them we may see a repeat of the difficulties in building IHR core capacities, with poorer countries lacking the resources to develop their national capacities.

24.6 Understandings and Definitions

Understandings of the mechanisms of development of AMR as well as definitions of key terms relating to infection and antimicrobial use are critical to effective policy response for AMR. Even the term 'antimicrobial resistance' may be poorly understood (Mendelson et al. 2017). The two pillars of the health policy response to antimicrobial resistance relate to control of its spread ('infection control') and to reduction of inappropriate antibiotic use that promotes it (antimicrobial stewardship) (Davies et al. 2013; Laxminarayan et al. 2013).

Measures of success and failure of public health policy are heavily contingent on definitions. The definition of 'inappropriate prescribing' remains poorly informed by research into antibiotic effects and is therefore regarded as a public policy research priority in the area of antibiotic resistance (Spellberg et al. 2011). Similarly, the definition of 'severe infection' or 'sepsis' is subject to major revision (Fullerton et al. 2017) and new definitions may alter the cohort included in this definition, especially at initial point of care where health policy must drive effective immediate responses (Fullerton et al. 2017).

These definitional questions are of key importance to policy initiatives such as antibiotic stewardship. AMR is an example of the tragedy of the commons in which overconsumption of a common resource harms the common good (Hollis and Maybarduk 2015; Giubilini and Savulescu 2020). For this reason, stewardship and the preservation of key antimicrobials is an important strategy in addressing AMR (Laxminarayan et al. 2013). Yet the development of policy responses for AMR is a particular challenge as there is a need to limit inappropriate use of antibiotics, (which assumes well-developed understandings of 'inappropriate prescribing'), while simultaneously ensuring equitable access to medicines (Laxminarayan et al. 2013, 2016; Padiyara et al. 2018; Selgelid 2007). An antibiotic that is harmful to the public health in terms of driving resistance in the longer term may be thought to have specific benefit to the individual for whom it is prescribed. This may lead to direct conflict between public health policy needs and the needs of an individual and means in turn that antimicrobial stewardship may be cast in a policing role.

These challenges may be described as ones of 'understandings' and of 'definitions.' To address these challenges additional funding is required to support research on the mechanisms for acquiring resistance to further our understandings of this important area. Agreed international standards for data collection on AMR (World Health Organization 2014) are also vital if surveillance is to be comprehensive and effective, making definitional issues ones of critical importance.

24.7   AMR and the Sustainable Development Goals (SDGs)

When the Sustainable Development Goals (SDGs) (United Nations 2015) were adopted by the United Nations General Assembly in September 2015, AMR was not listed amongst the 17 goals, although it is mentioned in the SDG declaration (World Health Organization 2015d: 103). Nor was there the inclusion of a broader SDG goal on global health security, which might have also provided a focus for action on AMR (Kickbusch et al. 2015). However, this is not to suggest that the SDGs will be irrelevant for global responses to AMR. For example, addressing antimicrobial resistance will clearly be important to achieving SDG3 ('Ensure healthy lives and promote well-being for all at all ages'), while the goal of universal health coverage in SDG3.8 will help to ensure the existence of strong health systems (Gostin 2017: 194). International trade of animals or products, the complexity of addressing both overconsumption and lack of access to medicines, and the relevance of trade agreements and intellectual property arrangements to research and development for new antimicrobials, all highlight the importance of considering the relevance of trade to the development of solutions to antimicrobial resistance (Hanefeld et al. 2017). Access to water and sanitation were also included as SDG6 (United Nations 2015). As the UK's Review on Antimicrobial Resistance has noted, improved water and sanitation can help to reduce antibiotic consumption by helping to prevent infections that may then be treated with antibiotics (Review on Antimicrobial Resistance 2016a: 21-23;

2016b). The association between poor sanitation, infections and antibiotic use shows that broader public health goals and improvements can play an important role in global strategies to prevent AMR (Review on Antimicrobial Resistance 2016a; 2016b).

24.8   Conclusion

Developing effective mechanisms to address the growing threat of AMR will be essential to safeguarding public health into the future. To achieve this will require, amongst other things, strong mechanisms for global health governance to support coordinated programs and building of capacity at the national, regional and global levels. The commitments to the development of national plans for AMR along with suggestions for new financing mechanisms to support capacity building and the funding of global public goods reflect a promising prioritisation of AMR. Whether we can develop practical and effective means of addressing AMR remains to be seen.

Acknowledgements Belinda Bennett gratefully acknowledges helpful discussions with Sara E Davies in the development of this chapter and Elizabeth Dallaston for her research assistance.

References

Adhikari, N.K.J., R.A. Fowler, S. Bhagwanjee, and G.D. Rubenfeld. 2010. Critical care and the global burden of critical illness in adults. *The Lancet* 375: 1339–1346.

Anomaly, J. 2010. Combating resistance: The case for a global antibiotics treaty. *Public Health Ethics* 3 (1): 13–22.

Årdal, C., K. Outterson, S.J. Hoffman, A. Ghafur, M. Sharland, N. Ranganathan, et al. 2016. International cooperation to improve access to and sustain effectiveness of antimicrobials. *Lancet* 387: 296–307.

Bennett, B., and T. Carney. 2017. Public health emergencies of international concern: Global, regional and local responses to risk. *Medical Law Review* 25 (2): 223–239.

Davies, S.C., T. Fowler, J. Watson, D.M. Livermore, and D. Walker. 2013. Annual report of the Chief Medical Officer: Infection and the rise of antimicrobial resistance. *Lancet* 381 (9878): 1606–1609.

Davies, S.E., A. Kamradt-Scott, and S. Rushton. 2015. *Disease diplomacy: International norms and global health security*. Baltimore: Johns Hopkins University Press.

Fidler, D.P., and L.O. Gostin. 2006. The new International Health Regulations: An historic development for international law and public health. *Journal of Law, Medicine & Ethics*. 34: 85–94.

Fullerton, J.N., K. Thompson, A. Shetty, J.R. Iredell, et al. 2017. New sepsis definition changes incidence of sepsis in the intensive care unit. *Critical Care and Resuscitation*. 19 (1): 9–13.

Giubilini, A. and J. Savulescu. 2020. Moral responsibility and the justification of policies to preserve antimicrobial effectiveness. In *Ethics and drug resistance: Collective responsibility for global public health*, ed. E. Jamrozik and M. Selgelid. Cham: Springer.

Goff, D.A., R. Kullar, E.J. Goldstein, M. Gilchrist, D. Nathwani, A.C. Cheng, et al. 2017. A global call from five countries to collaborate in antibiotic stewardship: United we succeed, divided we might fail. *Lancet Infectious Diseases* 17: e56–e63.

Gostin, L.O. 2017. Our shared vulnerability to dangerous pathogens. *Medical Law Review* 25 (2): 185–199.

Gostin, L.O., and R. Katz. 2016. The International Health Regulations: The governing framework for global health security. *The Milbank Quarterly* 94 (2): 264–313.

Gostin, L.O., M. DeBartolo, and E. Friedman. 2015. The International Health Regulations 10 years on: The governing framework for global health security. *Lancet* 386: 2222–2226.

Gostin, L.O., O. Tomori, S. Wibulpolprasert, A.K. Jha, J. Frenk, S. Moon, et al. 2016. Toward a common secure future: Four global commissions in the wake of Ebola. *PLoS Medicine* 13 (5): e1002042. https://doi.org/10.1371/journal.pmed.1002042.

Hanefeld, J., M. Khan, G. Tomson, and R. Smith. 2017. Trade is central to achieving the Sustainable Development Goals: A case study of antimicrobial resistance. *BMJ* 358: j3505. https://doi.org/10.1136/bmj.j3505.

Ho, C.W.L., and T.-L. Lee. 2020. Global governance of antimicrobial resistance: A legal and regulatory toolkit. In *Ethics and drug resistance: Collective responsibility for global public health*, ed. E. Jamrozik and M. Selgelid. Cham: Springer.

Hoffman, S., K. Outterson, J.-A. Røttingen, O. Cars, C. Clift, Z. Rizvi, et al. 2015. Editorial: An international legal framework to address antimicrobial resistance. *Bulletin of the World Health Organization* 93: 66.

Hoffman, S.J., J.-A. Røttingen, and J. Frenk. 2015. International law has a role to play in addressing antibiotic resistance. *Journal of Law, Medicine & Ethics* 43 (3 Suppl): 65–67.

Hoffman, S.J., G.M. Caleo, N. Daulaire, S. Elbe, P. Matsoso, E. Mossialos, et al. 2015. Strategies for achieving global collective action on antimicrobial resistance. *Bulletin of the World Health Organization* 93: 867–876.

Hollis, A., and P. Maybarduk. 2015. Antibiotic resistance is a tragedy of the commons that necessitates global cooperation. *Journal of Law, Medicine & Ethics* 43 (3 Suppl): 33–37.

Howard, S.J., M. Catchpole, J. Watson, and S.C. Davies. 2013. Antibiotic resistance: Global response needed. *Lancet Infectious Diseases* 13 (12): 1001–1003.

Kamradt-Scott, A. 2011. A public health emergency of international concern? Response to a proposal to apply the International Health Regulations to antimicrobial resistance. *PLoS Medicine* 8 (4): e1001021.

Kickbusch, I., J. Orbinski, T. Winkler, and A. Schnabel. 2015. We need a sustainable development goal 18 on global health security. *Lancet* 385: 1069.

Laxminarayan, R., A. Duse, C. Wattal, A. Zaidi, H. Wertheim, N. Sumpradit, et al. 2013. Antibiotic resistance – The need for global solutions. *Lancet Infectious Diseases* 13: 1057–1098.

Laxminarayan, R., P. Matsoso, S. Pant, C. Brower, J.-A. Røttingen, K. Klugman, et al. 2016. Access to effective antimicrobials: A worldwide challenge. *Lancet* 387: 168–175.

Littman, J., A.M. Viens, and D.S. Silva. 2020. The super-wicked problem of antimicrobial resistance. In *Ethics and drug resistance: Collective responsibility for global public health*, ed. E. Jamrozik and M. Selgelid. Cham: Springer.

Mendelson, M., O. Dar, S.J. Hoffman, R. Laxminarayan, M.M. Mpundu, and J.-A. Røttingen. 2016. A global antimicrobial conservation fund for low- and middle-income countries. *International Journal of Infectious Diseases* 51: 70–72.

Mendelson, M., M. Balasegaram, T. Jinks, C. Pulcini, and M. Sharland. 2017. Antibiotic resistance has a language problem. *Nature* 545: 23–25.

Moon, S., D. Sridhar, M.A. Pate, A.K. Jha, et al. 2015. Will Ebola change the game? Ten essential reforms before the next pandemic. The report of the Harvard-LSHTM independent panel on the global response to Ebola. *Lancet* 386: 2204–2221.

Moon, S., J.-A. Røttingen, and J. Frenk. 2017. Global public goods for health: Weaknesses and opportunities in the global health system. *Health Economics, Policy and Law* 12: 195–205.

Ottersen, T., S.J. Hoffman, and G. Groux. 2016. Ebola again shows the International Health Regulations are broken: What can be done differently to prepare for the next epidemic? *American Journal of Law & Medicine* 42: 356–392.

Ottersen, T., R. Elovainio, D.B. Evans, D. McCoy, D. McIntyre, F. Meheus, et al. 2017. Towards a coherent global framework for health financing: Recommendations and recent developments. *Health Economics, Policy and Law* 12: 285–296.

Padiyara, P., H. Inoue, and M. Sprenger. 2018. Global governance mechanisms to address antimicrobial resistance. *Infectious Diseases: Research and Treatment* 11: 1–4.

Piddock, L.J. 2012. The crisis of no new antibiotics – What is the way forward? *Lancet Infectious Diseases* 12 (3): 249–253.

Review on Antimicrobial Resistance. 2014. *Antimicrobial resistance: Tackling a crisis for the health and wealth of nations.* London.

———. 2015. *Tackling a global health crisis: Initial steps.* London.

———. 2016a. *Tackling drug-resistant infections globally: Final report and recommendations.* London.

———. 2016b. *Infection prevention, control and surveillance: Limiting the development and spread of drug resistance.* London.

Rochford, C., D. Sridhar, N. Woods, et al. 2018. Global governance of antimicrobial resistance. *Lancet* 391: 1976–1978.

Selgelid, M.J. 2007. Ethics and drug resistance. *Bioethics* 21 (4): 218–229.

Seymour, C.W., T.D. Rea, J.M. Kahn, A.J. Walkey, D.M. Yealy, and D.C. Angus. 2012. Severe sepsis in pre-hospital emergency care: Analysis of incidence, care, and outcome. *American Journal of Respiratory and Critical Care Medicine.* 186 (12): 1264–1271.

Spellberg, B., R. Guidos, D. Gilbert, J. Bradley, H.W. Boucher, W.M. Scheld, et al. 2008. The epidemic of antibiotic-resistant infections: A call to action for the medical community from the Infectious Diseases Society of America. *Clinical Infectious Diseases* 46 (2): 155–164.

Spellberg, B., M. Blaser, R.J. Guidos, H.W. Boucher, J.S. Bradley, B.I. Eisenstein, et al. 2011. Combating antimicrobial resistance: Policy recommendations to save lives. *Clinical Infectious Diseases* 52 (Suppl 5): S397–S428.

United Nations. 2015. *Sustainable Development Goals.* Available from: www.un.org/sustainabledevelopment/sustainable-development-goals.

United Nations General Assembly. 2016. *Political declaration of the high-level meeting of the General Assembly on antimicrobial resistance* (A/RES/71/3).

Viens, A.M., and J. Littman. 2015. Is antimicrobial resistance a slowly emerging disaster? *Public Health Ethics.* 8(3): 255-265.

Vincent, J.L., J. Rello, J. Marshall, E. Silva, A. Anzueto, C.D. Martin, et al. 2009. International study of the prevalence and outcomes of infection in intensive care units. *JAMA* 302 (21): 2323–2329.

Wernli, D., T. Haustein, J. Conly, Y. Carmeli, I. Kickbusch, and S. Harbarth. 2011. A call for action: The application of the International Health Regulations to the global threat of antimicrobial resistance. *PLoS Medicine* 8 (4): e1001022.

Wernli, D., P.S. Jørgensen, C.M. Morel, S. Carroll, S. Harbarth, N. Levrat, et al. 2017. Mapping global policy discourse on antimicrobial resistance. *BMJ Global Health* 2: e000378.

Woolhouse, M., and J. Farrar. 2014. An intergovernmental panel on antimicrobial resistance. *Nature* 509: 555–557.

World Health Assembly. 2014. *Antimicrobial resistance* (WHA67.25).

———. 2015. *Resolution WHA68.7 – Global action plan on antimicrobial resistance.*

World Health Organization. 2014. *Antimicrobial resistance: Global report on surveillance.* Geneva: World Health Organization.

———. 2015a. *Global action plan on antimicrobial resistance.* Geneva: World Health Organization.

———. 2015b. *Implementation of the International Health Regulations (2005): Report of the Review Committee on second extensions for establishing national public health capacities and on IHR implementation* (A68/22 Add.1).

———. 2015c. *Worldwide country situation analysis: Response to antimicrobial resistance.* Geneva: World Health Organization.

————. 2015d. *Health in 2015: From MDGs Millennium Development Goals to SDGs Sustainable Development Goals*. Geneva: World Health Organization.

————. 2016a. *International Health Regulations (2005)*. 3rd ed. Geneva.

————. 2016b. *Implementation of the International Health Regulations (2005): Annual report on the implementation of the International Health Regulations (2005). Report by the Director-General* (WHA A69/20).

World Health Organization, Food and Agricultural Organization of the United Nations, World Organisation for Animal Health. 2018. *Monitoring global progress on addressing antimicrobial resistance: Analysis report of the second round of results of AMR country self-assessment survey 2018.*

Chapter 25
Global Governance of Anti-microbial Resistance: A Legal and Regulatory Toolkit

Calvin W. L. Ho ⓘ and Tsung-Ling Lee

Abstract Recognizing that antimicrobial resistance (AMR) poses a serious threat to global public health, the World Health Organization (WHO) has adopted a Global Action Plan (GAP) at the May 2015 World Health Assembly. Underscoring that systematic misuse and overuse of drugs in human medicine and food production is a global public health concern, the GAP-AMR urges concerted efforts across governments and private sectors, including pharmaceutical industry, medical professionals, agricultural industry, among others. The GAP has a threefold aim: (1) to ensure a continuous use of effective and safe medicines for treatment and prevention of infectious diseases; (2) to encourage a responsible use of medicines; and (3) to engage countries to develop their national actions on AMR in keeping with the recommendations. While the GAP is a necessary step to enable multilateral actions, it must be supported by effective governance in order to realize the proposed aims.

This chapter has a threefold purpose: (1) To identify regulatory principles embedded in key WHO documents relating to AMR and the GAP-AMR; (2) To consider the legal and regulatory actions or interventions that countries could use to strengthen their regulatory lever for AMR containment; and (3) To highlight the crucial role of the regulatory lever in enabling other levers under a whole-of-system approach. Effective AMR containment requires a clearer understanding of how the regulatory lever could be implemented or enabled within health systems, as well as how it underscores and interacts with other levers within a whole-of-system approach.

C. W. L. Ho (✉)
Faculty of Law, The University of Hong Kong, Hong Kong S.A.R., People's Republic of China

Centre for Medical Ethics and Law, Li Ka Shing School of Medicine and Faculty of Law, The University of Hong Kong, Hong Kong S.A.R., People's Republic of China
e-mail: cwlho@hku.hk

T.-L. Lee
Taipei Medical University, Taipei, Taiwan

Shuang-Ho Hospital, New Taipei City, Taiwan
e-mail: tl265@georgetown.edu

© The Author(s) 2020
E. Jamrozik, M. Selgelid (eds.), *Ethics and Drug Resistance: Collective Responsibility for Global Public Health*, Public Health Ethics Analysis 5,
https://doi.org/10.1007/978-3-030-27874-8_25

Keywords Global action plan · Governance · Regulation · Law/legal · Collective action · Stewardship

25.1 Introduction

Antimicrobial resistance (AMR) is widely recognized as a public health threat, responsible for 700,000 deaths worldwide. If the spread of antimicrobial resistance is left unaddressed, it could lead to 10 million additional annual deaths by 2050, according to an estimation by the World Bank. Facilitated by inappropriate uses of medicines to control the spread of infection for human and animal health, antimicrobial resistance also poses long term threat to human development. The United Nations (UN) Secretary-General Ban Ki-Moon describes AMR as a "fundamental threat" to human development at a high-level UN meeting on drug-resistant bacteria (United Nations News Centre 2016). Likewise, recognizing the gravity of antimicrobial resistance on global health, the then Director-General for the World Health Organization (WHO) Margaret Chan characterizes the rise of AMR as a "slow-motion tsunami" (Leatherby 2017). Without an effective global containment strategy, the World Bank warns, the economic impact of AMR makes it unlikely for the world to reach the sustainable development goals set for 2030 (World Bank 2017).

Scientists have long known that microbes can become resistant to medicine. Alexander Fleming, the Nobel laureate for the discovery of penicillin, cautioned the world in his 1945 Nobel acceptance speech of the impending public health crisis (Fleming 1945, at p. 93): "… there is the danger that the ignorant man may easily under dose himself and by exposing his microbes to non-lethal quantities of the drug make them resistant." Since the 1950s the WHO has identified AMR as a global threat, but little progress has been made in improving access to antimicrobials and maintaining their appropriate consumption and effectiveness. Likewise, limited innovation in antimicrobials further compounds the challenge. For these reasons, a broad range of microorganisms have become more resistant to antimicrobials in all parts of the world. An emerging concern is AMR for diseases which affect low and middle income countries (LMICs) disproportionately, such as tuberculosis (TB), malaria and HIV (See Chaps. 2–4). Furthermore, with extensively drug-resistant to tuberculosis now identified in 105 countries, it further raises concerns of a future TB epidemic where limited treatment options are available.

Even though the direct consequences of AMR on human health were beginning to be scientifically well-understood several decades ago, international efforts to address this problem did not begin till the late 1990s and 2000s. The WHO played a key role in catalysing international actions on the issue. It convened a series of consultative groups and expert workshop to assess, evaluate, and develop a series of recommendation for effective containment interventions to garner international attention. This work culminated in the report *WHO Global Strategy for Containment of Antimicrobial Resistance* (World Health Organization 2001). Since this strategy was published, AMR has been discussed at several World Health Assembly meetings, resulting in the adoption of several resolutions such as WHA60.16 concerning the rational use of medicine and WHA62.15 on prevention and control of multidrug-resistant tuberculosis and extensively drug-resistant tuberculosis. At the 2015 World Health Assembly, member states endorsed a *Global Action Plan on Antimicrobial Resistance* (GAP-AMR; World Health Organization 2015a) – which calls for an effective One Health approach – and which was later endorsed through resolutions by the Food and Agriculture Organization of the United Nations (FAO) and World Organisation for Animal Health (OIE). In the same year, AMR was recognized as a threat to the world's sustainability and human development at the UN level. In a landmark UN resolution guiding the global development plan for the next 15 years entitled *Transforming our world: the 2030 Agenda for Sustainable Development* (United Nations 2015), AMR is mentioned but not explicitly set out as a Sustainable Development Goals (SDGs) target. Most recently, the G20 summit reaffirmed the commitment to combat antimicrobial resistance (G20 Leaders' Declaration 2017).

These commitments, both within and outside of the high-level political setting of UN organs, underscore the growing political interest in AMR. Two distinct but interrelated factors explain the recent high-level political attention. First, national governments have a strong self-interest in mitigating the negative impacts of AMR on public health: the estimated economic cost for failing to address the issue would be £66 trillion in lost productivity to the global economy (Public Health England 2015). Second, AMR transcends national borders and exposes a global vulnerability which necessitates collective action at the international level. The shared vulnerability underpinning AMR was recently acknowledged by Tedros Adhanom, the WHO Director-General, in his address to the G20 summit. Urging the world to act, he underscored the interdependence of the world, noting "… vulnerability for one is vulnerability for all of us" (World Health Organization 2017a). German Chancellor Angela Merkel echoed this concern, depicting AMR as akin to a global health security issue of global collective responsibility (Scheuber 2017).

This chapter has a threefold purpose: (1) To identify regulatory principles embedded in key WHO documents relating to AMR and the GAP-AMR; (2) To consider the legal and regulatory actions or interventions that countries could use to strengthen their regulatory lever for AMR containment; and (3) To highlight the crucial role of the regulatory lever in enabling other levers under a whole-of-system approach. In the section that follows, we consider how the WHO and other international bodies have systematically framed this global health issue as a collective action problem; initially by setting out the WHO GAP-AMR. But subsequently, it was quickly recognised that a global plan would not be self-enabling and therefore a global framework has since been proposed by the UN to facilitate implementation by all member states. We set out what we consider to be the core principles that are embedded in the GAP-AMR and the global framework built around it. We then consider the regulatory lever that member states need to establish and apply in order for these principles to effect change at the national level.

25.2 The WHO and AMR

25.2.1 Collective Action Problem

Ilona Kickbusch and David Gleicher (World Health Organization 2012a, b) define global health as health issues which transcend national boundaries and governments and call for actions on the global forces and global flows that determine the health of people. As microbes are capable of penetrating national borders, national efforts are contingent upon, and vulnerable to external actions and forces. It is widely recognised that a collective response is necessary to mitigate the negative consequences of AMR. As no country is capable of addressing the issue without some degree of mutual reliance on others to mount an effective response against AMR, the interdependency makes AMR containment a collective action problem.

Some scholars go further and argue that the containment of AMR is a public good: the benefits from effective containment are enjoyed by all and there is no rivalry in consumption (Smith and Coast 2002). If this is correct, it could present a free rider problem: individual states lack incentive to take the necessary actions and instead, rely on others to act. Arguably, the free rider problem can be addressed through international law (Wernli et al. 2011; Review on Antimicrobial Resistance 2014). Steven Hoffman and Asha Behdinan (2016), Reinl (2016), Susan Rogers Van Katwyk et al. (2016), Christine Årdal et al. (2016), and Asha Behdinan et al. (2015), for instance, argue that countries can be encouraged to act if international law embeds incentives. While many international institutions are involved in addressing the threat posed by AMR, as a starting point, we focus the discussion on the WHO, particularly for its instrumental role in developing the GAP-AMR.

25.2.2 Global Action Plan

Recognizing that systematic misuse and overuse of antimicrobial drugs in human medicine and food production puts every nation at risk, the overarching goal of the GAP-AMR is thus to ensure that the world is able to "treat and prevent infectious diseases with effective and safe medicines that are quality-assured, used in a responsible way, and accessible to all who need them" (World Health Organization 2015a, b, at p. 8). To achieve this goal, the GAP-AMR requires concurrent actions at the national and international levels. While the GAP-AMR is not technically binding, it seeks to harmonise practices across countries while affording regulatory flexibility. To assure policy coherence at the national and international levels, the GAP-AMR provides five objectives to guide and align national and international policy actions: (1) to improve awareness and understanding of AMR through effective communication, education and training; (2) to strengthen the knowledge and evidence base through surveillance and research; (3) to reduce the incidence of infection through effective sanitation, hygiene and infection prevention measures; (4) to optimize the use of antimicrobial medicines in human and animal health; and (5) to develop the economic case for sustainable investment that takes account of the needs of all countries, and increases investment in new medicines, diagnostic tools, vaccines and other interventions. Member states are urged to have national action plans that are aligned with the GAP-AMR within 2 years of the endorsement of the action plan by the World Health Assembly.

Moreover, because excessive human and animal use of antibiotics in multiple settings will have health, economic and security implications beyond national borders, the GAP-AMR embraces a *One Health* approach towards AMR. Defined as a collaborative, multi-sectoral and trans-disciplinary, the *One Health* approach recognizes interconnection between people, animals, plants and the shared environment. This approach calls for sectorial coordination involving human and veterinary medicine, agriculture, finance, environment, and consumers to optimal health outcomes. At the level of international health, horizontal coordination of different UN agencies occurs through the WHO. As the specialized public health agency within the UN agency, the WHO is charged with organizing international responses to shared health challenges, and this responsibility includes acting as "the directing and coordinating authority on international health work" (World Health Organization n. d.). The WHO works with the Strategic and Technical Advisory Group on AMR at the FAO and OIE to develop a framework for monitoring and evaluation of member states' national action plans. Likewise, the FAO, OIE and World Bank are encouraged to put in place and implement action plans in their respective fields. Notably, the regulatory functions bestowed on the WHO are broader in scope than any other international agency in the UN orbit.

As of 2017, more than one third of WHO member states have completed their national action plans on AMR, and a further 62 are in the process of doing so. These national action plans provide a basis for an assessment of the resource needs at national and international levels. The WHO is tasked with publishing biennial progress report on countries' progress in implementing their national action plans. The progress report will also include an assessment of progress made by the FAO, OIE and WHO.

25.2.3   *Limitations of the Global Action Plan*

To be sure, the GAP-AMR provides a good starting point, but the plan lacks concrete goals to compel action. Moreover, the WHO alone cannot be expected to solve the global AMR crisis. For instance, preserving antimicrobial medicines will require a global agreement as to what constitutes 'appropriate use'. Likewise, new financing mechanisms will be needed to incentivise global innovation in antimicrobial medicines. Thus, in the same resolution that endorsed the global action plan, the World Health Assembly (2015a, b) requested the Director-General to develop a global development and stewardship framework to support GAP-AMR in combating AMR. Specifically, the Health Assembly requires the Director-General to (World Health Assembly 2015a, b, Request 7):

develop, in consultation with Member States and relevant partners, options for establishing a global development and stewardship framework to support the development, control, distribution and appropriate use of new antimicrobial medicines, diagnostic tools, vaccines and other interventions, while preserving existing antimicrobial medicines, and promoting affordable access to existing and new antimicrobial medicines and diagnostic tools, taking into account the needs of all countries, and in line with the global action plan on antimicrobial resistance.

The global framework will build on the GAP-AMR, but with specific focuses on preservation of antimicrobial medicines and development of new antimicrobial medicines, diagnostic tools, vaccines and other interventions. According to a report issued by the FAO, OIE and WHO (Food and Agriculture Organization 2017, at p. 4), a global development and stewardship framework would have a threefold goal:

1. Stewardship: Preserving antimicrobial medicines through a stewardship framework covering control, distribution and appropriate use;
2. Research & Development: Developing of new health technologies for preventing and controlling antimicrobial resistance; and
3. Access: Promoting affordable access to existing and new antimicrobial medicines and diagnostic tools.

This framework further encapsulates key principles that have been expounded in earlier initiatives of the WHO. For the purposes of this chapter, we highlight three principles that are of especial pertinence to laws and regulations on pharmaceuticals, which are discussed in the section that follows (for a broader ethical discussion on AMR, see Haire, Chap. 3, this volume; Cheah et al., Chap. 4, this volume). Not necessarily in order of priority, these principles are:

1. Rational and responsible use of antimicrobials

At a practical level, the WHO defines rational use of medicines as patients receiving medications appropriate to their clinical needs, in doses that meet their own individual requirements, for an adequate period of time, and at the lowest cost to them and their community (World Health Organization 1985). This definition extends to the use of antimicrobials, where for instance, irrational use occurs when patients are prescribed and/or take antibiotics (intended to treat bacterial infection) when in fact they have a viral infection. As a matter of public policy, responsible use of antimicrobials is set out as a governing principle that should underpin national governments' efforts to curb AMR. This principle requires governments to ensure that existing activities, capabilities and resources of health system are aligned to ensure patients receive the right dosage of antimicrobials at the right time, use them appropriately and benefit from the usage (World Health Organization 2012a, b).

2. Equitable access to, and appropriate use of, existing and new antimicrobial medicines.

The GAP-AMR recognizes that all countries should have a national action plan on antimicrobial resistance that includes an assessment of resource needs. Moreover, recognizing the need to optimise the use of antimicrobial medicines in human and animal health, the WHO has updated its model list of essential medicine which categories antibiotics into three groups: access, watch and reserve. The access group include antibiotics (considered to have low resistance potential) recommended as first or second choice treatment options for common infections. This group of antibiotics should be widely available at an affordable cost and of assured quality. The watch group of antibiotics are those generally considered as to have higher resistance potential but are recommended as for first or second choice treatment for limited number of indicators. The reserve group consists of 'last-resort' options, or tailored to highly specific patients and setting, and when other alternatives have already failed. International efforts are required in monitoring, reporting the uses of reserve antibiotics to preserve their effectiveness (World Health Organization 2017b, c).

3. Transparency: Data Sharing, Collection and Evaluation

Data sharing, collection and evaluation have been emphasised in order to promote transparency and collaboration. A report by the WHO Secretariat on antimicrobial resistance sets these responsibilities out concisely as (World Health Organization 2015a, at p. 11): "Publishing biennial progress reports, including an assessment of countries and organizations that have plans in place, their progress in implementation, and the effectiveness of action at regional and global levels; and including an assessment of progress made by the FAO, OIE and WHO in implementing actions undertaken within the organizations' tripartite collaboration will also be included in these reports."

Significantly, the call for a global development and stewardship framework was later reiterated at a high-level meeting on antimicrobial resistance at the UN level in 2016 (United Nations 2016). It was the only fourth time that the UN General Assembly convened a high-level meeting on a health issue. Previous meetings – HIV/AIDS, Ebola and non-communicable diseases – catalysed and mobilized political actions at the international level. The UN Political Declaration on antimicrobial resistance was adopted by all 193 member states, signalling a global commitment to combat antimicrobial resistance.

In the discussion so far, we have considered the framing of AMR as a collective action problem the WHO and other international organisations. We have also briefly set out how the GAP-AMR and an enabling global framework have been constructed in response to this problem. More importantly, we have attempted to identify key principles that are embedded in the GAP-AMR, and in respect of which the global framework seeks to give expression to. For ease of reference, we refer to them generally as regulatory principles, because member states need to incorporate them – through laws and regulations – within their health systems. These laws and regulations collectively constitute a regulatory lever that will, at a basic level, enable member states to design, implement and manage policies (particularly pharmaceutical oriented ones) toward responsible use and good stewardship of antimicrobials. These laws and regulations are considered in the next section of the chapter. We then explain why a sound regulatory lever within a whole-of-system approach is critical to enable other levers (financial and information ones in particular) to operate effectively in meeting global AMR objectives.

25.3   Regulatory Leverage for Responsible Use and Good Stewardship of Antimicrobials

The prevalence of AMR is heavily influenced by the way that antimicrobials are consumed. It is now well established that overuse and underuse of antibiotics can lead to resistance (Jamrozik and Selgelid, Chap. 1, this volume). Even so, improper or imprudent use of antibiotics is deeply entrenched within health systems. In order for all member states of the WHO to meet their moral and political commitments set out under the GAP-AMR, it is critical for health systems to be strengthened on all fronts. Clearly the challenge of AMR is a complex one because it is influenced by many different factors and conditions. At the level of health systems, the relational dynamics between prescribers (or suppliers) of antibiotics and patients (or consumers), financial incentives, systemic commitments and characteristics, and the regulatory environment, are arguably the key contributors to AMR. For the purposes of this chapter, we focus on the role of regulation (broadly applied to refer to both legislative and regulatory actions) and examine how it could be used to support the containment of AMR. Many countries do not have a substantially clear and systematic legal and regulatory framework that is specifically directed at AMR (for example, Singh 2017). At a fundamental level, a clear regulatory position on responsible antimicrobial use that simultaneously prioritises effective antimicrobial stewardship is a pre-requisite to a coordinated response in policy decisions and actions across different domains within a health system. In addition, a variety of regulatory interventions should be considered to enable, as well as promote, the use of structural (delivery arrangement), information and financial levers to encourage appropriate use and stewardship of antimicrobials. It is further important for the regulatory environment to be sufficiently open in allowing a combination of top-down and bottom-up actions within a whole-of-system approach.

   Policy discussions on effective stewardship of pharmaceuticals in health systems as a response to the problem of AMR predate the GAP-AMR. In a WHO report (Bigdeli et al. 2014) published ahead of the 2015 World Health Assembly, four main policy objectives with respect to medicines (and antimicrobials) in health systems have been identified as: (1) widely available high-quality medicinal products; (2) equitable access; (3) appropriate and safe use; and (4) affordability. Different policy actions and conditions are matched to each of the four policy objectives in the following manner (Bigdeli et al. 2014, at p. 45):

1. Ensuring availability of quality generic and innovative products:

   • Monitoring product quality;
   • Prequalifying supplies and products;

- Negotiating prices, quality, volume, and supply-chain security;
- Promoting fair competition;
- Engaging in risk sharing agreements;
- Establishing patient access programmes;

2. Improving Equitable Access:

- Understanding utilization profiles;
- Assessing of care seeking behaviour and barriers to care;
- Expanding provider networks;
- Targeting policies and programmes to improve access for vulnerable populations;

3. Encouraging Appropriate Use:

- Implementing and updating standard treatment guidelines;
- Matching essential medicines and reimbursement lists to standard treatment guidelines;
- Assessing provider performance;
- Managing care comprehensively;
- Implementing and monitoring policies to encourage clinically appropriate and cost-effective use;

4. Keeping cost affordable:

- Monitoring routine medicines expenditures by therapeutic area;
- Evaluating health technologies and budget impact;
- Assessing household medicines expenditure;
- Implementing and monitoring policies and programmes to reduce waste and inappropriate use.

Not surprisingly, these policy objectives are closely aligned with what we have identified to be regulatory principles that underscore the GAP-AMR. For instance, rational and responsible use, equitable access and transparency are all necessary conditions to ensure the availability of quality antimicrobials within a health system. However, these policy objectives and their attending actions and conditions inevitably compete in many ways. Price pressures could limit investment in governance infrastructure, where such limitations are typically manifested in weak regulatory capacity, information imbalance, lack of coordination among different stakeholders and perverse incentives whereby irresponsible use is directly supported by direct financial gain. Over time, these practices not only strain health systems by increasing needless consumption, cost and inefficiencies, but also accentuate the global threat of AMR. Within the paradigm of value-based practice, the problem of AMR highlights the urgent need to shift current low-value practices to high-value ones (Porter 2010; Elshaug et al. 2017). A low-value practice is an intervention where evidence suggests that it confers no or very little benefit (to a patient for instance). It also depicts any practice where risk of harm exceeds probable benefit, or where added costs of the intervention do not provide proportional added benefits.

In contrast, a high-value practice is one where evidence suggests it confers benefit on the intervention subject, or probability of benefit exceeds probable harm, or where the added costs of the intervention provide proportional added benefits relative to alternatives. For instance, overprescribing that is incentivised, among other factors, by increased revenue for healthcare providers through greater pharmaceutical sales is a low-value practice that continues to be sustained in many health systems.

While it is beyond the scope of this chapter to address comprehensively the value implications of inappropriate use and poor stewardship of antimicrobials, our intent is to make explicit an implicit understanding that responding to the AMR challenge could be closely linked to addressing many on-going concerns relating to quality of care (World Health Organization 2016). The role and impact of regulation on quality of care have been a longstanding concern among a variety of scholars in different quarters. With limited exception however (notably in the work on refining the working definitions for substandard and falsified medical products (World Health Organization 2017c)), there has not been as much focus on the regulatory lever within health systems on containment of AMR as compared to the financial lever, for example. We hope to address this deficiency by proposing different tools and conditions that could make-up or compose the regulatory lever in relation to each of the four policy objectives identified by the WHO, as well as propose what we consider to be a sufficient open and responsive regulatory environment that could constructively resolve tensions by focusing on higher-value practice when these objectives come into conflict.

25.3.1   Ensuring Quality

A sufficiently robust and up-to-date legal and regulatory framework is necessary to control the quality, safety and efficacy of pharmaceuticals, including antimicrobials. Substandard or degraded antimicrobials, where dosage may be lower or less effective, contribute to therapeutic failure and could thereby encourage the development of drug-resistant strain of pathogens. Similarly, counterfeit antimicrobials could have an adverse effect if the active ingredients include other types of antibiotics (and/or other drugs). In order for such a legal and regulatory framework to be robust, regulatory actions must include accreditation, audit and inspection for the purposes of controlling quality and assessing the safety and efficacy of antimicrobials. For instance, health systems that manufacture antimicrobials must have a legal and regulatory framework to ensure that good manufacturing standards and practices are adhered to. It is not enough to only specify these requirements, but it is just as important for regulatory mechanisms to be in place that can effectively detect deficiencies or deviations from prescribed standards and practices.

Appropriate laws and regulations are also needed to legitimise, implement and sustain policies and programmes that are directed at rational and appropriate antimicrobial use. These policies and programmes generally relate to disease surveillance and management, and standard treatment guidelines. More recently, various measures have been introduced to incentivise the use of high value care through pay-for-performance programmes. These programmes reward healthcare providers for achieving quality, efficiency and "value" by increasing accessibility and appropriate use of drugs that are of proven efficacy. However, evidence of the effectiveness of such programmes are mixed in high-income countries and extremely limited in LMICs. From a regulatory standpoint, pay-for-performance programmes are not self-enabling everywhere but are likewise dependent on a supportive regulatory environment. Broadly speaking, appropriate legal or regulatory principles should be in place to ensure fair bargain, safety and (where appropriate) fair compensation, monitoring and data sharing.

In summary, the following legal and regulatory interventions should be considered in advancing the policy goal of ensuring availability of quality (generic and innovative) antimicrobials:

- Laws and regulations on standards and practices that ensure quality level is achieved (e.g. good manufacturing standards and practices);
- Legally sanctioned practices for monitoring product quality and prequalifying supplies and products (e.g. through licensing, accreditation, audit and inspection);
- Law and regulations that promote fair competition or that enable regulatory action to be taken against anti-competitive practices;
- Legally entrenching disease surveillance and management programmes; and
- Set out legal and regulatory baseline and principles for risk sharing agreements and patient access programmes.

25.3.2 *Improving Prescribing and Dispensing*

In many LMICs, antimicrobials are sold over-the-counter without a prescription or otherwise dispensed by individuals who lack professional training or authority. Even where there may be laws or regulations that proscribe such practices, they may be poorly or inadequately enforced (Singh 2017; World Health Organization 2015b). For instance, accreditation and professional licensing may not specifically target adherence to standard treatment guidelines. Consequently, the failure to adhere to guidance on responsible antimicrobial prescription would not render the healthcare provider professionally accountable or otherwise legally empower a professional body to take remedial action. Responsible prescribing and dispensing practices could also be hampered by weak healthcare infrastructure, expectations of patients and perverse financial incentives. Access to rapid or reliable diagnostic tests may be limited in a low resource health system. This could in turn encourage healthcare providers to veer towards prescribing an antibiotic in order to ensure that the prospect of a bacterial infection is addressed even if there is no reliable diagnosis to that effect. Such a conservative approach could even be a cost effective response (in the short term) if the cost of the antibiotic is lower than to order a laboratory test to validate a diagnosis (World Health Organization 2015b). For this and other reasons, providers may feel obligated to prescribe – while patients may feel entitled to use – antimicrobials, as a quick treatment option. Patients may not be aware of what appropriate use of antimicrobials means, particularly where duration of medical consultation is limited, and could consider themselves to have received substandard care if they have not been prescribed an antibiotic. At a systemic level, financial incentives may encourage overprescribing of antimicrobials. Where pharmaceutical sales generate revenue for healthcare providers and institutions, there would be a perverse financial incentive to overprescribe. Such a practice may be exacerbated where pharmaceutical companies themselves enter into profit-sharing arrangements with these providers or institutions. Additionally, it is currently impossible to determine the extent that antimicrobials are appropriately prescribed and consumed as there is a lack of reliable data across all health systems.

Laws and regulations are necessary to prohibit over-the-counter sale of antimicrobials while ensuring that patients continue to have access through appropriately trained and qualified healthcare professionals. In addition, requiring an appropriate amount of information to be indicated on packaging and to be shared as part of responsible prescribing practice could be given regulatory force. On the former, such a requirement could be taken up as a regulatory measure to ensure high-value or quality use of antimicrobials, particularly where full treatment courses are to be dispensed. Healthcare providers, institutions and professional associations have a crucial role to play in robust guideline development and implementation processes, filling evidence gaps with research, developing high-value practices, and leading or participating in efforts to shift from low-value to high-value practices. In many health systems, these stakeholders do not have sufficient or appropriate legal standing to contribute constructively to policy measure that are directed at improving prescribing and dispensing practices (Singh 2017; World Health Organization 2015b). As noted earlier, professional associations that have an interest in ensuring that standard treatment guidelines are observed by their members may not have any regulatory authority to monitor and improve such practices. Where professional organisations have the capability and motivation to improve professional practices, appropriate laws and regulations could be facilitative of this in a manner that is transparent and accountable. The challenge of perverse financial incentives is perhaps more difficult to surmount, particularly if the health system concerned is committed to particular structural arrangements or values. Legal or regulatory intervention could then be a platform for evaluation, discussion and change. For instance, legal and regulatory changes introduced by South Korea in 2000 to prohibit doctors from dispensing drugs have reduced inappropriate antibiotic prescribing (Kwon 2003; Park et al. 2005).

In summary, the following legal and regulatory interventions could be considered to improve prescribing and dispensing of antimicrobials:

- Prohibit over-the-counter sale of antimicrobials;
- Lend regulatory weight to standard treatment guidelines;

- Empower healthcare institutions and professional associations to improve prescribing and (where applicable) dispensing practices through means that include assessing provider performance;
- Introduce regulation to ensure that care is managed comprehensively;
- Laws and regulations that support implementing and monitoring policies to encourage clinically appropriate and cost-effective use of antimicrobials; and
- Evaluate, remove or manage perverse financial incentives through appropriate laws and regulations.

25.3.3   Ensuring Appropriate, Affordable and Equitable Access

The policy goals of ensuring appropriate, affordable and equitable access to pharmaceuticals (including antimicrobials) are aligned with the WHO's global health initiative on universal health coverage (UHC), broadly directed at promoting access for all to appropriate health services at affordable cost (World Health Assembly Resolution 2005; World Health Organization 2010a, b). Much discussion on UHC has focused on expanding populations covered by national payment systems, although comparatively little information exists on what pharmaceutical are provided, whether they meet the healthcare needs of the population, and how health systems manage pharmaceuticals so that patients receive high-value services at costs that households and systems can afford. When the types of pharmaceuticals provided do not meet population needs, risk protection is inadequate and this does not prevent household impoverishment (Yip and Hsiao 2009; Parry 2012; Wagner et al. 2008).

The economic burden of pharmaceuticals on households is high: they account for nearly half of household healthcare expenditures in 12 Asia-Pacific countries (Wagner and Ross-Degnan 2009) and for all healthcare expenses in four out of 10 households in 22 low- and 17 middle-income countries (Wagner et al. 2011). Pharmaceuticals also constitute a major source of inefficiencies in health systems. Of the ten leading sources of inefficiency in health systems identified in the 2010 World Health Report, pharmaceuticals account for the top three (World Health Organization 2010a). Underuse of generic products, use of substandard and counterfeit medicines, and inappropriate use of medicines waste scarce resources in systems. The World Health Organization estimates that more than half of all medicines globally are prescribed, dispensed, or sold inappropriately (World Health Organization 2009). For instance, overuse of antibiotics to treat acute respiratory tract infections wastes resources and leads to use of higher cost second and third line antibiotics for drug resistant infections. In many LMICs, access to a qualified healthcare professional may cost patients more time and money when compared with inappropriately or illegally obtaining antibiotics over-the-counter or from an unauthorised vendor. To promote appropriate and affordable access, national antibiotics policies and standard treatment guidelines must be supported by essential medicines lists or formularies that encourage rational and responsible use of antimicrobials. In addition, appropriate laws and regulations must be in place to ensure that the antimicrobial supply chain is secure in terms of their procurement, storage and sale (World Health Organization 2015b). These requirements extend to importation requirements and quality inspections for health systems that do not manufacture antimicrobials.

The use of antimicrobials in animals for food production or other purposes will also need to be carefully monitored and regulated. Whereas laws and regulations have conventionally been domain specific particularly in keeping regulations relating to humans distinct from nonhuman animals, the *One Health* approach endorsed in the GAP-AMR highlights the need for a more comprehensive and coordinated approach across the food, veterinary and health sectors. Many countries have yet to establish a regulatory mechanism to enforce requirements for appropriate use of antimicrobials in animals. In addition, there is inadequate infrastructure for monitoring and controlling the development of resistant pathogens in animals, their vertical transmission from one animal species to another, as well as zoonotic transmissions to humans.

Above all, WHO policy documents and guidance (2010a, 2014, 2016, 2017a) have consistently emphasised the importance of promoting equity through greater stakeholder engagement and prioritising the worst off (or otherwise the most vulnerable) in a given society. This could be especially important for decentralised health systems, where inequalities across regions may be great (see also Reid, Chap. 16, this volume). A related concern is that public awareness of appropriate antimicrobial use remains low in most, if not all, health systems. Even within healthcare institutions, infrastructure and human resources may not be adequately equipped to implement and manage infection prevention and control programmes. Equitable access as devised through paradigms such as "accountability for reasonableness" (Daniels and Sabin 2002) is arguably more likely to enable and encourage relevant stakeholders – particularly the broader community – to be interested and proactively involved in national antibiotics policies and related infection prevention and control programmes. As we shall elaborate on below, an equitable regulatory lever is crucial in support bottom-up approaches to promoting high-value use of antimicrobials (Tang et al. 2016).

In summary, the following legal and regulatory interventions could be considered to ensure appropriate, affordable and equitable access to antimicrobials:

- Introducing laws and regulations that implement and sustain infection prevention and control programmes, including appropriate surveillance mechanisms, to understand utilisation profiles, assess care seeking behaviour and barriers to care, and improve access for vulnerable populations;
- Regulation could be the basis of public awareness campaigns and continuing education for stakeholders;
- Laws and regulations may be needed to support monitoring routine medicines expenditures by therapeutic area, evaluating health technologies and budget impact, and assessing household medicines expenditure;
- Implementing and monitoring policies and programmes to reduce waste and inappropriate use through appropriate regulations;

- Requiring appropriate stakeholders' involvement or contribution through regulation; and
- Reduce the use of antibiotics as growth promoters in animals through laws and regulations.

25.4 Regulatory Lever Within a Whole-of-System Approach

In our discussion above, we have considered the different types of legal and regulatory actions or tools that could constitute the regulatory lever, taking into account pharmaceutical policy goals and the GAP-AMR regulatory principles. We have also noted that the regulatory lever is but one of other levers that are available to policy-makers, two of such levers being financial and information. In this section, we broadly explain why the regulatory lever underscores the effectiveness of these two other levers within a "whole-of-system" approach that is directed at AMR containment. By this approach, we adopt the WHO's emphasis that focus should not be limited to a particular component of a health system – broadly defined to mean "all organizations, people and actions whose primary intent is to promote, restore and maintain health"- but to recognise that different systemic components are interrelated and interact in ways that may be anticipated or unanticipated (World Health Organization 2010a; b, at p. 19).

25.4.1 Financial Lever

The financial lever could be thought of as being constituted by financial schemes that include budget controls, tax and incentive arrangements, and also the policies and actions of healthcare purchasers or payers, particularly social insurers (Bigdeli et al. 2014). Ideally, financing schemes should be designed to support decision-making through provision of information on demographic characteristics, healthcare needs and utilisation patterns of its users, and also of healthcare providers – particularly prescribing patterns and related costs. Implementers of financial schemes exert a degree of financial control over patients and healthcare providers in terms of what they pay for, and could shape patient demand through both financial and educational means. Additionally, financial incentives should encourage cost-effective prevention and care, while financial commitments should be directed at meeting infrastructural requirement, such as surveillance mechanisms that allow the use of international and local data on disease burden and utilisation patterns to signal potential inappropriate use patterns. In reality, financial interventions, like expenditure-focused policy instruments, tend to lack specificity and often have unintended effects. For instance, a cap on funding does not necessarily encourage clinically appropriate use or otherwise reduce wasteful spending as a result of over-treatment.

Within a whole-of-system approach, the financial lever should be applied together with the regulatory lever to support the establishment of a sound information environment, by making available evidence-based clinical guidelines and economic assessments that include health technology assessment and budget impact. As we have noted above, this is crucial in overcoming the current challenge that too little information on monitoring and evaluation activities is available, primarily due to lack of mechanisms in many health systems to monitor antimicrobial prescription and use. While some information is available on how medicines are financed in these health systems, there is little information to reliably determine equitable access and appropriate use.

25.4.2   Information Lever

The information lever of many health systems is disproportionately focused on price rather than on appropriate use and good stewardship. There is a need to review, revise and develop information systems to collect information that will enable policy-makers to target policies that improve prescribing practices, carry out audits and conduct education programmes. Such a system should also be able combine information from different parts of the healthcare system. Regular samples of paper-based facility prescribing and dispensing records can provide information on utilization to inform policy decisions (World Health Organization 2015b).

Policies that ensure appropriate use of antibiotics in a manner that is effective, safe, equitable and efficient depend on the availability of information including population demographics, disease epidemiology, treatment approaches, and political and economic environments. Health systems will need to be capable of generating routine, up-to-date information about the type of antibiotics that are needed by patients, which antibiotics are being used and how they are used across different patient populations within the health system, who prescribed them, whether they are clinically appropriate (such as in addressing the disease burden faced by the population) and the cost spent. Without such information, it will be difficult to determine if quality of care is provided. Ideally, information systems should capture details on the antibiotics use and expenditure, along with quality of care (such as the percentage of primary care patients receiving antibiotics), and details of misuse. However, these details are usually not captured by data collection systems. For instance, where providers are paid through bundled-payment arrangements (whether case or episode-based), the information system may not be designed to capture data on the type and amount of antibiotics prescribed, since payment does not depend on such information. As we have noted above, appropriate laws and regulations could help bring about changes to information systems that prioritises AMR containment.

Constant monitoring, feedback and evaluation are important to ensure that levers continue to achieve desired goals. Crucially, member states will need to ensure that the whole-of-system approach to AMR containment could be implemented bottom-up and top-down (Elshaug et al. 2017). Bottom-up actions require stakeholders who are not in any formal positions of authority to change practices that are within their sphere of influence. Such stakeholders could be patients, clinicians, cooperatives and agricultural producers (see Schwenkenbecher, Chap. 23, this volume and Oakley, Chap.8 in this volume). In contrast, top-down actions have wider impact and the drivers of change typically include governments, professional bodies and third party payers. The use of the regulatory lever, along with other levers, should enable as well as facilitate a combination of both bottom-up and top-down actions to improve policies and practices relating to antibiotics access, use and stewardship.

25.5 Conclusion

In this chapter, we have considered how AMR containment has become a collective action problem, perhaps most comprehensively mapped out in the GAP-AMR. While the GAP-AMR itself lacks legal or regulatory force, there are at least three regulatory principles that could be drawn from it. These regulatory principles in turn require countries to adopt a variety of legal and regulatory actions or interventions that may be necessary to strengthen their regulatory lever for AMR containment. We have attempted to explicate these legal and regulatory actions in terms of four pharmaceutical policy objectives that have been articulated in a number of WHO documents and initiatives. Finally, we highlighted the crucial role of the regulatory lever in implementing the GAP-AMR, and also in enabling other levers under a whole-of-system approach.

Policies on AMR in health systems need to be responsive to shifting contexts and goals. Such adaptations must be informed by the best available evidence of what works, for whom, how and why in a given situation. In addition, routine monitoring and periodic evaluations of the impacts are necessary, and they are further crucial to ensuring quality, appropriate use and good stewardship of antimicrobials. The regulatory lever could and arguably should be applied to introduce, guide, scale-up, adapt, adjust or terminate policies on AMR containment. While it is not disputed that the regulatory lever is generally recognised to be important, there has been relatively little attention as to what it means in terms of laws and regulations that could be directed at AMR containment. If the GAP-AMR is to be effectively enabled, this lacuna that we have highlighted will require greater attention.

References

Årdal, Christine, Kevin Outterson, Steven J. Hoffman, Abdul Ghafur, Mike Sharland, Nisha Ranganathan, Richard Smith, et al. 2016. International cooperation to improve access to and sustain effectiveness of antimicrobials. *The Lancet* 387 (10015): 296–307. https://doi.org/10.1016/S0140-6736(15)00470-5.

Behdinan, Asha, Steven J. Hoffman, and Mark Pearcey. 2015. Some global policies for antibiotic resistance depend on legally binding and enforceable commitments. *The Journal of Law, Medicine & Ethics* 43 (suppl. 3): 68–73. https://doi.org/10.1111/jlme.12277.

Bigdeli, Maryam, David H. Peters, and Anita K. Wagner, eds. 2014. *Medicines in health systems: Advancing access, affordability and appropriate use.* Geneva: World Health Organization. Available at: http://www.who.int/alliance-hpsr/resources/FR_webfinal_v1.pdf.

Cheah, Phaik-Yeong, Nicholas Day, and Michael Parker. 2020. Ethics and antimalarial drug resistance. In *Ethics and drug resistance: Collective responsibility for global public health*. Cham: Springer.

Daniels, Noman, and James E. Sabin. 2002. *Setting limits fairly: Can we learn to share medical resources?* New York/Oxford: Oxford University Press.

Elshaug, Adam G., Meredith B. Rosenthal, John N. Lavis, Shannon Brownlee, Harald Schmidt, Somil Nagpal, Peter Littlejohns, Divya Srivastava, Sean Tunis, and Vikas Saini. 2017. Levers for addressing medical underuse and overuse: Achieving high-value health care. *Lancet* 390: 191–202. https://doi.org/10.1016/S0140-6736(16)32586-7.

Fleming, Alexander. 1945. Penicillin: Nobel Lecture, December 11, pp. 83–93. Available at: https://www.nobelprize.org/uploads/2018/06/fleming-lecture.pdf

Food and Agriculture Organization of the United Nations, World Organisation for Animal Health, World Health Organization. 2017. *Global framework for development & stewardship to combat antimicrobial resistance*. WHO/EMP/IAU/2017.08. Available at: http://www.who.int/phi/implementation/research/WHA_BackgroundPaper-AGlobalFrameworkDevelopmentStewardship-Version2.pdf.

G20 Leaders' Declaration: Shaping an interconnected world. Hamburg, 7/8 July. Available at: https://www.g20.org/gipfeldokumente/G20-leaders-declaration.pdf.

Haire, Bridget. 2020. Antiretroviral resistance – Ethical issues. In *Ethics and drug resistance: Collective responsibility for global public health*. Cham: Springer.

Hoffman, Steven J., and Asha Behdinan. 2016. Towards an international treaty on antimicrobial resistance. *Ottawa Law Review* 47 (2): 503. Available at: https://commonlaw.uottawa.ca/ottawa-law-review/sites/commonlaw.uottawa.ca.ottawa-law-review/files/olr_47-2_09_hoffman_behdinan_final.pdf.

Jamrozik, Euzebiusz, and Michael Selgelid. 2020. Drug-resistant infection: Causes, consequences, and responses. In *Ethics and drug resistance: Collective responsibility for global public health*. Cham: Springer.

Kwon, Soonman. 2003. Pharmaceutical reform and physician strikes in Korea: Separation of drug prescribing and dispensing. *Social Science & Medicine* 57: 529–538. https://doi.org/10.1016/S0277-9536(02)00378-7.

Leatherby, Lauren. 2017. The rise of antibiotic-resistant infections threatens economies. *Financial Times* July 7. Available at: https://www.ft.com/content/1a3b06fa-57ff-11e7-80b6-9bfa4c1f83d2.

Michael E. Porter. 2010. What Is Value in Health Care?. *New England Journal of Medicine* 363 (26): 2477–2481

Park, Sylvia, Stephen B. Soumerai, Alyce S. Adams, Jonathan A. Finkelstein, Sunmee Jang, and Dennis Ross-Degnan. 2005. Antibiotic use following a Korean national policy to prohibit medication dispensing by physicians. *Health Policy and Planning* 20 (5): 302–309. https://doi.org/10.1093/heapol/czi033.

Parry, Jane. 2012. Nine of 10 Chinese are covered by medical insurance, but access to treatment remains a problem. *BMJ* 344: e248. https://doi.org/10.1136/bmj.e248.

Public Health England. 2015. *Health matters: Antimicrobial resistance*. Available at: https://www.gov.uk/government/publications/health-matters-antimicrobial-resistance/health-matters-antimicrobial-resistance.

Reid, Lynette. 2020. Antimicrobial resistance and social inequalities in health: Considerations of justice. In *Ethics and drug resistance: Collective responsibility for global public health*. Cham: Springer.

Reinl, James. 2016. UN declaration on antimicrobial resistance lacks targets. *The Lancet* 388 (10052): 1365. https://doi.org/10.1016/S0140-6736(16)31769-X.

Review on Antimicrobial Resistance. 2014. *Antimicrobial resistance: Tackling a crisis for the health and wealth of nations*. Available at: http://www.jpiamr.eu/wp-content/uploads/2014/12/AMR-Review-Paper-Tackling-a-crisis-for-the-health-and-wealth-of-nations_1-2.pdf.

Rogers Van Katwyk, Susan, Marie Evelyne Danik, Ioana Pantis, Rachel Smith, John-Arne Røttingen, and Steven J. Hoffman. 2016. Developing an approach to assessing the political feasibility of global collective action and an international agreement on antimicrobial resistance. *Global Health Research and Policy* 1 (1): 20. https://doi.org/10.1186/s41256-016-0020-9.

Scheuber, Andrew. 2017. *Angela Merkel discusses global antibiotic challenge with Imperial academic*. Imperial College London. Available at: http://www3.imperial.ac.uk/newsandeventspggrp/imperialcollege/newssummary/news_23-2-2017-15-42-19.

Singh, Poonam. 2017. One health approach to tackle antimicrobial resistance in South East Asia. *BMJ* 358: 3625. https://doi.org/10.1136/bmj.j3625.

Smith, Richard D., and Joanna Coast. 2002. Antimicrobial resistance: A global response. *Bulletin of the World Health Organization* 80 (2): 126–133. Available at: http://www.who.int/iris/handle/10665/71062.

Tang, Yuqing, Chaojie Liu, and Xinping Zhang. 2016. Public reporting as a prescriptions quality improvement measure in primary care settings in China: Variations in effects associated with diagnoses. *Scientific Reports* 6: 39361. https://doi.org/10.1038/srep39361.

United Nations. 2015. *Transforming our world: The 2030 agenda for sustainable development*. A/RES/70/1. Available at: https://sustainabledevelopment.un.org/content/documents/21252030%20Agenda%20for%20Sustainable%20Development%20web.pdf.

———. 2016. *Press release: High-level meeting on antimicrobial resistance*. Available at: http://www.un.org/pga/71/2016/09/21/press-release-hl-meeting-on-antimicrobial-resistance/.

United Nations News Centre. 2016. *At UN, global leaders commit to act on antimicrobial resistance*. September 21. Available at: http://www.un.org/apps/news/story.asp?NewsID=55011#.WcDFn9FpHb0.

Wagner, Anita K., and Dennis Ross-Degnan. 2009. The potential for insurance systems to increase access to and appropriate use of medicines in Asia-Pacific countries. In *Prescribing cultures and pharmaceutical policy in the Asia-Pacific*, ed. Karen Eggleston. Stanford: Walter H. Shorenstein Asia-Pacific Research Center.

Wagner, Anita K., Madeleine Valera, Amy J. Graves, Sheila Laviña, and Dennis Ross-Degnan. 2008. Costs of hospital care for hypertension in an insured population without an outpatient medicines benefit: An observational study in the Philippines. *BMC Health Services Research* 8: 161. https://doi.org/10.1186/1472-6963-8-161.

Wagner, Anita K., Amy J. Graves, Sheila K. Reiss, Robert LeCates, Fang Zhang, and Dennis Ross-Degnan. 2011. Access to care and medicines, burden of health care expenditures, and risk protection: Results from the World Health Survey. *Health Policy* 100 (2–3): 151–158. https://doi.org/10.1016/j.healthpol.2010.08.004.

Wernli, Didier, Thomas Haustein, John Conly, Yehuda Carmeli, Ilona Kickbusch, and Stephan Harbarth. 2011. A call for action: The application of the international health regulations to the global threat of antimicrobial resistance. *PLoS Medicine* 8 (4): e1001022. https://doi.org/10.1371/journal.pmed.1001022.

World Bank. 2017. *Drug-resistant infections: A threat to our economic future*. Washington, DC: World Bank. Available at: http://documents.worldbank.org/curated/en/455311493396671601/pdf/114679-REVISED-v1-Drug-Resistant-Infections-Executive-Summary.pdf.

World Health Assembly Resolution. 2005. *WHA58.33 – Sustainable health financing, universal coverage and social health insurance*. WHA Resolution: Fifty-eighth World Health Assembly. Available at: http://apps.who.int/medicinedocs/documents/s21475en/s21475en.pdf.

———. 2015a. *Antimicrobial resistance: Draft global action plan on antimicrobial resistance*. A68/20. Available at: http://apps.who.int/gb/ebwha/pdf_files/WHA68/A68_20-en.pdf.

———. 2015b. *WHA68.7 – Global action plan on antimicrobial resistance*. WHA Resolution: Sixty-eighth World Health Assembly. Available at: http://apps.who.int/medicinedocs/en/d/Js21889en/.

World Health Organization. 1985. *The rational use of drugs: Report of the conference of experts. Nairobi, 25-28 November 1985*. Geneva: World Health Organization. Available at: http://apps.who.int/medicinedocs/documents/s17054e/s17054e.pdf.

———. 2001. *WHO global strategy for containment of antimicrobial resistance*. Geneva: World Health Organization. Available at: http://www.who.int/drugresistance/WHO_Global_Strategy_English.pdf.

———. 2009. *Medicines use in primary care in developing and transitional countries*. Geneva: World Health Organization. Available at: http://www.who.int/medicines/publications/who_emp_2009.3/en/index.html.

———. 2010a. *The world health report. Health systems financing: The path to universal coverage*. Geneva: World Health Organization. Available at: http://www.who.int/whr/2010/en/index.html.

———. 2010b. *Western Pacific Regional strategy for health systems based on the values of primary health care*. Available at: http://www.wpro.who.int/topics/health_systems/wpro_strategy_health_systems_primary_health_care.pdf.

———. 2012a. *Governance for health in the 21st century*. Geneva: World Health Organization. Available at: http://www.euro.who.int/__data/assets/pdf_file/0019/171334/RC62BD01-Governance-for-Health-Web.pdf.

———. 2012b. *The pursuit of responsible use of medicines: Sharing and learning from country experiences*. Geneva: World Health Organization. Available at: http://apps.who.int/iris/bitstream/10665/75828/1/WHO_EMP_MAR_2012.3_eng.pdf.

———. 2014. *Making fair choices on the path to universal health coverage*. Geneva: World Health Organization. Available at: http://www.who.int/choice/documents/making_fair_choices/en/.

———. 2015a. *Global action plan on antimicrobial resistance*. Geneva: World Health Organization. Available at: http://www.who.int.libproxy1.nus.edu.sg/antimicrobial-resistance/publications/global-action-plan/en/.

———. 2015b. *Antimicrobial resistance in the Western Pacific Region: A review of surveillance and health systems response*. Geneva: World Health Organization. Available at: http://www.wpro.who.int/entity/drug_resistance/documents/amr_wpr.pdf.

———. 2016. *Health care without avoidable infections: The critical role of infection prevention and control*. Geneva: World Health Organization. Available at: http://apps.who.int.libproxy1.nus.edu.sg/iris/bitstream/10665/246235/1/WHO-HIS-SDS-2016.10-eng.pdf.

———. 2017a. *Health emergencies represent some of the greatest risks to the global economy and security: Remarks delivered by Dr Tedros Adhanom Ghebreyesus to G20 8 July 2017*. Available at: http://www.who.int/dg/speeches/2017/g20-summit/en/.

———. 2017b. *Report of the 21st WHO expert committee on the selection and use of essential medicines*. Geneva: World Health Organization. Available at: http://www.who.int/medicines/publications/essentialmedicines/EML_2017_ExecutiveSummary.pdf?ua=1.

———. 2017c. *WHO global surveillance and monitoring system for substandard and falsified medical products*. Geneva: World Health Organization. Available at: https://www.who.int/medicines/regulation/ssffc/publications/gsms-report-sf/en/.

———. n.d. *About WHO: What we do*. Available at: http://www.who.int/about/what-we-do/en/.

Yip, Winnie, and William C. Hsiao. 2009. Non-evidence-based policy: How effective is China's new cooperative medical scheme in reducing medical impoverishment? *Social Science and Medicine* 68: 201–209. https://doi.org/10.1016/j.socscimed.2008.09.066.

Chapter 26
The Super-Wicked Problem
of Antimicrobial Resistance

Jasper Littmann, A. M. Viens, and Diego S. Silva

Keywords Complex systems · Social policy · Bioethics · Political philosophy

Antimicrobial resistance (AMR) – the progressive process by which microbes, such as bacteria, through evolutionary, environmental and social factors develop the ability to become resistant to drugs that were once effective at treating them – is a threat from which no one can escape. It is one of the largest threats to clinical and global health in the twenty-first century – inflicting monumental health, economic and social consequences.[1] All persons locally and globally, and even all future persons yet to come into existence, all suffer the shared, interdependent vulnerability to this threat that will have a substantial impact on all aspects of our lives. For example,

---

[1] J O'Neill, et al., *Infection Prevention, Control and Surveillance: Limiting the Development and Spread of Drug Resistance* (London: Review on Antimicrobial Resistance, 2016) https://amr-review.org/sites/default/files/Health%20infrastructure%20and%20surveillance%20final%20version_LR_NO%20CROPS.pdf (accessed September 29, 2017); President's Council of Advisors on Science and Technology, *Report To The President On Combating Antimicrobial Resistance* (Washington: PCAST, 2014); World Health Organization, *The Evolving Threat of Antimicrobial Resistance: Options for Action* (Geneva: WHO Press, 2012); World Economic Forum, *Global Risks 2013 – Insight Report, Eighth Edition* (Geneva, 2013); S.C. Davies, J. Grant and M. Catchpole, *The Drugs Don't Work: A Global Threat* (London: Penguin, 2013); O.O. Adeyi, et al., *Drug-Resistant Infections: A Threat to Our Economic Future* (Washington, DC: World Bank, 2017).

J. Littmann (✉)
Norwegian Institute of Public Health, Oslo, Norway
e-mail: jasper.littmann@fhi.no

A. M. Viens
School of Global Health, York University, Toronto, Canada
e-mail: amviens@yorku.ca

D. S. Silva
Sydney Health Ethics, School of Public Health, University of Sydney,
Sydney, NSW, Australia
e-mail: diego.silva@sydney.edu.au

© The Author(s) 2020
E. Jamrozik, M. Selgelid (eds.), *Ethics and Drug Resistance: Collective Responsibility for Global Public Health*, Public Health Ethics Analysis 5,
https://doi.org/10.1007/978-3-030-27874-8_26

while reliable data are hard to find, the European Centre for Disease Prevention and Control (ECDC) has conservatively estimated that, in Europe alone, AMR causes additional annual cost to health care systems of at least €1.5 billion, and is responsible for around 25,000 deaths per year. Furthermore, AMR significantly increases the cost of treating bacterial infections with an increase in length of hospital stays and average number of re-consultations, as well as the resultant lost productivity from increased morbidity.[2] With a combined cost of up to $100 trillion to the global economy – pushing a further 28 million people into extreme poverty – this is one of the most pressing challenges facing the world.[3] Most troublingly, if we do not succeed in diminishing the progression of AMR, there is the very real potential for it to threaten common procedures and treatments of modern medicine, including the safety and efficacy of surgical procedures and immunosuppressing chemotherapy.[4] Some experts are warning that we may soon be ushering in a post-antibiotic area.[5]

26.1 Challenges in Responding to AMR

There exists a multitude of policy responses to AMR at the local, national and international levels.[6] Unfortunately, their success to date has only been limited.[7] This may partly be due to the fact that many microorganisms are highly adaptable and constantly evolving, thereby presenting a perpetual challenge to clinicians, researchers, public health professionals and policy-makers.[8] However, the formulation and

---

[2] L.L. Maragakis, E.N. Perencevich and S.E. Cosgrove, 'Clinical and Economic Burden of Antimicrobial Resistance,' *Expert Review of Anti-infective Therapy* 6(2008): 751–63; F. Alam, et al. 'The additional costs of antibiotics and re-consultations for antibiotic-resistant Escherichia coli urinary tract infections managed in general practice,' *International Journal of Antimicrobial Agents* 33(2009): 255–57; R. Smith and J. Coast, 'The true cost of antimicrobial resistance,' *BMJ* 2013; 346: f1493; Centers of Disease Control and Prevention, *Antibiotic Resistance Threats in the United States, 2013* (Atlanta: Centers of Disease Control and Prevention, 2013)

[3] J. O'Neill et al., *Tackling Drug-Resistant Infections Globally: Final Report and Recommendations* (London: Review on Antimicrobial Resistance, 2016), available at https://amr-review.org/sites/default/files/160518_Final%20paper_with%20cover.pdf (accessed September 29, 2017).

[4] A. Teillant, S. Gandra, D. Barter, D.J. Morgan, and R. Laxminarayan, 'Potential burden of antibiotic resistance on surgery and cancer chemotherapy antibiotic prophylaxis in the USA: a literature review and modelling study,' *Lancet Infectious Diseases* 15(2015): 1429–37.

[5] S.C. Davies, J. Grant and M. Catchpole, *The Drugs Don't Work: A Global Threat* (London: Penguin, 2013).

[6] World Health Organization, *Global Strategy for Containment of Antimicrobial Resistance* (Geneva: WHO, 2001); Infectious Diseases Society of America, 'The 10 × '20 Initiative: Pursuing a Global Commitment to Develop 10 New Antibacterial Drugs by 2020,' *Clinical Infectious Diseases* 8(2010): 1081–83; Bundesministerium für Gesundheit, *Deutsche Antibiotikaresistenz-Strategie* (Berlin, 2011); Department of Health, *UK Five Year Antimicrobial Resistance Strategy 2013 to 2018* (London, 2013)

[7] World Health Organization, *Antimicrobial Resistance – Global Report on Surveillance* (Geneva: WHO Press, 2014)

[8] A.S. Fauci and D.M. Morens, 'The Perpetual Challenge of Infectious Diseases,' *New England Journal of Medicine* 366(2012): 454–61.

implementation of effective polices are also compounded by the complexity of the challenge. This complexity pertains to the number of local, national and international stakeholders involved, the difficulty in establishing successful collaboration and coordination mechanisms across different policy areas, and the numerous inter-related drivers that make this a problem at the global scale.[9]

The number of stakeholders who contribute to the emergence of AMR is extraordinarily large.[10] Misuse and over-prescription of antibiotics, for example, are driven not only by health care professionals but also consumers – many of whom will not use antibiotics as instructed, often without being aware of the potential consequences of their actions.[11] Other patient groups entirely lack access to appropriate antimicrobial treatment, especially in poorer or less developed settings.[12] Where treatment is either unaffordable for many or infrastructure is lacking, people are more likely to self-medicate or buy counterfeit drugs through unofficial channels, increasing the chance that their drugs are less effective or unsuitable for treating the infection.[13] Beyond human usage, the pervasive use of antibiotics outside of medical settings also remains a key source of resistance. Indeed, it is estimated that more antibiotics are given to animals than consumed by humans – and, in many instances, for non-therapeutic use.[14] The wide use of antibiotics within agricultural, aquaculture, horticultural and animal farming industries all provide multiple opportunities for the increase incidence and prevalence of antibiotic-resistant bacteria.[15]

AMR does not, therefore, squarely fall into a single policy domain and any effective policy will require collaboration among a wide range of experts, such as clinicians, veterinarians, microbiologists, pharmacologists, epidemiologists, lawyers,

---

[9] R. Laximinarayan, et al., 'Antibiotic resistance - the need for global solutions,' *Lancet Infectious Diseases* 13(2013): 1057–98.

[10] W. Albrich, D. Monnet, S. Harbarth, 'Antibiotic Selection Pressure and Resistance in Streptococcus pneumoniae and Streptococcus pyogenes,' *Emerging Infectious Diseases* 10(2004): 514–17.

[11] C.C. Butler, et al., 'Understanding the culture of prescribing: qualitative study of general practitioners' and patients' perceptions of antibiotics for sore throat,' *BMJ* 317(1998): 637–42; I. Björkman, et al., 'Infectious disease management in primary care: perceptions of GPs.' BMC Fam Pract, 12:1 (2011); N. Britten, 'Patients' expectations of consultations', *BMJ*, 328 (2004):416–17

[12] M. Mendelson, et al. 'Maximising access to achieve appropriate human antimicrobial use in low-income and middle-income countries,' *Lancet* 387(2015):188–98; R. Laxminarayan, et al., 'Access to effective antimicrobials: a worldwide challenge,' *Lancet* 387(2016): 168–175.

[13] N. Gualde, *Resistance: The human struggle against infection* (New York: Dana Press, 2006); R. Laxminarayan and D.L. Heymann, 'Challenges of drug resistance in the developing world,' *BMJ* 344(2012): e1567.

[14] J. O'Neill, et al., *Antimicrobials in agriculture and the environment: reducing unnecessary use and waste* (London: Review on Antimicrobial Resistance, 2016) https://amr-review.org/sites/default/files/Antimicrobials%20in%20agriculture%20and%20the%20environment%20-%20Reducing%20unnecessary%20use%20and%20waste.pdf (accessed September 12, 2017).

[15] Food and Agriculture Organization of the United Nations, *Drivers, Dynamics and Epidemiology of Antimicrobial Resistance in Animal Production* (Geneva: WHO, 2016); R.W. Meek RW, H. Vyas and L.J.V. Piddock, 'Nonmedical Uses of Antibiotics: Time to Restrict Their Use?' *PLoS Biology* 13(2015): e1002266. https://doi.org/10.1371/journal.pbio.1002266.

philosophers, economists and public health professionals.[16] This requires high degrees of multi-sectorial integration, coordination and accountability mechanisms, for which existing policy-making structures are inappropriate or insufficient.[17] This is further compounded by the fact that AMR has numerous interdependent biological and social drivers that make it a problem on the global scale. Expanding international travel, tourism and trade will continue to perpetuate resistance beyond current levels.[18] As such, while national and regional policies can certainly impact on the prevalence of drug-resistant infections,[19] the global dissemination of resistant bacteria demonstrate why such localised efforts alone cannot be ultimately effective.[20]

As a result of all of these considerations, AMR also raises distinctive ethical issues, which must not only be accounted for in our policy and response activities, but will play an important role in supporting the numerous difficult choices involved in balancing the benefits and burdens associated with protecting antibiotic effectiveness and reducing the spread of drug-resistant infections.[21] Our traditional normative theories and principles, as developed in other infection prevention and control contexts, will be insufficient if mechanically applied as if it were just another problem of infectious disease ethics. It will require careful attention to the morally-relevant features of what makes AMR a distinctive problem, and due care and context-specificity in the application of moral guidance and regulation. The ethical analysis required to shape and guide our policy response to the problem of AMR

---

[16] S.B. Levy, *The Antibiotic Paradox* (Cambridge, MA: Perseus Publishing, 2002).

[17] R.D. Smith and J. Coast, 'Antimicrobial resistance: a global response,' *Bulletin of the World Health Organization* 80(2002): 126–133; R. Laxminarayan, et al., 'Antibiotic resistance-the need for global solutions,' *Lancet Infectious Diseases* 13(2013): 1057–98; C. Årdal, et al., 'International cooperation to improve access to and sustain effectiveness of antimicrobials,' *Lancet* 387(2015): 296–307; S. J. Hoffman and T. Ottersen, 'Addressing Antibiotic Resistance Requires Robust International Accountability Mechanisms,' *The Journal of Law, Medicine and Ethics* 43(2015): 53–64.

[18] D.W. MacPherson, B.D. Gushulak and W.B. Baine, 'Population Mobility, Globalization, and Antimicrobial Drug Resistance,' *Emerging Infectious Diseases* 17(2009): 1727–1732; Z. S. Ahammad, et al., 'Increased Waterborne blaNDM-1 Resistance Gene Abundances Associated with Seasonal Human Pilgrimages to the Upper Ganges River,' *Environmental Science & Technology* 48(2014): 3014–3020; A.H. Holmes, L.S. Moore and A. Sundsfjord, 'Understanding the mechanisms and drivers of antimicrobial resistance,' *Lancet* 387(2016): 176–87.

[19] For instance, as illustrated by the significant variation in the proportion of methicillin-resistant Staphylococcus aureus (MRSA) infections across European hospitals. See, e.g., European Centre for Disease Prevention and Control, *Annual epidemiological report on communicable diseases in Europe* (Stockholm: ECDC, 2011).

[20] I.N. Okeke and R. Edelman, 'Dissemination of Antibiotic-Resistant Bacteria across Geographic Borders,' *Clinical Infectious Diseases* 33(2001): 364–369; S.E. Majowicz, et al., 'The Global Burden of Nontyphoidal *Salmonella* Gastroenteritis,' *Clin Infect Dis* 50(2010): 882–89; L.S. Tzouvelekis, et al., 'Carbapenemases in Klebsiella pneumoniae and Other Enterobacteriaceae: an Evolving Crisis of Global Dimensions,' *Clinical Microbiology Reviews.* 25(2012): 682–707; Nicola K. Petty, et al. 'Global dissemination of a multidrug resistant Escherichia coli clone,' *PNAS* 111(2014): 5694–5699.

[21] J. Littmann and A.M. Viens, 'The Ethical Significance of Antimicrobial Resistance,' *Public Health Ethics* 8(2015): 209–224.

also necessitates thinking and theorizing that is able to incorporate and reflect the various biological, social, political and legal factors involved in the spread and control of drug-resistant infections.

In this paper, we will argue that the concept of super-wicked problems can provide an apt description of the current situation regarding AMR, and that it will help us better understand some of the complex ethical challenges associated with AMR. Furthermore, framing AMR as a super-wicked problem will help to explain why at first glance, AMR appears to be so similar to other large policy challenges, such as climate change. However, we will argue that the structural similarity should not lead us to conclude that attempts to mitigate one super-wicked problem can simply be transferred to another. Instead, we suggest that the crucial feature of super-wicked problems is the need to change *path dependency* – a term we will explain more fully below – and that this change will likely require unique approaches, methods and tools for each super-wicked problem. To develop this argument, we will first introduce the ideas of super-wicked problems as a way to frame AMR. Second, we will argue that conceptualizing AMR as a super-wicked problem can help inform policy making by highlighting how our efforts should be focused. Finally, in sections four and five, we provide a potential way to move forward, and highlight key ethical issues that arise in this context.

26.2   Framing AMR as a Super-Wicked Problem

The concept of wicked problems, which has its origins in Rittel and Webber's paper about planning theory, describes certain policy problems as complex challenges that do not respond to standard problem-solving mechanisms.[22] Rittel and Webber suggest that the success of social policy in the 19th and early twentieth century was essentially the picking of low-hanging fruits, and that the policy challenges that societies are now facing are much more difficult to address.[23] Inherent complexity (i.e., the overlapping and varied sources of influence or causes to a social problem that resist simple linear solutions),[24] inter-relatedness with other policy fields, and several conflicting goals that might each be reasonably pursued, make some policy problems "wicked". According to Rittel and Webber, these problems are not solvable by traditional instruments of policy making, especially not by cost-benefit or

---

[22] H.W.J. Rittel and M.M. Webber, 'Dilemmas in a General Theory of Planning,' *Policy Sciences* 4:2 (1973): 155–69.

[23] H.W.J. Rittel and M.M. Webber, 'Dilemmas in a General Theory of Planning,' *Policy Sciences* 4:2 (1973): 155–69.

[24] M.C. Jackson, *Systems Thinking: Creative Holism for Managers* (West Sussex, UK: John Wiley and Sons, Ltd., 2003).

system analysis, as these approaches cannot make sense of all dimensions of wicked problems.[25]

Wicked problems are characterized by the fact that proposed solutions cannot be judged as 'right' or 'wrong' but merely as more or less suitable, and they do not allow for a trial-and-error approach to policy-making. Instead, Rittel and Webber argue, wicked problems only allow policy-makers a single shot at solving the problem – if this fails, the unsuccessful policy will have changed the original problem to such an extent its initial alternatives may no longer be viable contenders.[26]

The concept of wicked problems has been applied to a number of areas as diverse as coastal governance,[27] liberal arts and design[28] and, more recently, climate change. It has been argued, however, that the latter no longer presents a wicked, but a 'super-wicked' problem.[29] Super-wicked problems provide an even greater challenge due to four additional complications that policy makers have to engage with. These criteria are:

(i)   Time for finding a solution to a policy challenge is running out;
(ii) Those seeking to solve the problem are part of the cause;
(iii) Central authorities to address the problem are either weak or non-existent; and
(iv) Policy responses discount the future irrationally.[30]

The following section illustrates how these criteria can be applied to the case of AMR, why it should be viewed as a super-wicked problem and how each aspect of the problem raises important ethical concerns that must be factored into how we will be justified in responding to AMR as a super-wicked problem.

26.2.1   Criterion 1: Time Is Running Out

Super-wicked problems require drastic and urgent responses if they are not to become unsolvable policy dilemmas. In the case of AMR, it is evident that increasing drug resistance will exacerbate the existing challenge, which has led to concerns

---

[25] H.W.J. Rittel and M.M. Webber, 'Dilemmas in a General Theory of Planning,' *Policy Sciences* 4:2 (1973): 155–69.

[26] H.W.J. Rittel and M.M. Webber, 'Dilemmas in a General Theory of Planning,' *Policy Sciences* 4:2 (1973): 155–69.

[27] S. Jentoft and R. Chuenpagdee, 'Fisheries and coastal governance as a wicked problem,' *Marine Policy* 33(2009): 553–60.

[28] R. Buchanan, 'Wicked Problems in Design Thinking,' *Design Issues* 8(1992): 5–21.

[29] R. Lazarus, 'Super wicked problems and climate change: restraining the present to liberate the future,' *Cornell Law Review* 94(2009): 1153–234; K. Levin, et al., 'Overcoming the tragedy of super wicked problems: constraining our future selves to ameliorate global climate change,' *Policy Sciences* 45:2 (2012): 123–52.

[30] K. Levin, et al., 'Overcoming the tragedy of super wicked problems: constraining our future selves to ameliorate global climate change,' *Policy Sciences* 45:2 (2012): 123–52.

about a "post-antibiotic era".[31] There is thus, in short, an observable trend towards more complex cases of AMR, which will either be very difficult to treat or no longer respond to antibiotics at all.[32] Indeed, for some infections, such as tuberculosis or gonorrhea, resistance to all common treatments can already be observed.[33] As multi- and extensively drug resistant infections are being registered more and more frequently and with a lack of new antibiotics under development, the prospect of total antimicrobial resistance is no longer an abstract worst-case scenario.[34] This not only signals the need to make exigent decisions about a large and serious risk of harm for individuals and populations, but also factoring in the ethics of acting under conditions of necessity, urgency and uncertainty – and how this may differ from policy responses taken under conditions of normalcy.[35]

26.2.2   Criterion 2: Those Seeking a Solution Are Part of the Problem

AMR is a multi-factorial problem and often those involved in combating it also contribute to, or exacerbate it, in some important way:

- *Health care professionals,* who in many countries act as gatekeepers for access to antibiotics, often prescribe more antibiotics than necessary, which can lead to higher prevalence of drug-resistant infections.[36] There in an increasing number of studies, which indicate that a high proportion of prescriptions of antibiotics do not comply with scientific guidelines.[37] Furthermore, appropriate infection

---

[31] N. Brown, 'Dawn of the post-antibiotic age?' *BMJ* 309(1994): 615; A.J. Alanis, 'Resistance to Antibiotics: Are We in the Post-Antibiotic Era?' *Arch Med Res* 36(2005): 687–705.

[32] European Centre for Disease Prevention and Control, *Antimicrobial resistance surveillance in Europe: Annual report of the European Antimicrobial Resistance Surveillance Network (EARS-Net) 2011* (Stockholm, 2012).

[33] E. Alirol, et al., 'Multidrug-resistant gonorrhea: A research and development roadmap to discover new medicines,' *PLoS Med* 14(2017): e1002366; World Health Organization, *Treatment guidelines for drug-resistant tuberculosis, 2016 update* (Geneva: WHO, October 2016 revision) http://apps.who.int/iris/bitstream/10665/250125/1/9789241549639-eng.pdf?ua=1&ua=1 (Accessed September 25, 2017).

[34] European Centre for Disease Prevention and Control, *The bacterial challenge: time to react* (Stockholm, 2009).

[35] See, e.g., A.M. Viens, 'Normative Uncertainty and the Ethics of Risk Regulation in Emergency,' in A. Alemanno (ed.), *Governing Disasters: The Challenges of Emergency Risk Regulation* (Edward Elgar Publishing, 2011), pp. 137–46; A.M. Viens, *In Extremis: Morality in Times of Emergency* (London: University of London PhD, 2012).

[36] R.G. Finch, 'Antibiotic resistance: a view from the prescriber,' *Nature Reviews Microbiology* 2(2004): 989–94; V.I. Enne, 'Reducing antimicrobial resistance in the community by restricting prescribing: can it be done?' *Journal of Antimicrobial Chemotherapy* 65(2010): 179–82.

[37] K.E. Fleming-Dutra, A.L. Hersh and D.J. Shapiro, 'Prevalence of Inappropriate Antibiotic Prescriptions Among US Ambulatory Care Visits, 2010–2011,' *JAMA* 315(2016): 1864–1873. doi:https://doi.org/10.1001/jama.2016.4151.

prevention and control can reduce the need for antibiotics in the first place – which is an area where improvements can be made too.[38]

- *Pharmaceutical companies* have steadily decreased research efforts into new antimicrobial drugs, despite mounting pressures on health care systems to introduce new and effective drugs. The lack of investment is, at least, partly caused by concerns over limited returns on investment due to advancing AMR.[39] Margaret Chan, former World Health Organization (WHO) Director-General, recently alluded to this problem, when she rhetorically asked: "[f]rom an industry perspective, why invest considerable sums of money to develop a new antimicrobial when irrational use will accelerate its ineffectiveness before the R&D investment can be recouped?"[40] Recent advances in clinical research show that it may be possible to develop entirely new classes of antibiotics.[41] However, the time frame for their development remains unclear and unless novel ways of financing are introduced, their prices are likely to be prohibitive for most patients in low and middle-income countries (LMICs). The industry perspective is that the investment into new antibiotics is only financially feasible, if considerable lump sum payments, so-called market entry rewards, are paid out. These are usually estimated to have to be greater than $ 1 billion.[42] Moreover, the antibiotic production process and discarded antibiotic-laced waste can also have a considerably negative environmental impact that further perpetuates resistance.[43]

---

[38] J. O'Neill et al., *Tackling Drug-Resistant Infections Globally: Final Report and Recommendations* (London: Review on Antimicrobial Resistance, 2016), available at https://amr-review.org/sites/default/files/160518_Final%20paper_with%20cover.pdf (accessed September 29, 2017).

[39] C. Morel and E. Mossialos, 'Stoking the antibiotic pipeline,' *BMJ* 340(2010): c2115.

[40] M. Chan, *Secondary Antimicrobial Resistance in the European Union and the World* (Geneva: World Health Organization, 2012), available at: http://www.who.int/dg/speeches/2012/amr_20120314/en/ (accessed September 29, 2017).

[41] L.L. Ling, et al., 'A new antibiotic kills pathogens without detectable resistance,' *Nature* 517(2015): 455–59.

[42] R. Laxminarayan, P. Matsoso, S Pant, C. Brower, J.A. Røttingen, K. Klugman and S. Davies, 'Access to effective antimicrobials: a worldwide challenge,' *Lancet* 387(2016): 168–175; K. Outterson, et al., 'Delinking Investment in Antibiotic Research and Development from Sales Revenues: The Challenges of Transforming a Promising Idea into Reality,' *PLOS Medicine* 13(2016): e1002043. https://doi.org/10.1371/journal.pmed.1002043.

[43] K.D. Brown, et al., 'Occurrence of antibiotics in hospital, residential, and dairy effluent, municipal wastewater, and the Rio Grande in New Mexico,' *Science of the Total Environment* 366(2006): 772–783; J. Fick, et al., 'Contamination of surface, ground, and drinking water from pharmaceutical production,' *Environ Toxicol Chem* 28(2009): 2522–7; D.G.J. Larsson, 'Pollution from drug manufacturing: review and perspectives. *Phil. Trans. R. Soc. B* 369(2014): 20130571; Changing Markets Foundation, *Superbugs in the Supply Chain: How Pollution from Antibiotics Factories in India and China is Fuelling the Global Rise of Drug-Resistant Infections* (Utrecht: Changing Markets Foundation, 2016); R. Laxminarayan and R.R. Chaudhury, 'Antibiotic Resistance in India: Drivers and Opportunities for Action,' *PLoS Medicine* 3(2016): e1001974; C. Lübbert, et al., 'Environmental pollution with antimicrobial agents from bulk drug manufacturing industries in Hyderabad, South India, is associated with dissemination of extended-spectrum beta-lactamase and carbapenemase-producing pathogens,' *Infection* 45(2017): 479–491.

- *Patients* who are prescribed antibiotics often fail to adhere to treatment recommendations. This problem is exacerbated in countries where antibiotics are available for purchase over the counter, and patients may be unable to afford or have access to a health care professional who can prescribe and dispense the appropriate drug.[44]
- *Food Producers and Sellers* must continue to take significant steps to reduce the use of antibiotics as growth promoters and adopt practices that reduce the extent to which the food chain continues to be a key pathway for the transmission of resistant pathogens.[45] We also need to ensure AMR surveillance systems include resistance levels in animal stocks and food systems.
- *Governments and intergovernmental organizations* still lack the requisite regulation and governance structures to address the complexity associated with AMR and remain focused on multilateral and bilateral agreements that were commonly used in the middle of the twentieth century.[46]

The various and complex interacting contributions to the spread of AMR make it difficult (though certainly not impossible) to attribute causal responsibility to past action, including a failure to act. This also raises important questions about how our policy response should hold individuals, groups and institutions morally responsible for their complicity or causal contribution to AMR in any accountability mechanisms developed going forward.

26.2.3   Criterion 3: Central Authorities to Address the Problem Are either Weak or Non-existent

There is currently a lack of institutional structures that can meet the challenge of AMR effectively and at all policy levels. While international organizations, such as WHO or ECDC, have developed guidelines for prudent use of antibiotics, as well as

---

[44] M.B. Planta, 'The Role of Poverty in Antimicrobial Resistance' *J Am Board Fam Med*, 20:6 (2007):533–39; I.N. Okeke, A. Lamikanra, R. Edelman, 'Socioeconomic and behavioral factors leading to acquired bacterial resistance to antibiotics in developing countries,' *Emerging Infectious Diseases* 5(1999): 18–27; D.J. Morgan, I.N. Okeke, R. Laxminarayan, E.N. Perencevich and S. Weisenberg, 'Non-prescription antimicrobial use worldwide: a systematic review,' *The Lancet Infectious Diseases* 11(2011): 692–701.

[45] C. Verraes, et al., 'Antimicrobial Resistance in the Food Chain: A Review,' *International Journal of Environmental Research and Public Health* 10(2013): 2643–2669; T. P. Van Boeckel, et al., 'Global trends in antimicrobial use in food animals,' *Proceedings of the National Academy of Sciences of the United States of America* 112(2015): 5649–5654; T. P. Van Boeckel, et al. 'Reducing antimicrobial use in food animals,' *Science* 357(2017): 1350–1352.

[46] A. C. Singer, et al. 'Review of Antimicrobial Resistance in the Environment and Its Relevance to Environmental Regulators,' *Frontiers in Microbiology* 7(2016):1728; Gabriel Birgand, et al., 'Comparison of governance approaches for the control of antimicrobial resistance: Analysis of three European countries,' *Antimicrobial Resistance & Infection Control* 7(2018): 28. More is written regarding the role of these stakeholders below, under criterion 3.

programs for co-operation and data sharing, their implementation lies with the respective national governments and adherence cannot be enforced.[47] Failure to follow these recommendations is not necessarily a matter of ignorance or antipathy – in many instances countries may simply lack the control mechanisms, infrastructure, expertise, or resources to meet all requirements and recommendations of best-practice guidelines. It is also being increasingly recognized that poor governance and political corruption make major contributions to our inability to effectively respond to AMR.[48] However, since drug resistance cannot be confined by national borders, these discrepancies and deficiencies in national health systems and policies inevitably threatens the effectiveness of antibiotics and control of drug-resistant infections, even in countries that strictly enforce measures to reduce and control the use of antibiotics and mitigate the spread of AMR. More recently, the United Nations has sought to address AMR more comprehensively across sectors and policy areas, both by establishing a temporary interagency coordination group, and by strengthening the tripartite, a coordinating forum comprised of WHO, FAO, OIE, and UN Environment.[49] Nevertheless, coordination between international mechanisms and national institutions remains a voluntary process, for which few dedicated resources are available.

This raises important questions about, for instance, who should take ultimate responsibility for antimicrobial stewardship when there are no central authorities or where jurisdictions lack the resources or infrastructure to undertake stewardship activities in accordance with international recommendations.[50] Indeed, the effect of the absence of authorities and resources in the face of the inherent complexity of global health governance has gained greater and greater traction in the global health arena. In particular, we are starting to see more of a role for formal and informal stakeholders (e.g., Wellcome Trust, Bill and Melinda Gates Foundation, World Bank, corporations) increasingly shaping and guiding global health activity through many channels – many of which are no longer transparent nor are the varied values and interests of these stakeholders always easy to fully comprehend.[51] As such, the super-wicked problem of AMR sits within a much broader complex challenge that

---

[47] See e.g. the World Health Organization's *Global Action Plan on Antimicrobial Resistance* (Geneva: WHO, 2015).

[48] P. Collignon, et al., 'Antimicrobial Resistance: The Major Contribution of Poor Governance and Corruption to this Growing Problem,' *PLoS One* 10(2015): e0116746; B. Rönnerstrand and V. Lapuente, 'Corruption and use of antibiotics in regions of Europe,' *Health Policy* 121(2017): 250–256.

[49] See World Health Organization, 'Interagency Coordination Group on Antimicrobial Resistance (IACG), https://www.who.int/antimicrobial-resistance/interagency-coordination-group/en/ (accessed October 10, 2018), and World Health Organization, *FAO, OIE, WHO – Tripartite Concept Note* (Geneva: WHO, 2010) http://www.who.int/foodsafety/areas_work/zoonose/concept-note/en/ (accessed October 10, 2018).

[50] B. Bennett and J. Iredell, "Global health governance and antimicrobial resistance" In *Ethics and Drug Resistance: Collective Responsibility for Global Public Health*. Springer, Cham.

[51] C. Clinton and D. Sridhar, *Governing Global Health* (Oxford: Oxford University Press, 2017); J. Youde. *Global Health Governance* (Cambridge: Polity Press, 2012); Nora Kenworthy, Ross

is making sense of, and impacting, global health governance, thus amplifying and multiplying the complexity of addressing AMR.

26.2.4   Criterion 4: Current Policies Discount the Future Irrationally

There is a notable disparity between the enormous significance that antibiotics have in today's health care systems and the lack of a comprehensive and realistically implementable global strategy to protect them as a resource for current and future generations.[52] Indeed, while the impact of AMR on health outcomes today and in the future is increasingly recognized, there remains a lack of coordination and sufficient funding to address the problem. When the health ministers of the G20 countries met for the first time in 2017, they declared "AMR has the potential to have a major negative impact on public health as well as on growth and global economic stability".[53] Other impact assessments have come to similar conclusions.[54] Since modern medicine relies so heavily on antibiotics as a prophylactic for standard surgical procedures, the further progression of AMR would be detrimental not only for infectious disease policy, it would also affect the outcome of surgical care or the survival chances of cancer patients. In short, AMR has become a wider health system issue.[55] It is therefore all the more surprising that current policies do not place greater emphasis on long-term strategies to preserve antibiotic effectiveness and best-case policy scenarios consist in the preservation of some level of antibiotic effectiveness in the short run, without a comprehensive replacement strategy for ineffective drugs.[56]

Due to the complexity of the drivers of AMR, its costs are extremely difficult to assess, even if they are only measured locally or in the short run.[57] A number of recent economic models have attempted to model the long-term costs, but due to the

---

MacKenzie, and Kelley Lee (eds.), *Case Studies on Corporations and Global Health Governance: Impacts, Influence and Accountability* (Rowman & Littlefield, 2016).

[52] M. Millar M, 'Can antibiotic use be both just and sustainable... or only more or less so?' *Journal of Medical Ethics* 37(2011): 153–7; L. Leibovici, M. Paul and O. Ezra, 'Ethical dilemmas in antibiotic treatment,' *Journal of Antimicrobial Chemotherapy* 67(2012): 12–16.

[53] Berlin Declaration of the G20 Health Ministers, Berlin 2017, p. 5 https://www.bundesgesundheitsministerium.de/fileadmin/Dateien/3_Downloads/G/G20-Gesundheitsministertreffen/G20_Health_Ministers_Declaration_engl.pdf (accessed on September 25, 2017).

[54] O. Adeyi. et al., Drug-resistant infections: A threat to our economic future, Volume 2 (Washington, D.C: World Bank, 2017), available at http://documents.worldbank.org/curated/en/323311493396993758/final-report

[55] R. Smith and J. Coast, 'The true cost of antimicrobial resistance,' *BMJ* 346(2013): f1493.

[56] A.J. Alanis, 'Resistance to Antibiotics: Are We in the Post-Antibiotic Era?' *Arch Med Res* 36(2005): 687–705.

[57] D.H. Howard, R.D. Scott, 'The Economic Burden of Drug Resistance,' *Clin Infect Dis* 41(2005): S283-S86; R. Smith and J. Coast, 'The true cost of antimicrobial resistance,' *BMJ* 346(2013): f1493.

complexity of AMR, they inevitably have to restrict their analysis to a select number of bacterial pathogens or consider a limited number of potential cost factors.[58] As a result, existing cost models likely underestimate the long-term effects of AMR. However, even these conservative cost models, which – by their own admission – systematically underestimate some of the costs associated with drug resistance, come to the conclusion that the cost of AMR (including resistance to antiviral drugs) may exceed $100 trillion in total GDP loss by 2050.[59] What this shows quite clearly is that current efforts to address AMR are simply not proportionate to the magnitude of the challenge at hand.

This also raises questions about the proportionality of our policy responses – in a moral and legal sense – in light of how it will be legitimate for us to act given the seriousness of the challenge faced. If the spread and magnitude of harm of AMR will be as predicted, what could be currently seen as a disproportionate response – for instance, what might be seen as overly paternalistic or coercive measures – may be seen to be justified as the problem worsens. Nevertheless, if the long game here is about keeping AMR at bay for as long as possible, should we actually have to wait until the problem gets much worse before we are justified in utilizing more paternalistic or coercive measures to achieve the same goal? The moral and legal calculus involved – and the extent to which the legitimacy of our responses should be determined in proportion to the risk of harm – could signal, for example, the development of new or more stringent stewardship obligations, the imposition of more risk on individuals (even without consent) and potential limitations on once established and uncontroversial entitlements and rights.[60] The nature of super-wicked problems are such that we must consider not only what norms and values should advance particular social, medical and public health goals, but also how such norms and values can set constraints or limits on these goals.

---

[58] J. Taylor, et al., *Estimating the economic costs of antimicrobial resistance: models and results* (Cambridge, UK: RAND Europe, 2014); KPMG, *The global economic impact of antimicrobial resistance* (London: KPMG: 2014).

[59] J. O'Neill et al., Tackling Drug-Resistant Infections Globally: Final Report and Recommendations. (London: Review on Antimicrobial Resistance, 2016), available at https://amr-review.org/sites/default/files/160518_Final%20paper_with%20cover.pdf (accessed September 29, 2017).

[60] Gro Harlem Brundtland, who served three terms as Prime Minister of Norway and as Director-General of the World Health Organization from 1998 to 2003, maintained that "AMR is just as much about human rights and justice as it is about health" [http://blogs.bmj.com/bmj/2017/12/06/gro-harlem-brundtland-new-who-guidelines-are-crucial-step-to-fighting-antimicrobial-resistance/]

26.3 How Conceptualizing AMR as a Super-Wicked Problem Can Inform Policy Making

The account of AMR as a super-wicked problem is not merely descriptive, but offers a framework for assessing the expected impact of policy making. Understanding AMR as a super-wicked problem underlines three important facts. First, it highlights the tendency of current policies to focus on preserving antibiotic effectiveness, while failing to reduce the future need for antibiotics. Second, it explains why traditional instruments, such as cost-benefit analysis (CBA), are unsuited to determining an appropriate policy response. Third, it helps bring to light how the complex interaction of ethical considerations will contribute to shaping which policy options will be viewed as acceptable.

Many of the existing policies that address AMR (and a considerable portion of the academic literature) emphasize the need for new antibiotics, as well as the cost-effective and prudent use of available resources.[61] The policy focus is thus on addressing problems on the supply side, and the creation of new resources, where a broad number of policy campaigns already exist.[62] However, if AMR is understood as a super-wicked problem, it becomes apparent that such approaches – while a necessary contribution – will ultimately and inevitably fall short of the goal of effectively controlling AMR in the long run. Consideration must take place of the relative value of focusing on upstream versus downstream determinants of drug-resistant infections, and the extent to which values other than efficiency or innovation (e.g., health equity) should guide both which drivers of resistance we focus on as well as which preventive and therapeutic responses we should pursue.

Framing AMR as a super-wicked problem should lead policy makers to place a much stronger emphasis on those policies, which – to paraphrase Levin et al. – generate a shift in path dependencies.[63] The concept of path dependencies explains current policies in light of their development of time, and as a result of earlier decision.[64] A given policy may therefore be more influenced by legacy decisions than by

---

[61] E. Mossialos, *et al, Policies and incentives for promoting innovation in antibiotic research* (Copenhagen: European Observatory on Health Systems and Policies, 2008); T. Groves, 'Back to basics with the three Rs,' *BMJ* 344(2012); R. Laxminarayan and G.M. Brown, *Economics of Antibiotic Resistance: A Theory of Optimal Use* (Washington D.C. Resources for the Future, 2001); C. Morel and E. Mossialos, 'Stoking the antibiotic pipeline,' *BMJ* 340(2010): c2115.

[62] L.D. Högberg, A. Heddini, and O. Cars, 'The global need for effective antibiotics: challenges and recent advances' *Trends Pharmacol Sci* 31(2010): 509–15.

[63] K. Levin, et al., 'Overcoming the tragedy of super wicked problems: constraining our future selves to ameliorate global climate change,' *Policy Sciences* 45(2012): 123–52.

[64] A more complete definition of path dependence is as follows: 'Path dependence is the idea that decisions we are faced with depend on past knowledge trajectory and decisions made, and are thus limited by the current competence base. In other words, history matters for current decision-making situations and has a strong influence on strategic planning... A well-known example is the QWERTY layout for typewriters. Despite the fact that different keyboard layouts in modern computer keyboards would allow faster typing, the QWERTY layout prevails.' *Financial Times Lexicon* http://lexicon.ft.com/Term?term=path-dependence (accessed September 30, 2017).

the current state of affairs and the latest available evidence. A shift in path dependencies is necessary once it has become apparent that the trajectory of earlier policy decisions is leading to an unsustainable outcome in the long run. In the case of climate change, for example, such path dependency is exemplified by the widespread and continued reliance on fossil fuels.[65] In the case of AMR, this path dependency is reflected by the reliance on antibiotics as not only a treatment against acute infection, but also as a tool for infection prevention in both clinical and veterinary settings. This dependency is replicated in most current policies that focus on pulling on the same levers, which aim at either increasing the availability of antibiotics or decreasing the use of antibiotics. What is crucially missing is sufficient emphasis on infection prevention and control measures that reduce the need for antibiotics in the first place.

Consequently, current policies do not offer a long-term fix to the problem of AMR and create, at best, a "faux paradigmatic change", in which the implementation of policy only makes small corrections to a previous policy failure temporarily.[66] The most obvious example for a faux paradigmatic shift is the reliance on future developments of new antibiotics, which are effective against resistant bacteria. While such a development will provide significant short-term improvements, past experience suggests that bacteria will ultimately develop resistance to new classes of antibiotics, as well. Consequently, as Spellberg has observed, "we will never truly defeat microbial resistance; we can only keep pace with it."[67] In the absence of a realistic option for true paradigmatic change (i.e., a technological method to avoid the further emergence and spread of AMR altogether), it appears advisable to abandon ambitions to outpace the adaptation of microbes to new antibiotics, and instead focus on the creation of incremental but transformative changes, which no longer follow the same policy trajectory and instead reduce the dependence on antibiotics. In particular, this new trajectory must have as a chief focus policy options that can be effectively implemented in LMICs.

One analogy that summarizes this situation, and which has been used repeatedly to describe the problem at hand, is that of a "leaky bucket".[68] If we think of antibiotic effectiveness as a resource contained in a bucket, the emergence of resistance is

---

[65] G. Unruh, 'Understanding Carbon Lock-In,' *Energy Policy* 28(2000): 817–30; P. Aghion, C. Hepburn, A. Teytelboym, and D. Zenghelis, *Path dependence, innovation and the economics of climate change* (London: Centre for Climate Change Economics and Policy Grantham Research Institute on Climate Change and the Environment, 2014) http://www.lse.ac.uk/GranthamInstitute/wp-content/uploads/2014/11/Aghion_et_al_policy_paper_Nov20141.pdf (accessed September 29, 2017).

[66] B. Cashore and M. Howlett, 'Punctuating Which Equilibrium? Understanding Thermostatic Policy Dynamics in Pacific Northwest Forestry,' *Am J Pol Sci* 51(2007): 532–51; K. Levin, et al. 'Overcoming the tragedy of super wicked problems: constraining our future selves to ameliorate global climate change,' *Policy Sciences* 45(2012): 123–52.

[67] H.W. Boucher, et al. 'Bad Bugs, No Drugs: No ESKAPE! An Update from the Infectious Diseases Society of America,' *Clin Infect Dis* 48(2009): 1–2.

[68] This analogy was originally proposed by Prof. Otto Cars, ReAct-Action on Antimicrobial Resistance, Uppsala University Sweden.

akin to holes in this bucket, slowly draining the effectiveness of available drugs. The development of a new drug would effectively add water to the leaking bucket. But, in the absence of a realistic option to develop antibiotics that avoid the subsequent emergence of resistance, this will only have a temporary effect and not address the underlying problem; namely, the holes in the bucket. Thus, not only are the antibiotics currently in clinical development not adequate to counter the increasing prevalence of AMR, it is unlike they ever could be.[69]

A second argument for framing AMR as a super-wicked problem is that this approach discourages a reliance on CBA in policy making. Super-wicked problems – by their nature – are not easily solvable with standard CBA tools because they describe scenarios in which the cost of inaction will be very high, yet occur at some point in the distant future. Consequently, CBA will usually recommend an insufficiently large commitment of resources to address super-wicked problems because costs incurred today are pitted against benefits at a later stage for which neither magnitude nor time frame are known.[70] As outlined earlier, existing cost-models for the assessment of the economic burden of AMR systematically underestimate the true cost of AMR because, as Smith and Coast argue, "[none] considered the bigger picture – a world in which there are no effective antibiotics for situations where they are currently used routinely".[71] And where it has been attempted to take these wider costs into consideration, the complexity of AMR has usually forced analysts to restrict their models to pathogens and geographical regions for which reliable data exists.[72] One response to this criticism of CBA in policy-making is to suggest that a bad estimate of cost is still better than no estimate at all. However, as Jamieson rightly points out, this response overlooks that whenever uncertainties about future developments are great, relying on the supposedly neutral judgment of a bad cost-estimate is a leap of faith – and it may prompt us to reach policy decisions that, in the long run, are far worse than the ones we might have considered had we not aimed for cost-efficiency based on unreliable or incomplete information.[73]

A final argument for framing AMR as a super-wicked problem is that this approach provides a greater prominence on the importance of the values and norms that should inform what would make particular policy options more or less

---

[69] World Health Organization, *Antibacterial Agents in Clinical Development: An Analysis of the Antibacterial Clinical Development Pipeline, including Tuberculosis* (Geneva: World Health Organization, 2017).

[70] R. Lazarus, 'Super wicked problems and climate change: restraining the present to liberate the future,' *Cornell Law Review* 94(2009): 1153–234.

[71] R.D. Smith and J. Coast, 'The true cost of antimicrobial resistance'. BMJ 2013; 346: f1493. Also cf. J. Coast and R.D. Smith, 'Distributional Considerations in Economic Responses to Antimicrobial Resistance,' *Public Health Ethics* 8(2015): 225–237.

[72] See e.g. the RAND corporations cost model for O'Neill's AMR review: https://www.rand.org/pubs/research_reports/RR911.html (accessed September 29, 2017).

[73] D. Jamieson, 'Ethics, Public Policy and Global Warming,' in S.M. Gardiner, et al. (eds.) *Climate Ethics* (Oxford: Oxford University Press, 2010), pp. 77–86. Also cf. A. Williams, 'Cost-effectiveness analysis: is it ethical?' *Journal of Medical Ethics* 18(1992): 7–11.

acceptable.[74] Through diminishing an analysis of the cost of AMR in primarily economic terms, it helps bring to light the non-economic costs and values involved in shaping which policy options will be viewed as acceptable. By emphasizing the need for shifts in path dependencies, and the resultant changes in our approaches and responses, it also emphasizes the need to re-evaluate the values and norms which underpinned our previous policies and activities. To put it another way, through framing AMR as a super-wicked problem, we not only acknowledge that previous approaches are unlikely to be sustainable in the long run, but we are also forced to ask what sort of values and norms could justify new policy options that would not only be effective but also ethical. This is all the more relevant because understanding AMR as more than a scientific or technical issue is a relatively new perspective. For most of the existence of antibiotics, their use has been primarily viewed as a medical or microbiological issue, and was governed accordingly. This means that value judgments were commonly only implicit and often incoherent. We have seen that these technical or medical matters do not exhaust all of the relevant considerations, and much of the decision-making and policy-making around AMR concerned matters that were inherently and inescapably ethical. Acting as an antimicrobial steward, for instance, often involves making moral judgments, promoting particular values and prioritising different aims – which are normative, and not merely technical, activities. In viewing AMR as a societal challenge, we can see the failings of earlier approaches that had not considered the normative significance this shift implies.

26.4 Incrementally Creating Transformative Shifts in Path Dependency: Alternative Strategies

Creating transformative changes in path dependency will require adjustments on all types and levels of antibiotic use, and is unlikely to come in the shape of a single intervention. While there are no formulaic strategies to address wicked and super-wicked problems, different kinds of policy response will be suited to different time frames and address different aspects of the challenge that AMR presents. Nancy Roberts distinguishes between three possible approaches to super-wicked problems: authoritative, competitive and collaborative strategies.[75]

---

[74] For different ways in which framing the problem of AMR can have ethical and political implications, see, for instance, A.M. Viens and Jasper Littmann, 'Is Antimicrobial Resistance a Slowly Emerging Disaster?' *Public Health Ethics* 8(2015): 255–265.

[75] N. Roberts, 'Wicked Problems and Network Approaches to Resolution,' *Int Public Manag Rev* 1(2000): 1–19.

26.4.1 Authoritative Strategies

These strategies involve a small number of decision-makers who develop policy solutions, which are then implemented by others. Such approaches require that decision-makers have the ability to enforce the implementation of their chosen strategy. The advantage of such a policy is that decision-making complexity is reduced, and policies can be implemented and adjusted relatively quickly. On the other hand, such approaches to solving super-wicked problems are likely to alienate a large proportion of stakeholders and they depend on the existence of power structures where they can be enforced.[76] To this end, recent proposals for international legal frameworks and an intergovernmental panel have been put forward.[77] However, the question how adherence at the local level could be controlled and, where necessary, enforced – let alone sufficiently resourced – make such centralized and costly proposals very difficult to implement. Authoritative strategies may therefore be of greater use at the national or regional level. At the global level, however, the current lack of adequate resource, governance and accountability structures, which could help to effectively implement them, is likely to limit their usefulness in addressing the challenge of AMR.

26.4.2 Competitive Strategies

These strategies let different stakeholders or corporations compete for the creation of (market-based) solutions and often lead to creative approaches to problem solving.[78] However, the commitment of stakeholders will largely depend on the strength of incentives to focus on a given policy area. In the case of AMR, the absence of sufficient market incentives has led to minimal investments into R&D for new antibiotics, which lies well below a socially optimal level.[79] Recent initiatives such as

---

[76] L. Briggs, 'Tackling Wicked Problems – A Public Policy Perspective,' in *Contemporary Government Challenges* (Australian Public Service Commission, 2007).

[77] J. Anomaly, 'Combating Resistance: The Case for a Global Antibiotics Treaty,' *Public Health Ethics* 3(2010): 13–22; M. Woolhouse and J. Farrar, 'Policy: An intergovernmental panel on antimicrobial resistance,' *Nature* 509(2014): 555–557; Steven J. Hoffman, 'An international legal framework to address antimicrobial resistance,' *Bulletin of the World Health Organization* 93(2015): 66; S.J. Hoffman, J-A. Røttingen and J. Frenk, 'International Law Has a Role to Play in Addressing Antibiotic Resistance,' *Journal of Law, Medicine & Ethics* 43(2015): 65–67; A.D So, et al., 'An integrated systems approach is needed to ensure the sustainability of antibiotic effectiveness for both humans and animals,' *J Law Med Ethics*. 2015:43(Suppl 3):38–45; N. Gulati, et al., *Using International Instruments to Address Antimicrobial Resistance* (Ottawa: University of Ottawa Global Strategy Lab, 2016).

[78] N. Roberts, 'Wicked Problems and Network Approaches to Resolution,' *Int Public Manag Rev* 1(2000): 1–19.

[79] E. Mossialos, et al., *Policies and incentives for promoting innovation in antibiotic research* (Copenhagen: European Observatory on Health Systems and Policies, 2008).

the US Generating New Antibiotic Incentives Now (GAIN) Act or the European Innovative Medicines Initiative (IMI) are trying to address this, but even if successful, their overall budget does not permit any kind of paradigmatic shift to current antibiotic use policy.[80] Other strategies which have been offered, such as antibiotic de-linkage schemes,[81] health impact funds,[82] public-private partnerships,[83] innovation prizes[84] and other mechanisms[85] have mostly sought to replicate the old approach of developing new antibiotics in the hopes of outpacing resistance. While some AMR funding streams now include calls for projects in behavioral science, communication and education, their funding remains a miniscule fraction of what is currently being invested into drug development. From an ethical perspective, this creates additional problems. New drugs are likely to disproportionately benefit high-income countries (HICs), despite LMICs having the greatest burden of drug-resistant infections. Behavioural science research is also predominantly focused on HICs. We currently lack sufficient behaviour change research into what can be effectively achieved in regions with high levels of resistance and limited resources, e.g. some African countries, India, or countries in South East Asia. Competitive strategies are unlikely to change this as they are currently structured and incentivized.

26.4.3 Collaborative Strategies

These strategies involve a large number of stakeholders, especially in cases where responsibility and decision-making power are widely dispersed. While more difficult to establish than top-down authoritative approaches, collaborative strategies are widely considered to be the most suitable approach to dealing with wicked and

---

[80] J.P. Roberts, 'Incentives aim to boost antibiotic development,' *Nature Biotechnology* 30:8 (2012):735.

[81] C. Clift, et al. (eds.), *Towards a New Global Business Model for Antibiotics: Delinking Revenues from Sales* (London: Chatham House, 2015).

[82] K. Outterson, T. Pogge and A. Hollis, 'Combating Antibiotic Resistance Through the Health Impact Fund,' in I. Glenn Cohen (ed.), *The Globalization of Health Care: Legal and Ethical Issues* (Oxford: Oxford University Press 2013), pp. 318–338.

[83] For example, DRIVE-AB (Driving reinvestment in research and development and responsible antibiotic use), available from: http://drive-ab.eu; CARB-X (Combating antibiotic resistant bacteria biopharmaceutical accelerator), available from: http://www.carb-x.org

[84] For example, the Longitude Prize, available at: https://longitudeprize.org; Antimicrobial Resistance Diagnostic Challenge, available at: https://dpcpsi.nih.gov/AMRChallenge. Also see J. Love and T. Hubbard, 'The Big Idea: Prizes to Stimulate R&D for New Medicines,' *Chicago-Kent Law Review* 82(2007): 1519–1554.

[85] K. Outterson, J.H. Powers, G.W. Daniel, and M.B. McClellan, 'Repairing the Broken Market for Antibiotic Innovation,' *Health Affairs* 34(2015): 277–285; M.J. Renwick, D.M. Brogan and E. Mossialos, 'A Systematic Review and Critical Assessment of Incentive Strategies for Discovery and Development of Novel Antibiotics,' *The Journal of Antibiotics* 69(2016): 73–88; R.D. Smith, and J. Coast. 'The economics of resistance through and ethical lens,' In *Ethics and Drug Resistance: Collective Responsibility for Global Public Health*. Springer, Cham.

super-wicked problems, especially in the absence of a strong global planning authority to address the challenge.[86] Collaborative strategies can and are already being used at different levels, from UN initiatives to regional local campaigns that focus on raising awareness or improve prescribing.[87]

These three strategies are not mutually exclusive and different approaches can and should be combined, where viable and appropriate. Viewing AMR as a super-wicked problem suggests that sustainable change is most likely to result from collaborative strategies focusing on multiple drivers, yet the urgency with which a response to AMR must be found may necessitate, for example, the inclusion of competitive strategies to develop new drugs and authoritative strategies at the national level to enforce strict prescribing guidelines or prohibitions on using antibiotics in farm animals as growth promoters. However, the critical reader will also notice that, for the most part, present efforts to address AMR can already be categorized as authoritative, competitive or collaborative strategies respectively. This begs a question as to whether framing AMR as a super-wicked problem can be of any further help in developing novel and effective strategies to combat drug resistance. In the paper's penultimate section, we canvass a few potential ways in which the super-wicked problem frame can impact on current AMR policy going forward.

26.5   Shifting Path Dependencies – The Way Forward

Understanding AMR as a super-wicked problem itself does not, unfortunately, generate a set of novel, easily implementable policy solutions. Yet, as outlined before, the complexity of AMR, as well as the countless factors that contribute to it, make one-stop solutions highly unlikely in the first place.

While it does not offer any immediate solution to the problem, the understanding of AMR as a super-wicked problem may, however, achieve another goal – namely, to prompt a reconsideration of the relative importance of different responses to drug resistance. One of the most important insights of framing AMR as a super-wicked problem is that there is no technological fix we can engineer, nor any simple market-based solution that will avoid the impending scarcity of effective antibiotics in the future. Any responses that will help us keep AMR at bay in a significant way will involve a delicate balancing of benefits and burdens that will require difficult choices and restrictions to be imposed on individuals and populations. This insight is at odds with current research funding in the area of AMR, which is heavily skewed towards drug development.[88] Moreover, given the global scale of the problem, success in one

---

[86] L. Briggs, 'Tackling Wicked Problems - A Public Policy Perspectiv,' in *Contemporary Government Challenges* (series) (Australian Public Service Commission, 2007).

[87] For a selection of local initiatives see for example the ReAct Toolbox, https://www.reactgroup. org/toolbox/

[88] R. Kelly, et al., 'Public funding for research on antibacterial resistance in the JPIAMR countries, the European Commission, and related European Union agencies: a systematic observational analysis,' *The Lancet Infectious Diseases* 16(2016): 431–440.

part of the world will likely only be temporary, given the drivers brought and accelerated by globalization. Framing AMR as such a fundamentally unsolvable policy challenge may appear to be defeatist. However, the point is not to admit defeat, but to focus on those interventions that may be of greatest use in the long run – even if we can never overcome the vicious cycle of bacterial resistance and antibiotic obsolescence.[89] It may come as good news at this stage that many potential candidates for such a strategy already exist, but have simply not have been implemented properly.

The most obvious example in this context is infection control and prevention and water, sanitation and hygiene (WASH). While ultimately decisive for the control of nosocomial infections, hospital and healthcare hygiene and infection prevention have often appeared to be more of an afterthought in recent policy discussion. However, they remain of crucial importance for the prevention of infections, in both high- and low-income settings.[90] Similarly, many of the ongoing efforts for the creation of broader public awareness for the problem and causes of AMR have not yet achieved their desired goals.[91] Education and hygiene measures are no silver bullets – and have their own issues with compliance and recalcitrance – but if AMR is understood as a super-wicked problem, these policies should receive much greater recognition as a crucial part of an effective AMR strategy. Indeed, these are just some of the many different behaviour change interventions – aimed at both professionals and the public – that should comprise a multipronged and diversified response to AMR.[92] Of course, the success of such behavior change interventions will itself be dependent upon fixing even more fundamental problems plaguing global health, including poverty and extreme income inequality, since many

---

[89] A.E. Aiello, N.B. King and B. Foxman, 'Ethical Conflicts in Public Health Research and Practice Antimicrobial Resistance and the Ethics of Drug Development,' *American Journal of Public Health*, 96(2006): 1910–1914; B. Spellberg, J.G. Bartlett and D.N. Gilbert, 'The Future of Antibiotics and Resistance,' *New England Journal of Medicine* 368(2013): 299–302.

[90] S. Harbarth, et al., 'Antimicrobial resistance: one world, one fight!' *Antimicrobial Resistance and Infection Control* 4(2015): 49 – https://doi.org/10.1186/s13756-015-0091-2; Peter Collingon, et al., 'Anthropological and socioeconomic factors contributing to global antimicrobial resistance: a univariate and multivariable analysis,' *The Lancet Planetary Health* 2018; 2: e398–405. Also World Health Organization, *Guidelines on Sanitation and Health* (Geneva: WHO, 2018).

[91] Special Eurobarometer, *Antimicrobial Resistance Report, May – June 2013* (Brussels: EU Directorate-General Communication, 2013).

[92] See, for example, T. Edgar, S.D. Boyd and M.J. Palamé, 'Sustainability for behaviour change in the fight against antibiotic resistance: a social marketing framework,' *Journal of Antimicrobial Chemotherapy* 63(2009): 230–237; C.S. Lundborg and A.J. Tamhankar, 'Understanding and changing human behaviour—antibiotic mainstreaming as an approach to facilitate modification of provider and consumer behaviour,' *Uppsala Journal of Medical Sciences* 119(2014):125–133; Pubic Health England, *Behaviour Change and Antibiotic Prescribing in Healthcare Settings: Literature Review and Behavioural Analysis* (London: Public Health England, 2015); S. Tonkin-Crine, A.S Walker and C.C Butler, 'Contribution of behavioural science to antibiotic stewardship,' *BMJ* 2015; 350:h3413; K. Chaintarli et al., 'Impact of a United Kingdom-wide campaign to tackle antimicrobial resistance on self-reported knowledge and behaviour change,' *BMC Public Health* 16(2016):393.

technical and behavioral interventions are limited by adverse social, economic, and political contexts.[93] Tackling AMR will require those working in global health to directly address injustices, however one conceives of justice. Here, for our purposes, any conception of justice will do since the problem of injustices in global health are theoretically overdetermined.[94] In other words, practically speaking, tackling the levers that make AMR a super-wicked problem will have to include making real efforts to address global injustices for the various ways in which they contribute to the level of drug-resistant pathogens around the world.

Finally, understanding AMR as a super-wicked problem and thereby as a global challenge that defies simple solutions by any one party suggests that a much greater part of our efforts must be directed towards increasing standards of access and quality of prescribing in those regions where resources continue to be limited. We have not yet managed to ensure the provision of adequate access to antibiotics in many regions of the world, where the price of drugs is often prohibitive for patients and where over-the-counter sales have led to an unregulated and uncontrolled use of antibiotics.[95] We are, therefore, faced with a situation in which we have to reduce the excessive use of antibiotics in some regions of the world while ensuring greater access in others.[96] As Daulaire et al. maintain, 'meaningful access is dependent on good stewardship and vice versa.'[97] Nevertheless, this will require us to successfully confront and find answers to difficult distributive questions about when access should be increased or limited, and how to maintain sustainable fair access in a way that attempts to diminish the rate of AMR as long as possible. Stewardship to protect the effectiveness of antibiotics presupposes a functioning healthcare and legal system with sufficient oversight to regulate antibiotic usage adequately. New drugs, especially against Gram-negative bacteria, are badly needed and if resistance against drugs of last resort (e.g., carbapenems, colistin) further increases, we are officially out of options for treatment. In many of the countries that require greater access, it is difficult to guarantee even the most basic stewardship mechanisms. As a result, there is a need to explore different ways of approaching antibiotic governance in a global setting as part of responding to this super-wicked problem that genuinely

---

[93] S. Gill and I. Bakker, 'The global crisis and global health,' in S. Benatar and G. Brock (eds.), *Global Health and Global Health Ethics* (Cambridge: Cambridge University Press, 2011), pp. 221–238.

[94] See, for example, L. Reid "Antimicrobial resistance and social inequalities in health: considerations of justice" In *Ethics and Drug Resistance: Collective Responsibility for Global Public Health*. Springer, Cham, where the author provides an egalitarian argument toward this conclusion.

[95] R. Laximinarayan, et al. 'Antibiotic resistance – the need for global solutions,' *Lancet Infect Dis*, 13:12 (2013): 1057–98.

[96] G. Heyman, O. Cars, M.T. Bejarano, and S. Peterson, 'Access, Excess, and Ethics—Towards a Sustainable Distribution Model for Antibiotics,' *Upsala Journal of Medical Sciences* 119(2014): 134–141.

[97] Nils Daulaire, et al., 'Universal Access to Effective Antibiotics is Essential for Tackling Antibiotic Resistance,' *The Journal of Law, Medicine and Ethics* 43(2015): 17–21 at 18.

empowers local and regional stakeholders to shape the path dependencies that guide our response.

26.6 Conclusion

While AMR is a complex and arduous challenge, understanding it as a super-wicked problem does not mean it is intractable or that the multitude of drivers and stakeholders defy making substantial progress. The discussion of potential initiators of path dependency in this paper serves as an illustration of possible scenarios rather than as a comprehensive list of recommendations, but it highlights that recognizing AMR as a super-wicked problem would indeed have policy implications and should lead us to reconsider our values and priorities in responding to AMR.

This will also include a discussion of ethical norms and standards that should be met in addressing the problem of drug resistance. Given that AMR has only recently moved from being a purely clinical (or veterinary) problem into the realm of global challenges that require a broader societal response, there exists currently little research, and even less agreement on what the most important ethical issues in AMR are, and what we should do about them.[98] Ethicists should weigh in on such complex and wicked problems,[99] but they must also be aware of the fact that this is not an abstract problem, but already a health challenge on a global scale.

Understanding AMR as a super-wicked problem is not merely a matter of categorization. Instead, it should lead us to reconsider current policy approaches in light of their expected usefulness and likely success of implementation. New approaches to tacking AMR should seek to avoid replicating earlier patterns and problems of jumping between 'one-best-way' approaches or 'one-size-fits-all' interventions – both empirically and ethically. If AMR does indeed present a super-wicked problem, policy efforts should be primarily directed at shifting path dependencies. As such, the aim is not 'solving' AMR but to make progress towards better mitigation and management through these shifts. Current policies, for instance, which promote infection prevention and control, antimicrobial stewardship and the development of new drugs, are a crucial contribution to curbing AMR because they prolong antibiotic effectiveness and prevent infections in the first place. However, they are clearly insufficient as answers in the medium to long run, and should therefore only constitute a first step in initiating more fundamental changes to public health policy to reduce future dependence on antibiotics and more general social policy affecting the drivers of drug resistance.

---

[98] J. Littmann J and A.M. Viens, 'The Ethical Significance of Antimicrobial Resistance,' *Public Health Ethics* 8(2015): 209–224.

[99] D.S. Silva, M.J. Smith MJ, and C.D. Norman, 'Systems thinking and ethics in public health: a necessary and mutually beneficial partnership,' *Monash Bioethics Review* 2018 Jun 13. doi: https://doi.org/10.1007/s40592-018-0082-1. [Epub ahead of print].

Framing AMR as a super-wicked problem also emphasizes the importance of ongoing trends towards more integrated collaborations across sectors and research disciplines. It should help in creating greater awareness for the true scope of the problem we are faced with and the urgency with which we must address it.

Acknowledgements The authors would like to thank James Wilson and Anthony Kessel for feedback on an earlier draft of this paper.

Index

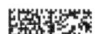